Diffuse Scattering of X-Rays and Neutrons by Fluctuations

Springer
Berlin
Heidelberg
New York
Barcelona
Budapest
Hong Kong
London
Milan
Paris
Santa Clara
Singapore
Tokyo

Mikhail A. Krivoglaz

Diffuse Scattering of X-Rays and Neutrons by Fluctuations

With 41 Figures

Springer

Professor Dr. Mikhail A. Krivoglaz †

Translator:

Dr. Oleg A. Glebov
Science and Life Foundation
Rozhdestvensky bul. 19, kv.15
103045 Moscow, Russia

Editors of the English Edition:

Prof. V. G. Baryakhtar
Institute of Metal Physics
Ukrainian Academy of Sciences
pr. Vernadskogo 36
252680 Kiev, Ukraine

Prof. S.C. Moss
Department of Physics
University of Houston
Houston, Texas 77204-5506, USA

Prof. M. A. Ivanov
Institute of Metal Physics
Ukrainian Academy of Sciences
pr. Vernadskogo 36
252680 Kiev, Ukraine

Prof. J. Peisl
Sektion Physik
Ludwig-Maximilians-Universität München
Geschwister-Scholl-Platz 1
D-80539 München, Germany

Title of the Russian original edition:
Diffuznoe rasseîanie rentgenovskikh lucheĭ i neĭtronov v na fluktuaŝionnykh neodnorodnosîakh v neideal 'nykh kristallakh.

Library of Congress Cataloging-in-Publication Data.
Krivoglaz, M. A. (Mikhail Aleksandrovich) [Diffuznoe rasseîanie rentgenovskikh lucheĭ i neĭtronov na fluktuaŝionnykh neodnorodnostîakh v neideal 'nykh kristallakh. English] Diffuse scattering of x-rays and neutrons by fluctuatios / Mikhail A. Krivoglaz. p. cm. Includes bibliographical referencs and index.
ISBN-13: 978-3-642-78767-6 e-ISBN-13: 978-3-642-78765-2
DOI: 10.1007/978-3-642-78765-2

1. Kirkendall effect. 2. Solids – Fluctuations. 3. X-rays – Diffraction. 4. X-rays – Scattering. 5. Thermal neutrons – Scattering.
QC176.8.D5K7513 1994 530.4'15 – dc20 94-3791

Cover design: Erich Kirchner
Typesetting: Camera-ready output produced from the translator's input files using a Springer TEX macro package.
SPIN: 10007876 54/3143 - 5 4 3 2 1 0 - Printed on acid-free paper

Foreword to the English Edition of Volume I

In 1969 the "Theory of X-ray and Thermal-Neutron Scattering by Real Crystals" by M.A. Krivoglaz appeared in English. While this book covered an extended series of publications by Professor Krivoglaz and his students and co-workers, much of it was new to western researchers (some of the work discussed, for example, had previously been available only in Ukrainian). Perhaps the key word in the original title was "Real"; for in that book, Krivoglaz laid out in systematic detail of ever increasing rigor, the influence of all manner of defects and disorder on the scattering patterns of real solids. For many of us this book was a rich lode indeed which we have been mining now for many years.

The logic of the Krivoglaz book was in the development of the basic principles of microscopic fluctuation theory and lattice statics (and dynamics) followed by Fourier representations of the relevant fields and leading naturally – in k-space – to scattering phenomena. Complexity was added gradually and deference was paid to the experimental situation although, as Krivoglaz himself noted at the time in his Preface, "the theory is treated as a branch of theoretical physics rather than 'crystallography' " so that "only in individual cases shall we consider the corresponding experimental results." However rewarding the task ultimately was, it was thus not always easy to chase down a particular experimental case. For example: were the defects in question dilute or concentrated, correlated or uncorrelated; were the displacements large and non-normally distributed or small and, perhaps, Gaussian distributed; were these bounded and of the first kind, producing Debye-Waller attenuation and diffuse scattering, or of the second kind, producing peak broadening, and how and when did these regimes overlap? One could also ask (and one did) how the often complex local atomic arrangements were related to the basic electronic structure and statistical mechanics of the solid in question. Analogies were thereby drawn between, say, Jahn-Teller-induced distortion fields in perovskites and the elastic dipolar fields in solid solutions. The electron theory of metals was also invoked to explain fine structure in the diffuse scattering from disordered alloys. One could go on but it suffices here to say that the 1969 monograph, carefully read, was a wonderful book.

Of course, since 1969 enormous advances have taken place both in the development of new laboratory X-ray sources, electronics and control systems and in wavelength-tuneable synchrotron radiation sources. These developments, along with continued interest within the neutron scattering community in larger crystals, have prompted studies of weaker diffuse scattering from crystals of (for X-rays) microscopic dimensions or from surfaces and very thin films. For example, monolayer films on crystals are now widely studied using synchrotron radiation since there are distinct surface correlations in the vicinity of bulk phase transitions. Defect concentrations in the range of 200 ppm can directly be measured in their influences both on Bragg peaks and the induced diffuse scattering, and weak scattering contrast can be enhanced by tuning to absorption edges. In short, the scattering enterprise is rather healthy.

Into this situation, at a truly propitious time, enter the new volumes of M.A. Krivoglaz. The present Volume I is a complete revision and up-date of those chapters of the early book dealing specifically with defects and defect-induced displacements with considerable attention now paid to the proliferation of experimental results since 1969. Correlated displacive defects are treated in detail but concentration – and related – fluctuations, phase transitions and similar topics are reserved for Volume II where fluctuation theory and lattice statics are joined. In the present book Krivoglaz has carefully re-examined the classification of defects and their displacement fields to present an increasingly complex hierarchy of diffraction effects associated with these fields. At one end, as before, is the isolated point defect and its $(1/r^2)$ displacement field. At the other end is the tangle of dislocations present in plastically deformed crystals which are often, but not always, separable into domains or sub-cells and walls. Throughout Krivoglaz notes how the details of peak shifts and line broadening evolve out of, or turn into, diffuse scattering and quasilines in diffraction patterns. All is systematically related to varying defect structures, sizes, strengths, symmetries, concentrations, correlations and dimensionalities along with the effects of crystal (elastic) anisotropy. To all of this Krivoglaz brings an uncanny physical sense of these varying influences, and the regimes in which they obtain, together with a complete mastery of the necessary mathematical tools and approximations to express it all. The Appendices, for example, are impressive in their detail. And finally, he is scrupulous in his appreciation and citations of the literature in this extremely busy field.

It is a particular irony and sadness to us that Professor Krivoglaz did not live to enjoy the openness between the former Soviet Union and the West. How he would have enjoyed visiting some of the scattering centers whose work figures so prominently in his papers; and how we, who have mined his books and papers for their treasures, would have loved to have him with us for a while! With these new books we shall have him with us at least in spirit; but we shall all have to be considerably cleverer now than before to cover the loss of our friend and colleague.

Houston and Munich
May, 1995

S.C. Moss
J.S. Peisl

Editors' Preface

Mikhail Alexandrovich Krivoglaz died unexpectedly when he was preparing the English edition of his two-volume monograph on diffraction and diffuse scattering of X-rays and neutrons in imperfect crystals. His death was a heavy blow to all who knew him, who had worked with him and to the world science community as a whole.

The application of the diffraction techniques for the study of imperfections of crystal structures was the major field of Krivoglaz' work throughout his career in science. He started working in the field in the mid-fifties and since then made fundamental contributions to the theory of real crystals. His results have largely determined the current level of knowledge in this field for more than thirty years. Until the very last days of his life, Krivoglaz continued active studies in the physics of diffraction effects in real crystals. His interest in the theory aided in the explanation of the rapidly advancing experimental studies.

The milestones marking important stages of his work were the first monograph on the theory of X-ray and neutron scattering in real crystals which was published in Russian in 1967 (a revised English edition in 1969), and the two-volume monograph published in Russian in 1983–84 (this edition is the revised translation of the latter).

The first volume of the monograph was completely rewritten by Krivoglaz in preparation for the English edition. He presented the latest experimental results and the newest theoretical concepts and made significant additions to the Russian text. Many sections were reworked completely and also new sections were added, such as on fluctuation waves in thin films, regular reflection intensities in imperfect crystals, diffuse scattering and correlation in defect positions, scattering by point defects in irradiated crystals and dilute solutions, the effects of finite defects in thin films and surface layers on X-ray scattering, as well as an additional appendix. As for the second volume, M.A. Krivoglaz had only time to set a general framework for the revision of the Russian text.

The co-workers and students of Mikhail Alexandrovich took it upon themselves to finish the preparation of his book for the English edition. The approach was to adhere as precisely as possible to the original author's framework. This was easier in the case of the first volume which had been fully completed by M.A. For the second volume we only had the brief notes left by him.

The preparation of the manuscript for the English edition was done by R.I. Barabash, D.A. Vul, I.M. Dubrovsky, B.V. Egorov, M.A. Ivanov, D.V. Lotsko, Yu. G. Pogorelow, K.P. Ryaboshapka, and B.B. Timofeev.

We are grateful to Profs. S.C. Moss (USA) and J. Peisl (Germany) for their kind agreement to edit the English text of the book. We all hope this edition may serve as a memorial to our late friend.

V.G. Bar'yakhtar
M.A. Ivanov

Preface

The techniques using diffuse scattering of X-rays or thermal neutrons are the most effective diffraction methods for extracting information on imperfect crystals. These techniques can be employed to study the types of various imperfections in crystals and their structure, the arrangement of atoms of different species and the static displacements on the scale of the interatomic spacing. Recent sophisticated experimental procedures have made these techniques more precise and easier to run and they are now increasingly applied to the analysis of defects in crystals, phase transformations in solids and alloy structures.

The imperfections in crystals may be produced by various nonequilibrium defects, for example, those generated by radiation exposure, straining or hardening of the crystal, or they can appear under thermal equilibrium. The nonequilibrium defects typically have a relatively low concentration. Therefore, the general methods developed for defects of an arbitrary type with a low concentration can be used for analyzing various diffraction effects including diffuse scattering caused by the defects. This analysis was conducted in the monograph *Diffraction of X-rays and Neutrons in Imperfect Crystals* [1.1] and this book is a sequel to it. The monograph [1.1] also treated some equilibrium imperfections, for instance, impurities in dilute solutions and thermal vacancies, but a detailed analysis of diffuse scattering was made only for small defect concentrations.

In many important cases, however, the equilibrium inhomogeneities in crystals are not just individual defects at a large distance from each other but occur throughout the bulk of the crystal and the imperfections may be highly correlated. These inhomogeneities are produced by the fluctuations of the concentration and various internal parameters. Such fluctuational inhomogeneities throughout the entire crystal appear in concentrated solid solutions, in systems with highly fluctuating internal parameters such as the ferroelectrics, in molecular crystals with orientational ordering or other ordered crystals with displacement or substitution disorder and so on. The fluctuational inhomogeneities may be especially significant in the vicinity of second-order phase transitions.

The diffuse scattering by the equilibrium inhomogeneities caused by fluctuations of the composition and the order parameters in a crystal is treated in this book. Such inhomogeneities can be conveniently described as a combination of the fluctuation waves of the concentration and the order parameters (the Fourier components of their fluctuational distributions in space). Then the diffuse scattering at a given angle may be regarded as produced by the respective fluctuation wave and the static displacement wave produced by it. This method of fluctuation waves gives a fairly simple unified framework for analyzing all effects of diffuse scattering (for instance, scattering by alloys, critical scattering, the contributions of the static distortions, the long-range forces, the fluctuations of the correlation parameters and so on). It is also useful for other problems in the theory of imperfect crystals. This is why the method is consistently employed throughout the book.

At first, some general aspects of the theory of fluctuations in solutions and ordered systems are discussed in considerable detail. It is easy then to analyze the peculiar features of diffuse scattering.

The first two chapters give the general theory of fluctuations of the concentration and the order parameters. Chapter 1 presents the phenomenological and microscopic theories of fluctuation waves in systems where the interaction between atoms rapidly becomes weaker with increasing distance and the effect of the long-range forces on the fluctuations can be ignored. The elastic, Coulomb or dipole–dipole long-range interaction in some systems may be fairly strong, however, as can the indirect interaction between atoms mediated by the conduction electrons of the metal, and they may significantly affect the character of the fluctuations, especially in the critical range, giving rise to qualitatively new effects. Sufficiently strong long-range forces also have a significant effect on the equilibrium state of solid solutions and systems with easily variable internal parameters. Therefore, metal and semiconductor systems may have a metastable state as well as a heterogeneous state at thermal equilibrium. Such structures may consist, for instance, of alternating ordered and disordered regions, as discussed in Chap. 2. This discussion is of interest not only for the theory of diffraction but also for some other applications and for the general theory of alloys and one-component ordered systems.

The theoretical results on fluctuations are employed in the next two chapters for analyzing diffuse scattering of X-rays and thermal neutrons in various systems. Chapter 3 treats scattering by fluctuations of the composition and the order parameters, in particular, in metal alloys and one-component ordered crystals and describes the diffraction techniques for studying the correlation parameters and the energy of interaction between atoms in solutions, the Fermi surfaces in alloys, the short-range order structures, and other characteristics of such systems. The critical scattering in systems near second-order phase transitions and critical points on decomposition diagrams is discussed in Chap. 4 which also presents the relevant information on phase transitions and critical fluctuations.

The author is grateful to M.N. Botvinko, D.A. Vul, V.B. Egorova, V.S. Lekhtsier, and L.N. Ryzhova for their assistance in the work on the manuscript.

Contents

1. Fluctuation Waves of the Composition and Order Parameters

The fluctuational inhomogeneities in solid solutions or one-component ordered systems mostly produce relatively small static displacements. Therefore, we can often retain only the linear terms in the expansions in powers of displacements when we calculate the amplitude of diffuse scattering of X-rays or thermal neutrons. Under such circumstances the scattering intensity for a given diffraction vector $\boldsymbol{Q}$ proves to be dependent only on a single Fourier component of the fluctuations and the static displacements related to them, that is, a single fluctuation wave whose wave vector is uniquely determined by $\boldsymbol{Q}$ (by analogy, the intensity of one-phonon thermal diffuse scattering depends on a single wave of thermal vibrations). The method of static fluctuation waves is thus convenient for the analysis of the scattering processes. In the framework of this method the intensity of diffuse scattering in weakly distorted crystals is determined by the probability of the fluctuation wave causing scattering in a specified direction of the diffraction vector. The conventional methods of the theory of fluctuations may be applied to the given thermodynamical or statistical model of the imperfect crystal to perform sufficiently simple calculations of this probability. The calculations must be considerably simpler than for the entire set of the correlation parameters which are needed for direct determinations of the scattering intensity when the scattering events by individual atoms are analyzed (in fact, the method of fluctuation waves also makes possible calculations of the correlation parameters).

The method of fluctuation waves was first employed for analyzing X-ray diffraction in the case of anomalous critical scattering by one-component crystals in the vicinities of the reciprocal lattice points [1.2]. This scattering is caused by the long-wavelength fluctuations of the long-range order which can be studied in the framework of the macroscopic theory of fluctuations. The long-range order parameters and the concentration also experience fluctuations which are not statistically independent in solid solutions and thus give rise to additional effects. The macroscopic theory of fluctuation waves may be used for analyzing diffuse scattering in the vicinities of the reciprocal lattice points in solid solutions when such effects are taken into consideration [1.3]. At large distances from the reciprocal lattice points, scattering is caused by short-wavelength fluctuation waves and diffuse scattering and fluctuation waves in such cases must be treated in the framework of the microscopic theory applied to specific statistical models of solutions [1.4–6]. The method of fluctuation waves is also a fairly simple tool for analyzing the static displacements in solid solutions and their effect on scattering [1.7–14], the effects of the long-range forces on fluctuations and scattering [1.15–19], the scattering caused by the short-range order fluctuations [1.20], and so on.

The fluctuation waves in this chapter are analyzed for circumstances where the effects of the long-range forces can be ignored (the latter will be treated in Chap. 2). Section 1.1 gives a general description of the long-range order and correlation in the ordered solid solutions and analyzes the relationship between the characteristics of the fluctuation waves and the correlation parameters. The macroscopic and microscopic theories of the fluctuation waves of the concentration and the long-range order parameters are given in Sects. 1.2, 3. The fluctuations of the correlation parameters (that occur even in the ideal solutions where the averages of these parameters vanish) are discussed in Sect. 1.4. The fluctuation waves of the concentration and the order parameters always produce waves of static displacements in real crystals which are discussed in Sect. 1.5.

1.1 Order in Solid Solutions and Fluctuations

As the temperature tends to absolute zero a solid solution which is in the thermodynamic equilibrium must tend to a state in which the arrangement of atoms at the crystal lattice sites is fully ordered. This result follows from the Nernst theorem since any macroscopic state which does not correspond to complete order can be realized with a large number of configurations of atoms of various species over the lattice sites. The number W of distinguishable permutations of atoms realizing such configurations and corresponding to the given macroscopic order and correlation parameters is very large while the configurational component $S = k_B \ln W$ of the entropy for such a state is nonvanishing which does not agree with the Nernst theorem for $T \to 0$ (here k_B is the Boltzmann constant). One of the consequences is that a nonstoichiometric solution cannot be in thermodynamic equilibrium at sufficiently low temperatures and it must decompose yielding at $T \to 0$ fully ordered stoichiometric solutions or pure elements. Such decomposition decreases the energy of the system. [This fact is not always taken into consideration in the analysis of the phase diagrams for the ordered solutions at low temperatures. According to the above discussion, the diagrams must include narrow ranges of the ordered phases near the stoichiometric compositions and wide two-phase ranges. For some special models of the interaction between atoms "frustration" occurs at certain concentrations and the energy of the disordered system is precisely equal to the energy of the ordered phases into which the system could have decomposed. This is the case, for instance, for the nearest-neighbour model applied to fcc solid solutions with the concentration $c = 3/8$ [1.21]. But when even a weak interaction at larger distances (between the crystal atoms occupying the lattice sites) is taken into account the above uncertainty is eliminated and a fully ordered structure with a large parameter becomes favored. A considerable number of various ordered phases with large parameters can thus be formed in systems with significant long-range forces.] The above discussion is, of course, applicable only to the equilibrium state while solutions with nonequilibrium order parameters may exist for a long time at low temperatures since the rate of the diffusion processes is low.

As the temperature increases some atoms migrate to "foreign" lattice sites and the solution is no longer fully ordered. A disorder of this kind produces a sharp increase in the number of distinguishable permutations of atoms and a decrease in the configurational entropy term $-k_B T \ln W$ in the free energy so that the solution which is not fully ordered proves to be in thermodynamic equilibrium. Disordered arrangements of atoms over the lattice sites is the most significant type of crystal imperfection for solid solutions and it will be discussed in this and the next chapters. It is primarily the very fact that the sites of the sublattice of a given species are occupied by different atoms (that is, having a difference between their atomic scattering factors) that produces the violation of regularity which causes diffuse scattering of X-rays or neutrons. In addition, the difference between the atomic radii of the solution components in the incompletely ordered solution produces the displacements of atoms from the lattice sites, that is static distortions (see Sect. 1.5).

Description of the Ordered State of a Solution. Let us consider at first the parameters which describe the ordered state of a solution. For the sake of simplicity let us limit the discussion to the case of binary solutions. We shall number the sites in the primitive unit cell with the subscript γ and the unit cells with the subscript s; then $\gamma = 1, 2, \ldots \nu$ and $s = 1, 2, \ldots, N$ where ν is the number of sites in the unit cell and N is the number of unit cells in the crystal. The arrangement of the atoms A and B over the lattice sites of the A-B solution is uniquely described in detail by the set of $N\nu$ variables $c_{s\gamma}$ each of which can have two values:

$$c_{s\gamma} = \begin{cases} 1 & \text{if the site } s\gamma \text{ is occupied by an atom } A \\ 0 & \text{if the site } s\gamma \text{ is occupied by an atom } B \end{cases} . \tag{1.1.1}$$

The subscript γ can be omitted for the disordered solutions with a single atom per unit cell ($\nu = 1$) so that the state of the solution can be described by the set of N numbers $c_s = 1, 0$.

It is clear, however, that not all numbers $c_{s\gamma}$ must be specified for the determination of any parameter of a macroscopic crystal which can be found in experiments. It is enough to know only the averages (over the crystal or the statistical ensemble) of some combinations of these parameters which determine the internal characteristics which describe the order and correlation in the solution. Such internal characteristics are determined primarily by the averages of the numbers $c_{s\gamma}$,

$$c_\gamma = \langle c_{s\gamma} \rangle , \tag{1.1.2}$$

which equal the probabilities that the sites of the sublattice γ are occupied by the atoms A. They uniquely determine the long-range order in the solution. The probabilities that the sites γ are occupied by the atoms B are then $1 - c_\gamma$.

In a disordered solution the quantities c_γ are, obviously, equal to the concentrations c of the atoms A in the solution. But in the ordered solution c_γ are the concentrations of the atoms A at the sites γ which are generally different for different sublattices. In the ordered systems containing more than two lattice sites per primitive unit cell some sites of the cell are reflected onto each other through the symmetry transformations of the crystal, for instance, rotation around a symmetry axis. Such lattice sites are equivalent and have the same probabilities c_γ. The number ν' of various types of the equivalent sites (which differ in the probabilities c_γ) is equal to or smaller than the number ν of atoms in the unit cell. The long-range order in a solution with ν' types of equivalent sites can, apparently, be described either by specifying the numbers c_γ for all sublattices or the mean concentration c and $\nu' - 1$ thermodynamical internal parameters (the long-range order parameters). They are linearly related to c_γ. For instance, if an ordered structure has two types of nonequivalent sites ($\nu' = 2$) and the unit cell contains ν_1 sites of the first type (for atoms A) and ν_2 sites of the second type for atoms B ($\nu_1 + \nu_2 = \nu$) then the probabilities c_1 and c_2 for such a solution can be expressed in terms of the concentration c and the long-range order parameter η:

$$c_1 = c + \frac{\nu_2}{\nu}\eta\,, \quad c_2 = c - \frac{\nu_1}{\nu}\eta\,. \tag{1.1.3}$$

The parameter η is selected to make $\eta = 1$ for the fully ordered solution of the stoichiometric composition $A\nu_1 B\nu_2$ (when $c_1 = 1$ and $c_2 = 0$). In the disordered solution (when $c_1 = c_2 = c$) we have $\eta = 0$.

One of the simplest ordered structures with the sites of two types is a beta-brass lattice (or $L2_0$ lattice, see Fig. 1.1a) in which the sites of the first type are at the centers of the cubic cells in the bcc lattice and the sites of the second type are at their vertices. The primitive unit cell has two sites, the numbers of which are identical ($\nu_1 = \nu_2 = 1$). In another common ordered structure of the type of Cu_3Au ($L1_2$) the sites of the first type are at the vertices of the cubic cells in the fcc lattice and the sites of the second type are at centers of their faces (see Fig. 1.1b). In this lattice the primitive cell has four sites. One is a site of the first type and the other three are equivalent sites of the second type ($\nu_1 = 1$, $\nu_2 = 3$). An example of an ordered structure with three types of lattice sites is the alloy Fe_3Al (see Fig. 1.2) where sites of the first type are at the vertices of the cubic cells in the bcc lattice while the sites of the second and third types are at the centers of the cells making up a lattice of the type of NaCl ($\nu_1 = 2$, $\nu_2 = \nu_3 = 1$). In this DO_3 structure the long-range order is described by two parameters. Other ordered structures, the results for the structures of specific solutions and the theoretical predictions are discussed in various review papers (see, for instance, [1.22–26]).

In some cases the process of ordering involves only the atoms in some of the sublattices while the probabilities that the sites of other sublattices are occupied by atoms of a given species are practically equal to unity. In such cases the unit cell of the disordered solution also contains more than one site. Such are the structures of some interstitial solutions in which the metal atoms at the lattice sites produce a perfectly ordered structure while the interstitial atoms (for instance, hydrogen

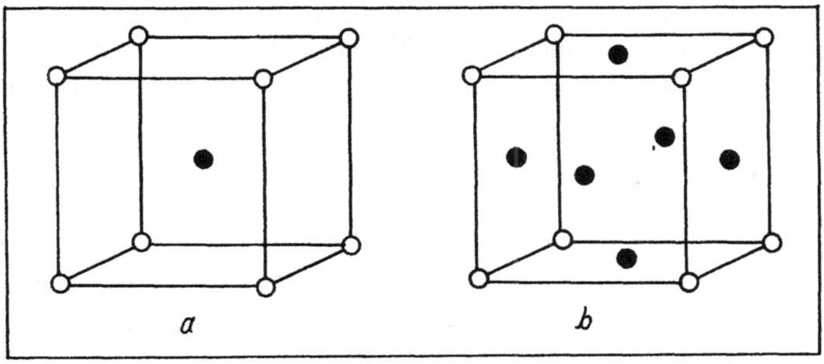

Fig. 1.1. Arrangement of lattice sites in the unit cell of an alloy of the beta brass type (**a**) and Cu_3Au (**b**). Open circles denote sites of the first type and solid circles of the second type

or carbon) are rearranged over the interstices and make up an ordered structure at low temperatures. It is sometimes more convenient to treat formally such processes in the interstitial solutions as ordering in binary solutions whose components are interstitial atoms and vacancies at the interstices that are not occupied by them. The atoms occupying the lattice sites just produce the external regular field and are responsible for the elastic interaction between the interstitial atoms.

An ordered structure is produced when the interaction between atoms of different species is stronger than the interaction between atoms of the same species (otherwise the solution decomposes). This energy factor is especially significant at low temperatures. As the temperature increases the contribution of the entropy term in the free energy grows and the degree of order in the solution diminishes (monotonically, as a rule), At a certain temperature of ordering the phase transition between the ordered state and the disordered state takes place and the long-range order vanishes so that the symmetry of the crystal is increased. But even in the

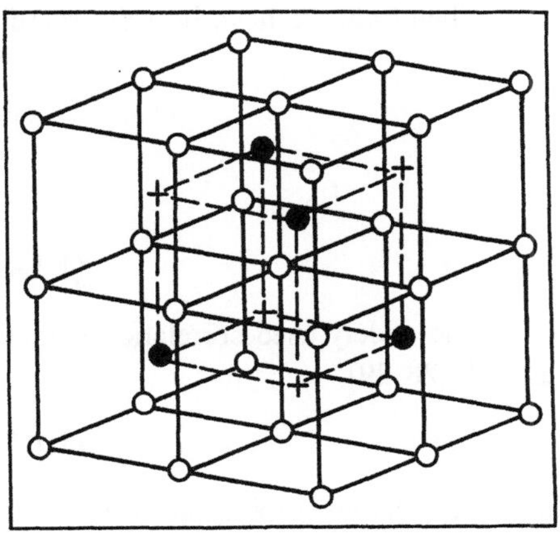

Fig. 1.2. Arrangement of lattice sites in the unit cell of an alloy of the Fe_3Al type. Open circles denote sites of the first type, solid circles sites of the of the second type, and crosses sites of the third type

absence of long-range order a given atom is preferentially surrounded with atoms of a different (or the same) species owing to the difference between the energies of interaction between different atoms. Thus, atoms are not randomly arranged over the lattice sites and there is a short-range order or correlation in the arrangement of atoms of different species over the lattice sites in a solution (a detailed review of theoretical and experimental results on the short-range order in alloys is given in the monograph [1.27]).

In a disordered solution such short-range order is described by the difference between the effective probabilities that a given site (site 1) is occupied by an atom of a given species under the condition that another site (site 2) is occupied by a certain atom, and the *a priori* probabilities for the occupancy of the site 1 which equal the concentrations of the atoms. (The concept of short-range order is applied by some authors only to the disordered solutions in which pairs of atoms of different species are preferentially produced. Here we shall regard short-range order as preferential production of the pairs of atoms of different and identical species.) While the long-range order is determined by the average numbers $c_{s\gamma}$, the correlation is described by the averages of their products.

Various parameters (differing in numerical factors) may be introduced to describe correlation in the arrangement of atoms in a solution. Let us first consider the correlation parameters $\varepsilon(\varrho)$ (in the disordered solution which describe the correlation in the occupancy of a pair of sites at a distance of the lattice vector ϱ. In the context of this discussion it is convenient to define these parameters as the difference between the actual probability $p_{AA}(\varrho)$ that this pair of sites is occupied by atoms A and the probability that their occupancy is in accordance with the random distribution of atoms over the lattice sites, that is, c^2. Since $c_s c_{s'}$ does not vanish only if the sites s and s' are occupied by atoms A we can write $p_{AA}(\varrho)$ as the average of this product $\langle c_s c_{s'}\rangle_\varrho$ where averaging is performed over the pairs of sites with distance of ϱ between them. When the arrangement of atoms is random the occupation probabilities for the sites s and s' are independent of each other and equal c while $\langle c_s c_{s'}\rangle_{(\varrho)} = c^2$. Therefore, we have in a disordered solution

$$\varepsilon(\varrho) = p_{AA}\varrho - c^2 = \langle c_s c_{s'}\rangle_\varrho - c^2 ,$$

or

$$\varepsilon(\varrho) = \langle (c_s - c)(c_{s'} - c)\rangle_\varrho . \tag{1.1.4}$$

The correlation parameters $\varepsilon(\varrho)$ are related to the widely used correlation parameters $\alpha(\varrho)$ introduced by Cowley and Warren [1.28, 29]:

$$\varepsilon(\varrho) = c(1-c)\alpha(\varrho) \quad (\varrho \neq 0) . \tag{1.1.5}$$

For the ordered solutions we can consider either the correlation in the occupancy for the pairs of sites of the given types γ and γ' which is described by the parameters $\varepsilon_{\gamma\gamma'}(\varrho)$ depending on ϱ, γ and γ', or the correlation in the occupancy for all sites in the solution at the distance of ϱ from each other which is described by the parameters $\varepsilon(\varrho)$ dependent only on ϱ. We can extend (1.1.4) and write the following equations for these correlation parameters in the ordered solution:

$$\begin{aligned}\varepsilon_{\gamma\gamma'}(\varrho) &= \langle (c_{s\gamma} - c_\gamma)(c_{s'\gamma'}) \rangle_{\varrho_{\gamma\gamma'}} ,\\ \varepsilon(\varrho) &= \langle (c_{s\gamma} - c_\gamma)(c_{s'\gamma'}) \rangle_{\varrho} .\end{aligned} \tag{1.1.6}$$

Here $\varrho = R_{s'\gamma'} - R_{s\gamma'}$ and $R_{s\gamma}$ is the radius-vector of the lattice site $s\gamma$. In the first equation (1.1.6) averaging is performed over the pairs of sites $s\gamma$ and $s'\gamma'$ at a distance of ϱ from each other in the given sublattices γ and γ' and in the second equation over the pairs of such sites from any sublattices n the crystal.

The correlation in solutions is often described only in terms of the binary parameters $\varepsilon_{\gamma\gamma'}(\varrho)$. One can easily see, however, that these parameters do not determine fully the short-range order in a solution. For instance, the short-range order in a disordered solution can be homogeneous throughout the crystal or the solution can consist of small regions with a high binary order arranged among regions with a low order. The short-range ordering of this type is known as "local order" [1.30]. In both cases of homogeneous and inhomogeneous short-range order the parameters $\varepsilon(\varrho)$ may, in principle, be identical but the states of the system are, apparently, quite different. Such heterogeneous states with a spatially inhomogeneous local order may be produced at thermodynamic equilibrium in systems with significant longrange forces (see Chap. 3) or in nonequilibrium systems at the initial stages of decomposition or other phase transformations.

To distinguish between states with homogeneous and significantly inhomogeneous short-range order and identical correlation parameters $\varepsilon(\varrho)$ (which are found by averaging over the entire crystal) we may specify the parameters which describe correlation in the occupancy not only of pairs but of larger numbers of sites. For instance, a disordered solution may be described with the correlation parameters

$$\begin{aligned}\varepsilon(\varrho_1, \varrho_2) &= \langle (c_s - c)(c_{s'} - c)(c_{s''} - c) \rangle_{\varrho_1, \varrho_2} ,\\ \varepsilon(\varrho_1, \varrho_2, \varrho_3) &= \langle (c_s - c)(c_{s'} - c)(c_{s''} - c)(c_{s'''} - c) \rangle_{\varrho_1, \varrho_2, \varrho_3} ,\end{aligned} \tag{1.1.7}$$

where $\varrho_1 = R_{s'} - R_s$, $\varrho_2 = R_{s''} - R_s$, $\varrho_3 = R_{s'''} - R_s$ and averaging is performed for constant values of these vectors. For large-scale inhomogeneities it may be more convenient to define the order and correlation parameters with the use of averaging over small crystal regions. Then these parameters will depend on the coordinates.

Static Fluctuation Waves of Composition and Order Parameters in Solutions. The order of atom arrangement in solutions is closely related to the fluctuations. For instance, if the short-range order in a disordered solution determines preferential production of the atom pairs of the same species then small regions with a higher concentration of the atoms A or B (in comparison with the average concentrations) appear in the solution, that is, the fluctuations of the composition occur in it. Long-range order fluctuations may be related to the preferential production of the atom pairs of different species.

The quantitative relationship between fluctuations and the correlation parameters can be conveniently found in terms of the Fourier representation. For instance, the parameters $c_{s\gamma}$ which describe uniquely the arrangement of atoms over the lattice sites can be replaced with their Fourier components:

$$c_{s\gamma} - c = \sum_{k} c_{k\gamma} e^{-i\boldsymbol{k}\cdot\boldsymbol{R}_{s\gamma}} , \quad c_{k\gamma} = \frac{1}{N} \sum_{s=1}^{N} (c_{s\gamma} - c_{\gamma}) e^{i\boldsymbol{k}\cdot\boldsymbol{R}_{s\gamma}} . \tag{1.1.8}$$

Since $\exp(-i\boldsymbol{k}\cdot\boldsymbol{R}_{s\gamma})$ remains unchanged under the replacement of $\boldsymbol{k}$ with $\boldsymbol{k}+\boldsymbol{G}$ where $\boldsymbol{G}$ is an arbitrary vector of the reciprocal lattice for the ordered solution we may assume that in (1.1.8) summation is performed over all vectors $\boldsymbol{k}$ which are in the first unit cell of reciprocal lattice and satisfy the cyclic conditions. Obviously, $c^{*}_{k\gamma} = c_{-k\gamma}$ and $c_{k\gamma} = 0$ for $k = 0$.

The sum $\sum_{s}(c_{s\gamma} - c)$ over s cells in a small region in the solution determines the fluctuational variation of the A atom concentration in the sites γ in this region. Therefore, it can be seen from (1.1.8) that for small $\boldsymbol{k}$ the parameters $c_{k\gamma}$ determine the amplitudes of the long fluctuation concentration waves for the sites γ. In the general case of arbitrary $\boldsymbol{k}$ values $c_{k\gamma}$ may also be treated as the Fourier components of the fluctuations of the A atom concentration at the sites γ.

Let us determine the averages $\langle c_{k\gamma} c*_{k\gamma'}\rangle$ of the squares or products of the parameters $c_{k\gamma}$ corresponding to the given $\boldsymbol{k}$. Equation (1.1.8) yields

$$\begin{aligned}\langle c_{k\gamma} c^{*}_{k\gamma'}\rangle &= \frac{1}{N^2} \sum_{s} [\langle (c_{s\gamma} - c_{\gamma})^2 \delta_{\gamma\gamma'} \\ &\quad + \langle (c_{s\gamma} - c_{\gamma})(c_{s'\gamma'} - c_{g'}\rangle e^{i\boldsymbol{k}\cdot(\boldsymbol{R}_{s\gamma} - \boldsymbol{R}_{s'\gamma'})} (1 - \delta_{\gamma\gamma'}, \delta_{ss'})] .\end{aligned} \tag{1.1.9}$$

According to (1.1.1) we have $c^2_{s\gamma} = c_{s\gamma}$, that is, $\langle c^2_{s\gamma}\rangle = c_{\gamma}$ and

$$\langle (c_{s\gamma} - c_{\gamma})^2 \rangle = c_{\gamma}(1 - c_{\gamma}) . \tag{1.1.10}$$

Using (1.1.6) we can rewrite (1.1.9) as

$$\langle c_{k\gamma} c^{*}_{k\gamma'}\rangle = \frac{1}{N} \left[c_{\gamma}(1 - c_{\gamma})\delta_{\gamma\gamma'} + \sum_{\varrho\neq 0} \varepsilon_{\gamma\gamma'}(\varrho) e^{i\boldsymbol{k}\cdot\boldsymbol{\varrho}} \right] . \tag{1.1.11}$$

We see that the mean squared Fourier components of the A atom concentrations in the lattice sites of different types (which determine the Fourier components of the composition and the long-range order parameters) are related by a simple equation to the Fourier components of the correlation parameters $\varepsilon_{\gamma\gamma'}(\varrho)$.

The Fourier series for $c_{s\gamma} - c_{\gamma}$ in (1.1.8) is written separately for each sublattice. In addition to the respective Fourier components $c_{k\gamma}$, we can also make use of the components c_{k} which are defined as the sums over all sites, rather than the sites of the given sublattice only:

$$c_{k} = \frac{1}{N\nu} = \sum_{s=1}^{N} \sum_{\gamma=1}^{\nu} (c_{s\gamma} - c_{\gamma}) e^{i\boldsymbol{k}\cdot\boldsymbol{R}_{s\gamma}} = \frac{1}{\nu} \sum_{\gamma=1}^{\nu} c_{k\gamma} . \tag{1.1.12}$$

Using (1.1.6) for $\varepsilon(\varrho)$ and (1.1.10) we can express the mean squared Fourier components $\langle|c_{\boldsymbol{k}}|^2\rangle$ in terms of the Fourier components of the correlation parameters $\varepsilon(\varrho)$:

$$\langle|c_{\boldsymbol{k}}|^2\rangle = \frac{1}{N\nu}\left[\frac{1}{\nu}\sum_{\gamma=1}^{\nu} c_\gamma(1-c_\gamma) + \sum_{\varrho\neq 0}\varepsilon(\varrho)\cos\boldsymbol{k}\cdot\varrho\right] . \tag{1.1.13}$$

We have used here the fact that $\varepsilon(-\varrho) = \varepsilon(\varrho)$.

For instance, for the disordered solutions with a single site per unit cell ($\nu = 1$, $c_{\boldsymbol{k}1} = c_{\boldsymbol{k}}$) we obtain

$$\langle|c_{\boldsymbol{k}}|^2 = \frac{1}{N}\left[c(1-c) + \sum_{\varrho\neq 0}\varepsilon(\varrho)\cos\boldsymbol{k}\cdot\varrho\right] . \tag{1.1.14}$$

If the arrangement of atoms over the lattice sites is not correlated ($\varepsilon(\varrho) = 0$), that is, for the ideal solution, we have

$$\langle|c_{\boldsymbol{k}}|^2\rangle = \frac{1}{N}c(1-c) \tag{1.1.15}$$

and $\langle|c_{\boldsymbol{k}}|^2\rangle$ does not depend on the wave vector $\boldsymbol{k}$. In the general case of nonideal solution, however, $\langle|c_{\boldsymbol{k}}|^2\rangle$ may strongly depend on $\boldsymbol{k}$.

Note that a simple integral relation (the summation rule) for $\langle|c_{\boldsymbol{k}}|^2\rangle$ is satisfied in this general case. Indeed, if we take sums of equations (1.1.13) or (1.1.14) over all $\boldsymbol{k}$ which are in any primitive unit cell of the reciprocal lattice of the disordered solution then we obtain

$$\sum_{\boldsymbol{k}}\langle|c_{\boldsymbol{k}}|^2\rangle = \frac{1}{\nu}\sum_{\gamma=1}^{\nu} c_\gamma(1-c_\gamma) = c(1-c) . \tag{1.1.16}$$

since $\sum_{\boldsymbol{k}}\cos\varrho\cdot\boldsymbol{k} = 0$ for $\varrho \neq 0$ and $\sum_{\boldsymbol{k}} 1 = N\nu$. The second equation (1.1.16) is written for the disordered solutions. We see that the mean squares $\langle|c_{\boldsymbol{k}}|^2\rangle$ for the cells of the reciprocal lattice are determined only by the long-range order parameters and the solution concentration.

Sometimes it is more convenient to consider, instead of the components $c_{\boldsymbol{k}}$ the Fourier components $c'_{\boldsymbol{k}}$ of the differences $c_{s\gamma} - c$ referred to the averaged solution concentration:

$$c'_{\boldsymbol{k}} = \frac{1}{N\nu}\sum_{s=1}^{N}\sum_{\gamma=1}^{\nu}(c_{s\gamma} - c)e^{i\boldsymbol{k}\cdot\boldsymbol{R}_{s\gamma}} . \tag{1.1.17}$$

Equations (1.1.12) and (1.1.17) differ from each other (in linear functions of the long-range order parameters) only in the ordered solutions for $\boldsymbol{k} = \boldsymbol{G}'$ where $\boldsymbol{G}'$ are the reciprocal lattice vectors produced by ordering. The (not necessarily mean) squared components (1.1.17) also satisfy a simple summation rule which is valid for arbitrary correlation and order parameters. Finding the square of (1.1.17) and performing summation over the primitive unit cell of the reciprocal lattice of the disordered solution we obtain

$$\sum_{k} |c'_{k}|^2 = c(1-c) . \tag{1.1.18}$$

The difference between (1.1.16) and (1.1.18) for the ordered solutions depends on the terms with $\boldsymbol{k} = \boldsymbol{G}'$.

The above results demonstrate that the correlation and the fluctuations of the concentration and the long-range order parameters are indeed definitely related. When we know all correlation parameters for the solution we can find the mean squares and products of various Fourier components of the fluctuations. On the contrary, when $\langle |c_{\boldsymbol{k}}|^2 \rangle \gg c^{*}_{\boldsymbol{k}\gamma'} \rangle$ are known for all wave vectors $\boldsymbol{k}$ which lie in the unit cell of the reciprocal lattice space then $\varepsilon_{\gamma\gamma'}(\boldsymbol{\varrho})$ for various $\boldsymbol{\varrho}$ values can be found from the Fourier transformation of equation (1.1.11). It is enough therefore to find only one set of parameters $\varepsilon_{\gamma\gamma'}(\boldsymbol{\varrho})$ or $\langle c_{\boldsymbol{k}\gamma} c^{*}_{\boldsymbol{k}\gamma'} \rangle$ (particularly, for the analysis of diffuse scattering whose intensity can be directly expressed in terms of these parameter for a given $\boldsymbol{k}$ value).

1.2 Thermodynamic Theory for Fluctuation Waves of Concentration and Long-Range Order Parameters in Undistorted Crystals

The probabilities of fluctuations in a small volume element or the probabilities of fluctuation waves can be calculated in the framework of the macroscopic or microscopic theories.

The macroscopic theory treats fluctuations in the volume elements which are small in comparison with the volume of the entire crystal but large in comparison with the atomic volumes. The surface effects can be ignored for such volume elements (if the long-range forces are not significant) and the specific thermodynamic functions in them do not differ from the respective functions in the massive bodies. The fluctuation probabilities then are calculated without any concrete model in the framework of the thermodynamic theory of fluctuations. These probabilities are expressed in terms of the derivatives of the thermodynamic potential which can be obtained from the independent experimental results in some cases.

An obvious advantage of the macroscopic theory of fluctuations consists in that it employs no simplified crystal models, is quite general and at the same time sufficiently uncomplicated. On the other hand, the results of this theory are valid only for volume elements which are much larger than the atomic volumes. The fluctuation waves in the macroscopic theory can accordingly be treated when the wavelength is much larger than the distance between atoms.

In some cases, however, for instance when we consider scattering of X-rays or thermal neutrons, in addition to the smoothly varying fluctuations which determine the intensity of diffuse scattering in the vicinities of the reciprocal lattice points, we must also take into account the fluctuations which sharply vary in space and correspond to the fluctuation waves with wavelengths of the order of the lattice parameter $\boldsymbol{d}$. The probabilities of such fluctuation waves can be calculated only in terms of the microscopic theory based on a specific model which is necessarily simplified. At the same time, with a specific model we can not only treat the short-wave fluctuations but also, in principle, calculate and analyze the temperature and concentration dependences for the thermodynamic parameters which determine the fluctuation probabilities in the macroscopic theory (and are not always directly obtainable from the independent experimental results).

We see that either the macroscopic or the microscopic theory is convenient for treating a specific problem and therefore we shall discuss both here.

This section will present the macroscopic theory of the fluctuation waves for those cases when we can ignore the longrange forces. First, we shall consider the concentration fluctuations and then the long-range order fluctuations. The microscopic theory of the fluctuation waves is given in Sect. 1.3.

1.2.1 Concentration Fluctuations

Let us consider the concentration fluctuations in a small volume element V' of a binary solid solution $A-B$ in which the lattice parameter is not significantly affected by the composition. Such fluctuations cause a change in the numbers of the atoms A and B while the total number V' of atoms in this volume remains constant. The remaining part of the crystal (with the volume $V \gg V'$) is then regarded as the environment which can exchange atoms with the given volume element V'.

The Boltzmann principle in the thermodynamic theory of fluctuations states that the probability w of a fluctuation being realized depends exponentially on the least work $\mathfrak{R}$ required for its production (see [1.31]):

$$w \sim \exp\left(-\frac{\mathfrak{R}}{k_{\mathrm{B}}T}\right). \tag{1.2.1}$$

A fluctuation involves a nonequilibrium state of the system and a transition to this state can be realized via different pathways. The external forces producing the fluctuation perform different work when different mechanisms are involved in realizing fluctuations of the given physical properties. The least work $\mathfrak{R}$ which determines the fluctuation probability according to (1.2.1) corresponds to the variation of the physical properties from the equilibrium values to the fluctuational values which is caused reversibly by an external source of work. It is significant here that the system is in the most complete state of equilibrium corresponding to these changed values of the fluctuating properties under consideration. For example, when we consider the concentration fluctuations the least work is calculated for the equilibrium values of the internal parameters (the long-range order parameters and correlation parameters) corresponding to the changed concentration.

If the temperature and volume, or the temperature and pressure, remain unchanged when a fluctuation occurs in a one-component solution then the least work equals respectively the difference between the free energies, or the thermodynamic potentials, of the states corresponding to the fluctuational and equilibrium values of the physical properties under consideration. (The meaning of the free energy of the nonequilibrium fluctuational state is discussed in detail by Leontovich; see Sect. 31 in [1.32].) The following equation can be written for the least work producing the fluctuational change in the solution concentration in the case when fluctuations do not cause distortions (see Sect. 115 in [1.31]):

$$\mathfrak{R} = \delta\Phi' - \mu_A \delta N_A - \mu_B \delta N_B = V'\delta\varphi - N'(\mu_A - \mu_B)\delta c\,. \tag{1.2.2}$$

Here $\delta\Phi' = \Phi'(c + \delta c) - \Phi'(c)$ is the variation of the thermodynamic potential of the volume element under consideration caused by the concentration variation of δc, μ_A and μ_B are the chemical potentials of the atoms A and B, and φ is the thermodynamic potential of the unit volume.

The definitions $\mu_A = \left(\dfrac{\partial\Phi}{\partial N_A}\right)_{PTN_B}$ and $\mu_B = \left(\dfrac{\partial\Phi}{\partial N_B}\right)_{PTN_A}$

for the chemical potentials (where Φ, N_A and N_B are given for the entire crystal) yield the obvious identity

$$\mu_A - \mu_B = \frac{1}{N_0}\left(\frac{\partial\Phi}{\partial c}\right)_{PTN_0} v_0\frac{d\varphi}{dc}\,, \tag{1.2.3}$$

where $N_0 = N\nu$ is the number of atoms in the crystal, $v_0 = V/N_0$ is the atomic volume and the derivative $d\varphi/dc$ is found for the constant T and P. If we expand $\delta\varphi$ in powers of δc and retain only the quadratic terms we find from (1.2.2, 3) that the least work

$$\mathfrak{R} = \frac{1}{2}V'\frac{d^2\varphi}{dc^2}(dc)^2 \tag{1.2.4}$$

is proportional to $(\delta c)^2$, that is, the probability distribution (1.2.1) of the concentration fluctuations has a Gaussian shape. Hence, the mean squared concentration fluctuations for a volume element are determined by the second derivative of the thermodynamic potential with respect to the concentration:

$$\langle(\delta c)^2\rangle = \frac{k_B T}{V'}\frac{1}{\frac{d^2\varphi}{dc^2}}\,. \tag{1.2.5}$$

To determine the fluctuation wave probability we must consider the fluctuational variation of the concentration throughout the entire crystal, rather than in the given volume element. In the macroscopic approximation a continuous function $\delta c(\boldsymbol{r})$ describes the concentration fluctuations at various points of the solution. The least work needed for producing the fluctuation distribution $\delta c(\boldsymbol{r})$ is now the functional $\delta c(\boldsymbol{r})$ which is determined by (1.2.2) integrated over the entire crystal volume

V. If the total numbers of the atoms A and B in the crystal are constant and fluctuations consist only in redistribution of the concentration between various parts of the crystal ($\int \delta c(\boldsymbol{r})d\boldsymbol{r} = 0$) then this integral is

$$\mathfrak{R} = \int \delta\varphi d\boldsymbol{r} \tag{1.2.6}$$

Since the concentration distribution is not homogeneous we must retain in the expansion of $\delta\varphi$ not only the terms with various powers of the function $\delta c(\boldsymbol{r})$ but also the terms with its derivatives with respect to the coordinates x_i. The terms with the first derivatives $\partial c/\partial x_i$ and such products as $[\delta c(\boldsymbol{r})](\partial c/\partial x_i) = (1/2)\{\partial[\delta c(\boldsymbol{r})]^2/\partial x_i\}$ will be eliminated after integration over the entire crystal (we shall introduce cyclic boundary conditions for the sake of simplicity). In the case of smoothly varying (long-wave) fluctuations discussed here we may ignore the terms with the derivatives of the third and higher order. Then only terms of the type of $(\partial c/\partial x_i)(\partial c/\partial x_j)$ out of the terms containing derivatives may be retained (in addition to the terms of the type of $\delta c(\boldsymbol{r})(\partial^2 c/\partial x_i \partial x_j)$ that can be reduced to the former after integration). Then we obtain the expansion

$$\delta\varphi = \frac{1}{2}\frac{d^2\varphi}{dc^2}(\delta c)^2 + \frac{1}{2}\beta_{ij}\frac{\partial c}{\partial x_i}\frac{\partial c}{\partial x_j} , \tag{1.2.7}$$

where β_{ij} is the symmetric tensor of the second rank, i and j denote the Cartesian coordinates, and summation is implied over repeated subscripts here and below.

According to the estimates made in Sect. 1.3, β_{ij} are of the order of $k_B T_0/r_0$ where T_0 is the temperature of ordering (or decomposition) of the solution and r_0 is the distance between atoms. The number of independent components of the tensor β_{ij} depends on the crystal symmetry. In cubic crystals it reduces to the scalar ($\beta_{ij} = \beta\delta_{ij}$) and the second term in (1.2.7) then has the simple form of $\beta(\nabla c)^2/2$.

The continuous concentration distribution $\delta c(\boldsymbol{r})$ in the crystal volume V can be expanded in the Fourier series:

$$\delta c(\boldsymbol{r}) = \sum_{\boldsymbol{k}} c_{\boldsymbol{k}} e^{-i\boldsymbol{k}\cdot\boldsymbol{r}}, \quad c_{\boldsymbol{k}} = c^*_{-\boldsymbol{k}} = \frac{1}{V}\int \delta c(\boldsymbol{r}) e^{i\boldsymbol{k}\cdot\boldsymbol{r}} d\boldsymbol{r} . \tag{1.2.8}$$

A comparison of (1.2.8) and (1.1.12) indicates that the expressions for $c_{\boldsymbol{k}}$ coincide for small $\boldsymbol{k}$ when the sum over s_γ in (1.1.12) can be replaced with the integral. If the distribution $\delta c(\boldsymbol{r})$ corresponds to a given fluctuation wave, that is, if

$$\delta c(\boldsymbol{r}) = c_{\boldsymbol{k}} \exp(-i\boldsymbol{k}\cdot\boldsymbol{r}) + c_{-\boldsymbol{k}} \exp(i\boldsymbol{k}\cdot\boldsymbol{r}) ,$$

then we obtain from (1.2.6, 7) the least work needed for producing this fluctuation:

$$\mathfrak{R} = V\left(\frac{d^2\varphi}{dc^2} + \beta_{ij}k_i k_j\right) c_{\boldsymbol{k}} c^*_{\boldsymbol{k}} . \tag{1.2.9}$$

If we write the complex quantities $c_{\boldsymbol{k}}$ as $c_{\boldsymbol{k}} = c'_{\boldsymbol{k}} + ic''_{\boldsymbol{k}}$ then we can see from (1.2.1, 9) that the fluctuation wave probabilities are Gaussian functions of the real and imaginary parts of their amplitudes. Then the mean squared kth Fourier component of the concentration fluctuation is given by

$$\langle |c_{\boldsymbol{k}}|^2 \rangle = \frac{k_B T}{V} \frac{1}{\frac{d^2\varphi}{dc^2} + \beta_{ij} k_i k_j} . \tag{1.2.10}$$

For instance, for undistorted cubic crystals we have

$$\langle |c_{\boldsymbol{k}}|^2 \rangle = \frac{k_B T}{V} \frac{1}{\frac{d^2\varphi}{dc^2} + \beta k^2} . \tag{1.2.11}$$

Equations (1.2.10, 11) have been derived from the macroscopic treatment of fluctuations and they are valid only for long fluctuation waves when

$$k r_0 \ll 1 \tag{1.2.12}$$

and the higher-order derivatives in the expansion (1.2.7) can be ignored.

The derivative $d^2\varphi/dc^2$ tends to zero as we approach the critical point of decomposition for the solution so that high-intensity critical scattering occurs near this point (see Sect. 4.2). The second term in the denominator of (1.2.10) or (1.2.11), $\beta_{ij} k_i k_j \sim k_B T_0 v_0^{-1} (k r_0)^2$, is considerably smaller than the first term outside the critical range for $k r_0 \ll 1$. Therefore, the second term can be ignored in the long-wave case and then the mean square fluctuations are fully dependent on the thermodynamic characteristic $d^2\varphi/dc^2$. As can be seen from (1.2.3), the latter can be found from the experimental thermodynamic results on the concentration dependence of the chemical potentials. Since $\partial\mu_A/\partial c$ and $\partial\mu_B/\partial c$ obey the Gibbs-Duhem equation it is, in fact, enough to know one of them only. The second term in the denominator of (1.2.10) or (1.2.11) is significant in the critical range (see Sect. 4.2).

Since expressions (1.2.10) and (1.1.13) describe fluctuations in terms of the thermodynamic characteristics and the correlation parameters, respectively, they clearly must be equivalent. Writing the equation for these two expressions for the limiting case of $k \to 0$ we obtain the general formula

$$\frac{1}{\nu} \sum_{\gamma=1}^{\nu} c_\gamma (1 - c_\gamma) + \sum_{\varrho \neq 0} \varepsilon(\varrho) = \frac{k_B T}{v_0} \left(\frac{d^2\varphi}{dc^2} \right)^{-1} , \tag{1.2.13}$$

which relates the purely thermodynamic quantity $d^2\varphi/dc^2$, on the one hand, and the parameters $\varepsilon(\varrho)$ describing the correlation in the arrangement of the neighbouring atoms, that is, the microscopic characteristics of the solution, on the other hand.

1.2.2 Fluctuations of Long-Range Order Parameters in One-Component Solutions

In Sect. 1.1 we discussed the long-range order in solid solutions associated with the difference between the occupation probabilities for the atoms (ions) of the given species in various sublattices. The long-range order may also exist in the one-component crystals which consist of "molecules" of only one type. Under certain conditions such long-range order is associated with the orientational ordering. For instance, in ammonium halides NH_4Cl or NH_4Br the ammonium ions may be aligned producing a structure of the ferromagnetic or antiferromagnetic type. In other cases a change in the long-range order involves displacement of ions. Such ordering occurs in perovskite-type crystals, for instance, the ferroelectric $BaTiO_3$.

The long-range order can be described by a single parameter η (as in alloys of the type of beta brass) or it can be complex in nature and described by a set of n parameters η_μ(where $\mu = 1, 2, \ldots, n$). The symmetry of a crystal is decreased by the phase transition to the ordered state involving the displacements of ions as well as the preferential occupation of sublattice sites by atoms of a given species and the orientational ordering. Under transformations of the symmetry group of the disordered phase the order parameters are transformed in one of its irreducible representations, The dimensionality of this representation gives the number of components n of the order parameters.

Typically a small number of the parameters η_μ is needed for describing the long-range order in a system in thermodynamic equilibrium. The variations of a very large number of the long-range order parameters must generally be specified for describing a fluctuational variation of the long-range order. The reason for this is that the fluctuations in small volume elements cause nonvanishing local variations $\delta\eta_\mu$ even for those order parameters for which the phase under consideration is not ordered and has the the mean equilibrium values $\eta_\mu = 0$.

In specific cases, however, it is sufficient to know the fluctuations of only some of the order parameters. For example, when we consider scattering of X-rays or neutrons we must take into account the fluctuations of only those parameters η_μ which make a significant contribution to the scattering amplitude for the given transferred momentum $\boldsymbol{Q}$ which is in the vicinity of the reciprocal lattice point (where the macroscopic treatment is valid; see Sect. 3.6). The number of such parameters, as well as the number of the equilibrium order parameters, is small.

As follows from the general equations of statistical physics (see, for instance, [1.31], the fluctuations of the above small group of the long-range order parameters can be treated independently, separately from the fluctuations of other order parameters and other inner parameters of the crystal system. We have only to calculate the least work needed for producing fluctuations of the given parameters η_μ that is, the second-order derivatives $\partial^2\varphi/\partial\eta_\mu\partial\eta_{\mu'}$ corresponding to the changed values of the other inner parameters under the incomplete equilibrium for the given fluctuating parameters.

Landau [1.2] analyzed the long-range order fluctuations by considering X-ray scattering near a second-order phase transition in the simplest case of one-component crystals described by a single long-range order parameter. At thermodynamic equilibrium the long-range order parameter for such a crystal has the mean value of η found from the equilibrium condition

$$\frac{\partial\varphi}{\partial\eta} \equiv \partial_\eta = 0\,. \tag{1.2.14}$$

Fluctuations cause deviations from this mean value which can be regarded as steadily varying functions $\delta\eta(\boldsymbol{r})$ of the coordinates in the macroscopic approximation. As in the case of the concentration fluctuations, the least work required for producing the variation $\delta\eta(\boldsymbol{r})$ is clearly given by (1.2.6) where, according to (1.2.14), we have

$$\delta\varphi = \frac{1}{2}\varphi_{\eta\eta}(\delta\eta)^2 + \frac{1}{2}\alpha_{ij}\frac{\partial\eta}{\partial x_i}\frac{\partial\eta}{\partial x_j}\,. \tag{1.2.15}$$

Here, $\varphi_{\eta\eta} = \partial^2\varphi/\partial\eta^2$ is the second derivative found, as noted above, for the inner parameters adjusted with respect to the variation of η. Since (1.2.5) differs from (1.2.7) only in the replacement of c with η' the probabilities of the fluctuation waves of the long-range order parameter,

$$\delta\eta(\boldsymbol{r}) = \eta_{\boldsymbol{k}}\,\exp(\mathrm{i}\boldsymbol{k}\cdot\boldsymbol{r}) + \eta_{-\boldsymbol{k}}\,\exp(\mathrm{i}\boldsymbol{k}\cdot\boldsymbol{r})$$

can be found from the treatment similar to that applied above to the concentration fluctuations. Then we can calculate the mean-squared $\boldsymbol{k}$th Fourier component of the fluctuations of the long-range order parameter:

$$\langle|\eta_{\boldsymbol{k}}|^2\rangle = \frac{k_{\mathrm{B}}T}{V}\,\frac{1}{\varphi_{\eta\eta} + \alpha_{ij}k_ik_j}\,. \tag{1.2.16}$$

For sufficiently small $\boldsymbol{k}$ the mean square (1.2.16) is determined by the thermodynamic characteristic $\varphi_{\eta\eta}$.

The results obtained by Landau can be easily extended to a more general case when the long-range order in the one-component crystals is described by several parameters η_μ (where $\mu = 1, 2, \ldots, n$). The variation $\delta\varphi$ of the thermodynamic potential caused by the long-range order fluctuations $\delta\eta_\mu$ under the equilibrium condition of $\partial\varphi/\partial\eta_\mu = 0$ reduces to the positive definite quadratic form of all variables $\delta\eta_\mu$:

$$\delta\varphi = \frac{1}{2}\sum_{\mu,\mu'}^{n}\left[\varphi_{\mu\mu'}\delta\eta_\mu\delta\eta_{\mu'} + \alpha_{\mu\mu' ij}\frac{\partial\eta_\mu}{\partial x_i}\frac{\partial\eta_{\mu'}}{\partial x_j}\right], \tag{1.2.17}$$

where

$$\varphi_{\mu\mu'} \equiv \frac{\partial^2\varphi}{\partial\eta_\mu\partial\eta_{\mu'}}\,.$$

Since $\delta\varphi$ is invariant under the symmetry transformations of the crystal with a single scalar order parameter the quantities $\alpha_{\mu\mu' ij} = \alpha_{ij}$ (see (1.2.15)) are the components of the symmetric second-rank,tensor which degenerates to a scalar in the cubic crystal. But when there are several order parameters the symmetry transformations relate to each other the quantities $\alpha_{\mu\mu' ij}$ with different μ and μ' (not only with different i and j so that $\alpha_{\mu\mu' ij}$ do not reduce to scalars even for cubic crystals.

If the variations $\delta\eta_\mu$ of the order parameter correspond to fluctuation waves with a given wave vector $\boldsymbol{k}$, that is, if

$$\delta\eta_\mu(\boldsymbol{r}) = \eta_{\mu \boldsymbol{k}} \exp(-i\boldsymbol{k}\cdot\boldsymbol{r} + c.c. ,$$

then, according to (1.2.1, 6, 17), the probabilities of the amplitudes of these waves are described by a Gaussian distribution

$$w \sim \exp\left[-\frac{V}{k_\mathrm{B}T}\sum_{\mu,\mu'=1}^{n} \varphi_{\boldsymbol{k}\mu\mu'}(\eta'_{\mu\boldsymbol{k}}\eta'_{\mu'\boldsymbol{k}} + \eta''_{\mu\boldsymbol{k}}\eta''_{\mu'\boldsymbol{k}})\right] , \tag{1.2.18}$$

where

$$\eta_{\mu\boldsymbol{k}} = \eta'_{\mu\boldsymbol{k}} + i\eta''_{\mu\boldsymbol{k}} \; ; \quad \varphi_{\boldsymbol{k}\mu\mu'} = \varphi_{\mu\mu'} + \alpha_{\mu\mu' ij}k_ik_j \; . \tag{1.2.19}$$

In the general case the derivatives $\varphi_{\mu\mu'}$ with $\mu \neq \mu'$ are nonvanishing, that is, the fluctuations $\eta_{\mu\boldsymbol{k}}$ of different order parameters generally are not statistically independent and the probability distribution (1.2.18) does not break down into a product of distributions corresponding to different μ'. It has been shown (see Sect. 111 in [1.31]) that in this general case the averages $\langle\eta'_{\mu\boldsymbol{k}}\eta'_{\mu'\boldsymbol{k}}\rangle$ (or $\langle\eta''_{\mu\boldsymbol{k}}\eta''_{\mu'\boldsymbol{k}}\rangle$) can be found by inverting the matrix of the nth order comprised of the coefficients at $\eta'_{\mu\boldsymbol{k}}\eta_{\mu'\boldsymbol{k}}$. We have

$$\frac{1}{2}\langle(\eta_{\mu\boldsymbol{k}}\eta^*_{\mu'\boldsymbol{k}} + \eta^*_{\mu\boldsymbol{k}}\eta_{\mu'\boldsymbol{k}})\rangle = \frac{k_\mathrm{B}T}{V}\varphi^{-1}_{\boldsymbol{k}\mu\mu'} \; . \tag{1.2.20}$$

Here $\varphi^{-1}_{\boldsymbol{k}\mu\mu'}$ are the elements of the inverted matrix $\varphi_{\boldsymbol{k}\mu\mu'}$. In the limiting case of $k \to 0$ the matrix $\varphi^{-1}_{\boldsymbol{k}\mu\mu'}$ coincides with the matrix $\varphi^{-1}_{\mu\mu'}$ which is the inverted matrix of the second-order derivatives $\varphi_{\mu\mu}$. (For crystals belonging to certain symmetry classes in which atoms are never centers of inversion the expansion in powers of the derivatives $\partial\eta_{\mu'}/\partial x_i$ may include the linear terms, in addition to the quadratic terms [1.31, 33]. The linear terms appear in the expansion of $\delta\varphi$ in the combinations

$$\frac{1}{2}\delta_{\mu\mu' i}\left(\delta\eta_\mu\frac{\partial\eta_{\mu'}}{\partial x_i} - \delta\eta_{\mu'}\frac{\partial\eta_\mu}{\partial x_i}\right)$$

(the symmetric combinations such as $\delta\eta_\mu\frac{\partial\eta_{\mu'}}{\partial x_i} + \delta\eta_{\mu'}\frac{\partial\eta_\mu}{\partial x_i} = \frac{\partial}{\partial x_i}(\delta\eta_\mu\delta\eta_{\mu'})$ will disappear after integration over the entire crystal). Then the exponential function (1.2.18) for the probability distribution will include the additional terms.

The averages $\langle\eta'_{\mu\boldsymbol{k}}\eta'_{\mu'\boldsymbol{k}}\rangle$ and $\langle\eta''_{\mu\boldsymbol{k}}\eta''_{\mu'\boldsymbol{k}}\rangle$ here are determined by inversion of the matrix of the order $2n$, rather than n as in the above case, and the averages

$\langle \eta'_{\mu k} \eta''_{\mu' k} \rangle = 0$ are proportional to $\boldsymbol{k}$, in contrast to the case when the terms linear in the derivatives vanish. We shall write here more complicated equations for this case. Note, however that in the limit of $k \to 0$ when $\varphi^{-1}_{k\mu\mu'}$ coincides with $\varphi^{-1}_{\mu\mu'}$ equation (1.2.20) remains valid.)

Generally, in contrast to $d^2\varphi/dc^2$ the second-order derivatives of $\varphi_{\mu\mu'}$ with respect to the order parameters cannot be expressed in terms of the quantities independently obtained from the experimental results but must be found theoretically. If the order is produced by a second-order phase transition then in the vicinity of the transition the temperature dependences of $\varphi_{\mu\mu'}$ are given by the well-known equations from the theory of phase transitions (see Sect. 4.1). In crystals where either the order is produced by a first-order phase transition or we are far from the second-order transition, the functions $\varphi_{\mu\mu'}$ may be approximately expressed in terms of the ordering energies in the given statistical ordering model (see Sect. 1.3).

In some cases of interest the derivatives $\varphi_{\mu\mu'}$ (or $\varphi_{k\mu\mu'}$) can be related to the quantities which are found directly from the independent experimental data (with the use of the generalized susceptibilities). Such relations can be derived for the proper ferroelectrics in which the long-range order parameters can be the components (all three or just one or two) of the spontaneous polarization vector $\boldsymbol{P}$, for the ferromagnetics where the long-range order parameters can be the components of the magnetization vector $\boldsymbol{M}$ (three components for the isotropic Heisenberg ferromagnetic or one component for the strongly anisotropic Ising ferromagnetic) and ferroelastics in which the long-range order is described by the components of the strain tensor u_{ij}. (For the crystals belonging to the large class of improper ferroelectrics the order parameters are not reduced to the components of the polarization vector but are described by another irreducible representation [1.34, 35]. For such structures $\varphi_{\mu\mu'}$ cannot be expressed in terms of the dielectric susceptibility.) Long-range dipole-dipole or elastic forces are effective in all such cases and they must be taken into consideration in the rigorous treatment of the fluctuations (see Sects. 2.1, 2). But in ferroelectrics with high concentrations of charges (conduction electrons or ions) the dipole–dipole forces are shielded (see Sect. 2.2) and the simpler equations derived in this section in the absence of long-range forces can be used for calculating the probabilities of the long-wave fluctuations. The relatively weak dipole–dipole interaction in ferromagnetics typically does not manifest itself outside a narrow range around the Curie point. We shall assume, therefore, that the long-range forces can be ignored, for instance, owing to shielding, and leave the discussion of the long-range forces until Chap. 2. Then we can easily derive, for instance, for the proper ferroelectrics the relation between the components $\varkappa_{ij}$ of the dielectric susceptibility tensor and the second-order derivatives $\partial^2\varphi/\partial P_i \partial P_j$ with respect to the long-range order parameters (which are the components P_i of the polarization vector) [1.10, 36].

Note that in the external electric field E the thermodynamic potential φ of the unit volume treated as a function of an internal parameter, namely, the polarization $\boldsymbol{P}$, and the field strength $\boldsymbol{E}$ can be written as

$$\varphi(T, \boldsymbol{E}, \boldsymbol{P}) = \varphi(\boldsymbol{P}) - \boldsymbol{P} \cdot \boldsymbol{E} - \frac{E^2}{8\pi} , \tag{1.2.21}$$

where $\varphi(\boldsymbol{p})$ do not depend on $\boldsymbol{E}$ (see Sect. 43 in [1.37] or Sect. 39 in [1.38]). The equilibrium condition $\partial\varphi/\partial\boldsymbol{P} = 0$ indicates that the equilibrium electric field corresponding to the given polarization $\boldsymbol{P}$ is

$$E_i = \frac{\partial\varphi(\boldsymbol{P})}{\partial P_i} . \tag{1.2.22}$$

Differentiating this equation with respect to P_j and using the definition of the electric susceptibility tensor, $\varkappa_{ij} = \partial P_i/\partial E_j$ we obtain a relation between $\partial^2\varphi/\partial P_i \partial P_j$ and $\varkappa_{ij}$:

$$\varphi_{ij} \equiv \frac{\partial^2\varphi}{\partial P_i \partial P_j} = \varkappa_{ij}^{-1} . \tag{1.2.23}$$

Here $\varkappa_{ij}^{-1}$ are the components of the reversed tensor $\varkappa_{ij}$. When we substitute (1.2.23) into (1.2.20) we find that in the case of complete shielding of the field E for $k \to 0$ the mean squares and products of the Fourier components of the polarization fluctuations are

$$\frac{1}{2}\langle P_{ki}P_{kj}^* + P_{ki}^*P_{kj}\rangle = \frac{k_B T}{V}\varkappa_{ij} , \tag{1.2.24}$$

that is, they indeed are expressed in terms of the quantities found from the independently obtained experimental data. (Since thermal fluctuations practically do not cause high-frequency electron polarization the meaning of the quantities $\varkappa_{ij}$ in (1.2.24) is changed when we replace the above classical treatment with the approach based on the quantum theory of fluctuations. These quantities then are the differences between the static dielectric susceptibility $\varkappa_{ij}$ and the high-frequency susceptibility $\varkappa_{\infty ij}(\boldsymbol{k})$; see Sect. 2.2.) Similar equations can be written for the magnetization fluctuations in the absence of long-range forces where $\varkappa_{ij}$ must be replaced with the magnetic susceptibility tensor χ_{ij}.

1.2.3 Fluctuations of Long-Range Order Parameters and Concentration in Solid Solutions

Typically, in solid solutions the fluctuations of the long-range order parameters occur together with the concentration fluctuations. The fluctuations of the order and composition are generally not statistically independent and their probabilities must be found by considering simultaneously the deviations of the order and composition from the equilibrium state. This gives rise to changes in the equations for fluctuations of the order parameters and some new features in comparison with the case of one-component crystals [1.3].

Assume that the long-range order in the solution depends only on a single parameter and in the macroscopic approximation the wave-like fluctuations of the composition and order are described by the smoothly varying functions

$$\delta c(\boldsymbol{r}) = c_k \exp(-i\boldsymbol{k}\cdot\boldsymbol{r}) + \text{c.c.}\,, \quad \delta\eta(\boldsymbol{r}) = \eta_k \exp(-i\boldsymbol{k}\cdot\boldsymbol{r}) + \text{c.c.}$$

Then the least work required for the reversible production of these fluctuation waves in a cubic crystal is given by the equation derived by simply extending (1.2.9):

$$\begin{aligned}\mathfrak{R}_k = V[&(\varphi_{\eta\eta} + \alpha k^2)|\eta_k|^2 + (\varphi_{cc} + \beta' k^2)|c_k|^2 \\ &+ 2(\varphi_{c\eta} + \gamma k^2)(\eta_k c_k^* + \eta_k^* c_k)]\,,\end{aligned} \tag{1.2.25}$$

Here $\varphi_{cc} \equiv \partial^2\varphi/\partial c^2$ and $\varphi_{c\eta} \equiv \partial^2\varphi/\partial c\partial\eta$. When we substitute (1.2.25) into (1.2.1) and invert the second-order matrix comprised of the coefficients in (1.2.25) we obtain

$$\langle|\eta_k|^2\rangle = \frac{k_B T}{V}\,\frac{\varphi_{cc} + \beta' k^2}{\varphi_{\eta\eta}\varphi_{cc} - \varphi_{c\eta}^2 + (\alpha\varphi_{cc} + \beta'\varphi_{\eta\eta} - 2\gamma\varphi_{c\eta})k^2}\,, \tag{1.2.26}$$

$$\langle|c_k|^2\rangle = \frac{k_B T}{V}\,\frac{\varphi_{\eta\eta} + \alpha k^2}{\varphi_{\eta\eta}\varphi_{cc} - \varphi_{c\eta}^2 + (\alpha\varphi_{cc} + \beta'\varphi_{\eta\eta} - 2\gamma\varphi_{c\eta})k^2}\,. \tag{1.2.27}$$

For noncubic crystals the terms αk^2 must be replaced with terms $\alpha_{ij}k_ik_j$ of course.

According to the definition of the partial derivatives $\partial/\partial c$ in the expressions for φ_{cc} and $\varphi_{c\eta}$ they correspond to the constant equilibrium value of η. In contrast, according to the principles of the thermodynamic theory of fluctuations, the derivative $d^2\varphi/dc^2$ is determined by the increase in the thermodynamic potential caused by the variation of the concentration accompanied by the variation of the order parameters. Therefore, the derivatives d/dc must be taken for η corresponding to the changed (fluctuational) concentrations (and the same temperature). Using the equilibrium condition (1.2.14) we obtain

$$\varphi_{cc} - \frac{\varphi_{c\eta}^2}{\varphi_{\eta\eta}} = \frac{d^2\varphi}{dc^2}\,. \tag{1.2.28}$$

Hence, the equations (1.2.27, 10) for $\langle c_k^2\rangle$ are equivalent as it must be expected, of course, since (1.2.10) is clearly a general equation determining the concentration fluctuations both in disordered and ordered solid solutions. (The condition (1.2.14), $\varphi_\eta = 0$, is satisfied for any concentration and therefore have

$$\frac{d\varphi_\eta}{dc} = \varphi_{c\eta} + \varphi_{\eta\eta}\frac{d\eta}{dc} = 0\,,$$

that is, $d\eta/dc = -\varphi_c/\varphi_{\eta\eta}$. When we substitute this expression into the equation (written for $\varphi_\eta = 0$)

$$\frac{d^2\varphi}{dc^2} = \varphi_{cc} + 2\varphi_{c\eta}\frac{d\eta}{dc} + \varphi_{\eta\eta}\left(\frac{d\eta}{dc}\right)^2,$$

we obtain (1.2.28). In a similar way, taking the derivative

$$\frac{d^2\varphi}{d\eta^2} = \varphi_{\eta\eta} + 2\varphi_{\eta c}\frac{dc}{d\eta} + \varphi_{cc}\left(\frac{dc}{d\eta}\right)^2$$

corresponding to the given chemical potentials of atoms (rather than given concentrations) and using the relation

$$\frac{d}{d\eta}\frac{\partial\varphi}{\partial c} = \varphi_{c\eta} + \varphi_{cc}\frac{dc}{d\eta} = 0\,,$$

derived from (1.2.3) we obtain

$$\frac{d^2\varphi}{d\eta^2} = \varphi_{\eta\eta} - \varphi_{c\eta}^2/\varphi_{cc}\,.$$

In disordered solutions when $\eta = 0$ for any concentration, that is, $d\eta/dc = 0$, the derivative $\varphi_{c\eta} = -\varphi_{\eta\eta}(d\eta/dc)$ also vanishes (here we have also $d^2\varphi/dc^2 = \varphi_{cc}$). Therefore, for small $\boldsymbol{k}$ the fluctuations of concentration and long-range order parameters in disordered solutions are statistically independent and the equation for $\langle|\eta_{\boldsymbol{k}}|^2\rangle$ has the same form in solid solutions and for one-component crystals. In ordered solutions we have, however, $\varphi_{c\eta} \neq 0$ (with the exceptions of solutions with certain concentrations) and thus the equations (1.2.26) and (1.2.16) for $\langle|\eta_{\boldsymbol{k}}|^2\rangle$ have different forms for solid solutions and one-component crystals. On the other hand, when we consider the derivative $d^2\varphi/d\eta^2$ for a system with the fixed chemical potentials of atoms, rather than fixed concentrations, then we can show (see the note after (1.2.28)) that $d^2\varphi/d\eta^2 = \varphi_{\eta\eta} - \varphi_{c\eta}^2/\varphi_{cc}$, that is, (1.2.26) may be written in the form of (1.2.16) (though under conventional experimental conditions the concentrations, rather than the chemical potentials, that is, $\varphi_{\eta\eta}$ and not $d^2\varphi/d\eta^2$ are constant).

If the long-range order in a solid solution is described by $n-1 > 1$ parameters η_μ then the fluctuations of these parameters and the concentration are not statistically independent, too, and thus they must be treated simultaneously. Under such conditions the mean squares and the products of the Fourier components can clearly be found from (1.2.20) if η_n in it is replaced with c.

We can find fluctuations from the above equations if we know the thermodynamic potentials as functions of c and η. These functions may be found from independent experimental data in some cases. (For ferroelectric solid solutions where the order parameters can be the components of the polarization vector $\boldsymbol{P}$, the derivatives $\partial^2\varphi/\partial P_i\partial P_j$ can still be expressed in terms of the susceptibilities via (1.2.23) and the derivative $\partial^2\varphi/\partial P_i\partial c$ may be found from the equation

$$\frac{d}{dc}\frac{\partial\varphi}{\partial P_i} = \frac{\partial^2\varphi}{\partial P_i\partial c} + \frac{\partial^2\varphi}{\partial P_i\partial P_j}\frac{dP_j}{dc} = 0\,,$$

if the concentration dependence of the spontaneous polarization is known while

$$\frac{d^2\varphi}{dc^2} = \varphi_{cc} + 2\frac{\partial^2\varphi}{\partial c\partial P_i}\frac{dP_i}{dc} + \frac{\partial^2\varphi}{\partial P_i\partial P_j}\frac{dP_i}{dc}\frac{dP_j}{dc} = \frac{1}{v_0}\frac{d(\mu_A - \mu_B)}{dc}$$

is determined by the concentration dependence of the chemical potentials.) Various thermodynamic or statistical solutions for φ can also be used for calculating the above functions and finding the explicit concentration and temperature dependences of the fluctuations. The most general equations are, clearly, the thermodynamic ones which have not been derived for any specific statistical model and are valid in various ranges of the phase diagram.

For dilute solutions in which the concentration of a component in a binary solution is low ($c \ll 1$) we can neglect the terms of the second order in c and write (see Sect. 88 in [1.31])

$$\varphi = \varphi_B + \frac{N}{V} k_B T c \ln c + c g(P,T) , \tag{1.2.29}$$

where φ_B is the thermodynamic potential of the unit volume of the pure component B and $g(P,T)$ is independent of the concentration c. When the concentration is sufficiently low the derivative of the thermodynamic potential (1.2.29), $\varphi_{cc} = N k_B T / V_c$ for any $\boldsymbol{k}$ (in the first cell of the reciprocal lattice is considerably larger than the second term in the denominator of (1.2.10). We have therefore in dilute solutions

$$\langle |c_{\boldsymbol{k}}|^2 \rangle = \frac{c}{N} (c \ll 1) . \tag{1.2.30}$$

At temperatures that are considerably higher than the temperature T_0 of ordering (or decomposition) the correlation is insignificant and the solution can be regarded as being ideal. Under such conditions we have a simple expression

$$\varphi = \varphi^0 + \frac{N}{V} k_B T [c \ln c + (1-c) \ln(1-c)] , \tag{1.2.31}$$

where φ^0 linearly depends on the concentration. Since we can ignore the dependence of φ on the gradient of c in the approximation of the ideal solution and take $\beta = 0$ then (1.2.30, 31) yield

$$\langle |c_{\boldsymbol{k}}|^2 \rangle = \frac{1}{N} c(1-c) \quad (T \gg T_0) . \tag{1.2.32}$$

for any $\boldsymbol{k}$ in the ideal solution. This result agrees with (1.1.15) derived with a different (probabilistic) approach.

A thermodynamic expression for φ can be obtained also for almost ordered solutions. These are the solutions with an almost stoichiometric composition at low temperatures $T \ll T_0$. In such solutions we may assume that the concentrations of atoms occupying "foreign" positions are small so that the thermodynamic calculations of φ can be performed. For instance, if the stoichiometric composition corresponds to $c = 1/2$ then for $c \cong 1/2$ the thermodynamic potential is given by [1.39]

$$\begin{aligned} \varphi = \varphi' &+ N\chi_1(1-c_1) + N\chi_2 c_2 + N k_B T [c_1 \ln c_1 + c_2 \ln c_2 + \\ &+ (1-c_1)\ln(1-c_1) + (1-c_2)\ln(1-c_2)] , \end{aligned} \tag{1.2.33}$$

where φ', χ_1 and χ_2 depend only on T and P but not on c and η. In the denominators of (1.2.26, 27) the terms independent of $\boldsymbol{k}$ grow faster than the terms of the second order in $\boldsymbol{k}$ when $\eta - 1$ and thus when we substitute (1.2.33) into these equations we obtain for the almost completely ordered solutions using (11.3) (for $\vartheta_1 = \vartheta_2 = \vartheta/2$)

$$\langle |c_{\boldsymbol{k}}|^2 \rangle = \frac{1}{4} \langle |\eta_{\boldsymbol{k}}|^2 \rangle = \frac{1}{2N} \left[c(1-c) - \frac{1}{4}\eta^2 \right] \quad (T \ll T_0) . \tag{1.2.34}$$

For instance, in the solid solutions of the stoichiometric composition ($c = 1/2$) we have

$$\langle |c_{\boldsymbol{k}}|^2 \rangle = \frac{1}{4} \langle |\eta_{\boldsymbol{k}}|^2 \rangle = \frac{1}{8N}(1-\eta^2) \approx \frac{1}{4N}(1-\eta) , \tag{1.2.35}$$

that is, the mean squared Fourier components of fluctuations are proportional to $1-\eta$. Since in such solutions η exponentially tends to unity with a decrease in temperature,

$$1 - \eta = 2 \exp\left(-\frac{\chi_1 + \chi_2}{2k_{\mathrm{B}}T} \right) ,$$

the mean squares $\langle |c_{\boldsymbol{k}}|^2 \rangle$ and $\langle |\eta_{\boldsymbol{k}}|^2 \rangle$ decrease exponentially.

The equation (1.2.34) for $\langle |c_{\boldsymbol{k}}|^2 \rangle$ coincides with (1.2.13) if we take $\varepsilon(\varrho) = 0$. We see that correlation is not significant in solid solutions not only at high temperatures but also at low temperatures near the state of complete order.

1.3 Microscopic Semiphenomenological Theory of Fluctuation Waves in Solid Solutions

The above macroscopic treatment of the fluctuation waves made it possible to express the mean-squared fluctuations in terms of thermodynamic characteristics without employing any specific model.These results are applicable, however, only for small k (long waves). The short-wave fluctuations corresponding to large k make a significant contribution to X-ray scattering. Such fluctuations cannot be treated in the framework of the macroscopic theory and a specific model of the solid solution must be employed for their analysis. Such a model makes it also possible to relate the fluctuation probabilities and hence the diffuse scattering intensity to the constants describing the interaction between atoms in the solution thus providing us with tools for analyzing these constants (and, in principle, for obtaining full information on the phase diagram and thermodynamics of the solution). It is possible at the same time to write the explicit expressions for the thermodynamic parameters as functions of the temperature and concentration which appear in the equations of the thermodynamic theory.

Mean-Square Amplitudes of the Fluctuation Waves in the Model of Pair Interaction. The simplified model of pair interaction between atoms is usually employed in the statistical theory of solutions [1.23–26]. It is assumed in this model that the solution energy E can be represented as a sum of the interaction energies for individual atom pairs:

$$E = \frac{1}{2} \sum_{s\gamma s'\gamma'} c_{s\gamma} c_{s'\gamma'} w_{\varrho} \qquad w(\varrho)|_{\varrho=0} = 0 \,. \tag{1.3.1}$$

Here E is determined to within the insignificant (independent of the atomic configuration in a solution of the given composition) terms of the zeroth and first order in the parameters $c_{s\gamma}$ (which have the values of 0 and 1, according to (1.1.1)). The ordering energies $w(\varrho)$ for the pairs of the lattice sites $s\gamma$ and $s'\gamma'$ connected by the vector ϱ are defined as the differences

$$w(\varrho) = 2v^{AB}(\varrho) - v^{AA}(\varrho) - v^{BB}(\varrho) \,, \tag{1.3.2}$$

where, for instance, $v^{AB}(\varrho)$ is the interaction energy for the pair of atoms A and B at the sites $s\gamma$ and $s'\gamma'$ taken with the reversed sign. It is assumed that $w(\varrho)$ are independent of the composition and the order of the solution. The energy (1.3.1) depends on the interaction between not only the neighbouring atoms but also more distant atoms. The solution model corresponding to (1.3.1) is equivalent to the Ising model in the magnetic ordering theory.

It should be born in mind that the equation for the energy must include also the "collective" terms associated with the energies of the conduction electrons and the vibrations of atoms in the solution, in addition to the energy of the direct interaction in the atom pairs. These terms may, however, also be written in the form of (1.3.1) if the deviation of the solid solution from the 'average' crystal can be regarded as a small perturbation which can be treated in the second-order approximation of the perturbation theory [1.40]. The pair interaction model can, therefore, be regarded essentially as a semiphenomenological general model which includes not only the direct pair interaction but also the indirect interaction via the conduction electrons or other quasiparticles. The energies $w(\varrho)$ for the simplest alloys can be calculated approximately with the pseudopotential method (see Sect. 2.4 and [1.41, 42]). Such calculations are more complicated for other crystals, particularly, those containing transition metals, since they must take into account higher-order terms in the perturbation theory and the effect of neighbouring atoms on the interaction energy for a given atom pair. The equations for the energy of the solution then contain terms with products of three or more parameters $c_{s\gamma}$ describing the many-body interaction and the energies $w(\varrho)$ in (1.3.1) depend on the composition and order of the solution.

Though the opportunities for calculating the constants of the interaction between atoms in solid solutions are increasingly improving, the accuracy of the quantitative results is not very high. It may be a reasonable approach, therefore, to regard $w(\varrho)$ as semiphenomenological parameters in the model of pair interaction described by (1.3.1) which are found by comparing theoretical predictions with experimental data. This model can be applied, rigorously speaking, only to the systems in which the many-body interaction is not strong. But some results of the semiphenomenological approach are apparently of a more general significance. At any rate, the method for calculating fluctuations given below can be readily extended to more complicated models of solid solutions (for instance, the many-body interaction must be included in order to account for the asymmetry of the phase diagrams [1.43]).

The model of pair interaction has been used to calculate the fluctuation wave probabilities in various approximations and with various techniques for the fluctuations in solid solutions [1.4, 5] and in the Ising-model magnetics (equivalent to the solution with $c = 1/2$) [1.44, 45]. One calculation approach [1.4] is based on breaking down the lattice of a given solid solution into a large number ν' of geometrically identical sublattices (comprising lattice sites of a given type) which are similarly arranged with respect to each other. The breakdown is performed in such a way that in this approximation each atom interacts only with the nearest neighbours in other sublattices and does not interact with atoms in its own sublattice. This technique concerns not only the direct interaction of atoms in the given sublattice but also the indirect "thermodynamic interaction" caused by the fact that each atom is surrounded with a cloud of atoms of a different species or a partially ordered cloud owing to the correlation between the intermediate atoms.

If the number ν' is sufficiently large the size of a cell consisting of ν' atoms is considerably larger than the range of the interaction between the atoms and the range of the correlation even when the atoms interact at relatively large distances. Then the thermodynamic potential Φ of the solid solution can be regarded as a function of the concentrations c_γ of the atoms on the sites of the sublattices γ which takes into account the (direct and indirect) interaction between atoms of different sublattices but ignores the interaction between atoms in the same sublattice. Though some of the different sublattices include sites of the same type and c_γ for them are identical these variables in the function $\Phi(c_\gamma)$ must be regarded as being formally different in the calculations of the fluctuation probabilities.

For the lattice with ν atoms per primitive unit cell the number ν' must be a multiple of ν (and can be considerably larger than ν. As in Sect. 1.1, the number of the sublattice will be denoted by the subscript γ which will have ν' values here corresponding to the sites of the larger cell rather than the unit cell as in Sect. 1.1. The Fourier component $c_{\boldsymbol{k}}$ of the concentration fluctuations in the solid solution may clearly be expressed in terms of the Fourier components $c_{\boldsymbol{k}\gamma}$ of the concentration fluctuations in individual sublattices with an equation similar to (1.1.12):

$$c_{\boldsymbol{k}} = \frac{1}{\nu'} \sum_{\gamma=1}^{\nu'} c_{\boldsymbol{k}\gamma}, \quad c_{\boldsymbol{k}\gamma} = \frac{1}{N^0} \sum_{s=1}^{N^0} (c_{s\gamma} - c_\gamma) \mathrm{e}^{\mathrm{i}\boldsymbol{k}\cdot\boldsymbol{R}_{s\gamma}}, \tag{1.3.3}$$

Here N^0 is the number of sites in a sublattice.

The Appendix presents the calculation procedures for the mean squares and products of the Fourier components $\langle c_{\boldsymbol{k}\gamma} c^*_{\boldsymbol{k}\gamma'} \rangle$ and $\langle |c_{\boldsymbol{k}}|^2 \rangle$. It can be seen from

A.6 that $\langle|c_{\boldsymbol{k}}|^2\rangle$ are determined by the matrix $a_{\gamma\gamma'}$ given by (A.6), that is, by the derivatives of the thermodynamic potential $\Phi c_\gamma)$ with respect to the concentrations c_γ in the sublattices (which are assumed to be independent variables even if the sites γ are equivalent). If the sum of the matrix elements in each row of the matrix $\|a_{\gamma\gamma'}\|$ is independent of the number of the row then, according to (A.9, 10), the expression for $\langle|c_{\boldsymbol{k}}|^2\rangle$ has the following simple form [1.4] (equations (1.3.4) can be derived in a different way [1.46]):

$$\langle|c_{\boldsymbol{k}}|^2\rangle = \frac{1}{N_0}\frac{1}{\chi_{\boldsymbol{k}}}\,, \quad \chi_{\boldsymbol{k}} = \sum_{\boldsymbol{\varrho}} \chi(\boldsymbol{\varrho}) \cos \boldsymbol{k}\cdot\boldsymbol{\varrho} = \frac{1}{k_{\mathrm{B}}T}\frac{\partial^2\Phi}{\partial c_{\boldsymbol{k}}\partial c_{\boldsymbol{k}}^*}\,, \tag{1.3.4}$$

where

$$\begin{aligned} &N_0 = N^0\nu'\,; \quad \chi(0) = \frac{1}{N^0 k_{\mathrm{B}}T}\frac{\partial^2\Phi}{\partial c_1^2} \\ &\chi(\boldsymbol{\varrho}) = \frac{1}{N^0 k_{\mathrm{B}}T z'}\frac{\partial^2\Phi}{\partial c_1 \partial c_{\boldsymbol{\varrho}}} \quad (\boldsymbol{\varrho} \neq 0)\,; \end{aligned} \tag{1.3.5}$$

and $c_{\boldsymbol{\varrho}}$ is the concentration at the sites of the sublattice displaced by the vector $\boldsymbol{\varrho}$ with respect to the first sublattice. After the derivatives in (1.3.5) have been found the concentrations c_γ must be replaced with the actual concentrations at the sites of the type under consideration which correspond to the equilibrium long-range order parameters. For instance, in disordered solutions c_γ must be replaced with the concentration c.

Equation (1.3.4) writes $(\langle c_{\boldsymbol{k}}^2\rangle)^{-1}$ as a Fourier series in terms of the second-order derivative of the thermodynamic potential with respect to the fluctuation wave amplitude. It is a rigorous equation (within the framework of the pair interaction described by (1.3.1)) and any errors it yields are caused by the approximations of the dependences $\Phi(c_\gamma)$. For instance, the expression (1.3.4) for $\langle|c_{\boldsymbol{k}}|^2\rangle$ must satisfy the exact equation (1.1.16) which makes it possible to write $\chi(0)$ in terms of $\chi(\boldsymbol{\varrho} \neq 0)$ and rewrite (1.3.4) in the equivalent form

$$\langle|c_{\boldsymbol{k}}|^2\rangle = \frac{C}{N_0}\frac{1}{\sum_{\boldsymbol{\varrho}\neq 0} y(\boldsymbol{\varrho}) \cos \boldsymbol{k}\cdot\boldsymbol{\varrho}}\,, \quad y(\boldsymbol{\varrho}) = \frac{\chi(\boldsymbol{\varrho})}{\chi(0)}\,, \tag{1.3.6}$$

where

$$C = \frac{N_0}{v}\sum_{\gamma} c_\gamma(1-c_\gamma)\left[\sum_{\boldsymbol{k}} \frac{1}{y(\boldsymbol{\varrho})\cos \boldsymbol{k}\cdot\boldsymbol{\varrho}}\right]^{-1} . \tag{1.3.7}$$

Disordered Solutions. Equations (1.3.4) or (1.3.6) make it possible, in particular, to analyze fluctuations in disordered solutions with Bravais-type lattices in which significant interaction occurs between different coordination spheres. Under such conditions the crystal can be broken down into the above sublattices if we choose a polyhedron which contains a large enough number of unit cells. Then we can ignore the interaction between the atom at the center of the polyhedron and the atoms outside the polyhedron so that the last term in (A.2) vanishes.Then if the sublattices are related to each of the ν' sites of the polyhedron the given atom does not interact with atoms in its own sublattice and interacts with not more than one atom in each other sublattice, so that the approach discussed in the Appendix can be used for calculating $\langle|c_{\boldsymbol{k}}|^2\rangle$. Since for any sublattice the number z' of sites neighbouring on a site in another sublattice is unity the sum (A.6) over $\mathfrak{Z}$ has only a single term. In a disordered solution the derivatives $\partial^2\Phi/\partial c_\gamma \partial c_{\gamma'}$ depend only on the vector ϱ between the nearest neighbouring sites γ and γ' but not on the numbers of these sublattices. Moreover, in a crystal lattice of the type under consideration all sites are equivalent and for any starting site the systems of vectors connecting the given site with all other sites are identical. All the diagonal matrix elements $a_{\gamma\gamma}$ are identical, therefore, and the same off-diagonal matrix elements occur on different rows of the matrix $\|a_{\gamma\gamma'}\|$, though in a different sequence. Then the sum of the matrix elements in each row is independent of the row number so that (1.3.4, 6) can be employed for calculating $\langle|c_{\boldsymbol{k}}|^2\rangle$ in the case of disordered solutions.

To use these equations we must know the behaviour of the thermodynamic potential as a function of the variables c_γ. We know that in the three-dimensional case no explicit exact expression has been derived for $\Phi(c_\gamma)$ and only approximate equations are available which are valid for various temperatures and solution compositions. For disordered solutions simple approximate equations have been derived for the case of high temperatures and arbitrary compositions in which the ratios $w(\varrho)/k_\mathrm{B}T$ can be regarded as small parameters, and in the case of small concentrations and arbitrary temperatures in which the concentration c is a small parameter.

When temperatures are sufficiently high (in comparison with the temperature T_0 of ordering or decomposition) we can calculate the thermodynamic potential in the self-consistent field approximation and ignore the correlation. In this approximation the energy is determined as the average of (1.3.1) and the correlation is ignored (that is in (1.3.1) $c_{s\gamma}c_{s'\gamma'}$ is replaced with $c_\gamma c_{\gamma'}$ and the conventional equation of the zeroth-order approximation of ordering theory is used for $S = k_\mathrm{B}T \ln W$ (where W is defined as the number of rearrangements of the atoms A and B over the lattice sites for the given probabilities c_γ; see, for instance, Sect. 16 in [1.23]). Then we have

$$\Phi = \Phi' + \frac{N^0}{2} \sum_{\substack{\gamma,\gamma'=1 \\ (\gamma\neq\gamma')}}^{\nu'} c_\gamma c_{\gamma'} w(\varrho)$$

$$+ N^0 k_\mathrm{B}T \sum_{\gamma=1}^{\nu'} [c_\gamma \ln c_\gamma + (1-c_\gamma)\ln(1-c_\gamma)] , \tag{1.3.8}$$

where Φ' is a linear function of c_γ and the vectors pass between the nearest sites of the sublattices γ and γ'. Equations (1.3.5, 6, 8) yield the following expressions for $\chi(\varrho)$ and $y(\varrho)$ in (1.3.4, 6) for disordered solutions at high temperatures:

$$\chi(0) = \frac{1}{c/1-c)}\,, \quad \chi(\varrho) = \frac{w(\varrho)}{k_\mathrm{B}T} \quad \text{for } \varrho \neq 0\,,$$
$$y(\varrho) = c(1-c)\chi(\varrho) \quad (|w(\varrho)| \ll k_\mathrm{B}T)\,. \tag{1.3.9}$$

Using (1.3.4, 6, 9) we obtain a simple expression for $\langle |c_{\boldsymbol{k}}|^2 \rangle$ in terms of the energy $w(\varrho)$ of ordering for the solution at high temperatures:

$$\langle |c_{\boldsymbol{k}}|^2 \rangle \approx \frac{C}{N} \frac{1}{1+c(1-c)\frac{w(\boldsymbol{k})}{k_\mathrm{B}T}} \approx \frac{1}{N} \frac{c(1-c)}{1+c(1-c)\frac{w(\boldsymbol{k})}{k_\mathrm{B}T}}\,,$$
$$w(\boldsymbol{k}) = \sum_{\varrho \neq 0} w(\varrho) \cos \boldsymbol{k} \cdot \boldsymbol{\varrho} \; (|w(\varrho)| \ll k_\mathrm{B}T)\,. \tag{1.3.10}$$

Though (1.3.8) used here ignores correlation it is automatically taken into account to some approximation in (1.3.10) where in both equations the terms linear in $w(\varrho)/k_\mathrm{B}T$ are retained in $\chi(\varrho)$. The first equation (1.3.10) includes the temperature-dependent constant C given by (1.3.7) and is more accurate as it rigorously takes into account also the terms proportional to $[w(\varrho)/k_\mathrm{B}T]^2$ in $\chi(0)$ but not in $\chi(\varrho)$.

Similar equations for the mean squares and products of the amplitudes of fluctuation waves have been derived also for the three- and multicomponent solutions in the approximation of the molecular field [1.47–49].

Simple equations for $\chi(\varrho)$, $y(\varrho)$ and, hence, for $\langle c_{\boldsymbol{k}}^2 \rangle$ can be derived also for solutions in which the concentration of a component is very low ($c \ll 1$). Such a solid solution must be in the disordered state at any temperature in the single-phase range of the phase diagram. The following expression can be derived for the thermodynamic potential in the pair interaction approximation when the terms of the higher order in c are ignored [1.50]:

$$\Phi = \Phi' + \frac{N^0}{2} k_\mathrm{B}T \sum_{\substack{\gamma,\gamma'=1 \\ (\gamma\neq\gamma')}}^{\nu'} c_\gamma c_{\gamma'} (1 - e^{-\frac{w(\varrho)}{k_\mathrm{B}T}})$$
$$+ k_\mathrm{B}TN^0 \sum_{\gamma=1}^{\nu'} [c_\gamma \ln c_\gamma + (1-c_\gamma) \ln(1-c_\gamma)]\,. \tag{1.3.11}$$

Hence we obtain for this case

$$\chi(0) = \frac{1}{c(1-c)}\,, \quad \chi(\varrho) = 1 - \exp\left(-\frac{w(\varrho)}{k_\mathrm{B}T}\right) \quad \text{for } \varrho \neq 0\,,$$
$$y(\varrho) = c(1-c)\chi(\varrho) \quad (c \ll 1)\,, \tag{1.3.12}$$

that is, the Fourier component $w(\boldsymbol{k})$ of the ordering energy is replaced with the Fourier component of the parameters $\{1 - \exp[-w(\varrho)/k_\mathrm{B}T]\}k_\mathrm{B}T$.

If we assume that the ordering energies $w(\varrho)$ vanish then, according to (1.3.9, 12), $\chi(\varrho) = 0$ for $\varrho \neq 0$ and (1.3.4) coincides with (1.1.5) for $\langle |c_{\boldsymbol{k}}|^2 \rangle$ in the ideal solution. It is a natural result since when $w(\varrho) = 0$ no correlation occurs in the solution and the latter is ideal. In the real solution $w(\varrho) \neq 0$, we have nonvanishing $\chi(\varrho)$ and the terms with $\varrho \neq 0$ in (1.3.4) account for nonideality of the solution causing the correlation. These terms determine the dependence of $\langle|c_{\boldsymbol{k}}|^2\rangle$ on $\boldsymbol{k}$ which is not a constant as is the case for the ideal solution. Equations (1.3.4, 9, 12) reveal that when temperature is increased or concentration is decreased the contribution of the correlation diminishes and the expression for $\langle|c_{\boldsymbol{k}}|^2\rangle$ reduces to (1.1.15).

According to (1.3.10), the behaviour of the solution becomes increasingly different from that of the ideal solution when the temperature is decreased. When the temperature reaches the point T_0' corresponding to

$$k_B T_0' = -c(1-c)w(\boldsymbol{k}_m) = -c(1-c)\sum_{\varrho\neq 0} w(\varrho)\cos \boldsymbol{k}_m \cdot \varrho\,, \tag{1.3.13}$$

where the point $\boldsymbol{k}_m$ gives the absolute minimum of the function $w(\boldsymbol{k})$ (which is negative) the second-order derivative $\partial^2\Phi/\partial c_{\boldsymbol{k}}\partial c^*_{\boldsymbol{k}}$ found in the self-consistent field approximation vanishes for $\boldsymbol{k} = \boldsymbol{k}_m$. Hence the solution becomes unstable while $\langle|c_{\boldsymbol{k}}|^2\rangle$ tends to infinity for $T \to T_0'$ and $\boldsymbol{k} \to \boldsymbol{k}_m$ according to (1.3.10). In the solutions where $w(\boldsymbol{k})$ has the minimum at $\boldsymbol{k}_m = 0$ decomposition occurs below the instability temperature while for $\boldsymbol{k}_m \neq 0$ ordering of the solution occurs at $\lesssim T_0'$ if it takes place as a second order phase transition. For the first order phase transition (or when the long-range forces are significant, see Chap. 2) the transition point T_0 (that is, the temperature at which the thermodynamic potentials of both phases are identical) is above the instability point T_0'.

The condition for applicability of the self-consistent field approximation, $|w(\varrho)| \ll k_B T$, is not satisfied, however, near the temperature T_0' and the thermodynamical properties of the solution can no longer be accurately described with this approximation. The errors in the calculated transition temperature are relatively small. For instance, in the nearest-neighbour approximation in which the only nonvanishing energy is $w(\varrho) \equiv w_1$ for atoms in the first coordination sphere, the decomposition temperature found from (1.3.13), $k_B T_0 = -c(1-c)zw_1$ (where z is the coordination number and $w_1 < 0$), is overestimated by 33% for the simple cubic lattice ($z = 6$), by 26% for the bcc lattice ($z = 8$), and by 22% for the fcc lattice ($z = 12$) (the error can be seen to decrease with an increase in z). The same error was obtained for the ordering temperature in alloys of the beta-brass type ($z = 8$) though the error is 70% for Cu_3Au with the effective $z = 4$. It is a significant fact, however, that the self-consistent field approximation (and other approximate statistical theories of ordering) yields an erroneous temperature dependence $\partial^2\Phi/\partial c_{\boldsymbol{k}}\partial c^*_{\boldsymbol{k}} \sim (T - T_0')$ near T_0' and therefore does not describe correctly the critical $\langle|c_{\boldsymbol{k}}|^2\rangle$ fluctuations which must be thus treated within the framework of a different model (see Sect. 4.1).

If the main contribution is made only by the interaction with the neighbouring atoms the self-consistent field equation (1.3.10) gives $\langle|c_{\boldsymbol{k}}|^2\rangle$ at $T > 2T_0'$ with a sufficient accuracy (the error is equal to or smaller than 10%). This follows, for instance, from the equations (1.3.15). The applicability range of the self-consistent field approximation becomes, however, much wider and covers the main range in which fluctuations grow critically if the interaction between atoms is significant in a large number of coordination spheres. It has been shown by the detailed analysis [1.51], for instance, that in the case of the energies $w(\varrho) < 0$ which have the constant sign and the same order of magnitude at large distances $r_0 \gg d_0$ (here d_0 is the distance between atoms) the self-consistent field approximation is inapplicable only in a narrow temperature range near the decomposition temperature T_0:

$$\tau = \frac{T - T_0}{T_0} \lesssim \frac{d_0^6}{r_0^6}, \quad r_0^2 = \frac{\sum_\varrho \varrho^2 w(\varrho)}{\sum\limits_\varrho w(\varrho)}. \tag{1.3.14}$$

To cover a wider temperature range in the case of shortrange interaction between atoms we must move outside the framework of the self-consistent field approximation and obtain equations for $\chi(\varrho)$ which must be more precise than (1.3.9) and take into account the higher-order terms in the expansion in powers of $w(\varrho)/k_\mathrm{B}T$ (for low concentrations such equations are given by (1.3.12)). This means that we must take into consideration the correlation when we calculate the thermodynamic potential of the solution and replace (1.3.8) derived in the zero-order approximation with a more rigorous result. We can use, for instance, the expansions of $\varPhi$ in powers of $w(\varrho)/k_\mathrm{B}T$ for this purpose. Such an expansion has been derived [1.52] (see also Sect. 17 in [1.23]) in the nearest-neighbour approximation for solutions of an arbitrary composition with the simple cubic or bcc lattice giving $\varPhi$ as a function of the concentrations c_1 and c_2 at the sites of two sublattices (including the vertices and centres of the cubes in the bcc crystals). If we ignore the terms proportional to $w_1(w_1/k_\mathrm{B}T)^3$ in the expansion of $\varPhi$ (it can be done at high enough temperatures) we can neglect both direct and indirect interactions between atoms in each sublattice and calculate $\langle|c_{\boldsymbol{k}}|^2\rangle$ from (1.3.4, 5) using the above procedure of breaking down the crystal into two sublattices. In this approximation $\chi(\varrho) = \chi_1$ do not vanish only for the first coordination sphere and the expressions for $\chi(0) = \chi_0$ and χ_1 have the following form [1.4]:

$$\begin{aligned} \chi_0 &= \frac{1}{Nk_\mathrm{B}T}\frac{\partial^2\varPhi}{\partial c_1^2} = \frac{1}{f} + zf\left(\frac{w_1}{k_\mathrm{B}T}\right)^2 - zf(1-2c)^2\left(\frac{w_1}{k_\mathrm{B}T}\right)^3, \\ \chi_1 &= \frac{1}{zNk_\mathrm{B}T}\frac{\partial^2\varPhi}{\partial c_1 \partial c_2} = \frac{w_1}{k_\mathrm{B}T} - \frac{1}{2}(1-2c)^2\left(\frac{w_1}{k_\mathrm{B}T}\right)^2 \\ &\quad + \frac{1}{6}(1-6f)^2\left(\frac{w_1}{k_\mathrm{B}T}\right)^3, \quad f = c(1-c). \end{aligned} \tag{1.3.15}$$

It can be seen that in the nearest-neighbour approximation the denominator in (1.3.4) includes only the sum over the sites of the first coordination sphere at high temperatures, that is $\langle|c_{\boldsymbol{k}}|^2\rangle$ has a simple dependence on $\boldsymbol{k}$ (though the temperature dependence is more complicated than (1.3.10) derived in the self-consistent field approximation). But when we take into account the terms proportional to $w_1(w_1/k_\mathrm{B}T)^3$ in the expansion of $\varPhi$ we obtain indirect interaction between more distant atoms and additional terms in the denominator in (1.3.4) corresponding to the vectors ϱ ending at the lattice sites in the second coordination sphere.

In the more compact fcc lattice the indirect interaction between atoms corresponds to the terms proportional to $w_1(w_1/k_BT)^3$ in the expansion of Φ. If we retain only the terms proportional to w^2/k_BT in Φ we can derive expressions for $\chi(\varrho)$ including the terms proportional to $w^2(\varrho)/(k_BT)^2$ for an arbitrary range of interaction between atoms [1.17];

$$\chi(0) = \frac{1}{f} + f\sum_{\varrho} \frac{w^2(\varrho)}{(k_BT)^2},$$

$$\chi(\varrho) = \frac{w(\varrho)}{k_BT} - \frac{1}{2}(1-2c)^2\frac{w^2(\varrho)}{(k_BT)^2} \quad (\varrho \neq 0). \tag{1.3.16}$$

When we substitute (1.3.15, 16) into (1.3.4) or (1.3.6) we can describe the dependence of $\langle|c_{\boldsymbol{k}}|^2\rangle$ on T, c and $\boldsymbol{k}$ in a wider temperature range than we can do in the self-consistent field approximation. In particular, (1.3.13, 16) show that an increase in the number of neighbours effectively interacting with the given atom contributes to improved convergence of the expansions in powers of T_0/T (and extends the applicability range of the self-consistent field approximation). Equations for $y(\varrho)$ that are more accurate than (1.3.16) and include terms proportional to $[w(\varrho)/k_BT]^3$ have been derived for the fcc solutions in [1.53] and high-temperature expansions with a large number of terms are given in [1.54] for the case of $c = 1/2$. Apart from the high-temperature expansions, other approximations can be used for the thermodynamic potential and the functions $\chi(\varrho)$. For instance, a special approximation of the entropy expression analyzed in [1.551 yields more accurate results than the self-consistent field approximation (though they are not more accurate than those obtained from the expansions (1.3.15, 16) for $\chi(\varrho)$. In another approach [1.56] calculations of $\langle c_s c_{s'}\rangle$ have been made by unlinking the chain of equations for the sequence of the correlation functions $\langle c_s c_{s'} c_{s''}, \ldots\rangle$. A simplified approximation for $\langle|c_{\boldsymbol{k}}|^2\rangle$ has been suggested in [1.57]. Various kinds of cluster expansions make possible a sufficiently accurate description of the thermodynamic properties of solutions (see [1.26, 58–63]). Various statistical characteristics of solutions may also be found with computer simulation, Monte-Carlo and molecular dynamics techniques (see, for instance, [1.21]). In principle, these techniques yield $\langle|c_{\boldsymbol{k}}|^2\rangle$ with a desired accuracy in a wide range of temperatures and compositions.

When we know the function $\langle|c_{\boldsymbol{k}}|^2\rangle$ for all $\boldsymbol{k}$ we can find the correlation parameters $\varepsilon(\varrho)$ for various coordination spheres from the Fourier transformation of (1.1.14)

$$\varepsilon(\varrho) = \frac{vN}{8\pi^3}\int \langle|c_{\boldsymbol{k}}|^2\rangle \cos(\boldsymbol{k}\cdot\varrho)d\boldsymbol{k}. \tag{1.3.17}$$

For instance, we can obtain the high-temperature decompositions for $\varepsilon(\varrho)$ from (1.3.4, 16) (without recurring to the nearest-neighbour approximation) [1.17, 56]:

$$\varepsilon(\varrho) = -c^2(1-c)^2\frac{w(\varrho)}{k_BT}\left[1-\frac{1}{2}(1-2c)^2\frac{w(\varrho)}{k_BT}\right]$$
$$+\,c^3(1-c)^3\frac{1}{(k_BT)^2}\sum_{s_1\neq s,s'} w_{ss_1}, w_{s's_1}\,, \qquad w_{ss'}\equiv w(\varrho=\boldsymbol{R}_{ss'})\,. \quad (1.3.18)$$

When we take into consideration the interactions of the type of $w_{ss's''}c_sc_{s'}c_{s''}$ for the triplets of atoms s, s', s'' (where $s=s'=s''$), in addition to the pair interaction $w_{ss'}c_sc_{s'}$ given by (1.3.1), we obtain the additional equation

$$\frac{c^4(1-c)^4}{2(k_BT)^2}\sum_{s_1s_2} w_{ss_1s_2}w_{s's_1s_2}+\frac{c^3(1-c)^3}{((k_BT)^4)}\sum_{s_1}(w_{ss_1}+w_{s's_1})w_{ss's_1}\,. \quad (1.3.18a)$$

Equations have also been derived for $\varepsilon(\varrho)$ in multicomponent solutions including the terms proportional to $(w/k_BT)^3$ [1.64].

If we compare the equations for $\langle|c_{\boldsymbol{k}}|^2\rangle$ derived with the microscopic and macroscopic theories we can express the quantities $d^2\varphi/dc^2$, $\varphi_{\eta\eta}$, β_{ij}, and α_{ij} appearing in the latter theory in terms of the statistical model of the solution. The function $\chi_{\boldsymbol{k}}$ in (1.3.4) exhibits an extremum for $\boldsymbol{k}=0$. If we expand it near this point in powers of $\boldsymbol{k}$ and compare the result with (1.2.11) we find that for cubic crystals

$$v\frac{d^2\varphi}{dc^2}=k_BT\sum_{\varrho}\chi(\varrho)\approx\frac{k_BT}{c(1-c)}+\sum_{\varrho\neq0}w(\varrho)-\frac{1}{2}(1-6c+6c^2)\sum_{\varrho\neq0}\frac{w^2(\varrho)}{k_BT}\,,$$

$$\beta=\frac{k_BT}{6v}\sum_{\varrho}\varrho^2\chi(\varrho)\approx\frac{1}{6v}\sum_{\varrho}\varrho^2w(\varrho)-\frac{(1-2c)^2}{12v}\sum_{\varrho}\frac{\varrho^2w^2(\varrho)}{k_BT}\,. \quad (1.3.19)$$

The above approximate expressions take into account the expansions (1.3.16) for $\chi(\varrho)$.

According to the definition (1.3.4), $\chi_{\boldsymbol{k}}$ is a periodic function in the reciprocal lattice space. Therefore, it must have extremum points inside each unit cell of the reciprocal lattice corresponding to some $\boldsymbol{k}=\boldsymbol{k}_m$ [1.65, 66]. The expression for $\chi_{\boldsymbol{k}}$ (and hence for $\langle|c_{\boldsymbol{k}}|^2\rangle$, according to (1.3.4)) can be rewritten in a simpler form near the extremum points:

$$\chi_{\boldsymbol{k}}=\chi_m+\beta_{mij}(k_i-k_{mi})(k_j-k_{mj})\,,\quad \chi_m=\sum_{\varrho}\chi(\varrho)\cos\boldsymbol{k}_m\cdot\varrho\,,$$

$$\beta_{mij}=-\frac{1}{2}\sum_{\varrho}\chi(\varrho)\varrho_i\varrho_j\cos\boldsymbol{k}_m\cdot\varrho\,. \quad (1.3.20)$$

In particular, the function $\chi_{\boldsymbol{k}}$ it must have extrema at some high-symmetry points of the reciprocal lattice space. When the coordinate axes are aligned along the high-symmetry directions the tensor β_{mij} undergoes transformation to principal axes and becomes diagonalized (see [1.26]). The point symmetry for the point $\boldsymbol{k}_m$

is generally lower than the point symmetry of the crystal and hence even in cubic crystals β_{mij} does not reduce to a scalar (with the exception of the cases when $\boldsymbol{k}_m$ coincide with the points of cubic symmetry). In the vicinities of these high-symmetry points we can relate $c_{\boldsymbol{k}}$ to the Fourier components $\eta_{\mu\boldsymbol{k}}$ of the order parameters and $\chi(\boldsymbol{\varrho})$ to the parameters $\varphi_{\mu\mu'}$ and $\alpha_{\mu\mu' ij}$ discussed in Sect. 1.2. One should bear in mind that in this section $c_{\boldsymbol{k}}$ are given for the reciprocal lattice cell of the disordered solution while $\eta_{\mu\boldsymbol{k}}$ are given in the reciprocal lattice cell of the ordered solution which is smaller by a factor of ν. In such small cell $\eta_{\mu\boldsymbol{k}}$ and $c_{\boldsymbol{k}}$ are linearly related to $c_{\boldsymbol{k}}$ at various points $\boldsymbol{k}+\boldsymbol{G}'$ of the large cell ($\boldsymbol{G}'$ are the new reciprocal lattice vectors produced by ordering).

Ordered Solutions. The above procedure for calculating $\langle|c_{\boldsymbol{k}}|^2\rangle$ becomes more complicated when applied to ordered solutions because the sums of the matrix elements in different rows of the matrix (A.6) are generally different. Therefore, we cannot start with (1.3.4) in such calculations but must use more general equations (A.6, 8). The simple equation (1.3.4) is, however, applicable to stoichiometric solutions AB in which the sites of the first type are surrounded with the sites of the second and first types in the same way as the sites of the second type are surrounded with the sites of the first and second types (a more general case is discussed in [1.4]). Crystals of the type of NaCl, beta-brass, AuCu, etc. have such structures. Their crystal lattices can be broken up into ν' sublattices as it was done above for the disordered solutions. Then $\nu'/2$ resulting sublattices include sites of the first type ($\gamma = 1, \ldots, \nu'/2$) and the same number of sublattices consist of the sites of the second type ($\gamma = \nu'/2+1, \ldots, \nu'$). For the solutions of the stoichiometric composition AB the derivatives $\partial^2\Phi/\partial c_\gamma^2$ are identical for various γ (corresponding to the sites of the first and second type) within the framework of the statistical model of the solution used here. In addition, the sum of the matrix elements of a row in the matrix (A.6) is independent of the row number for the crystal structures discussed here. Therefore, in this case $\langle|c_{\boldsymbol{k}}|^2\rangle$ can also be found from (1.3.4) as it was done for disordered solutions.

Under almost complete order we can derive a sufficiently accurate expression for the thermodynamic potential of the ordered solution. For the almost fully ordered solutions of the stoichiometric composition we have [1.50]

$$\begin{aligned}\Phi &= \Phi' + k_{\mathrm{B}}TN^0\sum_{\gamma=1}^{\nu'}[c_\gamma \ln c_\gamma + (1-c_\gamma)\ln(1-c_\gamma)] \\ &+ \frac{1}{2}k_{\mathrm{B}}TN^0\left\{\sum_{\gamma,\gamma'=1}^{\nu'}(1-c_\gamma)(1-c_{\gamma'})\left[1-\exp\left(-\frac{w(\boldsymbol{\varrho})}{k_{\mathrm{B}}T}\right)\right]\right. \\ &+ \sum_{\gamma,\gamma'=\frac{\nu'}{2}+1}^{\nu'} c_\gamma c_{\gamma'}\left[1-\exp\left(-\frac{w(\boldsymbol{\varrho})}{k_{\mathrm{B}}T}\right)\right] - 2\sum_{\gamma=1}^{\nu'/2}\sum_{\gamma'=\frac{\nu'}{2}+1}^{\nu'}(1-c_\gamma)c_{\gamma'} \\ &\times\left.\left[\exp\left(\frac{w(\boldsymbol{\varrho})}{k_{\mathrm{B}}T}\right)-1\right]\right\}. \qquad (1.3.21)\end{aligned}$$

When we substitute (1.3.21) into (1.3.5) and use $c_1 = (1+\eta)/2$ we can can find $\chi(\varrho)$ in (1.3.4):

$$\chi(0) = \frac{4}{1-\eta^2}\ ; \quad \chi(\varrho) = 1 - \exp\left(-\frac{w(\varrho)}{k_B T}\right)\ , \tag{1.3.22}$$

if the vector ϱ lies between sites of the same type ($\varrho \neq 0$), and

$$\chi(\varrho) = \exp\left(\frac{w(\varrho)}{k_B T}\right) - 1\ , \tag{1.3.23}$$

if the vector ϱ passes between sites of the first and second types.

For an almost fully ordered solution AB in which each lattice site is surrounded only by sites of another type (such crystals as NaCl or beta-brass) we obtain from (1.3.4, 22, 23) in the nearest-neighbour approximation

$$\langle |c_{\boldsymbol{k}}|^2 \rangle = \frac{1}{4N_0} \frac{1-\eta^2}{1 + \frac{1}{4}(1-\eta^2)\left[\exp\frac{w_1}{k_B T} - 1\right] \sum_{s=1}^{z} \cos \boldsymbol{k}\cdot\boldsymbol{\varrho}_s}\ . \tag{1.3.24}$$

Since for $T \to 0$ the parameter η exponentially tends to unity [1.50] according to $1 - \eta \approx 2\exp(-zw_1/2k_B T)$ the second term in the denominator in (1.3.24) describing correlation decreases as $\exp[-(z-2)w_1/2k_B T]$, that is, correlation is insignificant at low enough temperatures as in the case of high temperatures. When this term is ignored the expression (1.3.24) coincides with (1.2.34) derived in the macroscopic theory of fluctuations for almost fully ordered solutions ($N_0 = 2N$) and $\langle |c_{\boldsymbol{k}}|^2 \rangle$ no longer depends on $\boldsymbol{k}$. In the nearest-neighbour approximation the equilibrium value of $\langle |c_{\boldsymbol{k}}|^2 \rangle$ decreases as $\exp(-zw_1/2k_B T)$ with a decrease in temperature and in the fully ordered solution we obtain clearly $\langle |c_{\boldsymbol{k}}|^2 \rangle = 0$.

The fluctuations of composition and long-range order analyzed above for solutions are related to the static fluctuation waves (of infinitesimal amplitude) corresponding to various $\boldsymbol{k} \neq \boldsymbol{G}$ lying in the reciprocal lattice cell. The long-range ordering of the solution by itself can be regarded as the occurrence of the static concentration waves with the wave vector $\boldsymbol{k}$ equal to the reciprocal lattic vectors $\boldsymbol{G}$. It is very useful to describe the long-range order in terms of the Fourier representation in the thermodynamic theory for analyzing the probabilities of the second-order phase transitions in crystals of a given symmetry [1.67–70] (see Sect. 4.1 and [1.32]). The static concentration waves have been treated in several interesting studies [11.25, 26, 71–74] in the statistical theory of ordering using the above model of the pair interaction between atoms, They included, for instance, the analysis of the ordered structures that may be produced by the coherent phase transition (of the second or first order) from the disordered phase of a given structure, the analysis of the relationship between the stability ranges for these structures (in particular, at low temperatures) and the behaviour of the ordering energies $w(\varrho)$ and their Fourier components at various distances, and the analysis of the ordering kinetics.

1.4 Fluctuations of the Order Parameters in Solid Solutions

The fluctuations of the concentration and long-range order parameters discussed above produce inhomogeneities of various crystal parameters, for instance, the structure amplitudes or the light refraction index and give rise to physical phenomena, in particular, scattering of X-rays or light. But inhomogeneities in disordered solutions can be produced not only by the concentration fluctuations but also the fluctuations of the internal parameters describing the short-range order or correlation in the solution, that is, the correlation parameters $\varepsilon(\varrho)$ given by (1.1.4). Even when the solution concentration is fixed such fluctuations cause a difference in the numbers of the pairs of atoms of the same species at a given distance ϱ in different volume elements (of the same shape and size) which gives rise to additional inhomogeneities.

The occurrence of the fluctuations of internal parameters by itself does not affect the equations for the concentration fluctuations δc, such as (1.2.5, 10), derived with the thermodynamic theory of fluctuations. The only condition is that the derivatives $d^2\varphi/dc^2$ must take into account the fact that the equilibrium values of the internal parameters vary with variation of the concentration (this condition is always met). It was shown in Sect. 1.3 in the case of the fluctuations of the concentration and the long-range order that the simultaneous treatment of the statistical fluctuations of the concentration and internal parameters yields identical equations for $\langle(\delta c)^2\rangle$ if the relations between the total and partial derivatives of the thermodynamic potential are considered.

The fluctuations $\delta\varepsilon(\varrho)$ of the correlation parameters produce, however, additional inhomogeneities, namely, additional static distortions of the crystal, variations of the light refraction index, etc. which occur besides the inhomogeneities produced by the concentration fluctuations δc. They must cause additional scattering of X-rays, light or other waves which is often ignored but whose intensity is in some cases comparable to that of scattering by the concentration fluctuations. In this connection the effects associated with the fluctuations of the correlation parameters must be analyzed. (Similarly, when we consider scattering by the density fluctuations in one-component liquids, in principle, we must also take into account the fluctuations of the correlation parameters which describe in this case the probabilities of the relative distances between atom pairs.)

For the sake of simplicity we shall limit the treatment to the fluctuations of the correlation parameters in disordered solutions with a single atom per unit cell. As in the case of the concentration fluctuations, we may analyze here either the fluctuations of $\varepsilon(\varrho)$ in small volume elements or the fluctuations of their Fourier components $\varepsilon_{\boldsymbol{k}}(\varrho)$ in the entire crystal. (In the former analysis, averaging in (1.1.4) is performed not over the statistical ensemble (or the entire crystal) but over the pairs of atoms linked by the vector ϱ in the volume element V'. Since different atom distributions may occur with certain probabilities the averages calculated in this way may exhibit fluctuations dependent on the volume V'. According to (1.1.4), the $\boldsymbol{k}$th Fourier component of the $\varepsilon(\varrho)$ fluctuations in the disordered solution is given by

$$\varepsilon_{\boldsymbol{k}}(\boldsymbol{\varrho}) = \frac{1}{N} \sum_{s}^{(\boldsymbol{\varrho})} (c_s - c)(c_{s'} - c) \exp(\mathrm{i}\boldsymbol{k} \cdot \boldsymbol{R}_s) \,. \tag{1.4.1}$$

Here in summation over s the number s' of the site is definitely related to by the equation $\boldsymbol{R}_{s'} = \boldsymbol{R}_s - \boldsymbol{\varrho}$ where $\boldsymbol{\varrho} \neq 0$ (for $\boldsymbol{\varrho} = 0$ and $\boldsymbol{k} \neq 0$ the component $\varepsilon_{\boldsymbol{k}}(\boldsymbol{\varrho})$ reduces to $c_{\boldsymbol{k}}$.

It can be seen from (1.4.1) and (1.8.1) that the Fourier components of the correlation parameters and the concentration are determined by the same quantities c_s and therefore they must be related. Indeed, when we substitute the expansion (1.1.8) for $c_s - c$ into (1.4.1) we can express $\varepsilon_{\boldsymbol{k}}(\boldsymbol{\varrho})$ in terms of $c_{\boldsymbol{k}}$:

$$\varepsilon_{\boldsymbol{k}}(\boldsymbol{\varrho}) = \sum_{\boldsymbol{k}'} c_{\boldsymbol{k}'} c_{\boldsymbol{k}-\boldsymbol{k}'} \exp(\mathrm{i}\boldsymbol{k}' \cdot \boldsymbol{\varrho}) \,. \tag{1.4.2}$$

We see that the set of the Fourier components $\varepsilon_{\boldsymbol{k}}(\boldsymbol{\varrho})$ for various $\boldsymbol{\varrho}$ is definitely related to the set of the products $c_{\boldsymbol{k}'} c_{\boldsymbol{k}-\boldsymbol{k}'}$ for various $\boldsymbol{k}'$. It can also be seen from (1.4.2) that the $\boldsymbol{k}$th Fourier component of the squared concentration fluctuation $[\delta c(\boldsymbol{r})]^2$ is unambigously expressed in terms of the $\boldsymbol{k}$th Fourier components $\varepsilon_{\boldsymbol{k}}(\boldsymbol{\varrho})$ of the correlation parameters for various $\boldsymbol{\varrho}$. (It should be noted that $\langle \varepsilon_{\boldsymbol{k}}(\boldsymbol{\varrho})^2 \rangle$ does not reduce to the sum of the products of the mean squares $\langle |c'_{\boldsymbol{k}}|^2 \rangle \langle |c_{\boldsymbol{k}-\boldsymbol{k}'}|^2 \rangle$ as one might think because of the form of (1.4.2). Indeed, though the means of the four-factor products, $\langle c_{\boldsymbol{k}'} c_{\boldsymbol{k}-\boldsymbol{k}'} c^*_{\boldsymbol{k}''} c^*_{\boldsymbol{k}-\boldsymbol{k}''} \rangle$ are of the order of N^{-3}, while the product of the mean squares is of the order of N^{-2}, the number of such fours is of the order of N^2 and their contribution to $\langle |\varepsilon_{\boldsymbol{k}}(\boldsymbol{\varrho})|^2 \rangle$ is of the same order of magnitude as that of the terms $\langle |c_{\boldsymbol{k}'}|^2 \rangle \langle |c_{\boldsymbol{k}-\boldsymbol{k}'}|^2 \rangle$.)

The mean squares and the products of the Fourier components of the correlation parameters and the concentration, $\langle \varepsilon_{\boldsymbol{k}}(\boldsymbol{\varrho}) \varepsilon^*_{\boldsymbol{k}}(\boldsymbol{\varrho}') \rangle$ and $\langle \varepsilon_{\boldsymbol{k}}(\boldsymbol{\varrho}) c^*_{\boldsymbol{k}} \rangle$, can be readily calculated for the ideal solutions when atoms are randomly arranged over lattice sites and the means of the correlation parameters vanish. Though the means of $\varepsilon(\boldsymbol{\varrho})$ vanish their fluctuations of course still occur since the numbers of the pairs of atoms of the same species at a distance $\boldsymbol{\varrho}$ can be higher or lower than the mean numbers simply because of the random distribution of atoms in various small volume elements even when no preferential repulsion or attraction occurs between atoms of the same species. This fact causes the fluctuations of $\varepsilon(\boldsymbol{\varrho})$. Since for ideal solutions $\langle c_s^2 \rangle = \langle c_s \rangle = c$ and $\langle c_s c_{s'} \rangle = c^2$ for $s \neq s'$ we obtain from (1.1.8) and (1.4.1) (for $\boldsymbol{\varrho} \neq 0$)

$$\begin{gathered}\langle |\varepsilon_{\boldsymbol{k}}(\boldsymbol{\varrho})|^2 \rangle = \frac{1}{N} c^2 (1-c)^2 \,, \quad \langle \varepsilon_{\boldsymbol{k}}(\boldsymbol{\varrho}) \varepsilon^*_{\boldsymbol{k}}(-\boldsymbol{\varrho}) \rangle = \frac{1}{N} c^2 (1-c)^2 \exp(\mathrm{i}\boldsymbol{k} \cdot \boldsymbol{\varrho}) \,, \\ \langle c_{\boldsymbol{k}} \varepsilon^*_{\boldsymbol{k}}(\boldsymbol{\varrho}) \rangle = 0 \,, \quad \langle \varepsilon_{\boldsymbol{k}}(\boldsymbol{\varrho}) \varepsilon^*_{\boldsymbol{k}}(\boldsymbol{\varrho}') \rangle = 0 \quad \text{for } \boldsymbol{\varrho}' \neq \pm \boldsymbol{\varrho} \,. \end{gathered} \tag{1.4.3}$$

It can be concluded then that in ideal solutions the fluctuations of the concentration and the correlation parameters are statistically independent and for low concentrations the relative contribution made by the fluctuations of the correlation parameters is insignificant in comparison with that made by the concentration fluctuations (see (1.1.15)).

Finding the square of (1.4.1) and averaging the result we can easily see that in nonideal solutions $\langle c_k \varepsilon_k^*(\varrho)\rangle$ and $\langle \varepsilon_k(\varrho)\varepsilon_k^*(\varrho')\rangle$ are expressed in terms of the correlation parameters $\varepsilon(\varrho_1, \varrho_2)$ and $\varepsilon(\varrho_1, \varrho_2, \varrho_3)$ for a large number of sites given by (1.1.7). Then the variables c_k and $\varepsilon_k^*(\varrho)$ (as well as $\varepsilon_k(\varrho)$ and $\varepsilon_k^*(\varrho')$ with $\varrho' \neq \pm\varrho$) are generally statistically dependent. For small $\boldsymbol{k}$ the respective means can be found from the macroscopic theory of fluctuations but the dependence of the thermodynamic potential on $\varepsilon(\varrho)$ must be known for that.

The fluctuations of the correlation parameters in solution can be analyzed within the framework of the model of pair interactions discussed in Sect. 1.3. For instance, in the nearest-neighbour approximation the mean number N_{AA} of the pairs of the neighbouring atoms A and the fluctuations of this number can be simply expressed in terms of the derivatives of the thermodynamic potential Φ with respect to the ordering energy w_1 for the nearest neighbours:

$$\langle N_{AA}\rangle = \frac{\partial \Phi}{\partial w_1}, \quad \langle (N_{AA} - \langle N_{AA}\rangle)^2\rangle = \frac{\partial^2 \Phi}{\partial w_1^2}. \tag{1.4.4}$$

Since $N_{AA} = \frac{1}{2}Nzc^2 + \frac{1}{2}N|\sum_{\varrho}{}' \varepsilon(\varrho)$ where summation is performed over the nearest sites, we can find from (1.4.4) the fluctuations of $\varepsilon(\varrho)$ in a volume element containing $N = N'$ atoms.

Taking an expansion of Φ in powers of $w_1/k_B T$ (see, for instance, Sect. 17 in [1.23]) and ignoring the terms proportional to $w_1(w_1/k_B T)^3$ we can demonstrate that for solutions with the simple cubic and bcc lattices only the following terms do not vanish:

$$\begin{aligned}\langle[\varepsilon(\varrho) - \varepsilon]^2\rangle &= \langle[\varepsilon(\varrho) - \bar\varepsilon][\varepsilon(-\varrho) - \bar\varepsilon]\rangle \\ &= \frac{1}{N}c^2(1-c)^2\left[1 - (1-2c)^2\frac{w_1}{k_B T}\right], \quad \bar\varepsilon = \langle\varepsilon(\varrho)\rangle.\end{aligned} \tag{1.4.5}$$

Here the vector ϱ lies between the neighbouring sites. In a higher-order approximation in $w_1/k_B T$ products of other fluctuations of the correlation parameters also do not vanish. Thus when we take into consideration nonideality of the solution in this approximation we obtain an additional factor $[1 - (1-2c)^2 w_1/k_B T]$ in equations (1.4.3) for small $\boldsymbol{k}$. It should be noted that, according to (1.4.4), the fluctuations of the correlation parameters for the nearest neighbours must exhibit the same behaviour as the specific heat when we approach the critical point (that is, the fluctuations must be much weaker than the concentration fluctuations).

The contribution of the correlation parameters may be especially significant for heterogeneous systems in thermodynamic equilibrium which are produced owing to strong long-range forces (see Sect. 2.5). Large fluctuations of the many-body and pair correlation parameters are large in systems possessing such heterogeneous local order (cf. Sect. 1.1).

Relatively large fluctuations of the long-range and short-range order parameters and the concentration (in solid solutions) can occur also near the first-order phase transitions if the interphase surface energy is small. Since the specific thermodynamic potentials of both phases are identical, as noted by *Frenkel* [1.75, 76] the least work required for the fluctuation production of a new-phase particle within the matrix is sharply decreased at the transition point. The probability of such heterophase fluctuations near the transition point is limited only by the interphase interfacial energy. This probability is considerable even for particles of relatively large sizes in systems with a very small interphase energy per atom which is lower by several orders of magnitude than $k_B T$ (for instance, owing to accidental reasons). On the other hand, production of such particles of a new phase with a different structure can clearly be regarded also as large fluctuations of the longrange and short-range order. (Note a qualitative difference between the fluctuations near phase transitions of the first and second order). Continuous branched fluctuations $\delta\eta$ of the long-range order parameter exhibiting complicated variations in space occur in a large volume region near a second-order phase transition and for large N' the fluctuation probability monotonically decreases with an increase in $\delta\eta$ for large N. On the other hand, near a first-order phase transition the heterophase fluctuations correspond to a finite variation $\delta\eta$ (which is not proportional to $1/\sqrt{N'}$) but only in some fairly strictly limited crystal regions [1.75, 76]. Under such conditions a particularly large contribution must naturally be made by the fluctuations of the correlation parameters which accompany such local fluctuations of the long-range order parameters and the concentration.) If in solid solutions the new-phase particles have a changed composition large concentration fluctuations also occur. In most cases the interfacial energy seems to be sufficiently large so that intense production of the heterophase fluctuations is inhibited (in the absence of long-range forces) and fewer reasons can be given for describing them in macroscopic terms.

1.5 Static Displacements Caused by Fluctuation Waves of Composition, Order and Correlation Parameters

Although the analysis of fluctuations in this chapter ignores the accompanying elastic distortions, the fluctuations of concentrations and internal parameters in real crystals always produce some distortions. We shall discuss here only the static displacements caused by a certain given distribution of fluctuations (the effects of the respective elastic energy on the fluctuation probability will be discussed in Sect. 2.1).

It has been noted above that the static displacements can be calculated with a much simpler procedure using the Fourier representation and the method of fluctuation waves [1.7–14]. This method has been applied to the analysis of static displacements produced by finite defects of an arbitrary type (see Sects. 2.2–24 in [1.1]). The results derived for crystals with defects are clearly applicable directly to low-concentration substitutional or interstitial solutions since the impurity atoms can be regarded as defects.

In high-concentration solutions the static displacements exhibit some additional features caused, for instance, by the fluctuations of the order and correlation parameters. The latter produce static displacements also in one-component ordered crystals. The method of fluctuation waves can be also conveniently used for treating these more complicated systems [1.8, 10, 11, 20].

The static displacements caused by the fluctuations of concentration, the correlation parameters and the long-range order parameters in high-concentration ordered solid solutions and one-component crystals will be first treated in Sect. 1.5.1 in the limit of long waves (smoothly varying fluctuations). In these conditions we can use the macroscopic approach which simplifies the results and in some cases relates the displacement characteristics to the parameters found from independent experiments. Later in Sect. 1.5 we shall apply the microscopic treatment to the disordered high-concentration solutions including the case of short wave fluctuations (with $k \sim \pi/d$). In particular, we shall pay a closer attention to the effects caused by the fluctuations of the correlation parameters and analyze the conditions under which they can become significant.

1.5.1 Long-Wavelength Limiting Case

Disordered Substitutional Solutions. In disordered solutions with a single atom per unit cell the Fourier series for the static displacements can be written as

$$\boldsymbol{u}_s = \mathrm{i}\sum_{\boldsymbol{k}} \boldsymbol{u}_{\boldsymbol{k}} \exp(-\mathrm{i}\boldsymbol{k}\cdot\boldsymbol{R}_s)\,, \quad \boldsymbol{u}_{-\boldsymbol{k}} = -\boldsymbol{u}^*_{\boldsymbol{k}}\,, \quad \mathrm{i} = \sqrt{-1}\,. \tag{1.5.1}$$

Here (and below) the displacements $\boldsymbol{u}_s$ are taken from the lattice sites of the "averaged crystal". The static displacements are determined by the distributions of the atoms A and B over the lattice sites of the binary solution $A-B$, that is, by the occupation numbers c_s, and they can be written as linear functions of the variables $c_s, c_{s'}, c_{s''}, \ldots$ where the coefficients depend only on the vector differences $\boldsymbol{R}_s - \boldsymbol{R}_{s'}, \boldsymbol{R}_s - \boldsymbol{R}_{s''}, \ldots$. The Fourier coefficients $\boldsymbol{u}_{\boldsymbol{k}}$ are expressed accordingly in terms of the Fourier components $c_{\boldsymbol{k}}$ of the concentration fluctuations with the same $\boldsymbol{k}$ and in terms of the products $c_{\boldsymbol{k}'}, c_{\boldsymbol{k}-\boldsymbol{k}'}, \ldots$ with different $\boldsymbol{k}'$ but a given $\boldsymbol{k}$. As shown by (1.4.2) such products $c_{\boldsymbol{k}'}, c_{\boldsymbol{k}-\boldsymbol{k}'}$ can be written as a linear function of the $\boldsymbol{k}$th Fourier components $\varepsilon_{\boldsymbol{k}}(\varrho)$ of the correlation parameters for different distances ϱ. Therefore $\boldsymbol{u}_{\boldsymbol{k}}$ are linearly related to the Fourier components of the fluctuations of the concentration and the correlation parameters which correspond to the same wave vector $\boldsymbol{k}$:

$$\boldsymbol{u}_{\boldsymbol{k}} = \boldsymbol{A}_{\boldsymbol{k}} c_{\boldsymbol{k}} + \sum_{\varrho\neq 0} \boldsymbol{A}_{\boldsymbol{k}}(\varrho)\varepsilon_{\boldsymbol{k}}(\varrho) + \ldots \tag{1.5.2}$$

We have not written here explicitly the often insignificant terms $\boldsymbol{A}_{\boldsymbol{k}}(\varrho_1, \varrho_2)\varepsilon_{\boldsymbol{k}}(\varrho_1 \varrho_2), \ldots$ which describe the fluctuations of the correlation parameters for three or more lattice sites.

The equilibrium correlation parameters vary when the solution concentration is changed. A convenient approach is to identify in $\varepsilon_{\boldsymbol{k}}(\varrho)$ the component

$[d\varepsilon_k(\varrho)/dc_k]c_k$ which is related by equilibrium equations to the Fourier components c_k of the concentration fluctuations. Then the coefficients $\boldsymbol{A}_k$ in (1.5.2) are altered and this equation can be rewritten as

$$\begin{aligned} \boldsymbol{u}_k &= \boldsymbol{A}_k c_k + \sum_{\varrho\neq 0} \boldsymbol{A}_k(\varrho)\tilde{\varepsilon}_k(\varrho) + \dots , \\ \boldsymbol{A}_k &= \boldsymbol{A}_k^c + \sum_{\varrho\neq 0} \frac{\partial \varepsilon_k(\varrho)}{\partial c_k} \boldsymbol{A}_k(\varrho) + \dots , \\ \tilde{\varepsilon}_k(\varrho) &= \varepsilon_k(\varrho) - \frac{d\varepsilon_k(\varrho)}{dc_k} c_k . \end{aligned} \tag{1.5.3}$$

The transition from $\varepsilon_k(\varrho)$ to $\tilde{\varepsilon}_k(\varrho)$ is convenient, in particular, because the variables c_k and $\varepsilon_k(\varrho)$ are statistically independent. It follows from the fact that for any concentration we have $\partial\Phi/\partial\varepsilon_k(\varrho) = 0$, that is,

$$\frac{\partial^2\Phi}{\partial c_k^* \partial\varepsilon_k(\varrho)} + \frac{\partial^2\Phi}{\partial\varepsilon_k(\varrho)\partial\varepsilon_k^*(\varrho)} \frac{d\varepsilon_k(\varrho)}{dc_k} = 0$$

(see the derivation of (1.2.28)). It is precisely this expression that appears as the coefficient $c_k^*\tilde{\varepsilon}_k(\varrho)$ in the formula for $\mathfrak{N}_k$ (see (1.1.25)). When nonideality of the solution is ignored we have $d\varepsilon_k(\varrho)/dc_k = 0$ that is, $\tilde{\varepsilon}_k(\varrho)$ and $\varepsilon_k(\varrho)$ (as well as $\boldsymbol{A}_k^c$ and $\boldsymbol{A}_k$ coincide.

For small $\boldsymbol{k}$ we can calculate the coefficients $\boldsymbol{A}_k$ and $\boldsymbol{A}_k(\varrho)$ in equation (1.5.3) for the static displacements with the use of the macroscopic elasticity theory in the same way as the amplitudes of the static displacement waves caused by the defect concentration fluctuations are calculated in Sect. 2.2 in [1.1]. We should bear in mind that in this case the concentration stresses are produced not only by the concentration fluctuations but also by the the fluctuations of the correlation parameters. Using the arguments of Sect. 2.2 in [1.1] we obtain

$$\boldsymbol{A}_k = \hat{C}_k^{-1}\boldsymbol{P}_k , \quad \boldsymbol{A}_k(\varrho) = \hat{C}_k^{-1}\boldsymbol{P}_k(\varrho) . \tag{1.5.4}$$

Here $\hat{C}_k^{-1}$ is a reciprocal of the tensor $\hat{C}_k$. It follows from (2.2.17) in [1.1] that the tensor components C_{kij} and the components of the vectors $\boldsymbol{P}_k$ and $\boldsymbol{P}_k(\varrho)$ are expressed in terms of the elastic moduli $\hat{c}$ and the parameters $\hat{L}$ and $\hat{L}(\varrho)$ which describe the crystal strain u caused by a uniform variation of the concentration or the parameters $\varepsilon(\varrho)$:

$$\begin{aligned} C_{kij} &= c_{ijlm}n_i n_l , \quad P_{ki} = \frac{1}{k} c_{ijlm} n_j L_{lm} , \\ P_{ki}(\varrho) &= \frac{1}{k} c_{ijlm} n_j L_{lm}(\varrho) , \quad \boldsymbol{n} = \frac{\boldsymbol{k}}{k} , \end{aligned} \tag{1.5.5}$$

where

$$L_{ij} = \frac{du_{ij}}{dc} ; \quad L_{ij}(\varrho) = \frac{\partial u_{ij}}{\partial\varepsilon(\varrho)} . \tag{1.5.6}$$

According to (1.5.3), A_k accounts also for the distortions caused by the component in the variation of $\varepsilon(\varrho)$ which is adjusted to the concentration variation at equilibrium and therefore (1.5.6) includes the total derivative d/dc accounting for the dependences of $\varepsilon(\varrho)$ on c and of u_{ij} on $\varepsilon(\varrho)$ while the partial derivative $(\frac{\partial u_{ij}}{\partial c})_{\varepsilon(\varrho)}$ enters into the respective equation for A_k^c. It is the total derivative that is found directly from the experimental results on the behaviour of the equilibrium size and shape of the unit cell as functions of the solution concentration. The derivative with respect to $\varepsilon(\varrho)$ in (1.5.6) is calculated for a constant concentration and constant other parameters $\varepsilon(\varrho')$.

For high-symmetry crystals the expressions (1.5.4, 6) have much simpler forms (see Sect. 2.2.4 and Appendix B in [1.1]). In our case of substitutional solutions the symmetry of the tensor L_{lm} is determined by the crystal system (the symmetry of the Bravais lattice of the crystal). For instance, the matrix L_{lm} has the diagonal form for all crystals with the exception of the triclinic and monoclinic ones:

$$L_{ij} = \frac{1}{d_i}\frac{d(d_i)}{dc}\delta_{ij}\ . \tag{1.5.6a}$$

In the above equation d_i is the length of a segment between two lattice points parallel to the ith axis and δ_{ij} is the Kroneker delta.

The solution of the system of equations (1.5.4) for A_k can be obtained in the following explicit form for cubic crystals [1.77]:

$$A_{kx} = \frac{1}{3}\frac{c_{11}+2c_{12}}{kD(n)}(1+\xi n_y^2)(1+\xi n_z^2)\frac{1}{v}\frac{dv}{dc}n_x\ . \tag{1.5.7}$$

Here v is the unit cell volume, and c_{11}, c_{12}, and c_{44} are the elastic moduli,

$$\begin{aligned} D(n) &= c_{11} + \xi(c_{11}+c_{12})(n_x^2n_y^2+n_x^2n_z^2+n_y^2n_z^2) \\ &\quad + \xi^2(c_{11}+2c_{12}+c_{44})n_x^2n_y^2n_z^2\ , \\ \xi &= \frac{c_{11}-c_{12}-2c_{44})}{c_{44}}\ . \end{aligned} \tag{(1.5.7a)}$$

The expressions for A_{ky} and A_{kz} are obtained by cyclic replacement of the subscripts in (1.5.7). It can be seen from (1.5.7) that A_k is inversely proportional to k and depends on the direction of the unit vector $n = k/k$.

For elastically isotropic crystals the anisotropy parameter $\xi = 0$ and A_k is given by the simple expression

$$A_k = \frac{1+\sigma}{3(1-\sigma)}\frac{1}{v}\frac{dv}{dc}\frac{k}{k^2} \tag{1.5.7b}$$

where σ is the Poisson ratio.

The tensor $L_{ij}(\varrho)$ is associated with the preferred direction ϱ and therefore it has a lower symmetry than L_{ij}. For instance, for the bcc lattice the components of the tensor $L_{ij}(\varrho)$ for the pairs of the nearest neighbouring sites can be written as

$$L_{xx}(\varrho) = L_{yy}(\varrho) = L_{zz}(\varrho) = \tfrac{1}{3}[2L_1(\varrho) + L_3(\varrho)] \,,$$

$$L_{xy}(\varrho) = L_{xz}(\varrho) = L_{yz}(\varrho) = \tfrac{1}{3}[L_3(\varrho) - L_1(\varrho)] \,.$$

Here $L_3(\varrho)$ and $L_1(\varrho)$ are the principal values of the tensor $\hat{L}(\varrho)$ which can be seen from (1.5.6) to determine the strains along the axis $\langle 111 \rangle \| \varrho$ and in the perpendicular direction caused by fluctuations of $\varepsilon(\varrho)$. Under such conditions $\hat{L}(\varrho)$ is described by two parameters. For substitutional fcc solutions $\hat{L}(\varrho)$ is dependent on three parameters $L_1(\varrho)$, $L_2(\varrho)$, and $L_3(\varrho)$ which determine the strain along the axes [110], [110], and [001] for $\varrho \| [110]$:

$$L_{xx}(\varrho) = L_{yy}(\varrho) = \tfrac{1}{2}[L_1(\varrho) + L_2(\varrho)] \,, \quad L_{zz}(\varrho) = L_3(\varrho) \,,$$

$$L_{xy}(\varrho) = \tfrac{1}{2}[L_1(\varrho) - L_2(\varrho)] \,, \quad L_{xz}(\varrho) = L_{yz}(\varrho) = 0 \,.$$

For such $L_{ij}(\varrho)$ one can find $\boldsymbol{A}_k(\varrho)$ from (2.2.4) in [1.1].

The experimental data on the effect produced by annealing (giving rise to the short-range order) on the lattice parameters enable us to find only a single combination of the tensor components $L_{ij}(\varrho)$ even if we can ignore correlation outside the first coordination sphere and assume the variations of $\varepsilon(\varrho)$ caused by annealing to be known. Therefore, special experiments must be conducted with external forces violating the cubic symmetry in order to determine fully the tensors $L_{ij}(\varrho)$. One may use, for instance, experimental results on the variation of the lattice parameter caused by the short-range order produced in crystals by elastic stresses or a magnetic field (in ferromagnetics) or the results on internal friction associated with the short-range order. In many alloy systems the lattice parameters exhibit a much weaker dependence on the correlation parameters than on the composition and hence distortions are chiefly produced by the concentration fluctuations. The appropriate conditions are given in Sect. 1.5.2.

Distortions Associated with the Long-Range-Order Parameters. In single-component crystals inhomogeneities are associated with the fluctuations of the long-range-order parameters η_μ and the correlation parameters $\varepsilon_{\mu\mu'}(\varrho)$. In the limit of long waves we can ignore the difference between the displacements of atoms in various sublattices. Then the Fourier components $\boldsymbol{u}_k$ of the static displacements caused by fluctuations of the long-range order parameters and the correlations are given by an equation similar to (1.5.3):

$$\boldsymbol{u}_k = \sum_{\mu} \boldsymbol{A}_{\mu k} \eta_{\mu k} + \sum_{\mu\mu'} \sum_{\varrho \neq 0} \boldsymbol{A}_{\mu\mu' k}(\varrho) \tilde{\varepsilon}_{\mu\mu' k}(\varrho) + \dots \qquad (1.5.8)$$

Here $\tilde{\varepsilon}_{\mu\mu' k}(\varrho)$ are related to the components of the fluctuations of $\varepsilon_{\mu\mu'}(\varrho)$ which do not follow the fluctuations of η_μ. The respective coefficients $\boldsymbol{A}_{\mu\mu' k}(\varrho)$ are given by (1.5.4–6) where the subscripts $\mu\mu'$ must be added to $\boldsymbol{A}_k(\varrho)$, $L_{ij}(\varrho)$, and $\varepsilon(\varrho)$. The coefficients $\boldsymbol{A}_{\mu k}$ take into account the contributions to distortions made by the components of the correlation parameters which follow the fluctuations of the order parameters:

$$A_{\mu k} = \hat{C}_k^{-1} P_k(\hat{L}_\mu) , \quad L_{\mu ij} = \frac{\partial u_{ij}}{\partial \eta_\mu} , \tag{1.5.9}$$

Here the relationship between P_k and $\hat{L} = \hat{L}_\mu$ is given by (1.5.5) and the derivatives with respect to η_μ are calculated for constant other parameters $\eta_{\mu'}$ but for the correlation parameters $\varepsilon_{\mu\mu'}(\varrho)$ which adjust to the fluctuations of the order parameters.

If the long-range order is described by a single parameter $\eta(\mu = 1)$ then $L_{\mu ij}$ (and, hence, $A_{\mu k}$) can be found directly from the results on the variation of the lattice parameter caused by ordering. In the case of intrinsic ferroelectrics when the components P_l of the polarization vector may be taken as the order parameters η_μ the tensor components $L_{\mu ij}$ are found from the data on the behaviour of the lattice parameters as functions of P (or of the external electric field). It is more difficult to determine $L_{\mu ij}$ in the general case of a multicomponent order parameter.

The lattice parameters are even functions of η_μ in the crystals which have a center of inversion. Hence, in disordered crystals we have $L_{\mu ij} = 0$ and the fluctuations of the long-range order parameter do not make a contribution to the long-wave fluctuations of the static displacements so that $A_{\mu k} \neq 0$ only in the ordered phase. When the crystal does not have a center of inversion, however, in some cases (for instance, if a crystal is piezoelectric when it is not in the ferroelectric phase) a fluctuation $\delta\eta_\mu$ of the long-range order parameter in the disordered phase produces strains which are linear in $\delta\eta_\mu$ and $L_{\mu ij} = 0$. In such crystals fluctuations of the long-range order parameter cause static displacements both in the ordered and disordered phases.

In the ordered solid solutions the displacements are caused by the fluctuations of concentration and the fluctuations of the parameters of long-range order and correlation. Thus u_k is given by a sum of (1.5.3) and (1.5.8). Under such conditions in order to find L_{ij} and $L_{\mu ij}$ we must know the dependence of the shape and size of the unit cell on the composition and on the order parameters.

The experimental results demonstrate that in many solid solutions the derivatives of the lattice parameter d with respect to the order parameter are considerably smaller than the derivative with respect to the concentration. Under such conditions we may take into consideration only the distortions $u_k = A_k c_k$ in (1.5.3) caused by the fluctuations of the composition. It should be noted that the dependence of d on η_μ is generally described by the same small parameters as the dependence on $\varepsilon(\varrho)$. Therefore, it would be inconsistent in the general case to take into consideration the effects caused by the fluctuations of the order parameters η_μ and ignore those caused by the fluctuations of the correlation parameters. The fluctuations of η_μ may, however, make a major contribution in some cases, for instance, in the vicinities of phase transitions.

For the ordered solid solutions the Fourier components of the static displacements of atoms occupying sites in the sublattice γ can be expressed in terms of the Fourier components of the concentration fluctuations in various sublattices γ' by the following equation which is valid for any k:

$$u_{k\gamma} = \sum_{\gamma'=1}^{\nu} A_{k\gamma\gamma'} c_{k\gamma'} \; . \qquad (1.5.10)$$

In the limit of small k the coefficients $A_{k\gamma\gamma'}$ are independent of γ and given by the equations (1.5.4–6) for A_k where additional summation over γ' must be performed.

These equations can also be applied to the interstitial solid solutions if the interstices are regarded as sublattice sites occupied by interstitials and vacancies. It should be born in mind here that if the interstitial atoms occupy interstices of various types, for instance, octahedral interstices with the orientations X, Y, Z in bcc crystals then the symmetry of the respective tensors $L_{\mu ij}$ is lower than the symmetry of L_{ij} for the substitutional solutions. When the concentration of interstitials is low they can be regarded as single defects and the static displacements can be calculated from the equations derived in Chap. 2 of [1.1].

1.5.2 Fluctuation Waves of Static Displacements in Concentrated Solutions

When we analyze the intensities of scattering and other related phenomena we must know the amplitudes of the static displacement waves not only in the long-wave limit but also in the case of large $\boldsymbol{k}$. These amplitudes can be found only in the framework of the microscopic theory using a specific model of an imperfect crystal. Such a model has been treated in Sect. 2.3 of [1.1] for the case of low defect concentration under the assumption that the force constants in the free energy expansion remain unchanged in the presence of defects. This model yields a linear dependence of $\boldsymbol{u}_k$ on c_k which is reasonable for low defect concentrations.

In high-concentration solutions, however, a significant effect on the static displacements may be produced by higher-order terms in the expansion of $\boldsymbol{u}_k$ in powers of c_k or (what is equivalent) by the fluctuations of correlation. A more general model of imperfect crystals is applicable to concentrated solutions and it must be employed for analyzing these effects and finding the applicablity conditions for the simpler model of Sect. 2.3 in [1.1]. Such a model has been used for analyzing the static displacements in solutions [1.13, 20] (it was also used for analyzing the static and thermal displacements in alloys though with a somewhat different approach based on the equation for the hamiltonian rather than the free energy [1.78]).

Let us apply the above general model to the disordered solid solutions with a single atom per unit cell. In accordance with the Born theory of vibrations we shall regard the free energy F as a function of the static displacements $\boldsymbol{u}_s$ and the occupation numbers c_s (averaged over thermal vibrations for the given $\boldsymbol{u}_s$ and c_s) and expand it in powers of $\boldsymbol{u}_s$ retaining only the quadratic terms. Since the force constants $W_{ss'}$ and $V_{ss'}$ depend on the species of atoms occupying the sites s and s' we can write

$$F = F_0 - \sum_{ss'} W_{ss'i}^{aa'} u_{si} + \frac{1}{2} \sum_{ss'} V_{ss'ij}^{aa'} u_{si} u_{s'j} \; . \qquad (1.5.11)$$

The constants W and V generally depend on the composition of the solution, the correlation parameters and the temperature. It is taken into account in (1.5, 11) that the force with which atom s' acts on atom s depends on the displacement of the former and the species of both atoms. The fact that this force can depend on the species of all other neighbouring atoms is ignored, however. It can be shown that this dependence does not produce qualitative changes in the results (it gives rise only to small quantitative changes). The model described by (1.5.11) for the free energy (in contrast to the simpler model in Sect. 2.3 of [1.1]) makes it possible to describe the dependence of the elastic moduli on the concentration c, to take into account deviations from Vegard's rule (a linear dependence of the lattice parameters on c), and to analyze the effect of the fluctuations of the correlation parameters on the static distortions.

The constants $W^{\alpha\alpha'}_{ss'i}$ have the values $W^{AA}_{ss'i}$, $W^{AB}_{ss'i}$, and $W^{BB}_{ss'i}$. They can easily be expressed in terms of the numbers $c_s = 1, 0$ with the use of (1.1.1):

$$W^{\alpha\alpha'}_{ss'i} = W^0_{ss'i} + W_{ss'i}(c_s + c_{s'} - 2c) + W'_{ss'i}(c_s - c)(c_{s'} - c)\,, \qquad (1.5.12)$$

Here

$$\begin{aligned} &W'_{ss'i} = W^{AA}_{ss'i} + W^{BB}_{ss'i} - 2W^{AB}_{ss'i}\,; W_{ss'i} = W^{AB}_{ss'i} - W^{BB}_{ss'i} + cW'_{ss'i}\,;\\ &W^0_{ss'i} = W^{BB}_{ss'i} + 2cW_{ss'i} - c^2 W'_{ss'i}\,. \end{aligned} \qquad (1.5.13)$$

Similar equations can be written for $V^{\alpha\alpha'}_{ss'ij}$ for $s' \neq s$ (for $s = s'$ we have the condition $\sum_{s'} V^{\alpha\alpha'}_{ss'ij} = 0$).

The static displacements of atoms in the solid solution are determined by the equilibrium conditions according to which the derivatives of the free energy (1.5.11) with respect to u_{si} must vanish. When we express $W^{\alpha\alpha'}_{ss'}$ and $V^{\alpha\alpha'}_{ss'}$ in terms of c_s these conditions can be written as the following system of equations:

$$\begin{aligned} &\sum_{s'} [V_{ss'ij} + V'_{ss'ij}(c_s + c_{s'} - 2c) + V''_{ss'ij}(c_s - c)(c_{s'} - c)](u_{s'j} - u_{sj})\\ &\quad = \sum_{s'} [W_{ss'i}(c_s + c_{s'} - 2c) + W'_{ss'i}(c_s - c)(c_{s'} - c)]\,. \end{aligned} \qquad (1.5.14)$$

The sum $\sum_{s'} W^0_{ss'i}$ on the right-hand side of (1.5.14) is omitted because it vanishes since we have $W^{\alpha\alpha'}_{ss'i} = -W^{\alpha'\alpha}_{s'si} = -W^{\alpha\alpha'}_{s'si}$.

We shall assume below that $V'_{ss'}$ and $V''_{ss'}$ are considerably smaller than $V_{ss'}$ and we shall find the solution of the system of equations (1.5.14) by successive approximations. This assumption is valid when the differences between the elastic moduli of solid solutions with different compositions are much smaller than these moduli (this is the case for many binary systems). In the zero approximation we take $V'_{ss'} = V''_{ss'} = 0$. The solution for the resulting system of equations can be easily found with Fourier transformations as it was done for a similar system of equations (2.35) in [1.1]. In contrast to the latter system, however, the right-hand side of (1.5.14) includes, apart from the differences $c_s - c$ describing the concentration fluctuations, the products $(c_s - c)(c_{s'} - c)$ whose deviations from the mean $\varepsilon(\varrho)$ determine the fluctuations of the correlation parameters ($\varrho = R_s - R_{s'}$). After multiplying (1.5.14) by $\exp(i\boldsymbol{k}\cdot\boldsymbol{R}_s$, adding up the results and making use of (1.1.8), (1.4.1) and (1.5.1) for the Fourier expansions of $\boldsymbol{u}_s$, $c_s - c$ and $(c_s - c)(c_{s'} - c)$ we obtain a system of three linear equations which express the Fourier components $\boldsymbol{u_k}$ of the static displacements in terms of the Fourier components $c_{\boldsymbol{k}}$ and $\varepsilon_{\boldsymbol{k}}(\varrho)$ of the concentration and correlation parameters:

$$\begin{aligned}
&V_{\boldsymbol{k}ij}u_{\boldsymbol{k}j} = W_{\boldsymbol{k}i}c_{\boldsymbol{k}} + \sum_{\varrho} W'_{\boldsymbol{k}i}(\varrho)\varepsilon_{\boldsymbol{k}}(\varrho)\,,\\
&W_{\boldsymbol{k}i} = -\sqrt{-1}\sum_{s'} W_{ss'i}\exp(i\boldsymbol{k}\cdot\boldsymbol{R}_{ss'})\,,\\
&W'_{\boldsymbol{k}i}(\varrho) = W'_i(\varrho) = -\sqrt{-1}W'_{ss'i}\,,\\
&V_{\boldsymbol{k}ij} = \sum_{s'} V_{ss'ij}\exp(i\boldsymbol{k}\cdot\boldsymbol{R}_{ss'})\,.
\end{aligned} \tag{1.5.15}$$

We can see from (1.5.15) that in this model the Fourier components of the static displacements are linear functions of not only the Fourier components of the concentration but also of the Fourier components of the correlation parameters $\varepsilon_{\boldsymbol{k}}(\varrho)$.

A solution of the system of equations (1.5.15) can clearly be written as

$$\boldsymbol{u_k} = \boldsymbol{A}^{c}_{\boldsymbol{k}}c_{\boldsymbol{k}} + \sum_{\varrho\neq 0} \boldsymbol{A}_{\boldsymbol{k}}(\varrho)\varepsilon_{\boldsymbol{k}}(\varrho)\,, \tag{1.5.16}$$

where $A^{c}_{\boldsymbol{k}i}$ and $A_{\boldsymbol{k}i}(\varrho)$ are given by the equations

$$\begin{aligned}
&V_{\boldsymbol{k}ij}A^{c}_{\boldsymbol{k}i} = W_{\boldsymbol{k}i} \quad (i = 1,2,3)\,,\\
&V_{\boldsymbol{k}ij}\boldsymbol{A}_{\boldsymbol{k}j}(\varrho) = W'_i(\varrho) \quad (i = 1,2,3)\,.
\end{aligned} \tag{1.5.17}$$

Equation (1.5.16) can be also written in the form of (1.5.3) if we make the transition from $\varepsilon_{\boldsymbol{k}}(\varrho)$ to $\tilde{\varepsilon}_{\boldsymbol{k}}(\varrho)$ and from $\boldsymbol{A}^{c}_{\boldsymbol{k}}$ to $\boldsymbol{A}_{\boldsymbol{k}}$.

In order to find $\boldsymbol{u_k}$ with the iteration method in the first-order approximation in $V'_{ss'}$ and $V''_{ss'}$ we must first determine the displacements $\boldsymbol{u}^0_s$ in the zero approximation from (1.5.1, 16), substitute the results into the terms with $V'_{ss'}$ and $V''_{ss'}$ in equations (1.5.14) and regard the resulting expressions as additional terms on the right-hand sides of these equations. Then we can use the Fourier transformation to find the solution of the system of equations for $\boldsymbol{u}_s$ as was done in the zeroth approximation and obtain the first-order approximation for $\boldsymbol{u_k}$. The coefficients $W_{\boldsymbol{k}i}$ and $W'_{\boldsymbol{k}i}(\varrho)$ in (1.5.15) are changed as a result. For instance, in low-concentration solutions we can use (2.4.33) in [1.1] for renormalizing the constants $W_{ss'i}$ in the expressions for $W_{\boldsymbol{k}i}$ under the change V' of the force constants. We obtain the replacement $W_{ss'i} \rightarrow W_{ss'i} - V'_{ss_1s'il}u_{s_1l}$ where for small V' we have $u_{s_1l} \approx u^0_{s_1l}$ and obtain them directly from (2.4.33) in [1.1] if V' are not small.

When we take into account the terms with V' and V'' in addition to renormalization of the coefficients, on the right-hand side of the equations (1.5.15) we obtain, besides the terms proportional to c_k and $\varepsilon_k(\varrho)$, new terms proportional to the Fourier components of the fluctuations of the correlation parameters (1.1.7), $\varepsilon_k(\varrho_1, \varrho_2)$ and $\varepsilon_k(\varrho_1, \varrho_2, \varrho_3)$ which corresponds to triplets and quadruplets of lattice sites, rather than pairs. The fluctuations of such complex correlation functions do not change the results qualitatively but produce small quantitative corrections if the dependence of the elastic moduli on the composition is weak ($V'_{ss'}$ and $V''_{ss'}$ are small) and especially if the concentration is low (in this case we have $\langle \varepsilon_k(\varrho_1, \varrho_2)|^2\rangle \sim c^3$). Therefore, we shall limit the discussion to the fluctuations of only binary correlation parameters and assume that the appropriate changes have been already made in the coefficients $W_{ss'i}$ and $W'_{ss'}$ in W_{ki} and $W'_i(\varrho)$.

According to (1.5.2), A_k^c and $A_k(\varrho)$ describe the Fourier components of the static displacements. They can be found by solving the system of linear equations (1.5.17) if one knows the force constants $V_{ss'ij}$, $W_{ss'j}$ and $W_{ss'i}$ which determine the coefficients in (1.5.17), according to (1.5.15). In principle, these force constants can be calculated with the microscopic models of interatomic interaction in crystals. In many cases, however, it is more convenient to make use of a semiphenomenological model in which only several force constants describing the interaction between the neighbouring atoms do not vanish. The parameters $V_{ss'ij}$ definitively determine the elastic moduli c_{pr} of the crystal, $W_{ss'i}$ determine the components of the tensor L_{ij} which describes the concentration dependence of the lattice parameters according to (1.5.6) (see, for instance, (2.3.9, 11) in [1.1]), and $W_{ss'i}$ determine the components of the tensor which describes the dependence of the lattice parameters on $\varepsilon(\varrho)$. Thus, if we select a model in which the number of the independent parameters $V_{ss'ij}$ is equal to the number of the independent elastic moduli and the number of the parameters $W_{ss'i}$ and $W'_{ss'i}$ is equal to the number of the independent components of the tensors L_{ij} and $L_{ij}(\varrho)$ we can express the force constants in terms of parameters which can be found in independent experiments.

For instance, the model of noncentral interaction between nearest neighbours may be applied to the fcc substitutional solutions. If we express the three independent parameters $V_{ss'ij}$ of this model in terms of the elastic moduli c_{11}, c_{12}, and c_{44} of a cubic crystal and the independent parameter $W_{ss'i}$ in terms of the derivative dv/dc (where v is the unit cell volume) we can find expressions for the coefficients V_{kij} and W_{ki} in (1.5.17) ([1.9], see Sect. 2.3 and Appendix C in [1.1]):

$$\begin{aligned} V_{kxx} &= dc_{11}\left[2 - \cos\frac{kdn_x}{2}\left(\cos\frac{kdn_y}{2} + \cos\frac{kdn_z}{2}\right)\right] \\ &\quad + d(2c_{44} - c_{11})\left(1 - \cos\frac{kdn_y}{2}\cos\frac{kdn_z}{2}\right), \end{aligned}$$

$$
\begin{aligned}
V_{kxy} &= V_{kyx} = d(c_{12}+c_{44})\sin\frac{kdn_x}{2}\sin\frac{kdn_y}{2} \,, \\
W_{kx} &= \frac{1}{12}d^2(c_{11}+2c_{12})\sin\frac{kdn_x}{2} \\
&\quad\times\left(\cos\frac{kdn_y}{2}+\cos\frac{kdn_z}{2}\right)\frac{1}{v}\frac{dv}{dc} \,.
\end{aligned}
\tag{1.5.18}
$$

Cyclic replacement of the subscripts x, y, z in (11.5.18) yields expressions for other coefficients.

The derivative dv/dc in the equation for W_{ki} is taken for constant correlation parameters. Appendix C in [1.1] gives similar equations for the substitutional bcc solutions and the interstitial solutions.

In the nearest-neighbour approximation the coefficients $W'_{ss'i}$ also do not vanish only for the nearest-neighbour pairs and are described by only one independent parameter. For an arbitrary pair of nearest neighbours at a distance ϱ we can write $W_{ss'i} \equiv W'_j(\varrho)$ in the form

$$
W'_i(\varrho) = \frac{z}{d}\varrho_i W' \,.
$$

The only parameter W' in this equation is expressed in terms of the derivative of the lattice parameter d with respect to the correlation parameter. To write this expression we must go over to the limit of $k \to 0$ in the equations (1.5.17) of the microscopic theory and compare the results with the respective equations for $A_k(\varrho)$ in the macroscopic theory, $\hat{C}_k A_k(\varrho) = P_k(\varrho)$ (see (1.5.4–6)). We can easily see [1.20] that

$$
W' = \frac{1}{24}d^2(c_{11}+2c_{12})\frac{1}{v}\frac{\partial v}{\partial \varepsilon} \,,
\tag{1.5.19}
$$

where the derivative $\partial v/\partial\varepsilon$ is determined by the variation of the atomic volume v caused by identical variations of the correlation parameters for all pairs of neighbours (rather than a single pair as in the case of the derivative $\partial v/\partial\varepsilon(\varrho)$). This derivative can be found from the experimental results on the changes in the lattice parameter produced by annealing. Then the equations (1.5.17) for $A_k(\varrho)$ for the fcc crystals can be rewritten in the nearest-neighbour approximation in the following form:

$$
V_{kij}A_{kj}(\varrho) = -\sqrt{-1}\frac{d}{12}(c_{11}+2c_{12})\frac{1}{v}\frac{\partial v}{\partial\varepsilon}\varrho_i \,,
\tag{1.5.20}
$$

where V_{kij} are given by (11.5.18). In the nearest-neighbour approximation the only nonvanishing $A_k(\varrho)$ correspond to the vectors ϱ passing through the lattice sites in the first coordination sphere and $W'_i(\varrho)$ and $A_k(\varrho)$ vanish for all other vectors ϱ.

The relative contributions of the effects caused by the fluctuations of the correlation parameters can be estimated by finding the mean squares and products of the components of the vector u_k and comparing the contributions made by the fluctuations of the concentration and the correlation parameters. The easiest estimates can be made for the ideal solutions in which the squared Fourier components for the fluctuations of the concentration and the correlation parameters are given by the simple equations (1.1.15) and (1.4.3) and these fluctuations are statistically independent, according to (1.4.3). Denote by $\boldsymbol{u}_{\boldsymbol{k}}^{(1)}$ and $\boldsymbol{u}_{\boldsymbol{k}}^{(2)}$ the Fourier components for the displacements caused by the fluctuations of the concentration and the correlation parameters and write, according to (1.1.15), (1.4.3) and (1.5.16),

$$\begin{aligned} &\langle u_{ki}^{(1)} u_{kj}^{(1)*} \rangle = \frac{1}{N} c(1-c) A_{ki} A_{kj} \,, \quad \langle u_{ki}^{(2)} u_{kj}^{(2)*} \rangle = \frac{1}{N} c^2 (1-c)^2 \frac{1}{k^2} f_{ij} \,, \\ &\frac{1}{k^2} f_{ij} = {\sum_{\varrho}}' A_{ki}(\varrho) A_{kj}(\varrho)(1 - \cos \boldsymbol{k} \cdot \boldsymbol{\varrho}) \,. \end{aligned} \tag{1.5.21}$$

In the nearest-neighbour approximation here summation over ϱ is performed only over the lattice sites in the first coordination sphere.

The coefficients $A_{ki}(\varrho)$ are found from (1.5.20). For small k we have $V_{kij} \sim k^2$ and $A_{ki}(\varrho) m \sim k^{-2}$, that is, $f_{ij} \equiv f_{ij}(\boldsymbol{n})$ are dependent only on the direction of the vector $\boldsymbol{k} = k\boldsymbol{n}$. In the general case of elastically anisotropic crystals the equations for f_{ij} prove to be too cumbersome even for small k (they are given in the explicit form in [1.20]). We shall write these equations here therefore only for elasticallv-isotropic fcc crystals:

$$\begin{aligned} f_{xx}(\boldsymbol{n}) &= \frac{1}{72}\left(\frac{1+\sigma}{1-\sigma}\right)^2 \left(\frac{1}{v}\frac{\partial v}{\partial \varepsilon}\right)^2 \{1 + n_x^2 + 2\nu'(1-n_x^2) \\ &\quad + \nu'^2[(1-n_x^2)^3 n_x^4 (1-n_x^2) + 2n_x^2 n_y^2 n_z^2]\} \,, \\ f_{xy}(\boldsymbol{n}) &= \frac{1}{72}\left(\frac{1+\sigma}{1-\sigma}\right)^2 \left(\frac{1}{v}\frac{\partial v}{\partial \varepsilon}\right)^2 n_x n_y [2 - \nu'(1+n_z^2) \\ &\quad - \nu'^2(1 - n_z^2 + 2n_z^4 - 2n_x^2 n_y^2)] \,, \quad \nu' = \frac{1}{1-2\sigma} \,. \end{aligned} \tag{1.5.22}$$

Let us compare, for instance, the mean squares of the static displacements $\boldsymbol{u}_{\boldsymbol{k}}^{(1)}$ and $\boldsymbol{u}_{\boldsymbol{k}}^{(2)}$ caused by the fluctuations of the concentrations and the correlation parameters in the case of the long-wave fluctuation waves with the wave vector $\boldsymbol{k}$ parallel to the axis [100] (the displacements are also parallel to this axis). Finding $f_{xx}(\boldsymbol{n})$ from (1.5.22) and A_{kx} from

$$\langle |\boldsymbol{u}_{\boldsymbol{k}}^{(2)}|^2 \rangle \ll \langle |\boldsymbol{u}_{\boldsymbol{k}}^{(1)}|^2 \rangle$$

$$\text{for} \quad \left(\frac{\partial v}{\partial c}\right)_{\varepsilon}^2 \gg \frac{1}{8}(1+\nu'^2) c(1-c) \left(\frac{\partial v}{\partial \varepsilon}\right)^2 . \tag{1.5.23}$$

Only for few solid solations we have information on the derivative $\partial v / \partial \varepsilon$, that is, on the variation of the crystal volume caused by annealing which gives rise to the short-range order. It is more convenient, therefore, to transform the condition

(1.5.23) and replace $\partial v/\partial\varepsilon$ with other parameters which are easier to determine from the experimental data. For instance, for small $(1/v)(\partial v/\partial c)$ when we ignore the dependence of $W'_{ss'i}$ on the composition we can easily express W' in terms of the second derivative $\partial^2 v/\partial c^2$ and then to relate the first derivative of v with respect to the correlation parameter ε with the second derivate with respect to the concentration: $2(\partial v/\partial\varepsilon) \approx \partial^2 v/\partial c^2$. Another (and, apparently, more accurate) estimate is $\partial v/\partial\varepsilon \approx \delta v/\delta\varepsilon$ where δv is the variation of the atomic volume caused by the transition from a fully ordered solution to the disordered solution with a random arrangement of atoms while the variation $\delta\varepsilon$ of the short-range order parameter is $1/16$ for such a transition in solid solutions as Cu_3Au. Using these estimates we can rewrite the condition (1.5.23) as

$$\left(\frac{\partial v}{\partial c}\right)^2 \gg c(1-c)\left(\frac{\partial^2 v}{\partial c^2}\right)^2 \quad \text{or} \quad \left(\frac{\partial v}{\partial c}\right)^2 \gg c(1-c)\left(\frac{\delta v}{\delta\varepsilon}\right)^2, \qquad (1.5.24)$$

where we have used the fact that $(1+\nu'^2)/32 < 1$.

The results discussed here indicate that if $W'_{ss'}$, $\hat{V}'_{ss'}$ and $\hat{V}''_{ss'}$ vanish, that is, if Vegard's rule is satisfied and the elastic moduli are independent of the concentration then the fluctuations of the correlation parameters do not affect the static displacements and $\boldsymbol{u}_{\boldsymbol{k}} = \boldsymbol{A}_{\boldsymbol{k}} c_{\boldsymbol{k}}$ linearly depend on $c_{\boldsymbol{k}}$ according to (1.5.15, 16). Though these conditions are not satisfied precisely this simple ($\boldsymbol{u}_{\boldsymbol{k}} = \boldsymbol{A}_{\boldsymbol{k}} c_{\boldsymbol{k}}$) approximation, in fact, is shown by (1.5.24) to give small errors if the lattice parameter exhibits only a weak dependence on the correlation parameters or (with a poorer accuracy) if the concentration dependence of the lattice parameters is close to being linear (and the elastic moduli dependence on the concentration is weak). It should be noted that the relative contribution of the effects caused by the fluctuations of the correlation parameters in nonideal solutions can be greater than in ideal solutions (see [1.20]).

2. Effects of Long-Range Forces on Fluctuations in Crystals and the Production of Heterogeneous States

Chapter 1 presented the thermodynamic theory of the fluctuations of the concentration and the order parameters in which the interaction between atoms is assumed to diminish rapidly with increasing distance. In this case the mean squared fluctuation of the concentration in a volume element V' is given by the conventional equation (1.2.5) of the thermodynamic theory of fluctuations while the mean squared Fourier component is given by (1.2.10). When long-range interactions are considered, however, the results obtained in Chap. 1 must be modified.

Equation (1.2.5) was derived for the case when the volume element V' under consideration was sufficiently large. Thus when we determined the variation $\delta\Phi'$ of the thermodynamic potential of this volume we could neglect the surface effects regarding it as a subsystem weakly interacting with the remaining solid solution. This approach is valid, however, only for a short-range interaction between atoms. In the case of long-range interaction the potential variation $\delta\Phi'$ depends on the concentration variation not only in the volume V' but also in other parts of the solid solution, that is, $\delta\Phi'$ is a functional of the function $\delta c(\boldsymbol{r})$ for the entire solution. The mean squared $\langle(\delta c)^2\rangle$ must now be found with a more complicated averaging procedure (over various distributions $\delta c(\boldsymbol{r})$ in the entire solution rather than in the volume V' than that used for (1.2.5) so that the latter equation is no longer valid. The equation for $\langle|c_{\boldsymbol{k}}|^2\rangle$ is also changed because (1.2.10) does not take into account the interaction at distances comparable with the wavelength, or larger than it, which now becomes significant even for long fluctuation waves. In the case of anisotropic long-range forces we obtain, for instance, a dependence of $\langle|c_{\boldsymbol{k}}|^2\rangle$ on the direction of $\boldsymbol{n} = \boldsymbol{k}/k$ for small k, that is, $\langle|c_{\boldsymbol{k}}|^2\rangle$ is nonanalytical for $k \to 0$ (though $k > L^{-1}$ where L is the size of the crystal.

The effect of long-range interaction is significant if the contribution to $\delta\Phi'$ produced by the variation of the concentration or an internal parameter outside the given volume V' is proportional to V' rather than the interface area. This contribution can be easily shown to be proportional to the surface area of the volume V' if the interaction energy decreases exponentially or as r^{-n} where $n > 4$ while for $n = 3$ the contribution is proportional to the volume V'. It is quite common for the interaction forces in crystals to decrease with distance as $1/r^3$ or at an even slower rate. For instance, the elastic interaction energy in elastically anisotropic crystals varies as $1/r^3$ for large r. The long-range elastic interaction between atoms in a solid solution which does not decrease at distances much smaller than the crystal size is caused by the mirror image forces (and in elastically isotropic crystals; see Sect. 2.4.3 in [1.1]). Another illustration of the interaction energy varying as $1/r^3$ is the dipole-dipole interaction energy for the fluctuations of polarization or magnetization. The interaction energy for impurity ions or other charged defects varies at an even slower rate according to the Coulomb law.

If the long-range forces make a significant contribution to $\delta\Phi'$ then the calculations for the fluctuations must be made not with such equations as (1.2.5, 10) but with the generalized equations which take into account these forces. Such equations have been derived in [1.15] for the fluctuations of the charged defects in crystals with Coulomb forces acting between them. The effect of the elastic and dipole-dipole interactions on the fluctuations has been treated in [1.16; 2.1] (and for some special cases the effects produced by the elastic energy on the composition fluctuations have been discussed in [2.2]).

We shall discuss in Sect. 2.1.1 how the equations of the theory of fluctuations are changed when we take into consideration the elastic interaction which is the most common type of long-range interaction in solids. This interaction is closely associated with the static distortions which appear, for instance, in solid solutions when the atomic radii of the components are different. This means that this interaction occurs in all systems to a varying degree. The effects caused by the dipole-dipole and Coulomb interactions are discussed in Sects. 2.2.3.

These long-range forces vary with distance as r^{-n} and affect the probabilities of the fluctuation waves determining the nonanalytic behaviour of their mean squared amplitudes $\langle|c_{\boldsymbol{k}}|^2\rangle$ for $k \to 0$ (which depend on the direction of $\boldsymbol{k}$ for $n = 3$ while for $n > 3$ their derivatives with respect to $\boldsymbol{k}$ exhibit singularities). Long-range interaction of a different kind is also effective in metal solutions and generally it gives rise to blurred, but in some cases quite sharply defined, singular features of $\langle|c_{\boldsymbol{k}}|^2\rangle$ for large $\boldsymbol{k}$ on some surfaces in the $\boldsymbol{k}$ space. These features are caused by a discontinuity occurring on the Fermi surface in the distribution of the conduction electrons in the metal. This produces Friedel oscillations of the electron density which decay comparatively slowly and an associated indirect interaction between atoms in the alloy. Since the power-function variation of the indirect interaction is modulated by the rapid oscillations (whose period is determined by the diameter of the Fermi surface) the related singular features of $\langle|c_{\boldsymbol{k}}|^2\rangle$ in [1.17] (which are largely similar to Cahn's singular features in the oscillation frequencies) occur for finite $\boldsymbol{k}$, rather than for $\boldsymbol{k} = 0$. The effect of the indirect interaction between atoms on the fluctuations in alloys is discussed in Sect. 2.4.

The long-range forces can not only produce a significant effect on the fluctuations; in some cases they radically alter the state of the crystal giving rise to metastable and equilibrium heterogeneous states. Crystals in such states consist of regions of a relatively large size (in comparison with the distance between atoms) where an internal parameter has different values.

The best-known heterogeneous structure is the domain structure in ferromagnets which produces a decrease in the dipole-dipole interaction energy. In a similar way the elastic energy in coherent nonequilibrium two-phase systems is decreased as the particles of the second phase acquire a special shape and align to form what may be called elastic domains (see Sect. 2.12).

The long-range interaction associated with indirect interaction caused by the conduction electrons can also give rise to ordered-state systems in which the order parameter phase varies regularly at relatively large distances, such as occurs in alloys with large spatial periods or charge density waves. In some circumstances the indirect interaction may make it thermodynamically favourable to produce equilibrium heterogeneous states consisting of alternating ordered and disordered regions (see Sect. 2.5).

In semiconductor crystals with readily variable parameters the Coulomb long-range forces can give rise to equilibrium heterogeneous states of a different kind. They are made up of alternating small regions (whose size is of the order of the shielding length) with a significantly higher electron density and a changed internal parameter (or composition) and regions with a low electron density. This state is also treated in Sect. 2.5.

2.1 Effects Caused by Elastic Interaction

2.1.1 Effect of the Elastic Interaction on Fluctuations of Concentration and Order Parameters in Crystals

Long-Wave Concentration Fluctuations. Consider first the effect of the long-range elastic forces on the fluctuation waves of the concentration in solid solutions. The energy of the elastic interaction between atoms in the volume V' and the atoms outside this volume can be found by directly adding up the interaction energies for atom pairs which have a component varying as $1/r^3$ and a component smoothly varying at distances of the order of the crystal size (see Sect. 2.4.3 in [1.1]). It is more convenient, however, not to consider individual atoms outside the volume V' but to describe the total effect of such atoms by the strains they produce inside V'. Since long-range effects must be most noticeable for long waves we shall consider first long-wave fluctuations when the wavelength is much larger than the distance d_0 between atoms and much smaller than the crystal size L, that is, when

$$kd_0 \ll 1\,, \quad kL \gg 1 \tag{2.1.1}$$

Here we may employ the macroscopic approximation in which the fluctuational variation of the concentration is described by a smoothly varying function $\delta c(\boldsymbol{r})$ and the crystal distortions by the strain tensor u_{ij} and the stress tensor δ_{ij}. We shall assume at first that the distortions are produced only by the concentration fluctuations and that the strains produced by the fluctuations of the order parameters and the correlation can be ignored. The strains produced by the concentration fluctuations are described by the tensor L_{ij} given by (1.5.6).

In the absence of stresses the least work $\mathfrak{R}$ required for reversibly producing the concentration fluctuations δc in the region of the crystal with a constant total number N' of the atoms A and B is given by (1.2.4) and expressed in terms of the second-order derivative of the thermodynamic potential density φ taken at a constant, for instance zero, pressure. The product φV is equal to the conventional thermodynamic potential of the crystal (found using standard experimental techniques) and it takes into account all types of interactions between atoms (including the elastic interactions). Therefore, (1.2.4) essentially takes into account the elastic energy associated with the homogeneous strain $L_{ij}\delta c$ of the crystal which is caused by the variation of the concentration in the volume V' under consideration (that is, caused by the elastic interaction between atoms in this volume). The elastic energy associated with the variation of the concentration outside the volume V' is not taken into account here. This concentration variation produces some additional elastic strain $u_{ij} - L_{ij}\delta c$ in V' which is added to the strain $L_{ij}\delta c$ caused by the concentration fluctuations δc inside the volume. Hence, additional stresses are generated. We must add then to the expression for the work $\mathfrak{R}$ a term describing the work needed for producing this additional strain (and stresses) in each volume element. Then we obtain the following equation for the work needed for producing an arbitrary inhomogeneous smoothly varying concentration distribution:

$$\mathfrak{R} = \frac{1}{2}\int \left[\frac{d^2\varphi}{dc^2}(\delta a)^2 + c_{ijlm}(u_{ij} - L_{ij}\delta c)u_{lm} - L_{lm}\delta c)\right] d\boldsymbol{r} \ . \tag{2.1.2}$$

If the concentration distribution corresponds to the $\boldsymbol{k}$th fluctuation wave then (1.2.8) and (1.5.1, 3) give

$$\delta c = c_{\boldsymbol{k}}\mathrm{e}^{-\mathrm{i}\boldsymbol{k}\cdot\boldsymbol{r}} + \text{c.c.}\,, \quad u_{ij} = \tfrac{1}{2}(n_i \boldsymbol{A}_{\boldsymbol{k}j} + n_j \boldsymbol{A}_{\boldsymbol{k}i} k c_{\boldsymbol{k}}\mathrm{e}^{-\mathrm{i}\boldsymbol{k}\cdot\boldsymbol{r}} + \text{c.c.}\,, \tag{2.1.3}$$

where the proportionality coefficient $\boldsymbol{A}_{\boldsymbol{k}}$ between the waves of static displacements and concentration are found from (1.5.4, 5). When we substitute (2.1.3) into (2.1.2) we find that in the limit of small k the work needed for production of the $\boldsymbol{k}$th fluctuation wave is

$$\mathfrak{R}_{\boldsymbol{k}} = V\left[\frac{d^2\varphi}{dc^2} + c_{ijlm}(kn_i\boldsymbol{A}_{\boldsymbol{k}j} - L_{ij})(kn_l\boldsymbol{A}_{\boldsymbol{k}m} - L_{lm})\right]|c_{\boldsymbol{k}}|^2 \,. \tag{2.1.4}$$

This equation differs from (1.2.9) (written for the limiting case of $k \to 0$) in the last term which describes the elastic energy. When we substitute (2.1.4) into (1.2.1) we obtain a probability distribution for fluctuations of $c_{\boldsymbol{k}}$ which has the Gaussian shape and then we obtain [1.16, 2.1]

$$\begin{gathered}\langle|c_{\boldsymbol{k}}|^2\rangle = \frac{k_\mathrm{B}T}{V\varphi_{cc}(\boldsymbol{k})}\,, \quad \varphi_{cc}(\boldsymbol{k}) = \frac{d^2\varphi}{dc^2} + \varphi_{cc}^{\mathrm{el}}\,, \\ \varphi_{cc}^{\mathrm{el}} = c_{ijlm}(kn_i\boldsymbol{A}_{\boldsymbol{k}j} - L_{ij})(kn_l\boldsymbol{A}_{\boldsymbol{k}m} - L_{lm}) = c_{ijlm}L_{ij}(L_{lm} - kn_l\boldsymbol{A}_{\boldsymbol{k}m})\,.\end{gathered} \tag{2.1.5}$$

The second term in the above equation for $\varphi_{cc}(\boldsymbol{k})$ describes the elastic long-range forces and makes (2.1.5.) different from the conventional formula (1.2.10) derived for the undistorted solid solutions. Significantly, for small k the function $\varphi_{cc}^{\mathrm{el}}$ and hence $\langle|c_{\boldsymbol{k}}|^2\rangle$ depend on $\boldsymbol{n}$ (on the orientation of the vector $\boldsymbol{k}$ with respect to the crystallographic axes). Thus, in the infinite crystal $\langle |c_{\boldsymbol{k}}|^2 \rangle$ is a nonanalytic function of $\boldsymbol{k}$. It should be noted that for $k \lesssim L^{-1}$ the boundary conditions which must be satisfied on the crystal surface distort the strain field (2.1.3) found for the infinite crystal so that (2.1.5) is no longer applicable [1.13]. If the fluctuations in the crystal are homogeneous then $u_{ij} = L_{ij}\delta c$, the second term in (2.1.2) and φ_{cc}^{el} in (2.1.5) vanish and $\langle (\delta c)^2 \rangle$ or $\langle |c_{\boldsymbol{k}}|^2 \rangle$ for $\boldsymbol{k} = 0$ are determined by the conventional equations (1.2.5, 10). The probability of fluctuations noticeably varying at distances of the order of the crystal size L (we have $k \sim L^{-1}$ depends significantly on the shape of the crystal [2.3, 4].

When k is not very small we must also include the term $\beta_{ij}j(\boldsymbol{k})k_i k_j$ in the expression for $\varphi_{cc}(\boldsymbol{k})$ as in (1.2.10). In contrast to the case of the short-range forces treated in Sect. 1.2 when β in cubic crystals is a constant, in the case of elastic interactionl $\beta(\boldsymbol{k})$ depends on $\boldsymbol{n}$ even in cubic crystals. We can easily find this result by treating the fluctuation waves for small k and describing the elastic energy with the equations derived in Sect. 1.5.

For instance, we can use (1.5.6, 7) and rewrite (2.1.5) as

$$\begin{aligned} \varphi_{cc}^{\text{el}} &= \frac{1}{3} K \left(\frac{1}{v} \frac{dv}{dc} \right)^2 \chi(\boldsymbol{n}) \,, \\ \chi(\boldsymbol{n}) &= 3 - \frac{c_{11} + 2c_{12}}{D} \left[1 + 2\xi (n_x^2 n_y^2 + n_x^2 n_z^2 + n_y^2 n_z^2) + 3\xi^2 n_x^2 n_y^2 n_z^2 \right] \,. \end{aligned} \tag{2.1.6}$$

Here $K = (c_{11} + 2c_{12})/3$ is the bulk modulus, D and ξ are expressed in terms of $\boldsymbol{n} = \boldsymbol{k}/k$ and the elastic moduli via equations (1.5.7a). Thus we see that even in cubic crystals $\varphi_{cc}(\boldsymbol{k})$ and $\langle |c_{\boldsymbol{k}}|^2 \rangle$ depend on the direction of the vector $\boldsymbol{k}$. For instance, we have in the high-symmetry directions

$$\begin{aligned} \chi(\boldsymbol{n}) &= 2\frac{c_{11} - c_{12}}{c_{11}} \quad (\boldsymbol{n} \| \langle 100 \rangle) \,, \\ \chi(\boldsymbol{n}) &= \frac{c_{11} - c_{12} + 6c_{44}}{c_{11} + c_{12} + 2c_{44}} \quad (\boldsymbol{n} \| \langle 110 \rangle) \,, \\ \chi(\boldsymbol{n}) &= \frac{12c_{44}}{c_{11} + 2c_{12} + 4c_{44}} \quad (\boldsymbol{n} \| \langle 111 \rangle) \,. \end{aligned} \tag{2.1.7}$$

For most crystals we have $\xi < 0$ and the equation (2.1.6) for φ_{cc}^{el} has a minimum for the directions of $\boldsymbol{n} \| \langle 100 \rangle$. If $\xi > 0$ then φ_{cc}^{el} has a minimum for $\boldsymbol{n} \| \langle 111 \rangle$. We will always have then a $\min \varphi_{cc}^{\text{el}} > 0$.

In hexagonal crystals we can find φ_{cc}^{el} from (2.1.5) just by solving a system of two linear equations determining $\boldsymbol{A}_{\boldsymbol{k}}$ (see equations (B.8) in [1.1]). The resulting φ_{cc}^{el} depends only on the angle between $\boldsymbol{k}$ and the hexagonal axis $\boldsymbol{Z}$. In particular, for the directions of $\boldsymbol{k}$ parallel or perpendicular to the axis Z equations (B.9, 10) in [1.1] give $\boldsymbol{A}_{\boldsymbol{k}}$ and we obtain

$$\varphi_{cc}^{\mathrm{el}} = 2\frac{c_{11}c_{33} + c_{12}c_{33} + 2c_{12}^2}{c_{33}}\left(\frac{1}{d_1}\frac{\partial d_1}{\partial c}\right)^2 \quad (\boldsymbol{k}\| Z)\,,$$

$$\varphi_{cc}^{\mathrm{el}} = \frac{c_{11}^2 - c_{12}^2}{c_{11}}\left(\frac{1}{d_1}\frac{\partial d_1}{\partial c}\right)^2 + 2\frac{(c_{11} - c_{12})c_{13}}{c_{11}}\frac{1}{d_1}\frac{\partial d_1}{\partial c}\frac{1}{d_z}\frac{\partial d_z}{\partial c}$$
$$+ \frac{c_{11}c_{33} - c_{13}^2}{c_{11}}\left(\frac{1}{d_z}\frac{\partial d_z}{\partial c}\right)^2 \quad (\boldsymbol{k} \perp Z)\,, \tag{2.1.8}$$

where d_z and d_1 are the lengths of the segments parallel or perpendicular to the axis Z.

For elastically isotropic crystals we obtain the following simple expression from (1.5.7b) and (2.1.5):

$$\varphi_{cc}^{\mathrm{el}} = \frac{2}{3}\frac{1 - 2\sigma}{1 - \sigma}K\left(\frac{1}{v}\frac{dv}{dc}\right)^2\,, \tag{2.1.9}$$

This means that $\varphi_{cc}(\boldsymbol{k})$ and $\langle|c_{\boldsymbol{k}}|^2\rangle$ are independent of the direction of the vector $\boldsymbol{k}$. But here we have $\varphi_{cc}(\boldsymbol{k}) \neq d^2\varphi/dc^2$, as is the case in elastically anisotropic crystals and the equations for $\langle|c_{\boldsymbol{k}}|^2\rangle$ are changed when the elastic interaction is taken into account. In a liquid we have $\sigma = 1/2$ and $\varphi_{cc}^{\mathrm{el}} = 0$. The elastic energy is only produced by shear strains typical of solids.

In the approximation of the elastic isotropic continuum we can clearly illustrate the fact that in the presence of elastic forces $\langle|c_{\boldsymbol{k}}|^2\rangle$ is not determined only by $d^2\varphi/dc^2$ as it is the case for the short-range forces. The long-range elastic interaction between the impurity atoms which varies as $1/r^3$ vanishes in this approximation (see Sect. 2.4.3 in [1.1]). Therefore, the elastic long-range interaction remains the same in the elastically isotropic solid solution under fluctuational rearrangement of the concentration at small distances ($kL \ll 1$), and in the absence of other types of interaction the fluctuations would be the same as in the ideal solution. At the same time introduction of impurity atoms into the crystal changes the elastic energy which is determined by the mirror-image forces and is not altered when the impurity atom is displaced at a distance considerably smaller than the crystal size (see Sect. 2.4.3 in [1.1]). This elastic energy which is almost independent of the impurity positions (and hence does not affect the fluctuations) makes a contribution to φ equalling

$$\frac{1}{3}K\frac{1 - 2\sigma}{1 - \sigma}\left(\frac{1}{v}\frac{dv}{dc}\right)^2 c(1 - c)$$

(see (2.4.53) in [1.1]). This contribution of the second order in concentration determines the nonideality of the solution and its decomposition at low temperatures; production of two crystals without a coherent link reduces the total elastic energy generated by the mirror-image forces in each of the crystals. But the term in $d^2\varphi/dc^2$ corresponding to this contribution will be cancelled by (2.1.9) since, as noted above, the elastic energy remains the same under coherent rearrangement of the composition in the given crystal at distances much smaller than L and in the absence of other interactions the solution behaves with respect to the fluctuations as an ideal solution.

To evaluate the relative contribution of the term φ_{cc}^{el} determined by the elastic energy we shall take, for instance, $c_{11} = 2.5$, $c_{12} = 1.5$, and $c_{44}, = 1.2 \cdot 10^{12}$ erg/cm^3 (elastic moduli of nickel). Then for $(1/v)(dv/dc) = 0.1$ we obtain $\varphi_{cc}^{el} = 5 \cdot 10^9$ erg/cm^3 for the vectors $\boldsymbol{k}$ parallel to the axes $\langle 100 \rangle$ and $8 \cdot 10^9$ erg/cm^3 for $\boldsymbol{k}$ parallel to the axes $\langle 111 \rangle$. Note for the sake of comparison that in the ideal solutions at 700 K for $v = 10^{23}$ cm^3 and $c = 1/2$ equation (1.2.13) yields $d^2\varphi/dc^2 = 4 \cdot 10^{10}$ erg/cm^3. In some cases the relative contribution of φ_{cc}^{el} may be significantly higher since in some substitutional solutions $(1/v)(dv/dc)$ can be about 0.4 while in interstitial solutions it equals approximately unity. On the other hand, for low concentrations, (when $d^2\varphi/dc^2 \sim 1/c$) and in ordered solutions with a high short-range order $d^2\varphi/dc^2$ is considerably larger than in the above estimate and the relative contribution of φ_{cc}^{el} decreases (especially in solutions with small $(1/v)(dv/dc)$).

The effects caused by the long-range elastic forces are particularly significant near the critical point on the decomposition curve. Since we have $d^2\varphi/dc^2 = 0$ at the critical point the probabilities of the long-wave fluctuations (where $kd_0 \ll 1$) in the critical region become anomalously high in the undistorted solutions. But in distorted solutions we obtain $\varphi_{cc}^{el} \neq 0$ if we take into account the long-range elastic forces. It can be seen then from (2.1.5) that $\varphi_{cc}(\boldsymbol{k})$ does not vanish at the critical point so that the fluctuations at the critical point are finite for $L^{-1} \ll k \ll d^{-1}$ though they can grow considerably for small elastic energies (for small L_{ij}). In the presence of distortions anomalous growth is exhibited only by those fluctuations whose wavelengths are comparable with the crystal size ($k \sim L^{-1}$) and for which φ_{cc}^{el} decreases (and vanishes for $\boldsymbol{k} = 0$). This suppression of long-range fluctuations by the long-range elastic forces must change the behaviour of the thermodynamic parameters near the critical points on the decomposition curves (see Chap. 4).

Though elastic forces suppress anomalous growth of the fluctuations with $L^{-1} \ll k \ll d^{-1}$ near the critical point this anomalous growth may occur at lower temperatures in supercooled solid solutions when a homogeneous undecomposed solution is at a nonequilibrium state. If a crystal has almost no defects and new-phase nuclei are produced and grow at a low enough rate, the solution can be in the undecomposed state for a considerable period which is sufficiently long for setting a quasi-equilibrium distribution of fluctuations (with not too small $\boldsymbol{k}$) corresponding to the given temperature. Here $\langle |c_{\boldsymbol{k}}|^2 \rangle$ is also given by (2.1.5) in which, however, we have $d^2\varphi/dc^2 < 0$ at temperatures below the decomposition point. As the temperature decreases $\varphi_{cc}(\boldsymbol{k})$ will decrease and $\langle |c_{\boldsymbol{k}}|^2 \rangle$ will grow until $\varphi_{cc}(\boldsymbol{k})$ vanishes for certain directions $\boldsymbol{n}_m = \boldsymbol{k}_m/k_m$ of the vector $\boldsymbol{n}$ at a (spinodal) temperature $T_0'(c)$. In this case $\langle |c_{\boldsymbol{k}}|^2 \rangle$ becomes anomalously large for such directions of $\boldsymbol{n}$ (note that $|c_{\boldsymbol{k}}|^2$ does not tend to infinity if terms of the order of k^2 in $\varphi_{cc}(\boldsymbol{k})$ are taken into account). In elastically anisotropic crystals $\varphi_{cc}(\boldsymbol{k})$ does not vanish for $T = T_0'(c)$ and $k \to 0$ for other directions of $\boldsymbol{n}$, that is, $\langle |c_{\boldsymbol{k}}|^2 \rangle$ must exhibit a sharp dependence on $\boldsymbol{n}$ near this temperature.

At the temperature $T_0'(c)$ the solid solution becomes unstable with respect to long-wave inhomogeneities of an arbitrarily small amplitude. For $T < T_0'(c)$ such inhomogeneities grow continuously with time and the solution decomposes via the spinodal mechanism. Many features of the decomposition kinetics and solution ordering can be understood from the analysis [2.5] of the spinodal mechanism responsible for decomposition of solution and the important suggestion made by *Cahn* [2.6] about the significant and often decisive contribution made by the elastic forces to the process (see the reviews [1.25, 26; 2.7–9]). (The long-range elastic forces determine also some qualitatively different kinetic effects in solids. For instance, the chemical diffusion coefficient in a cubic crystal has been found to depend on direction, that is, the crystal is anisotropic with respect to diffusion [2.9]. If the characteristic size of the concentration inhomogeneity is comparable with the crystal size the effective diffusion coefficient depends on the crystal shape [2.4]. The elastic forces also affect significantly the relaxation processes in solids and the viscosity properties [2.10].)

One should bear in mind that the above treatment of fluctuations assumed the absence of defects. However, crystals always contain nonequilibrium defects (dislocations, boundaries of grains and blocks, boundaries of antiphase domains in ordered crystals and so on) as well as the equilibrium defects. For instance, vacancies appear in crystals at high temperatures while complexies can be produced in solutions where components differ significantly in their radii (see Sect. 3.1.3 in [1.1]). Defects cause partial relaxation of the coherent elastic stresses discussed above and a decrease in the contribution of the elastic energy. If the concentrations of the equilibrium and nonequilibrium defects are small, however, then the effects caused by them may be ignored in the calculations of $\langle |c_{\boldsymbol{k}}|^2 \rangle$.

Short-Wave Concentration Fluctuations and the Energy of Elastic Interaction Between Atoms in Solution. The general macroscopic theory can describe the effects of the elastic interaction in the case of the steadily varying long-wave fluctuations discussed above. The microscopic theory must be applied to the short-wave fluctuations rapidly varying in space for analyzing the effect of the elastic interaction on the fluctuations and the ordering process in solid solutions. In contrast to the conventional theoretical model of ordering which ignores distortions and the elastic energy (see Sect. 1.3), such general theory must take into consideration the static displacements of atoms.

This can be done with the equation (1.5.11) for the free energy of the disordered solution [1.8, 9, 20], which was used for calculating the static displacements in Sect. 1.5. This equation is derived from the expansion in powers of the displacements $\boldsymbol{u}_s$ from the sites of the perfect crystal consisting of the "average" atoms. The free energy of such "average" crystal is denoted by F_0 and the expansion coefficients W and V depend on the occupation numbers c_s according to (1.5.12) and a similar equation for V. For the sake of simplicity we shall consider only the crystals in which the elastic moduli have a weak dependence on the composition and we shall assume that the coefficients $V_{ss'}^{\alpha\alpha'} = V_{ss'}$ in (1.5.11) are independent of the species α, α' of atoms occupying the sites s and s'. Then we can use (1.5.12, 13), the Fourier components u_k c_k and $\varepsilon_k(\varrho)$ of u_s, $c_s - c$ and $(c_s - c)(c_{s'} - c)$ and equations (1.1.8), (1.4.1) and (1.5.1) to write the free energy of the solution as

$$F = F_0 - N \sum_k \left[W_{ki} c_k + \sum_{\varrho \neq 0} W'_{ki}(\varrho) \varepsilon_k(\varrho) - \frac{1}{2} V_{kij} u_{kj} \right] u^*_{ki} , \qquad (2.1.10)$$

where $W_k, W'_k(\varrho)$, and $\hat{V}_k$ are given by (1.5.15).

Using (1.5.17) and the condition for the minimum of F with respect to u_k we obtain (1.5.16). Substitution of the latter to (2.1.10) yields

$$\begin{aligned} F &= F_0 - \frac{1}{2} N \sum_k \left[W_{ki} c_k + \sum_{\varrho \neq 0} W'_{ki}(\varrho) \varepsilon_k(\varrho) \right] V^{-1}_{kij} \\ &\times \left[W_{kj} c_k + \sum_{\varrho \neq 0} W'_{ki}(\varrho) \varepsilon_k(\varrho) \right]^* . \end{aligned} \qquad (2.1.11)$$

If the constants W' are small in comparison with W then we can ignore the terms including the Fourier components of the correlation parameters $\varepsilon_k(\varrho)$ in (2.1.11). Then we obtain the following simple equation for the free energy:

$$\begin{aligned} F &= \tilde{F}_0 - \frac{1}{2} N \sum_k [\mathscr{E}(k) - \bar{\mathscr{E}}] |c_k|^2 \\ &= \hat{F}_0 - \frac{1}{2} \sum_{\substack{s,s' \\ (s \neq s')}} \mathscr{E}(\varrho)(c_s - c)c_{s'} - c) . \end{aligned} \qquad (2.1.12)$$

Here

$$\begin{aligned} &\tilde{F}_0 = F_0 - \frac{1}{2} N c(1-c) \bar{\mathscr{E}} , \quad \bar{\mathscr{E}} = \frac{1}{N} \sum_k \mathscr{E}(k) , \\ &\mathscr{E}(k)) W_k A_k = W_{ki} V^{-1}_{kij} W^*_{kj} , \quad \mathscr{E}(\varrho) = \frac{1}{N} \sum_k (\mathscr{E}(k) - \bar{\mathscr{E}}) \exp(ik \cdot \varrho) \end{aligned} \qquad (2.1.13)$$

and the equation (1.1.18) was taken into account. In (2.1.12) we take the difference between the elastic energy $\mathscr{E}(k)$ of the kth wave and its average $\bar{\mathscr{E}}$ over the reciprocal lattice cell in order to eliminate the self-action term (with $\varrho = 0$) from the expression for the pair energy of elastic interaction between atoms. Then the energy of the elastic interaction between atoms is given by the conventional expression (1.3.1), that is, the term $-\mathscr{E}(\varrho)$ describing the relaxation drop in the energy caused by displacements of atoms from the sites of the average lattice is added to the energies $w(\varrho)$ found for undistorted (average) crystal and included in F_0.

In addition to the energy $\mathscr{E}(\varrho)$ which decreases with distances at a fairly fast rate as $1/\varrho^3$ the elastic interaction produces also a contribution to F which depends on the concentration and is determined by the interaction at distances of the order of the crystal size L (associated with the mirror-image forces at the crystal surface). This energy remains unchanged, however, under fluctuational rearrangements of atoms at distances of the order of $k^{-1} \ll L$. Therefore, when we calculate fluctuations in the microscopic model with $W' = 0$ and $V' = V'' = 0$ and take into consideration the static distortions, the constants $w(\varrho)$ in the equations derived in Sect. 1.3 are just replaced with $w(\varrho) - \mathscr{E}(\varrho)$. For instance, at high temperatures we then find from (1.3.4, 10) the term

$$\delta\chi_{\boldsymbol{k}} = -\frac{\mathscr{E}(\boldsymbol{k}) - \bar{\mathscr{E}}}{k_{\mathrm{B}}T} \tag{2.1.14}$$

to be added to $\chi_{\boldsymbol{k}}$ in the denominator of (1.3.4). (Note that since we take the difference between $\mathscr{E}(\boldsymbol{k})$ and the average $\bar{\mathscr{E}}$ in (2.1.12) we can ignore the condition (1.1.18) and the change in $|c_{\boldsymbol{k}'}|$ with $\boldsymbol{k}' \neq \boldsymbol{k}$ caused by variation of F with respect to $|c_{\boldsymbol{k}}|^2$.)

Using (2.1.5, 13) and the limiting equation for $W_{\boldsymbol{k}i}$ see (2.2.17) and (2.3.10, 11) in [1.1]) we obtain for small k

$$\mathscr{E}(\boldsymbol{k}) = vkc_{ijlm}n_jL_{lm}A_{\boldsymbol{k}i} = vc_{ijlm}L_{ij}L_{lm} - v\varphi_{cc}^{\mathrm{el}} \,. \tag{2.1.15}$$

This energy depends on the direction of the vector $\boldsymbol{n} = \boldsymbol{k}/k$ and, as discussed above, causes nonanalytic behaviour of $\langle|c_{\boldsymbol{k}}|^2\rangle$ for $k \to 0$ according to (1.3.4) and (2.1.14). Since the nonanalytic behaviour is determined by the contribution of $\mathscr{E}(\varrho)$ in the range of large ϱ where $\mathscr{E}(\varrho)$ are small the above expression for the nonanalytic contribution to $\chi_{\boldsymbol{k}}$ is valid at arbitrary temperatures.

To find $d^2\varphi/dc^2$ in distorted crystals we must take into account the fact that a homogeneous variation of the composition produces homogeneous strain changing the elastic energy of the homogeneous strain (in particular, that due to the mirror-image forces) included in F_0. When we write

$$\frac{dF_0}{dc} = \left(\frac{\partial F_0}{\partial c}\right)_{u_{ij}} + \left(\frac{\partial F_0}{\partial u_{ij}}\right)_c \frac{du_{ij}}{dc}$$

$\partial F_0/\partial u_{ij} = 0$ as follows from the definition of the "average" crystal in the absence of stresses and

$$\frac{d}{dc}\left(\frac{\partial F_0}{\partial u_{ij}}\right)_c = \frac{\partial^2 F_0}{\partial c\partial u_{ij}} + \frac{\partial^2 F_0}{\partial u_{ij}\partial u_{lm}}\frac{du_{lm}}{dc} = 0 \,,$$

we obtain then

$$\frac{d^2F_0}{dc^2} = \left(\frac{\partial^2 F_0}{\partial c^2}\right)_{u_{ij}} - Vc_{ijlm}L_{ij}L_{lm} \,. \tag{2.1.16}$$

The equation (2.1.13) for F_0 and (2.1.16) show that the total contribution of the elastic energy to $d^2\varphi/dc^2 = (1/V)(d^2F_0/dc^2$ is $-c_{ijlm}L_{ij}L_{lm} + \bar{\mathscr{E}}/v$. Using it and the equation (2.1.5) for $\varphi_{cc}^{\mathrm{el}}$ we obtain the same change in the denominator $\varphi_{cc}(\boldsymbol{k})$

in the eqation (2.1.5) for $\langle |c_k|^2 \rangle$ for small k as that given by (2.1.14, 15), that is, in the limit of long waves the results obtained in the microscopic and macroscopic theories coincide.

If $w(\varrho)$ are treated as semi-phenomenological parameters found from the experimental data then for small ϱ the energies $\mathscr{E}(\varrho)$ in (2.1.12) may be simply included in $w(\varrho)$. Thus we shall take into consideration explicitly only the elastic interaction at large distances which gives rise to singularities in the behaviour of $\langle |c_k|^2 \rangle$ and affects the processes of ordering and decomposition of a solid solution (for instance, producing a shift of the spinodal with respect to the decomposition curve). In crystal systems with very large distortions, for instance, in the interstitial solutions, the elastic interaction can make a major contribution to the energy of interaction between atoms [2.11, 12]. If we know the force constants $W_{ss'}$ and $\hat{V}_{ss'}$ that is, the parameters W_k and $\hat{V}_k$ which determine the static displacements according to equations derived in Sect. 1.5 then for such solid solutions we can directly calculate the energies $w(\varrho) \approx -\mathscr{E}(\varrho)$ and their Fourier representations [1.25; 2.3, 13–15, 25, 26], that is, obtain full data on the phase diagrams and thermodynamic functions.

In the general case, the Vegard rule is not satisfied, the elastic moduli of a solid solution depend on the composition and $W'_{ki}(\varrho)$ in (2.1.11) do not vanish (as well as the constants V' and V'' which cause renormalization of W' and give rise to new terms of the type of $\varepsilon_k(\varrho_1, \varrho_2), \ldots$ in (2.1.11); see Sect. 1.5). As shown by (1.4.1) and (2.1.11), the elastic interaction then gives rise to contributions not only to the energies of atom pairs (as in (2.1.12)) but also to the energies of triplets and quadruplets of atoms (see also [1.78]). If the interaction between more than two atoms is strong it significantly affects the thermodynamic properties of a solution [1.43].

It can be convenient to identify in $\varepsilon_k(\varrho)$ a component which has an equilibrium relation to c_k, that is, to make a transition from $\varepsilon_k(\varrho)$ to $\tilde{\varepsilon}_k(\varrho)$ according to (1.5.3). Then $\varepsilon_k(\varrho)$ in (2.1.11) is replaced with $\tilde{\varepsilon}_k(\varrho)$ and W_{ki} with $\tilde{W}_{ki} = W_{ki} + \sum_\varrho (\partial \varepsilon_k(\varrho)/\partial c_k) W'_{ki}$ and they will depend on the concentration and the order parameters. Under a certain approximation we can eliminate the terms with $\tilde{\varepsilon}_k(\varrho)$ and employ the model described by (2.1.12, 13) with such variable coefficients W_k (and $\hat{V}_k$). It was essentially this model that served as a basis for an interesting study [2.16] of the effects produced by the elastic interaction on the ordering process which demonstrated the cause of the first-order phase transition occurring in CuPt alloy upon ordering. The terms containing $\tilde{\varepsilon}_k(\varrho)$ in F may, strictly speaking, give rise to corrections of a comparable order of magnitude.

Long-Wave Fluctuations of the Long-Range Order Parameters. The above macroscopic results on the long-wave concentration fluctuations can be readily applied to the fluctuations of various internal parameters. For instance, if the long-range order in a one-component crystal is described by a single parameter $\eta_\mu = \eta$ then the equations describing the long- range order fluctuation waves can be derived from the appropriate equations for the concentration fluctuation waves simply by replacing c with η According to (2.1.5), when we take into account the elastic interaction we can write the following expression for $\langle|\eta_{\boldsymbol{k}}|^2\rangle$ for small k instead of (1.2.16) [1.16]:

$$\langle|\eta_{\boldsymbol{k}}|^2\rangle = \frac{k_{\mathrm{B}}T}{V\varphi_{\eta\eta}(\boldsymbol{k})}\,, \quad \varphi_{\eta\eta}(\boldsymbol{k}) = \varphi_{\eta\eta} + \varphi_{\eta\eta}^{\mathrm{el}}\,, \quad \varphi_{\eta\eta} = \frac{\partial^2\varphi}{\partial\eta^2}\,, \qquad (2.1.17)$$
$$\varphi_{\eta\eta}^{\mathrm{el}} = c_{ijlm}L_{\eta ij}(L_{\eta lm} - kn_l A_{\eta km})\,.$$

Here $L_{\eta ij}$ describes the self-strain in the crystal produced by the uniform variation of η and $A_{\eta km}$ are given by (1.5.9) where $\mu \to \eta$.

In crystals with an inversion center the lattice parameters are even functions of η and in the disordered phase (where $\eta = 0$) we have $L_{\eta ij} = 0$ that is, (2.1.17) shows that the elastic long-range forces do not affect the long-wave fluctuations of the long-range order. When η are small we have in this case $L_{\eta ij} \sim \eta$. This means that in the equation (2.1.17) for $\varphi_{\eta\eta}(\boldsymbol{k})$ both the first term and the second term (due to the elastic energy) tend to zero as we approach a second-order phase transition in an ordered crystal and $\langle|\eta_{\boldsymbol{k}}|^2\rangle$ becomes anomalously large for small k. But if a crystal lacks a center of inversion then in a disordered phase in some cases (for instance, if a crystal in the nonferroelectric phase is a piezocrystal or for ferroelastic crystals) a variation $\delta\eta$ of the long-range order produces strains proportional to this variation (the lattice parameters depend linearly on η) and $L_{\eta ij} \neq 0$. Under such conditions the elastic forces affect significantly the long-wave fluctuations of η both in the ordered and disordered phases. In the vicinity of a second-order phase transition $\langle|\eta_{\boldsymbol{k}}|^2\rangle$ remains finite for small k (though for small $L_{\eta ij}$ it may grow significantly) with the exception of the range of very small $k \sim L^{-1}$.

In such crystals with $L_{\eta ij} \neq 0$ for $\eta = 0$ (such as ferroelectric crystals exhibiting piezoeffect in the paraphase) the term $\varphi_{\eta\eta}^{\mathrm{el}}$ in (2.1.17) accounting for the elastic interaction causes a significant suppression of fluctuations which is especially strong in the vicinity of a second-order phase transition where $\varphi_{\eta\eta} = 0$ and $\varphi_{\eta\eta}^{\mathrm{el}}$ generally does not vanish [1.13, 16]. The difference in symmetry between the tensors L_{ij} and $L_{\eta ij}$ may produce a difference between the effects of the elastic interaction on the fluctuations of the concentration and the order parameters. In the first instance we have $\varphi_{cc}^{\mathrm{el}} > 0$ for all directions of $\boldsymbol{n}$ (see (2.16–19)). In the case of fluctuations of η and a certain symmetry of $L_{\eta ij}$ we find that $\varphi_{\eta\eta}^{\mathrm{el}}$ vanish but only for special directions of $\boldsymbol{n}$. For instance, when only one component $L_{\eta xy}$ of the tensor $L_{\eta ij}$ is nonvanishing and the elastic moduli c_{ijlm} correspond to the approximation of an elastic isotropic continuum [2.17, 18] we obtain from (1.5.4, 5) and (2.1.17)

$$\varphi_{\eta\eta}^{\mathrm{el}} = 4\mu L_{\eta xy}^2\left(n_z^2 + \frac{2}{1-\sigma}n_x^2 n_y^2\right) \qquad (2.1.18)$$

where μ is the shear modulus and σ is the Poisson ratio. This function vanishes for $\boldsymbol{k}$ parallel to the axes X or Y. For all other directions of $\boldsymbol{n}$ (2.1.18) does not vanish, that is, the elastic interaction suppresses fluctuations.

In elastically anisotropic crystals of orthorhombic, tetragonal or cubic symmetry the expressions for $\varphi^{\mathrm{el}}_{\eta\eta}$ can be written explicitly if the vector $\boldsymbol{k}$ is in the $k_x k_y$ plane or parallel to the direction k_z. When only $L_{\eta xy} \neq 0$ we obtain from (2.1.17) and the equations (B.2, 13, 14) in [1.1] which determine $\boldsymbol{A_k}$

$$\varphi^{\mathrm{el}}_{\eta\eta} = 4c_{66}L^2_{\eta xy}\left[1 - \frac{c_{66}(c_{11}n_x^4 + c_{22}n_x^4 - 2c_{12}n_x^2 n_y^2)}{c_{66}(c_{11}n_x^4 + c_{22}n_y^4) + (c_{11}c_{22} - c_{12}^2 - 2c_{12}c_{44})n_x^2 n_y^2}\right]$$

$$(\boldsymbol{k} \perp k_z) \qquad (2.1.19)$$

$$\varphi^{\mathrm{el}}_{\eta\eta} = 4c_{66}L^2_{\eta xy} \quad (\boldsymbol{k} \| k_z) .$$

If the long-range order in one-component crystals is described by several parameters η_μ (where $\mu = 1, 2, \ldots, n$) then we can determine the least work needed for producing the fluctuation $\delta\eta_\mu$ in the same way as it was done above. If we consider now the Fourier components and take into account the elastic energy we can easily obtain equations for the mean squares and products of the Fourier components replacing (1.2.20) derived for the case without long-range forces [1.16]:

$$\begin{aligned} &\frac{1}{2}\langle(\eta_{\mu\boldsymbol{k}}\eta^*_{\mu'\boldsymbol{k}} + \eta^*_{\mu\boldsymbol{k}}\eta_{\mu'\boldsymbol{k}})\rangle = \frac{k_{\mathrm{B}}T}{V}\varphi^{-1}_{\mu\mu'}(\boldsymbol{k}) , \\ &\varphi_{\mu\mu'}(\boldsymbol{k}) = \varphi_{\mu\mu'} + \varphi^{\mathrm{el}}_{\mu\mu'} , \quad \varphi_{\mu\mu'} = \frac{\partial^2\varphi}{\partial\eta_\mu\partial\eta_{\mu'}} , \\ &\varphi^{\mathrm{el}}_{\mu\mu'} = c_{ijlm}L_{\mu ij}(L_{\mu' lm} - kn_l A_{\mu' km}) . \end{aligned} \qquad (2.1.20)$$

Here $L_{\mu ij}$ and $A_{\mu km}$ are given by (1.5.9). We can also calculate from (2.1.20) the mean squares and products of the Fourier components of the concentration and the order parameters in the ordered solutions if we denote by η_μ the long-range order parameters (where $\mu = 1, 2, \ldots, n-1$) and by $\eta_n \equiv c$ the concentration.

According to the above calculations $\varphi_{\mu\mu'}$ in (2.1.20) are the derivatives with respect to η_μ taken for constant stresses (it is precisely these parameters that are found in conventional thermodynamic experiments). The elastic moduli $\hat{c} = \hat{c}^\infty$ are determined for the unchanged (not adjusted to the strain) parameters η_μ (this is done, for instance, by high-frequency measurements in which the period is much shorter than the relaxation time for η_μ. Equations (1.5.8) and (2.1.20) give the fluctuations $\langle|\boldsymbol{u}^\eta_{\boldsymbol{k}}|^2\rangle$ of the displacements caused by the fluctuations $\delta\eta_\mu$ of the long-range order parameters. These "static" displacements in fact vary with time with $\delta\eta_\mu$ (though fairly slowly in some crystal systems) and for instance in ferroelectrics the fluctuations of η_μ are related to oscillations. Thermal motion of atoms produces also additional displacement fluctuations which are not related to the fluctuations of η_μ. The probabilities of the displacement fluctuations are expressed in terms of the elastic moduli $\hat{c}^\infty$ (see the relevant expressions in Sect. 126 in [1.38]). Using these expressions, (1.5.8) and (2.1.20) we obtain the following equation for the total mean squared fluctuations:

$$\langle u_{\boldsymbol{k}i}u^*_{\boldsymbol{k}j}\rangle = \frac{k_{\mathrm{B}}T}{V}\left[\sum_{\mu,\mu'} A_{\mu\boldsymbol{k}i}A^*_{\mu'\boldsymbol{k}j}\varphi^{-1}_{\mu\mu'}(\boldsymbol{k}) + k^{-2}C^{-1}_{\boldsymbol{k}ij}\right] . \qquad (2.1.21)$$

Here $\hat{C}_k^{-1}$ is the inverted tensor $\hat{C}_k$ given by (1.5.5) (with the elastic moduli $\hat{c} = \hat{c}^\infty$). The last term in (2.1.21) describes the displacements which are not directly related to the order fluctuations and it is relatively small if the elastic moduli are not anomalously small while the first term always increases sharply near a phase transition (where $\varphi^{-1}_{\mu\mu'} \to 0$) if $\hat{L}_\mu \neq 0$.

We may rewrite (2.1.21) as

$$\langle u_{ki} u^*_{kj} \rangle = \frac{k_B T}{V k^2} C^{0-1}_{kij} , \quad C^0_{kim} = c^0_{ijlm} n_j n_l , \tag{2.1.21a}$$

where $\hat{c}^0$ are the low-frequency elastic moduli which take into account relaxation of the order parameters η_μ under strain (equations (2.1.21, 21a) describe the relation between $\hat{c}^0$ and $\hat{c}^\infty$ or $\hat{C}^0$ and $\hat{C}^\infty$. For ferroelastic crystals in which the components of the strain tensor are the long-range order parameters (2.1.21a) describes the mean squared Fourier components for the fluctuations of the displacements and the long-range order parameters. Fluctuations in ferroelastic crystals are treated in [2.19, 20].

Scattering of X-rays (see Chap. 4) and light [1.16; 2.22, 23] near second-order phase transitions in crystals exhibits characteristic features determined by the significant effects produced by the elastic long-range forces on fluctuations and described by (2.1.20–21a).

2.1.2 Effect of Elastic Interaction on the Shape of Second-Phase Particles and Metastable Heterogeneous Structures in Two-Phase Systems

Elastic Energy of Inhomogeneous Coherent Crystal. The general equations (2.1.4, 5, 17, 20) for the elastic component of the free energy associated with the cosinusoidal distribution (2.1.3) of the concentration or the long-range order parameter were derived in the study of the effects produced by the elastic interaction on the fluctuations [1.16; 2.1]. But they can be used also for calculating the elastic energy for an arbitrary steadily varying distribution of the concentration $\delta c(\boldsymbol{r})$ and the order parameters. Indeed, the component E_{el} of the elastic free energy determined by the difference between the strain u_{ij} and the self-strain $L_{ij}\delta c$ for the concentration fluctuation $\delta c(\boldsymbol{r})$ is given by the second term in (2.1.2) if we ignore the concentration dependence of the elastic moduli and deviations from the Vegard rule. If we substitute the expansion (1.2.8) for δc and a similar expansion for the strains into this term and make use of (2.1.3) relating u_{kij} and c_k we find that E_{el} is broken down into a sum of terms equal to the second term in (2.1.4) and corresponding to individual Fourier components:

$$E_{\text{el}} = \frac{V^2}{8\pi^3} \int d\boldsymbol{k} E_{\text{el}}(\boldsymbol{n}) |c_{\boldsymbol{k}}|^2 , \quad E_{\text{el}}(\boldsymbol{n}) = \frac{1}{2} \varphi^{\text{el}}_{cc}(\boldsymbol{k}) . \tag{2.1.22}$$

Here $\varphi^{\text{el}}_{cc}(\boldsymbol{k})$ is given by the general equation (2.1.5) or formulas (2.1.6–9). Equation (2.1.22) gives E_{el} in terms of the Fourier components $c_{\boldsymbol{k}}$ of the deviations of the concentration from the mean concentration (see (1.2.8)).

Since $E_{\rm el}(\boldsymbol{n})$ depends only on the direction of the wave vector $\boldsymbol{k}$ but not on its magnitude, for one-dimensional distributions $\delta c(\boldsymbol{r})$ we can take $E_{\rm el}(\boldsymbol{n})$ outside the integral in (2.1.22) and use (1.2.8) to obtain the following simple expression:

$$E_{\rm el} = E_{\rm el}(\boldsymbol{n}) \int [\delta c(\boldsymbol{r})]^2 d\boldsymbol{r} \, . \tag{2.1.23}$$

In elastically isotropic crystals this expression is valid for an arbitrary distribution $\delta c(\boldsymbol{r})$ (which is not necessarily one-dimensional).

We should bear in mind that $E_{\rm el}$ determines only a part of the variation of the elastic energy in a crystal determined by the difference between the strain $u_{ij}(\boldsymbol{r})$ in the inhomogeneous solid solution and the self-strain $L_{ij}\delta c(\boldsymbol{r})$. As noted above, under a uniform concentration variation the self-strain $L_{ij}\delta c(\boldsymbol{r})$ produces a variation of $-c_{ijlm}L_{ij}L_{lm} + \mathscr{E}/v$ in $d^2\varphi/dc^2$ caused by the elastic interaction. In the approximation of the elastic isotropic continuum the respective contribution to the variation of the thermodynamic potential, $1/2(d^2\varphi/dc^2)\int[\delta c(\boldsymbol{r})]^2 d\boldsymbol{r}$ will be cancelled by (2.1.23), as can be seen from (2.1.15, 22). This means that, as should be expected, under this approximation a coherent rearrangement of the atoms in a solid solution at distances much smaller than the crystal size does not change its elastic energy. When we consider the thermodynamic characteristics of various processes in crystals, however, it is convenient to use the functions $\varphi(c)$ which take into account self-strain and elastic interaction between atoms in homogeneous crystals (it is precisely these functions $\varphi(c)$ that are found from the conventional data on the equilibrium activities) without identifying the elastic components in them. It is convenient to consider separately only the additional elastic energy (2.1.22, 23) determined by the difference between $u_{ij}(\boldsymbol{r})$ and $L_{ij}\delta c(\boldsymbol{r})$.

An equation similar to (2.1.22) can be derived for the elastic energy determined by an arbitrary steadily varying inhomogeneous distribution $\eta_\mu(\boldsymbol{r}) = \langle\eta_\mu(\boldsymbol{r})\rangle + \delta\eta_\mu(\boldsymbol{r})$ of the long-range order parameters in the crystal:

$$E_{\rm el} = \frac{V^2}{8\pi^3}\int d\boldsymbol{k} \sum_{\mu,\mu'} E_{\mu\mu'{\rm el}}(\boldsymbol{n})\eta_{\mu\boldsymbol{k}}\eta^*_{\mu'\boldsymbol{k}} \, , \quad E_{\mu\mu'{\rm el}}(\boldsymbol{n}) = \frac{1}{2}\varphi^{\rm el}_{\mu\mu'} \, . \tag{2.1.24}$$

Here $\varphi^{\rm el}_{\mu\mu'}$ is determined by the equations (2.1.17–20).

The elastic interaction affects significantly, for instance, phase transformations in solids since it is a major factor determining the transformation rate and the shapes and mutual arrangement of the new-phase particles. This is why the elastic energy of the new-phase particles was calculated in many studies including the first results obtained by *Nabarro* [2.24] and *Eshelby* [2.25] for the new-phase particles coherently linked to the matrix (see the reviews [2.26–2.28]). When phases in the crystal are coherently linked it is convenient to consider the elastic interaction using the above equations for $E_{\rm el}$ written in the $\boldsymbol{k}$ representation. When the difference between the elastic moduli of the phases can be ignored they give a fairly simple approach to the treatment of the general case in which new-phase particles of an arbitrary shape are produced in an elastically anisotropic crystal [1.25; 2.29, 30].

As it is described above for crystals with point defects, the method of the static displacement waves gives a good description of new-phase particles or other macroscopic inclusions [2.29–32] with the use of the continuum-theory approximation for the defects (see Sect. 2.23 in [1.1]). In this method (which is applicable if the inclusion size is large in comparison with the distance between atoms) the inclusions of a given type (or orientation) α are described by the self-strain tensor $L^0_{\alpha ij}(\boldsymbol{r})$. The tensor describes the strain generated by the inclusions α produced by the phase transformation of the surrounding matrix.

If the inclusions are produced in the process of continuous diffusion-generated variation $\delta c(\boldsymbol{r})$ of the concentration in a binary solution (which is generally inhomogeneous both in the inclusion and in the matrix), if they have the same structure as the matrix and are coherently linked to it then the self-strain tensor

$$L^0_{\alpha ij}(\boldsymbol{r}) = L^0_{ij}(\boldsymbol{r}) = L_{ij}\delta c(\boldsymbol{r})$$

is proportional to the local variation $\delta c(\boldsymbol{r})$ and the proportionality coefficient L_{ij} given by (1.5.6). Under such conditions the elastic energy associated with production of inclusions is expressed in terms of the concentration inhomogeneities according to (2.1.22, 23).

In some cases inclusions differ from the matrix in their structure (and generally in the composition) and can be described in the framework of a model in which the inclusion with homogeneous structure and composition is separated from the matrix with a sharply defined interface. In such a model the functions $L^0_{\alpha ij}(\boldsymbol{r})$ are expressed in terms of the shape function $s_\alpha(\boldsymbol{r})$ and the self-strain tensor $L^0_{\alpha ij}$ corresponding to the total phase transformation of the crystal:

$$L^0_{\alpha ij}(\boldsymbol{r}) = L^0_{\alpha ij} s_\alpha(\boldsymbol{r}) \, .$$

Here the function $s_\alpha(\boldsymbol{r})$ equals unity inside the inclusions with the orientation α and vanishes outside them. If we refer strains to the average crystal with the self-strain

$$\bar{L}^0_{ij} = \sum_\alpha p_\alpha L^0_{\alpha ij}$$

(where p_α are the volume fractions of the inclusions of the type α we can write the differences in the equation for the elastic energy density in the form

$$u_{ij}(\boldsymbol{r}) - L^0_{ij}(\boldsymbol{r}) = u_{ij}(\boldsymbol{r}) - \bar{L}^0_{ij} - \sum_\alpha L^0_{\alpha ij}[s_\alpha(\boldsymbol{r}) - p_\alpha] \, . \tag{2.1.25}$$

The elastic energy E_{el} of the crystal containing arbitrary concentrations of inclusions of the type α with sharp boundaries can be found directly from (2.1.24) if an inclusion of the type α corresponds to a crystal region described by the order parameter $\eta_\mu = \eta_\alpha = 1$ and the tensor $\hat{L}_\mu = \hat{L}^0_\alpha$:

$$E_{\rm el} = \frac{1}{8\pi^3} \sum_{\alpha\alpha'} \int d\boldsymbol{k} E_{\rm el\alpha\alpha'}(\boldsymbol{n}) s_\alpha(\boldsymbol{k}) s_{\alpha'}(\boldsymbol{k}) \, ,$$

$$s_\alpha(\boldsymbol{k}) \int [s_\alpha(\boldsymbol{r}) - p_\alpha] \mathrm{e}^{-i\boldsymbol{k}\cdot\boldsymbol{r}} d\boldsymbol{r} \, , \tag{2.1.26}$$

$$E_{\rm el\alpha\alpha'}(\boldsymbol{n}) = \tfrac{1}{2} c_{ijlm} L^0_{\alpha ij} (L^0_{\alpha' lm} - k n_l A_{\alpha' km}) \, ;$$

Here $A_{\alpha' km}$ are coefficients in the relations between the Fourier components of $u_{ij}(\boldsymbol{r}) - L^0_{ij}$ and $s_\alpha(\boldsymbol{r}) - p_\alpha$ which are given by (1.5.9) where $\hat{L}_\mu$ are replaced with $\hat{L}^0_\alpha$.

For instance, if the matrix contains inclusions of only one orientation we can omit the subscripts α and α' and write for such two-phase system

$$E_{\rm el} = \frac{1}{8\pi^3} \int d\boldsymbol{k} E_{\rm el}(\boldsymbol{n}) |s(\boldsymbol{k})|^2 \, . \tag{2.1.27}$$

Using (2.1.9, 22) we can rewrite (2.1.27) for defects of cubic symmetry in elastically isotropic crystals as

$$E_{\rm el} = \frac{1}{3} \frac{1 - 2\sigma}{1 - \sigma} K (L^0_{ii})^2 V p (1 - p) \, . \tag{2.1.28}$$

In this case the elastic energy depends only on the concentration of the new-phase inclusions. In elastically anisotropic crystals or for defects of a lower symmetry $E_{\rm el}$ depends also on the distance between them and the energy of interaction between defects can be calculated from (2.1.27) by integration in the $\boldsymbol{k}$ space [1.25, 26].

Shape and Orientation of Single New-Phase Inclusions. We can find the optimal shape and arrangement of new-phase inclusions corresponding to the minimum of the elastic energy by analysing the above equations for $E_{\rm el}$. When the concentration of inclusions is low the interaction between them is insignificant and it is sufficient to consider the elastic energy of a single inclusion coherently linked to the matrix. This energy is given by (2.1.22) or (2.1.27) in the models with diffuse or sharply defined inclusion boundaries. In the latter equation $s(\boldsymbol{k})$ is the Fourier component of the shape function for a single inclusion (for small p the term $pV\delta_{\boldsymbol{k}0}$ in $s(\boldsymbol{k})$ is insignificant). Since the function $|s(\boldsymbol{k})|^2$ is normalized according to the condition $\int |s(\boldsymbol{k})|^2 d\boldsymbol{k} = 8\pi^3 V$ (see Sect. 1.3 in [1.1]) the elastic energy of the inclusion has the following minimum for the given volume:

$$E_{\rm el\,min} = V E_{\rm el}(\boldsymbol{n}_{\rm min}) \, , \tag{2.1.29}$$

if $s(\boldsymbol{k})$ are large only near certain directions of the vectors $\boldsymbol{k} = k\boldsymbol{n}$ which are near the minimum $\boldsymbol{n} = \boldsymbol{n}_{\rm min}$ of the positive function $E_{\rm el}(\boldsymbol{n})$. This rod-like distribution of $|s(\boldsymbol{k})|^2$ along the direction of $\boldsymbol{n}_{\rm min}$ is exhibited by the plate-like inclusions aligned perpendicular to $\boldsymbol{n}_{\rm min}$ in which the thickness $D_\parallel$ (in the direction parallel to $\boldsymbol{n}_{\rm min}$) is considerably smaller than its characteristic size $D_\perp$ in its plane (see Sect. 1.3 in [1.1]).

For instance, if inclusions of cubic symmetry are produced in a cubic crystal matrix, that is, $L^0_{ij} = L^0_{ij}\delta$, then we obtain from (2.1.5,6,26)

$$E_{\text{el}}(\boldsymbol{n}) = \frac{3}{2}KL^{02}\chi(\boldsymbol{n}) \,. \tag{2.1.30}$$

In elastically isotropic crystals $\xi = 0$) the function $\chi(\boldsymbol{n})$ is independent of $\boldsymbol{n}$, as can be seen from (2.1.6), and the elastic energy (2.1.29) depends only on the volume of the inclusion (but not on its shape which is determined under these circumstances by the condition that the surface energy be minimal). In real elastically anisotropic crystals, however, the function $\chi(\boldsymbol{n})$ depends on $\boldsymbol{n}$. For most crystals the anisotropy parameter $\xi < 0$ and $\chi(\boldsymbol{n})$ has the minimum for $\boldsymbol{n}\|\langle 100\rangle$, that is, the plate-like new-phase inclusions are in the $\{100\}$ planes. For the few crystals with $\xi > 0$ the production of plates in the $\{111\}$ planes is more favourable energetically. The minimum of $E_{\text{el}}(\boldsymbol{n})$ can be found even for more general forms of the tensors L_{ij} and c_{pq} from the general equation for $\varphi^{\text{el}}_{cc} = 2E_{\text{el}}(\boldsymbol{n})$ (where L_{ij} is replaced with L^0_{ij}. The specific expressions for the elastic energies of plate-like inclusions are given for some crystals in [1.25; 2.27].

In the limit of the thin inclusion ($D_\| \ll D_\perp$) the incoherence between the inclusion and the matrix relaxes at the expense of the elastic strain of the inclusion while the matrix remains unstrained in this limit. Indeed, for a plate-like inclusion $s(\boldsymbol{k})$ depends only on a single variable k (we have $\boldsymbol{k} \parallel \boldsymbol{n}$). This is why in the expression for the Fourier components of the distortion tensor,

$$u_{ij\boldsymbol{k}} = k_j u_{\boldsymbol{k}i} = C^{-1}_{\boldsymbol{k}ii'} P_{i'} n_j s(\boldsymbol{k}) \,,$$
$$P_{i'} = \frac{1}{V} c_{i'j'lm} n_{j'} L^0_{lm}$$

(see (2.2.36) in [1.1]), $C_{\boldsymbol{k}ii'}$ is given by (1.5.5)) the coefficient before $c(\boldsymbol{k})$ can be regarded as being constant so that the distortion $u_{ij}(\boldsymbol{r})$ is proportional to the shape function in the limit of $D_\| D_\perp^{-1} \to 0$.

$$u_{ij}(\boldsymbol{r}) = a_{ki} n_j s(\boldsymbol{r}) \,, \quad a_{ki} = C^{-1}_{\boldsymbol{k}ii'} c_{i'j'lm} n_{j'} L^0_{lm} \tag{2.1.31}$$

and does not vanish only inside the inclusion. Equation (2.1.31) yields also the total (intrinsic and elastic) strain u_{ij} and the angle of rotation Φ of the inclusion lattice with respect to the matrix lattice:

$$\begin{aligned} u_{ij} &= \tfrac{1}{2}(a_{ki}n_j + a_{kj}n_i)s(\boldsymbol{r}) \,, \\ \Phi_i &= \tfrac{1}{2}\delta_{ijl}(u_{j,l} - u_{l,j}) = \tfrac{1}{2}[\boldsymbol{a}\cdot\boldsymbol{n}]_i \,. \end{aligned} \tag{2.1.32}$$

In the important special case of phase transformations in which the plane normal to $\boldsymbol{n}$ remains invariant we obtain

$$L^0_{ij} = \tfrac{1}{2}(a^0_i n_j + a^0_j n_i) \,, \tag{2.1.33}$$

where $\boldsymbol{a}^0$ is a vector. When we substitute (2.1.33) into (2.1.31) and use the equation (1.5.5) for $C_{\boldsymbol{k}ij}$ we find that in this case $a_i = a^0_i$ and u_{ij} coincides with the self strain described by L^0_{ij}, that is, no additional elastic strain is produced and the elastic energy vanishes. Under nonisomorphic phase transformations a decrease in the elastic energy is favoured by the production of polydomain plate-shaped inclusions consisting of new-phase regions with different orientations, that is, the elastic domains for which the average tensor $\hat{L}^0$ is described by (2.1.33) (see below).

The elastic energy of an inclusion reduces to the minimum (2.1.29) which depends on its volume only in the limit of $D_\perp/D_\parallel \to \infty$. For finite ratios $D_\perp/D_\parallel$ one must bear in mind that at the interface of an inclusion the matrix is displaced by a constant vector of the order of $L^0 D_\parallel$ and thus the same strains are generated in it as the strains around a dislocation loop with the Burgers vector of the order of $L^0 D_\parallel$. The energy of the equivalent dislocation loop is of the order of $K|\xi|(L^0)^2 D^2_\parallel D_\perp \ln(D_\parallel/D_\perp)$ [2.33], it depends on the shape of the inclusion, and for the given volume $V \sim D_\parallel/D^2_\perp$ decreases with an increase in the ratio $D_\perp/D_\parallel$ so that the total energy tends to (2.1.29). This increase in the ratio $D_\perp/D_\parallel$ for the given volume is hindered, however, by an increase in the surface energy which is of the order of $\gamma D^2_\perp$ (where γ is the interfacial surface tension). The sum of these two terms for the given volume V has a minimum for $D_\parallel \sim (\gamma D_\perp/KL^0|\xi|)^{1/2}$. Therefore, a decisive energetic factor affecting the orientation and shape of the inclusion is the elastic energy including the energy of the dislocation loop (producing plate-shaped inclusions) when the following condition is satisfied [1.25]:

$$\gamma \ll K(L^0)^2 D_\parallel |\xi| \,. \tag{2.1.34}$$

When the reversed condition is satisfied the elastic energy has only a weak effect on the shape of the inclusions which become almost equaxial.

It should be noted that though the above energetic factors are quite important they are not the only ones affecting the inclusion shape. Kinetic factors can also play a significant (sometimes decisive) part. In the presence of the elastic interaction the kinetic treatment typically produces the same qualitative results as the energetic treatment.

Heterogeneous States in Coherent Two-Phase Systems (Elastic Domains). When the concentration of the second-phase particles is considerable the elastic interaction between them may play a significant part and the condition that the elastic energy (2.1.22) or (2.1.26) be minimal must be considered for the Fourier components $c_{\boldsymbol{k}}$ or $s_\alpha(\boldsymbol{k})$ for the total particle distribution, rather than for a single particle as in the case of low concentration discussed above. The elastic energy here may be decreased considerably if the second-phase particles form an ordered heterogeneous metastable structure.

Let us consider, for instance, such a heterogeneous structure produced by coherent isomorphic decomposition of a binary solution giving rise to a phase which differs from the matrix only in composition (but not in structure) [2.34]. If the deviations from the Vegard rule and the difference between the elastic moduli of the phases are small then the elastic energy of the resulting two-phase system with coherently linked phases is given by (2.1.22). The normalization condition (1.18) for the coefficients c_k shows that E_{el} has a minimum if c_k do not vanish only for $\boldsymbol{k}$ parallel to the vectors $\boldsymbol{n}_{min}$ for which $E_{el}(\boldsymbol{n})$ is minimal. This condition is satisfied, for instance, for the one-dimensional concentration distribution $c(\boldsymbol{r})$ along a direction $\boldsymbol{n}_{min}$ (for instance, in cubic crystals with $\xi < 0$ we have $\boldsymbol{n}_{min} \parallel \langle 100 \rangle$ as noted above). The elastic energy for this concentration distribution is given by (2.1.23). Therefore, the total thermodynamic potential of the coherent two-phase system under consideration is given by

$$\Phi = \int \{\varphi c(\boldsymbol{r})] + E_{el}(\boldsymbol{n}_{min})[c(\boldsymbol{r}) - c_0]^2\} d\boldsymbol{r} \, , \quad E_{el}(\boldsymbol{n}) = \tfrac{1}{2}\varphi_{cc}(\boldsymbol{n}) \, , \qquad (2.1.35)$$

where $\varphi(c)$ is the thermodynmic potential density in the absence of stresses and c_0 is the mean concentration (equal to the initial concentration of the solution).

As follows from the general concepts of thermodynamics of solutions, the integrand in (2.1.35) has a minimum for certain concentrations c_1 and c_2. If the system conserves coherence decomposition must give rise to regions with precisely these concentrations. Their relative fractions are found from the conventional lever rule. Since the distribution $c(\boldsymbol{r})$ is one-dimensional the regions with different compositions must have a plate-like shape, that is, the resulting heterogeneous structure consists of alternating particles with the concentrations c_1 and c_2. As shown by (1.5.3–6), under such a one-dimensional concentration distribution the lattice distortions are described by the distortion tensor $u_{ij}(\boldsymbol{r})$ proportional to $c(\boldsymbol{r})$ at the given point:

$$u_{ij}(\boldsymbol{r}) = a_{ki} n_j c(\boldsymbol{r}) \, , \qquad (2.1.36)$$

where a_{ki} is given by (2.1.31) where L^0_{lm} is replaced with L_{lm}. If, for instance, in a cubic crystal a regular structure is formed in the direction of $\boldsymbol{n} \parallel [100]$ then regions of each phase exhibit tetragonal distortions in this directions (in contrast to the case of low concentrations of the new-phase particles when such distortions described by (2.1.31) occur only in the second-phase particles).

These colonies of plate-shaped particles can be aligned along any of the possible directions of $\boldsymbol{n}_{min}$, for instance, $\langle 100 \rangle$ in the cubic matrix. The particle widths $D_\perp$ may be much smaller than the crystal size but they must be much larger than the parameter D of the regular structure under the condition that the system is quasi one-dimensional. The distortion fields cease to be one-dimensional in the regions of the width about D at the side surface of the colony and therefore the elastic energy density e is increased by $K|\xi|(D/D_\perp)[(1/v)(dv/dc)]^2$ (see [2.34]) while the resulting surface energy per unit volume is of the order of γ/D. The sum of these two terms has the minimum

$$e_{min} \sim \left(\frac{K|\xi|\gamma}{D_\perp}\right)^{1/2} \frac{1}{v}\left|\frac{dv}{dc}\right| \quad \text{for} \quad D \sim \left(\frac{\gamma D_\perp}{K|\xi|}\right)^{1/2} \left|\frac{1}{v}\frac{dv}{dc}\right|^{-1} . \qquad (2.1.37)$$

Two-dimensional and three-dimensional metastable modulated structures may also be produced but they are less favourable thermodynamically than the one-dimensional structure [1.25]. The heterogeneous structures occurring in real crystals are affected also by the kinetic factors, in addition to the energetic ones.

An interesting illustration of an ordered heterogeneous structure is a three-dimensional lattice of pores produced by coalescence of vacancies in irradiated materials which was reported by a number of authors starting with [2.35] (see the review [2.36]). This lattice may also be produced owing to the elastic interaction between the pores [2.36, 37]. But even in crystals where the elastic interaction is not significant such a lattice can be formed owing to the kinetic processes giving rise to a diffusion interaction or a peculiar diffusive-elastic interaction between the pores producing a dissipative structure which remains stable under radiation exposure [2.36, 38].

Decomposition of a solution giving rise to an isomorphic second phase causes a decrease in the elastic energy as a heterogeneous phase is produced but it remains quite large. If the second phase has a lower symmetry than the matrix then the elastic energy can be decreased to a negligibly low level by producing a heterogeneous structure consisting of alternating thin second-phase particles (elastic domains) with different orientations and taking the limit of large particle widths. A good illustration is the formation of a tetragonal phase in the cubic matrix which was discussed in detail in [2.39, 40] and phenomenologically treated in [2.41–43]. Here we can identify a polydomain colony of alternating plate-shaped particles with the tetragonal axes parallel to the axes [100] and [010] of the cubic phase. The respective self-strain tensors $\hat{L}^0$ for the two cases I and II have the following forms:

$$\hat{L}^{0(\mathrm{I})} = \begin{pmatrix} L' & 0 & 0 \\ 0 & L'' & 0 \\ 0 & 0 & L'' \end{pmatrix}, \quad \hat{L}^{0(\mathrm{II})} = \begin{pmatrix} L'' & 0 & 0 \\ 0 & L' & 0 \\ 0 & 0 & L'' \end{pmatrix}. \tag{2.1.38}$$

Only the components $\Delta L^0_{xx} = -\Delta L^0_{yy}$ of the tensor difference $\Delta\hat{L}^0 = \hat{L}^{0(\mathrm{I})} - \hat{L}^{0(\mathrm{II})}$ do not vanish. If the normal to the particles is parallel to [110] such tensor $\Delta\hat{L}^0$ can be written in the form (2.1.33) by taking the vector $\boldsymbol{a}^0$ parallel to [1$\bar{1}$0]. This means that a transition from the case I to the case II is a phase transformation with the invariant plane which does not produce elastic energy. This is a natural result for an interface in the (110) plane in which atoms have the identical arrangement in both tetragonal regions and elastic strain is not required for achieving coherence of the lattices.

We can make the average tensor $\bar{\hat{L}}^0 = p_1\hat{L}^{0(\mathrm{I})} + p_2\hat{L}^{0(\mathrm{II})}$ acquire the form (2.1.33) by selecting the concentrations p_1 and p_2 of the particles with the orientations I and II in a colony shaped as a polydomain slab and taking a special direction of the normal $\tilde{\boldsymbol{n}}$ to the plane of this slab. When $\hat{L}^{0(\mathrm{I})}$ and $\hat{L}^{0(\mathrm{II})}$ are given by (2.1.38) and and L' and L'' have different signs and $p_2 = 1 - p_1 = L'/L' - L''$, the nonvanishing components of the tensor $\bar{L}^0$ are $\bar{L}^0_{yy} = L' + L''$ and $\bar{L}^0_{zz} = L''$. Such a tensor has the form (2.1.33) if

$$\tilde{\boldsymbol{n}} = \boldsymbol{n}_{1,2} = \left(\frac{L' + L''}{L'}\right)^{1/2} \boldsymbol{e}_y \pm \left(-\frac{L''}{L'}\right)^{1/2} \boldsymbol{e}_z\,, \quad \boldsymbol{a}^0_{1,2} \parallel \tilde{\boldsymbol{n}}_{2,1}\,, \tag{2.1.39}$$

where the unit vectors e_y and e_z are parallel to the axes Y and Z and $|L''| < |L'|$. The direction of $\tilde{n}$ differs from the direction of $n_{\min}$ for a one-domain particle.

When a polydomain slab (colony), consisting of thin alternating plate-like particles of the phases with the tetragonal axes parallel to [100] and [010], has such an orientation the elastic energy proportional to the volume of the colony vanishes. The only nonvanishing elastic energy corresponds to the strain in the region at the slab surface with the width of the order of the parameter D of the regular structure. The elastic energy density $e_{\min}$ (and the parameter D) are determined by (2.1.37) where $(1/v)|\partial v/\partial c|$ is replaced with $|L'|$ for $|L'| \sim |L''|$. When the width $D_\perp$ of the particles is large enough the energy density $e_{\min}$ is small and the production of a polydomain slab is energetically favourable. At the initial stage of phase transformation, however, when $D_\perp$ is small the additional phase interfaces make a polydomain slab less favourable than a single-domain one which is aligned along the direction of $n_{\min}$ (see details in the reviews [2.26, 27] where more complicated polydomain structures are also discussed).

It should be noted that the heterogeneous states which are produced in coherent systems owing to the elastic interaction correspond to the metastable, rather than full equilibrium, in contrast to the largely similar magnetic domains in ferromagnetics and the heterogeneous states caused by indirect interaction (see Sect. 2.5). Under stable equilibrium the system must break down into two massive unstressed phases in which the surface energy and the energy of elastic stresses both vanish.

The elastic interaction described by the above equations may play a decisive part in the production of the equilibrium one-dimensional regular structures with large parameters in systems with flat defects. Such defects are produced, for instance, in nonstoichiometric oxides of the type of TiO_{2-x} or WO_{3-x} by removal of some atomic planes of oxygen which is accompanied by shearing strains, and by introduction of impurities between the atomic planes of carbon in intercalated graphite compounds. The significant elastic interaction between such planes at large distances gives rise to their regular arrangement [2.44–50].

2.2 Effects of Dipole-Dipole Interaction on Polarization Fluctuations

General Equations for the Amplitudes of the Polarization Fluctuation Waves. Similarly to the elastic interaction, the dipole-dipole interaction varies as $1/r^3$ and when it is taken into account the equations for the fluctuation probability may change their form. The electric diple-dipole forces affect significantly the polarization fluctuations in dielectric crystals while the magnetic forces affect the magnetization fluctuations.

Let us illustrate the effects produced by these forces with the case of the polarization fluctuations [1.16]. We shall consider the long-wave fluctuations for which the long-range interactions are particularly significant. The polarization of the crystal can therefore be described by a macroscopic parameter, namely, the polarization vector $\boldsymbol{P}(\boldsymbol{r})$. In proper ferroelectrics the components of this vector are long-range order parameters.

In the presence of a long-range dipole-dipole interaction the work required for producing polarization fluctuations in the volume V' under consideration depends not only on the state of the crystal in this volume but also on the polarization of the rest of the crystal. As in the case of elastic forces the energy of the interaction between the dipoles in the volume V' and the dipoles outside this volume can be conveniently found not by directly calculating the respective sums but by considering the electric field $\boldsymbol{E}$ produced inside V' by the dipoles located outside V'. For a nonequilibrium crystal we can obtain from (1.2.21) the thermodynamic potential φ of the unit volume regarded as a function of the internal parameter $\boldsymbol{P}$ and the external parameter $\boldsymbol{E}$ (the electric field). We must also make use of the relation (1.2.23) between the second-order derivatives of φ with respect to $\boldsymbol{P}$ and the dielectric susceptibility tensor $\varkappa_{ij}$ as well as the condition (1.2.22), $\boldsymbol{E} = \partial\varphi/\partial\boldsymbol{P}$, for the equilibrium field. Then the least work required for producing the polarization distribution $\delta\boldsymbol{P}$ varying in space and the electric field variation generated by this polarization, $\delta\boldsymbol{E}$ (which is zero on the average in the crystal), can be written as

$$\mathfrak{R} = \int \left[\tfrac{1}{2}\varkappa_{ij}^{-1}\delta P_i \delta P_j - \delta P_i \delta E_i\right] d\boldsymbol{r} \,. \tag{2.2.1}$$

The electric field variation $\delta\boldsymbol{E}$ caused by the static polarization fluctuation $\delta\boldsymbol{P}$ is given by the equations of electrostatics:

$$\operatorname{div} \delta\boldsymbol{E} = -4\pi \operatorname{div} \delta\boldsymbol{P} \,, \quad \operatorname{rot} \delta\boldsymbol{E} = 0 \tag{2.2.2}$$

(for the dielectric crystal in the absence of free charges). The second equation (2.2.2) demonstrates that for the $\boldsymbol{k}$th fluctuation wave the field amplitude $\boldsymbol{E}_{\boldsymbol{k}}$ is parallel to $\boldsymbol{k}$ (that is, $|\boldsymbol{k} \cdot \boldsymbol{E}_{\boldsymbol{k}}| = 0$) while the first equation yields a relationship between the amplitudes $\boldsymbol{E}_{\boldsymbol{k}}$ and $\boldsymbol{P}_{\boldsymbol{k}}$. Then we can write for the polarization and the electric field in this fluctuation wave

$$\delta\boldsymbol{P} = \boldsymbol{P}_{\boldsymbol{k}} \mathrm{e}^{-\mathrm{i}\boldsymbol{k}\cdot\boldsymbol{r}} + \text{c.c.} \,, \quad \delta\boldsymbol{E} = \boldsymbol{E}_{\boldsymbol{k}} \mathrm{e}^{\mathrm{i}\boldsymbol{k}\cdot\boldsymbol{r}} + \text{c.c.} \,, \quad \boldsymbol{E}_{\boldsymbol{k}} = -4\pi(\boldsymbol{P}_{\boldsymbol{k}}\boldsymbol{n})\boldsymbol{n} \,. \tag{2.2.3}$$

When we substitute these equations into (2.2.1) we obtain the least work required for producing the $\boldsymbol{k}$th fluctuation wave:

$$\mathfrak{R}_{\boldsymbol{k}} = \tfrac{1}{2} V \varkappa_{ij}(\boldsymbol{k})(P_{\boldsymbol{k}i} P_{\boldsymbol{k}j}^* + P_{\boldsymbol{k}i}^* P_{\boldsymbol{k}j}) \,, \tag{2.2.4}$$

where

$$\varkappa_{ij}^{-1}(\boldsymbol{k}) = \varkappa_{ij}^{-1} + 4\pi n_i n_j \,, \quad \boldsymbol{n} = \frac{\boldsymbol{k}}{k} \,. \tag{2.2.5}$$

The last term in (2.2.5) describing the effect of the long-range dipole-dipole forces changes significantly the least work.

In the classical (nonquantum) theory of fluctuations the expression (2.2.4) for the least work and the Gaussian distribution (1.2.1) of the fluctuation probabilities yield

$$\frac{1}{2}\langle (P_{ki}P^*_{kj} + P^*_{ki}P_{kj})\rangle = \frac{k_BT}{V}\varkappa_{ij}(\boldsymbol{k}) , \tag{2.2.6}$$

where $\varkappa(\boldsymbol{k})$ is the inverted tensor $\varkappa^{-1}(\boldsymbol{k})$ given by (2.2.5).

The classical theory is applicable, however, only if the condition $\hbar\omega_i \ll k_BT$ is satisfied for all characteristic frequencies ω_i of the polarization oscillations (and for all effective frequencies of the diffusion motion of the dipoles). Conditions of this type are typically satisfied to a high accuracy for the fluctuations of the concentration and the order parameters in the solid solutions in which very slow diffusion processes determine the variation of fluctuations with time. On the other hand, when we consider the polarization fluctuations the condition $\hbar\omega_i \ll k_BT$ is satisfied only for the fluctuation variations associated with relatively slow oscillations of ions. The reversed condition $\hbar\omega_i \gg k_BT$ is practically always satisfied for the polarization component associated with the electronic, rather than ionic, excitations. The contribution of these high-frequency fluctuations must be treated in the framework of the quantum theory.

In the calculations of the mean squared polarization fluctuations, a fraction of which have the quantum character, we can write the mean products of the fluctuations given by (2.2.6) in the classical theory of the "static" fluctuations in the form of the integrals

$$\tfrac{1}{2}\int_{-\infty}^{\infty}\langle[P_{ki}(\omega)P^*_{kj}(\omega) + P^*_{ki}(\omega)P_{kj}(\omega)]d\omega . \tag{2.2.7}$$

Here we have the mean products of the Fourier components

$$P_{ki}(\omega) = \frac{1}{2\pi}\int_{-\infty}^{\infty} P_{ki}(t)\exp(\mathrm{i}\omega t)dt ,$$

which can also be calculated in the classical theory. This result follows from the theory describing correlation of the fluctuations with time (see, for instance [1.31]). But we cannot use the equations of the classical theory of fluctuations for calculating the probabilities of the high-frequency fluctuation harmonics $\boldsymbol{P}_{\boldsymbol{k}}(\omega)$ for which $\hbar\omega_i \gg k_BT$. These harmonics are not excited and their contribution to the integral (2.2.7) may be ignored (if the quantum zero-point vibrations are neglected). To find the average products of the actual polarization, fluctuations we must remove from (2.2.6) or (2.2.7) the contribution produced by such unexcited harmonics.

Consider, for instance, the case when $k_BT \sim \hbar\omega'$ where ω' is much higher than the ionic frequencies and much lower than the electronic frequencies. Denote by $\varkappa_{\infty ij}$ the dielectric susceptibility tensor for the frequencies of the order of ω' corresponding to the electronic polarization. Then we can calculate the contribution of the high frequencies $\omega > \omega'$ to the integral (2.2.7) obtained with the classical treatment by replacing $\varkappa(\boldsymbol{k})$ in (2.2.6) with the tensor $\varkappa_\infty(\boldsymbol{k})$ which is related to $\varkappa_{\infty i-j}$ by (2.2.5). Finally, we deduct from (2.2.6) the contribution of the high-frequency fluctuations which are not excited at the given temperature and obtain [1.16]

$$\frac{1}{2}\langle(P_{ki}P^*_{kj} + P^*_{ki}P_{kj})\rangle = \frac{k_B T}{V}[\varkappa_{ij}(\boldsymbol{k}) - \varkappa_{\infty ij}(\boldsymbol{k})] . \tag{2.2.8}$$

At lower temperatures when the electronic oscillation frequencies and some ionic oscillation branches are not excited we can also make use of (2.2.8) if the tensor $\varkappa_\infty$ takes into account both the electronic and high-frequency ionic polarizations. When we derived (2.2.8) we assumed that the contribution of the low-frequency fluctuations under consideration was not anomalously small so that we could ignore the quantum (zero-point) high-frequency fluctuations.

One can see from (2.2.5) that all parameters which describe the mean products (2.2.8) of the polarization fluctuations can be found from the experimental results on the static dielectric susceptibility $\varkappa_{ij}$, and the refraction indexes for the infrared light which determine $\varkappa_{\infty ij}$. Formula (2.2.8), in similarity to (2.1.5), generally exhibits a singularity for $\varkappa \to 0$ (but when $k \gg L^{-1}$) since it follows from (2.2.5) that $\varkappa(\boldsymbol{k})$ and $\varkappa_\infty(\boldsymbol{k})$ are generally nonanalytic functions of $\boldsymbol{k}$ (for $k \to 0$ the limiting value of $\boldsymbol{k}$ depends on the direction of $\boldsymbol{k}$).

Equations (2.2.5, 8) are valid in the limiting case of long waves when $k \to 0$. Terms of the type of $\alpha_{ijlm}k_l k_m$ must be added to (2.2.5) in the case of shorter waves. The coefficients α_{ijlm} also depend on $\boldsymbol{n}$ if the quadrupole-quadrupole and dipole-octupole interactions are taken into consideration. (This effect must occur also for the fluctuations of the long-range order parameters in antiferroelectric crystals.)

Equations of the type of (2.2.8) are clearly applicable also to the magnetization fluctuations when we must take into consideration the magnetic dipole-dipole interaction (the dielectric susceptibility must be replaced with the magnetic susceptibility in these equations). Since for all materials the magnetic permittivity differs little from unity the effects caused by the long-range forces in this case are much weaker than those for the polarization fluctuations and typically they are observed only near the Curie point.

Crystals of Different Symmetries. In the special case of cubic crystals we have $\varkappa_{ij} = (\varepsilon - 1)\delta_{ij}/4\pi$ (where ε is the static permittivity) and $\varkappa_{\infty ij} = (\varepsilon_\infty - 1)\delta_{ij}/4\pi$. If we take take the axis Z parallel to $\boldsymbol{k}$ and substitute the above expressions into (2.2.5, 8) we find that (2.2.8) vanishes for $i \neq j$ while for $i = j$ we obtain

$$\begin{aligned} \langle|P_{kx}|^2\rangle &= \langle|P_{ky}|^2\rangle = \frac{k_B T}{4\pi V}(\varepsilon - \varepsilon_\infty) , \\ \langle|P_{kz}|^2\rangle &= \frac{k_B T}{4\pi V}\frac{\varepsilon - \varepsilon_\infty}{\varepsilon\varepsilon_\infty} . \end{aligned} \tag{2.2.9}$$

We see that the probabilities of production of the longitudinal and transverse fluctuation waves may be quite different. This difference is fully determined by the long-range dipole-dipole interaction which does not affect $\langle|P_{kx}|^2\rangle$ and decreases the probabilities of the longitudinal fluctuations and $\langle|P_{kz}|^2\rangle$. In contrast to the transverse polarization fluctuations, the longitudinal fluctuations are accompanied with the generation of an electric field which increases their energy (see (2.2.3)).

The effects caused by the dipole-dipole interaction must be especially noticeable in the proper ferroelectrics at the temperature corresponding to the second-order phase transition to the ferroelectric state when some components of the tensor $\varkappa_{ij}$ tend to infinity and the polarization fluctuations exhibit anomalous growth. The long-range interaction can significantly suppress such growth. For instance, it can be seen from (2.2.9) that at temperatures above the phase transition point in cubic crystals for $\varepsilon \to \infty$ the fluctuations of the transverse components of the polarization tensor in the plane perpendicular to $\boldsymbol{k}$ become anomalously large as it is found in theoretical models without the long-range forces. In contrast to theoretical predictions, however, the fluctuations of the polarization component parallel to $\boldsymbol{k}$ remain finite. After the transition of a cubic crystal to the ferroelectric state its symmetry becomes lower, the tensor components $\varkappa_{ii}$ are no longer identical and, according to the general equation (2.2.8), the expressions for the polarization fluctuations become more complicated than (2.2.9). In this case, however, the result on the anomalous growth of the polarization fluctuations perpendicular to $\boldsymbol{k}$ (but not those parallel to $\boldsymbol{k}$) is also applicable.

In noncubic (in the nonferroelectric phase) crystals the probability of the fluctuation waves depends strongly on the orientation of the vector $\boldsymbol{k}$ in a uniaxial ferroelectric crystal (possessing an axis of easy polarization) in which only one component $\varkappa_{zz}$ of the tensor $\varkappa_{ij}$ exhibits anomalous growth near the transition temperature ($\varkappa_{zz} \gg 1$). Under such conditions anomalously high polarization fluctuations occur only if $\boldsymbol{k}$ is almost perpendicular to the axis Z (if $n_z \ll 1$). For $\varkappa_{yy} = \varkappa_{xx}$ and $\varkappa_{ij} = 0$ for $i \neq j$ we obtain in this case [1.16] from (2.2.5, 8) if we ignore terms of the order of unity in comparison with $\varkappa_{zz}$ or n_z^{-2}

$$\langle |P_{\boldsymbol{k}z}|^2 \rangle = \frac{k_B T}{V} \frac{(1 + 4\pi\varkappa_{xx})\varkappa_{zz}}{1 + 4\pi\varkappa_{xx} + 4\pi\varkappa_{zz} n_z^2} , \tag{2.2.10}$$

that is $\langle |P_{\boldsymbol{k}z}|^2 \rangle$ reaches the maximum proportional to $\varkappa_{zz}$ for $n_z = 0$ and varies as $1/n_z^2$ for $4\pi\varkappa_{zz} n^2 \gg 1 + 4\pi\varkappa_{xx}$. The squares and products of other components of the vector $\boldsymbol{P}_{\boldsymbol{k}}$ in this case do not exhibit anomalous growth.

In ferroelectric crystals with a plane of easy polarization containing two axes of spontaneous polarization, X and Y, $\varkappa_{xx}$ and $\varkappa_{yy}$ grow as the temperature approaches the transition point. According to (2.2.5, 8), only the components of the vector $\boldsymbol{P}_{\boldsymbol{k}}$ in the plane XY perpendicular to $\boldsymbol{k}$ exhibit anomalous growth as in the case of cubic crystals [2.51].

The matrix $\varkappa_{ij}^{-1}(\boldsymbol{k})$ given by (2.2.5) can also be inverted in the general case of arbitrary tensors $\varkappa_{ij}$ and arbitrary directions of $\boldsymbol{n}$. Using the frame of reference in which $\varkappa_{ij} = \varkappa_i \delta_{ij}$ has the diagonal form we can write [2.52]

$$\varkappa_{ij}(\boldsymbol{k}) = \varkappa_i \delta_{ij} - \frac{4\pi\varkappa_i\varkappa_j n_i n_j}{1 + 4\pi \sum\limits_l \varkappa_l n_l^2} . \tag{2.2.10a}$$

A similar expression can be written for $\varkappa_{ij}^{\infty}(\boldsymbol{k})$ (in the frame of reference in which the tensor $\varkappa_{ij}^{\infty}$ has the diagonal form). These expressions can be used to calculate the polarization fluctuations in the general case according to (2.2.6, 8).

The suppression of the fluctuations caused by the dipole-dipole interaction producing a sharp anisotropy in their distribution gives rise to characteristic features exhibited by scattered X-rays, neutrons (see Chap. 4) and light [1.16; 2.23] as well as by the relaxation effects in ferroelectric [2.51, 53, 54] and ferromagnetic crystals [2.55, 56].

The fluctuations which occur near the second-order phase transitions determine the behavior of the thermodynamic parameters in the vicinity of the transition. As noted in [1.13], therefore, the character of this behaviour must be changed owing to the suppression of the critical fluctuations by the long-range elastic and dipole forces. For instance, the long-range effects must have a strong influence on the phase transitions in uniaxial ferroelectric crystals in which, as can be seen from (2.2.10), the fluctuations are suppressed for all $\boldsymbol{k}$ with the exception of the vicinity of the plane of $k \perp Z$. The additional terms due to fluctuations for the thermodynamic results derived from the Landau theory for second-order phase transitions were estimated with the perturbation theory [1.51; 2.57]. Detailed analysis [2.58, 59] demonstrates that the changes are not very significant and give rise only to additional logarithmic terms in the equation for the susceptibility. Similar changes are produced by the elastic interaction in ferroelectric crystals exhibiting a piezoeffect in the symmetric phase and in ferroelastic crystals (see Chap. 4).

Effects Caused by Screening of Fields and Piezoelectric Strain. The polarization fluctuations were calculated above for the dielectric crystals. The conduction electrons or charged defects in the crystal cause screening of the fields at distances of the order of the Debye length r_{D}. This is why the equations (2.2.8–10) are valid only for not too long fluctuation waves (shorter than r_{D}) for which $k^2 r_{\mathrm{D}} \gg \varepsilon$. For smaller k we must analyze the fluctuations of the polarization and the charge carrier density at the same time and then the contribution of the dipole-dipole forces becomes much smaller owing to the screening of the field produced by dipoles at large distances. For such small k in (2.2.8) $\varkappa_{ij}(\boldsymbol{k})$ coincides with $\varkappa_{ij}$. On the other hand, for high frequencies corresponding to $\varkappa_\infty$ the electric charges typically have no time for rearrangement required for screening of the electric fields produced by the highfrequency polarization fluctuations so that $\varkappa_{\infty ij}(\boldsymbol{k})$ remains unchanged. Using (1.2.24) which was derived for the case without long-range forces we can transform (2.2.8) into

$$\frac{1}{2}\langle(P_{ki}P^*_{kj} + P^*_{ki}P_{kj})\rangle = \frac{k_{\mathrm{B}}T}{V}[\varkappa_{ij} - \varkappa_{\infty ij}(\boldsymbol{k})] \,. \tag{2.2.11}$$

If there were enough time for relaxation of the high-frequency fields then $\varkappa_{\infty ij}(\boldsymbol{k})$ in (2.2.11) would have to be replaced with $\varkappa_{\infty ij}$.

The effects produced by screening on the dependence of the mean squared fluctuations on k for $k \sim \sqrt{\varepsilon} r_{\mathrm{D}}$ can be illustrated, for instance, with the results for the simplest case of the cubic crystal for which $\varepsilon_\infty = 1$. In this case the Debye screening does not change the form of the equations (2.2.9) (which were derived for $\varepsilon_\infty = 1$) giving $\langle |P_{kx}|^2 \rangle = \langle |P_{ky}|^2 \rangle$ while the third equation is transformed into

$$\langle |P_{kz}|^2 \rangle = \frac{k_B T}{4\pi V} \frac{\varepsilon - 1}{\varepsilon} \frac{\varepsilon + k^2 r_D^2}{1 + k^2 r_D^2} . \tag{2.2.12}$$

This equation coincides with (2.2.9) for $k r_D \gg \sqrt{\varepsilon}$ and then screening is not significant. When $k r_D \ll 1$ screening suppresses the effects of the long-range interaction and $\langle |P_{kz}|^2 \rangle$ coincides with $\langle |P_{kx}|^2 \rangle$.

If elastic strain is caused by the polarization fluctuations (piezoeffect) then we must take into account the long-range forces of two types, that is, the elastic and dipole-dipole forces. As it is the case for other internal parameters of long-range order, the inclusion of the elastic energy in the calculations of the polarization fluctuations gives rise to additional terms of the type of $\varphi^{\mathrm{el}}_{\mu\mu'}$ in the second-order derivatives of the thermodynamic potential in (2.1.20). Thus, we can take into account the elastic interaction by replacing $\partial^2 \varphi / \partial P_i \partial P_j \equiv \varkappa_{ij}$ with the tensor $\varkappa_{ij}$ in the equations derived in this section:

$$\varkappa'^{-1}_{ij} = \varkappa^{-1}_{ij} + c_{lmnp} L^i_{lm} (L^j_{np} - k_n A^j_{kp}) . \tag{2.2.13}$$

Here $\varkappa_{ij}$ is the susceptibility tensor for constant stresses, $\hat{c}$ is the tensor of the elastic moduli found for constant polarization, $L^i_{lm} = \delta u_{lm} / \delta P_i$ describes the strain δu_{lm} produced by the homogeneous variation of the ith polarization component by δP_i and the vector $\boldsymbol{A}^j_{\boldsymbol{k}}$ with the components A^j_{km} is determined by (1.5.9) where $L_{\mu lm}$ is replaced with L_{lm}. When the elastic energy is considerable, therefore, we must replace $\varkappa_{ij}(\boldsymbol{k})$ in the equation (2.2.8) for the fluctuations with the tensor $\varkappa'_{ij}(\boldsymbol{k})$ which is derived from $\varkappa'^{-1}_{ij}$ according to (2.2.5) in the same way as $\varkappa_{ij}(\boldsymbol{k})$ was derived from $\varkappa_{ij}$. When we calculate the second term in (2.2.8) associated with high-frequency fluctuations the strains that vary relatively slowly with time make no contribution. Thus, when we take into account the elastic interaction we obtain

$$\frac{1}{2} \langle (P_{ki} P^*_{kj} + P^*_{ki} P_{kj}) \rangle = \frac{k_B T}{V} [\varkappa'_{ij}(\boldsymbol{k}) - \varkappa_{\infty ij}(\boldsymbol{k})] . \tag{2.2.14}$$

When the wavelength of the fluctuation wave is larger than the Debye length ($k r_D \ll 1$) we must replace $\varkappa'_{ij}(\boldsymbol{k})$ with $\varkappa'_{ij}$ in (2.2.14), as follows from the above discussion.

2.3 Fluctuations of the Charged Defect Concentration

Some defects in crystals are charged. Such charged defects are vacancies and interstitials in ionic crystals, impurity ions with valencies different from the valencies of the matrix ions, for instance, Ca^{++} ions occupying the sites of Na^+ ions in NaCl crystals, ionized impurity atoms, etc. Strong Coulomb forces slowly decreasing with distance act between the charged defects. They have a significant effect on the fluctuations since the fluctuations of the concentration of charged defects with the same sign occuring in a large volume sharply increase the electrostatic energy and are energetically unfavourable. This is why the probability of such smoothly varying fluctuations is not large. On the other hand, the electrostatic energy is considerably lower under simultaneous fluctuations of the concentrations of positively and negatively charged defects so that the charge fluctuations are compensated. Thus we must obtain a strong statistical correlation between the concentration fluctuations of the charged defects of the opposite signs. It corresponds to a peculiar correlation between charged defects of the opposite signs at large distances, that is, the Debye screening caused by the Coulomb forces. As it is done for other problems of a similar type, a convenient approach to treating the correlation is to consider directly the fluctuation waves of the concentration for defects of the opposite signs [1.15].

Let us consider, for the sake of definiteness, a cubic ionic crystal containing positive and negative defects of the same valency. The mean concentrations of the defects must be identical according to the principle of electroneutrality. Assume that the unit cell has one position s_1 accessible for the positive defect and one position s_2 for the negative defect. As it was done above, the arrangement of defects over these positions can be described by the numbers c_{s1} and c_{s2} which equal unity if the position is occupied by a defect and vanish if the position is not occupied by a defect. We can also give their Fourier components c_{k1} and c_{k2} found from the expansion (1.1.8). It is convenient to make use of the sums and differences of these numbers:

$$2c_k^+ = c_{k1} + c_{k2}\,, \quad 2c_k^- = c_{k1} - c_{k2}\,. \tag{2.3.1}$$

The fluctuation wave for the difference (but not the sum) of the concentrations of the positive and negative defects with the amplitude c_k^- gives rise to a wave of electric charge,

$$\varrho_k = \frac{1}{V}\int \varrho(\boldsymbol{r})\exp(i\boldsymbol{k}\cdot\boldsymbol{r})d\boldsymbol{r} = \frac{2ez'}{v}c_k^- \tag{2.3.2}$$

(where z' is the defect charge and $e > 0$) and the associated electric flux density wave ($\mathrm{div}D = 4\pi\varrho$):

$$\boldsymbol{D}_k = \frac{1}{V}\int \boldsymbol{D}(\boldsymbol{r})\exp(i\boldsymbol{k}\cdot\boldsymbol{r})d\boldsymbol{r} = 4\pi i\varrho_k\frac{\boldsymbol{k}}{k^2} = \frac{8\pi iez'}{v}c_k^-\frac{\boldsymbol{k}}{k^2}\,. \tag{2.3.3}$$

The concentration of charged defects is always small and on the average they are at a considerable distance from each other which is much larger than the distance between atoms in the crystal. Therefore, when we calculate the thermodynamic potential we can ignore all short-range interaction between defects and take into consideration only the large electrostatic energy determined by the long-range Coulomb interaction. According to (2.3.3), the $\boldsymbol{k}$th fluctuation wave produces the following variation of this energy

$$\frac{1}{8\pi\varepsilon}\int D^2(r)dr = \frac{VD_kD_k^*}{4\pi\varepsilon} = \frac{16\pi e^2 z'^2 V}{\varepsilon v^2 k^2}|c_k^-|^2 . \tag{2.3.4}$$

In crystals with low defect concentration discussed here the configurational part of the entropy can be found in the simplest approximation which ignores the correlation between defects: $S = -Nk_B(c_1 \ln c_1 + c_2 \ln c_2)$. The respective contribution of the entropy to the variation of the thermodynamic potential produced by the $\boldsymbol{k}$th fluctuation wave is

$$Nk_BTc^{-1}\left[|c_{k1}|^2 + |c_{k2}|^2\right] = 2Nk_BTc^{-1}\left[|c_k^+|^2 + |c_k^-|^2\right] . \tag{2.3.5}$$

The least work $\mathfrak{R}_k$ required for producing the $\boldsymbol{k}$th fluctuation wave is the sum of (2.3.4) and (2.3.5):

$$\mathfrak{R}_k = \frac{16\pi e^2 z'^2 V}{\varepsilon v^2 k^2}|c_k^-|^2 + \frac{2Nk_BT}{c}(|c_k^+|^2 + |c_k^-|^2) . \tag{2.3.6}$$

The mean squared Fourier components of the fluctuations can be found by performing transformations of (1.2.1) and (2.3.6):

$$\langle|c_k^+|^2\rangle = \frac{c}{2N} , \quad \langle|c_k^-|^2\rangle = \frac{c}{2N}\frac{r_D^2k^2}{1+r_D^2k^2} , \quad \langle c_k^+ c_k^{-*}\rangle = 0 , \tag{2.3.7}$$

Here r_D is the Debye screening length:

$$r_D^2 = \frac{\varepsilon k_BTv}{8\pi e^2 z'^2 c} . \tag{2.3.8}$$

It can be seen from (2.3.7) that the Coulomb interaction does not affect the fluctuations of the total concentration of the positive and negative defects described by $c_k^+ = (c_{k1} + c_{k2})/2$ which conserve (on the average) electroneutrality. The Coulomb interaction does not affect also the fluctuations of the concentration difference described by $c_k^- = (c_{k1} - c_{k2})$ corresponding to short fluctuation waves whose length must be smaller than the Debye length ($kr_D \gg 1$). According to (2.3.4), however, the electrostatic energy increases as $1/k^2$. For sufficiently small $\boldsymbol{k}$ (when $kr_D \ll 1$) we find then that the Coulomb interaction suppresses the production of the fluctuation waves described by $\bar{c}_k$ which violate the electroneutrality and (2.3.7) demonstrates that long-wave fluctuations $\bar{c}_k$ have then a low probability.

The equations (2.3.7) were derived for the case when the field around the given defect was screened by other defects. But these results remain qualitatively the same for a semiconductor crystal where Debye screening is produced by the nondegenerate gas of the conduction electrons.

2.4 Effects of Long-Range Forces on Fluctuation Waves in Metal Alloys

The interaction between atoms in metal alloys is largely determined by the conduction electrons. The distribution of the electron density and the electron energy in alloys depend on the arrangement of atoms over the lattice sites causing a significant and sometimes decisive, contribution made by the conduction electrons to the configuration energy. The resulting indirect interaction between atoms in the alloy is similar to the indirect interaction between the spins of nuclei or atoms of the transition elements in metallic magnetic systems [2.60, 61].

The electron density in the fully degenerate gas of free electrons around a point defect is known to exhibit the Friedel oscillations with the period of π/k_{F} (where k_{F} is the wave vector of an electron on the Fermi surface) whose amplitude varies as $1/r^3$ (see, for instance, [2.62]). The indirect interaction between the impurity ions caused by the rearrangement of the conduction electrons in the field of the ions respectively varies as $1/r^3$ that is, it is a long-range interaction. This result is due to the sharp cut-off of the momentum distribution of electrons on the Fermi surface. In real alloys the conduction electrons are not free and their Fermi surface can differ significantly from the sphere so that the interaction may weaken with increasing distance at an even slower rate.

In alloys the slowly decreasing forces of the indirect interaction can produce such features in the energy of the concentration fluctuation waves and the probabilities of their generation regarded as functions of the wave vector [1.17] (see also Sect. 3 and [2.63] and [2.64]) which are largely similar to the Kohn features in the oscillation frequencies for metals [2.65] (see also [2.66]). In contrast to the elastic and dipole interaction forces, the indirect interaction forces depend on a rapidly oscillating factor and therefore the above features appear for large k, rather than $k \to 0$, on some surfaces in the $\boldsymbol{k}$ space associated with the Fermi surface.

To analyze the contribution of the indirect interaction we must find the energy of the conduction electrons as a function of the occupation numbers c_s for the lattice sites. The results on the electron energy (and other contributions to the alloy energy) calculated at the same time demonstrate the physical meaning of the ordering energy $w(\varrho)$ discussed in the semiphenomenological theory in Sect. 1.3 which can be found for some crystals.

Calculations of the electron energy E_e is much simpler for the alloys with high potential energies $V_A(\boldsymbol{r} - \boldsymbol{R}_s)$ and $V_B(\boldsymbol{r} - \boldsymbol{R}_s)$ for the interaction between the conduction electrons and the ions A and B. If the difference $V_A - V_B$ small then E_e can usually be found from the perturbation theory in the second-order approximation in the small parameter $V_A - V_B$. Let us consider, for example, the simplest case of the binary disordered solution in which the atomic radii of the components do not differ much and the static distortions can be ignored. Then the potential energy $V(\boldsymbol{r})$ of a conduction electron in the alloy can be written as

$$V(\boldsymbol{r}) = \bar{V}(\boldsymbol{r}) + \Delta V(\boldsymbol{r}) \,,$$
$$\Delta V(\boldsymbol{r}) = \sum_{s} (c_s - c)[V_A(\boldsymbol{r} - \boldsymbol{R}_s) - V_B(\boldsymbol{r} - \boldsymbol{R}_s)] \,. \tag{2.4.1}$$

Here $\bar{V}(\boldsymbol{r})$ is the potential energy of an electron in the perfect crystal consisting of the "average" atoms, and $\Delta V(\boldsymbol{r})$ describes the fluctuational deviation from the regularly varying potential energy. Using (1.1.8) to perform the transition from $c_s - c$ to $c_{\boldsymbol{k}}$ we can write $\Delta V(\boldsymbol{r})$ as a sum over a cell of the reciprocal lattice:

$$\Delta V(\boldsymbol{r}) = \sum_{\boldsymbol{k}} c_{\boldsymbol{k}} v_{\boldsymbol{k}}(\boldsymbol{r}) \mathrm{e}^{-\mathrm{i}\boldsymbol{k}\cdot\boldsymbol{r}} \,,$$
$$v_{\boldsymbol{k}}(\boldsymbol{r}) = \sum_{s} [V_A(\boldsymbol{r} - \boldsymbol{R}_s) - V_B(\boldsymbol{r} - \boldsymbol{R}_s)] \mathrm{e}^{\mathrm{i}\boldsymbol{k}(\boldsymbol{r}-\boldsymbol{R}_s)} \,. \tag{2.4.2}$$

When $\Delta V(\boldsymbol{r})$ is a small perturbation we can take the perfect crystal with the average potential energy $\bar{V}(\boldsymbol{r})$ as the zero-order approximation. If the interaction between electrons can be ignored the conduction electrons are then described by the Bloch wave functions and the electron energy $E_e^{(o)}$ is independent of the arrangement of the atoms A and B over the lattice sites (the numbers c_s). Statistical averaging results in vanishing of the additional term $E_e^{(1)}$ derived in the first-order approximation. In the second-order approximation we obtain the configuration part of the electron energy which depends on c_s. In the one-electron theory the respective contribution to the electron energy must be written in the following form (see [1.40]:

$$E_e^{(2)} = \frac{1}{2} N \sum_{\boldsymbol{k}} w_e(\boldsymbol{k}) |c_{\boldsymbol{k}}|^2 \tag{2.4.3}$$

It can also be written in the form corresponding to the general equation (1.3.1),

$$E_e^{(2)} = \frac{1}{2} \sum_{s,s'} w_e(\boldsymbol{\varrho})(c_s - c)(c_{s'} - c) \,,$$
$$w_e(\boldsymbol{\varrho}) = \frac{1}{N} \sum_{\boldsymbol{k}} w_e(\boldsymbol{k}) \mathrm{e}^{\mathrm{i}\boldsymbol{k}\cdot\boldsymbol{\varrho}} \,, \quad w_e(0) = 0 \,, \tag{2.4.4}$$

where

$$w_e(\boldsymbol{k}) = -\frac{2}{N} \sum_{\boldsymbol{\varkappa}\boldsymbol{G}} |v_{\boldsymbol{k}}(\boldsymbol{\varkappa}, \boldsymbol{\varkappa} + \boldsymbol{k} + \boldsymbol{G})|^2 \frac{n(E_{\boldsymbol{\varkappa}}) - n(E_{\boldsymbol{\varkappa}+\boldsymbol{k}+\boldsymbol{G}})}{E_{\boldsymbol{\varkappa}+\boldsymbol{k}+\boldsymbol{G}} - E_{\boldsymbol{\varkappa}}} \,; \tag{2.4.5}$$

$E_{\boldsymbol{\varkappa}}$ is the energy of an electron with the wave vector $\boldsymbol{\varkappa}$ in the scheme of broadened bands, $n(E_{\boldsymbol{\varkappa}})$ are the Fermi occupation numbers, $v_{\boldsymbol{k}}(\boldsymbol{\varkappa}, \boldsymbol{\varkappa} + \boldsymbol{k} + \boldsymbol{G})$ is a matrix element of the function $v_{\boldsymbol{k}}(\boldsymbol{r}) \exp(-\mathrm{i}\boldsymbol{k}\cdot\boldsymbol{r})$ produced with the Bloch wave functions $(\boldsymbol{\varkappa}|$ and $|\boldsymbol{\varkappa}+\boldsymbol{k}+\boldsymbol{G})$ corresponding to the potential energy $\bar{V}(\boldsymbol{r})$. The integrals over $\boldsymbol{\varkappa}$ corresponding to the sum (2.4.5) must be singular integrals in the Cauchy sense.

Approximate explicit expressions for the potential energies $V_A(\boldsymbol{r})$ and $V_B(\boldsymbol{r})$ have been derived in the theory of pseudopotentials (see [1.41,42; 2.62]). Using these expressions and the above formulas we can calculate numerically the parameters $w(\varrho)$ for those crystals where the theory of pseudopotentials gives accurate enough results (see, for example, [2.67-72]). We shall consider below, however, not the numerical results for $w(\varrho)$ but the behavior of $w_e(\boldsymbol{k})$ (and, hence, $\langle |c_{\boldsymbol{k}}|^2\rangle$ regarded as a function of $\boldsymbol{k}$.

This behaviour exhibits special features determined by the sharp (step-like) dependence of the distribution function $n(E_{\varkappa})$ for the degenerate gas (or liquid) of the Fermi particles. The terms for which $E_{\varkappa}$ and $E_{\varkappa+k+G}$ are similar, that is the denominator is small, make a significant contribution to the sum in (2.4.5). These terms correspond to the points on the equal-energy surface which are connected by the vector $\boldsymbol{k}+\boldsymbol{G}$. It can be seen, however, that the contributions to the sum in (2.4.5) are made only by such pairs of the electron states $\varkappa$ and $\varkappa+\boldsymbol{k}+\boldsymbol{G}$ in which one state has the energy above the Fermi level and the other below it. The contributions to (2.4.5) made by the terms with small denominators is significant, therefore, when the vector $\boldsymbol{k}+\boldsymbol{G}$ connects parallel parts of the Fermi surface (then the denominator is small in entire regions rather than individual points of the surface). The specific features of $w_e(\boldsymbol{k})$ are the more noticeable the larger are the parts of the Fermi surface which approximately coincide when one of them is displaced by the vector $\boldsymbol{k}+\boldsymbol{G}$ (that is, the larger is the range of summation in (2.4.5) for the terms with small denominators). The areas that approximately coincide are the larger, the larger are their curvature radii and thus the behavior of $w_e(\boldsymbol{k})$ depends significantly on the geometry of the Fermi surface.

We obtain from (2.4.5) for the simplest excitation described by $v_{\boldsymbol{k}}(\boldsymbol{r}) = v_{\boldsymbol{k},\boldsymbol{G}}\exp(-\mathrm{i}\boldsymbol{G}\cdot\boldsymbol{r})$ and the quadratic dispersion relation when $E_{\varkappa} = \hbar^2\varkappa^2/2m$ and $v(\varkappa,\varkappa+\boldsymbol{k}+\boldsymbol{G}) = v_{\boldsymbol{k},\boldsymbol{G}}$:

$$w_e(\boldsymbol{k}) = -\frac{mk_0 v|v_{\boldsymbol{k},\boldsymbol{G}}|^2}{2\pi^2\hbar^2}\left(1+\frac{4k_{\mathrm{F}}^2-k'^2}{4k_{\mathrm{F}}k'}\ln\left|\frac{2k_{\mathrm{F}}+k'}{2k_{\mathrm{F}}-k'}\right|\right). \tag{2.4.6}$$

Here $\boldsymbol{k}'=\boldsymbol{k}+\boldsymbol{G}$, k_{F} is the radius of the spherical Fermi surface, and $v=v_0$ is the atomic volume. The function (2.4.6) exhibits a singular feature for $k'\to 2k_{\mathrm{F}}$ that is, at the points where k' equals the diameter of the Fermi surface. This singular feature is not very sharp, however, reducing to the logarithmic divergence of the derivative.

In the general case of the Fermi surface of an arbitrary shape the function $w_e(\boldsymbol{k})$ exhibits singular features if the vector $\boldsymbol{k}+\boldsymbol{G}$ (where $\boldsymbol{G}$ is a reciprocal lattice vector) connects two points $\varkappa_1$ and $\varkappa_2=\varkappa_1+2\boldsymbol{k}_{\mathrm{F}}$ of the Fermi surface where the normals to the surface are antiparallel or parallel. Explicit expressions for these singularities have been derived in [1.17; 2.74, 75].

The function $w_e(\boldsymbol{k})$ exhibits sharper singular features if the parts of the Fermi surface near the points $\varkappa_1$ and $\varkappa_2$ have a shape which is nearly cylindrical or flat [2.75, 76]. This is the case for quasi-two-dimensional and quasi-one-dimensional systems. For a cylindrical part of the Fermi surface with the length $\mathcal{L}$ the singular term in w_e is [2.73]

$$w_e'(\boldsymbol{k}) = \frac{\sqrt{2}}{\pi^2} \frac{v\mathcal{L}v_{\boldsymbol{k}}^2}{|u_{z2} - u_{z1}|\sqrt{\hbar|a_{xx}|}} \sqrt{\delta k_z} \quad \text{for} \quad \delta k_z > 0\,,$$
$$w_e'(\boldsymbol{k}) = 0 \quad \text{for} \quad \delta k_z < 0\,, \tag{2.4.7}$$

where $v_{\boldsymbol{k}}^2$ is the mean value of $|v_{\boldsymbol{k}}(\boldsymbol{\varkappa}, \boldsymbol{\varkappa}+\boldsymbol{k}+\boldsymbol{G})|^2$ over the cylindrical part (over the flat part for (2.4.8)), $\delta \boldsymbol{k} = \boldsymbol{k} + \boldsymbol{G} - 2\boldsymbol{k}_\mathrm{F}$, the axis $\varkappa_z$ is perpendicular to the Fermi surface at the point $\boldsymbol{\varkappa}_1$, and u_{z1} and u_{z2} are the components of the electron velocity $\boldsymbol{u} = \hbar^{-1}\nabla_{\boldsymbol{\varkappa}} E_{\boldsymbol{\varkappa}}$ at the points $\boldsymbol{\varkappa}_1$ and $\boldsymbol{\varkappa}_2$ along the axis $\varkappa_z$, $a_{xx} = \hbar^4(m_{xx2}^{-1}u_{z2}^{-1} - m_{xx1}^{-1}u_{z1}^{-1})$. m_{xx} is a component of the effective mass tensor, and the axis $\varkappa_x$ is perpendicular to the axis of the cylinder.

For a flat part of the Fermi surface with the area $S_{\boldsymbol{k}}$ we obtain

$$w_e'(\boldsymbol{k}) = -\frac{1}{2\pi^3} v_{\boldsymbol{k}}^2 \frac{vS_{\boldsymbol{k}}}{\hbar|u_{z2} - u_{z1}|} \ln \frac{C}{|\delta k_z|}\,, \tag{2.4.8}$$

where the constant C is of the order of $E_\mathrm{F}/\hbar|u_z|$ (and E_F is the Fermi energy).

Under such conditions it is the function $w_e'(\boldsymbol{k})$, rather than its derivative, that exhibits a logarithmic singularity. It is not necessarily flat parts of the Fermi surface which produce such singularities. It is enough that the surface parts with parallel normals have the same curvature (in sign and magnitude), that is, they must coincide when one of them is displaced by a vector $k \to 2\boldsymbol{k}_\mathrm{F}$. This phenomenon known as nesting can occur when a multiply connected Fermi surface has both electron and hole regions. (As shown by (2.4.4), the singularities exhibited by the Fourier components (2.4.6–8) of the electron energy for the ellipsoidal, cylindrical and flat parts of the Fermi surface correspond to the functions $w_e(\varrho) \sim \varrho^{-3}$, ϱ^{-2} and ϱ^{-1} describing the amplitudes of the oscillating "electronic" components of the ordering energy. In accordance with (1.3.18), the correlation parameters $\varepsilon(\varrho)$ must exhibit the same behaviour for large ϱ. In a similar way, a singularity of the type of (2.4.7) must occur if the signs of the curvature of the parts of the Fermi surface which come into contact after displacement by $2\boldsymbol{k}_\mathrm{F}$ coincide in one direction and differ in another direction (as it happens when an ellipsoid comes into contact with a hyperboloid). In quasi-two-dimensional crystals the singularities exhibited by $w_e(\boldsymbol{k})$ may be strengthened becoming logarithmic (such as (2.4.8)) if the cylindrical parts of the Fermi surface are greatly flattened or if the Fermi surface has saddle points connected by the vector $2\boldsymbol{k}_\mathrm{F}$ [2.77].

One should bear in mind that, rigorously speaking, the Fermi surfaces in real three-dimensional crystals cannot have precisely cylindrical or flat surfaces. They must always have a "corrugated" shape with a characteristic (small) corrugation height $\tilde{k}$ and the extremum points corresponding to a finite (though small) curvature. This leads to violations of the equations (2.4.7, 8) for $|\delta k_z| < \tilde{k}$ where corrugation occurs (see, for instance, (2.5.4)). In addition, scattering of electrons by the static inhomogeneities and vibrations makes the lifetime of the electronic states finite and introduces uncertainty of the order of the decay parameter Γ into the determinations of the electron energy. As a result, the above singularities must be spread over the range of $|\delta k_z| \sim k_\mathrm{F}\Gamma/E_\mathrm{F}$ (and over the range of the order of

$k_F k_B/E_F$ because of the thermal spreading of the Fermi distribution step) [2.78]. Finally, if we go beyond the second-order approximation in the perturbation theory as it must be done for some values of $\varkappa$ and finite amplitudes $V_{\boldsymbol{k}} = v_{\boldsymbol{k}} c_{\boldsymbol{k}}$ of the wave of the potential energy regarded as a perturbation we obtain a significant change in the form of the singular part of $w_e(\boldsymbol{k})$ in the range where all the above factors cause spreading of the singularities $w_e'(\boldsymbol{k})$ described by the equations (2.4.7, 8) which remain applicable only for

$$|\delta k_z| \gg \tilde{k}\,, \quad k_F \frac{\Gamma}{E_F}\,, \quad k_F \frac{k_B T}{E_F}\,, \quad \frac{v_{\boldsymbol{k}}}{\hbar |u_z|}\,. \tag{2.4.9}$$

For instance, for almost flat (or coinciding) parts of the Fermi surface w_e accordingly does not tend to $-\infty$ but remains finite. But for small $\tilde{k}$ and Γ the dependence of w_e on $\boldsymbol{k}$ can exhibit a sharp minimum (a dip) for some systems (for instance, it has been observed for chromium, niobium, tungsten and some quasi-one-dimensional systems).

When we take into consideration the interaction between electrons the equations (2.4.6–8) for the singular component of the electron energy are changed and $w_e'(\boldsymbol{k})$ is replaced with a renormalized function $\tilde{w}_e'(\boldsymbol{k})$. We obtain for the homogeneous electron liquid with the permittivity $\varepsilon(\boldsymbol{k})$ in the constant phase approximation

$$\begin{aligned} \tilde{w}_e'(\boldsymbol{k}) &= \frac{w_e'(\boldsymbol{k})}{\varepsilon(\boldsymbol{k}+\boldsymbol{G})}\,, \quad \varepsilon(\boldsymbol{k}) = 1 + v_e(\boldsymbol{k}) P(\boldsymbol{k})\,, \\ v_e(\boldsymbol{k}) &= \frac{4\pi e^2}{v k^2}\,, \quad P(\boldsymbol{k}) = \frac{2}{N} \sum_{\varkappa} \frac{n(E_{\varkappa}) - n(E_{\varkappa+\boldsymbol{k}})}{E_{\varkappa+\boldsymbol{k}} - E_{\varkappa}}\,, \end{aligned} \tag{2.4.10}$$

where $\boldsymbol{G}' \approx 2\boldsymbol{k}_F - \boldsymbol{k}$ and $\varepsilon(\boldsymbol{k}+\boldsymbol{G}') \approx 1$.

In some cases the singularity $w_e'(\boldsymbol{k})$ is determined primarily by the numerator, rather than denominator, of (2.4.10). When we take into account inhomogeneity of the electron liquid and the exchange and correlation effects the expression for $\tilde{w}_e'(\boldsymbol{k})$ has a more complicated form (it depends on the matrix of the electron suceptibility) but the qualitative features of the electron energy described by (2.4.6–8) are preserved.

Figure 2.1 shows schematically the behaviour of $w_e(\boldsymbol{k})$ described by (2.4.6–8). If $\varepsilon(2\boldsymbol{k}_F)$ is not much larger than unity then $\tilde{w}_e'(\boldsymbol{k})$ have similar features, according to (2.4.10). In the limiting case of flat parts of the Fermi surface we obtain from (2.4.8, 10) logarithmic divergence of $\varepsilon(\boldsymbol{k})$, $P(\boldsymbol{k})$ and $W_e(\boldsymbol{k}')$ and a sharp growth of $\varepsilon(\boldsymbol{k})$ for $\varepsilon(\boldsymbol{k}) \gg 1$ almost entirely suppresses the singularity of $w_e(\boldsymbol{k})$ since the logarithmic factors in $w_e'(\boldsymbol{k}) \sim P(\boldsymbol{k})$ and $\varepsilon(\boldsymbol{k})$ are almost fully cancelled out, especially if $v_e(\boldsymbol{k})$ and the coefficient in the logarithmic term in $P(\boldsymbol{k})$ are large (see [2.79]). But even under such conditions we must observe the singular feature $\tilde{w}_e'(\boldsymbol{k}) \sim P/\varepsilon \sim (\varepsilon - 1)/\varepsilon = 1 - 1/\varepsilon$ though it remains finite (see the dashed curve in Fig. 2.1).

According to (1.3.10), at high temperatures $\langle |c_{\boldsymbol{k}}|^2 \rangle$ is directly expressed in terms of $w(\boldsymbol{k})$:

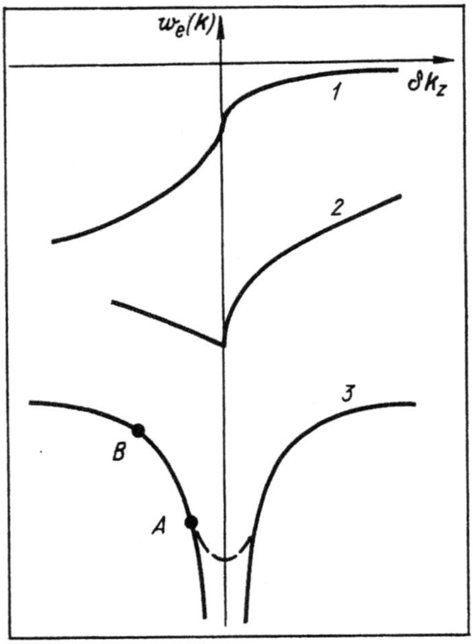

Fig. 2.1. Singularities of the coefficient $w_e(\boldsymbol{k})$ in the electron component of the energy of the concetration fluctuation wave. *1* – displacement by the vector $\boldsymbol{k}+\boldsymbol{G}=\delta\boldsymbol{k}+\varkappa_1-\varkappa_2$ brings into contact the ellipsoidal parts of the Fermi surface near the the points $\varkappa_1$ and $\varkappa_2$ with the antiparallel normals. 2 – Fermi surface sections with almost cylindrical shape. *3* – Fermi surface section with almost flat shapes or almost identical curvatures and parallel normals. The dashed line shows the spreading of the logarithmic singularity caused by violation of the condition (2.4.9) or by a significant contribution of the interaction between electrons

$$\langle|c_{\boldsymbol{k}}|^2\rangle=\frac{1}{N}c(1-c)\left\{1+\frac{c(1-c)}{k_{\mathrm{B}}T}\left[w_e(\boldsymbol{k})+w_0(\boldsymbol{k})-\frac{1}{N}\sum_{\boldsymbol{k}}w_e(\boldsymbol{k})\right]\right\}^{-1}. \tag{2.4.11}$$

Here $\hbar|u_z\delta k_z|<v_{\boldsymbol{k}}$ (see Sect. 2.5), $w_0(\boldsymbol{k})$ is the Fourier component for the part of the ordering energy $w_0(\boldsymbol{\varrho})$ which is not associated with the conduction electrons and depends on the direct interaction between ions, the energy of atom oscillations, etc. ($w_0(0)=0$).

Therefore, $\langle|c_{\boldsymbol{k}}|^2\rangle$ treated as a function of $\boldsymbol{k}$ must also exhibit a singular feature for the vector $\boldsymbol{k}+\boldsymbol{G}$ which connects two points of the Fermi surface at which the normals to the surface are antiparallel or parallel, We shall obtain the logarithmic singularity of the derivative of this function with respect to $\boldsymbol{k}$, a root singularity of the function or a spread-out logarithmic singularity when the parts of the Fermi surface near these points have finite curvature radii or have almost cylindrical or flat shapes, respectively [1.17]. In the latter case the dependence of $\langle|c_{\boldsymbol{k}}|^2\rangle$ on $\boldsymbol{k}$ must exhibit a sharp maximum. These singular features of $\langle|c_{\boldsymbol{k}}|^2\rangle$ give rise to special features found in the X-ray scattering intensity distributions (see Sect. 3.3). A more general analysis using the coherent potential approximation for calculating $w_e(\boldsymbol{k})$, instead of the perturbation theory, yields similar results [1.46].

Since the singularities of $P(\boldsymbol{k}+\boldsymbol{G})$ are spread out, the positions and behaviour of the singularities of $w_e(\boldsymbol{k})$ and $\langle|c_{\boldsymbol{k}}|^2\rangle$ can be significantly affected by the dependence of $v_{\boldsymbol{k}}^2$ on $\boldsymbol{k}$. In transition metals where the wave of concentration (or atom displacements) is accompanied by a redistribution of the d electron density which decreases the electron energy, the function $w_e(\boldsymbol{k})$ can exhibit characteristic minima which are due not only to $P(\boldsymbol{k}+\boldsymbol{G})$ but also to the Fourier components of the energy of interaction of electrons with the concentration waves and with each other regarded as functions of $\boldsymbol{k}$ [2.80–82]. If this minimum (which is not dependent directly on the diameter of the Fermi surface) is sufficiently sharp it also gives rise to an effective longrange interaction and singular features of $\langle |c_{\boldsymbol{k}}|^2 \rangle$.

The function $w_e(\boldsymbol{k})$ must exhibit the above characteristic features when the Fermi surface has small ellipsoidal cavities or thin bridges [1.17]. According to (2.4.5), the singular features under these conditions must occur in the range of $\boldsymbol{k}$ for which $\boldsymbol{k}+\boldsymbol{G}$ is close to the vectors connecting two different cavities or bridges. The singularity then corresponds to a dip of $w_e(\boldsymbol{k})$ in this range whose width is of the order of the cavity or bridge and the two Kohn singularities discussed above must also be found in this range. It follows from (2.4.11) then that $\langle |c_{\boldsymbol{k}}|^2 \rangle$ as a function of $\boldsymbol{k}$ must also exhibit dips in these ranges.

2.5 Equilibrium Heterogeneous States with Long-Range Forces

2.5.1 Crystal Structures with Large Parameters

Crystal structures with small numbers of atoms per unit cell are typically produced in systems with short-range forces. Crystal structures with large lattice parameters may appear in the presence of long-range forces, for instance, those due to the indirect interaction between ions in metallic systems. Under some conditions it is convenient to treat such structures as heterogeneous systems with the parameters of the long-range order (and sometimes composition) varying in space.

For instance, as discussed in Sect. 2.4, the Fourier components of the energy $w_e(\boldsymbol{k})$ of the indirect interaction exhibit singularities for $\boldsymbol{k}$ (or $\boldsymbol{k}+\boldsymbol{G}$) coinciding with the "diameter" of the Fermi surface, $2\boldsymbol{k}_{\mathrm{F}} = \varkappa_1 - \varkappa_2$ and these sigularities facilitate the production of regular structure with the wave vector $\boldsymbol{k} \approx 2\boldsymbol{k}_{\mathrm{F}} - \boldsymbol{G}$. Such structures are produced if the gain in the singular component of the electron energy is greater than the loss in the nonsingular part of the free energy due to the resulting inhomogeneity of the structure. For instance, if some "superlattice" vectors $\boldsymbol{G}'$ of the reciprocal lattice of the ordered alloy do not differ much from $2\boldsymbol{k}_{\mathrm{F}}$ and are in the range of singularity of $w_e(\boldsymbol{k})$ then the energetically favoured structure may be a large-parameter regular structure consisting of the alternating domains of the normal (small-parameter) ordered structure in which the order parameters differ in the sign or phase (see, for instance, [2.83–85]). Such a structure gives rise to "splitting" of the reciprocal lattice points $\boldsymbol{G}'$ and a decrease in the electron energy, especially in the cases corresponding to the curves 2 and 3 in Fig. 2.1. The structure produced in this way may be regarded as an equilibrium heterogeneous state of the alloy in which the order parameter $\eta(\boldsymbol{r})$ regularly varies in space.

In single-component crystals the singularities of the electron energy can give rise to bound static waves of charge density and atom displacement with a wavelength which is a either a multiple of the parameter of the original crystal lattice or incommensurate with it. Under certain conditions the respective order parameter also inhomogeneously varies in space. The charge density waves are produced when the gain in the electron energy is larger than the increase in the nonsingular component of the elastic energy.

Charge Density Waves. When we consider charge density waves for the sake of simplicity we shall limit the discussion to crystals with a single atom per unit cell. The wave of atom displacements $u = iu_k \exp(-i\boldsymbol{k}\cdot\boldsymbol{r})$+c.c. (compare with (1.5.1)) produces a change in the potential energy of the conduction electrons which is given by the following equation in the model of rigid ions:

$$\begin{aligned}
\Delta V(\boldsymbol{r}) &= \sum_{s}[V_A(\boldsymbol{r}-\boldsymbol{R}_s-\boldsymbol{u}_s)-V_A(\boldsymbol{r}-\boldsymbol{R}_s)] \\
&\approx u_k v_k'(\boldsymbol{r})e^{-i\boldsymbol{k}\cdot\boldsymbol{r}} + v_k^{(2)}u_k^2 + \text{c.c.}\,, \\
v_k'(\boldsymbol{r}) &= -i\sum_{s}\nabla V_A(\boldsymbol{r}-\boldsymbol{R}_s \exp[i\boldsymbol{k}(\boldsymbol{r}-\boldsymbol{R}_s] = \sum_{G} v'_{k+G}e^{-i\boldsymbol{G}\cdot\boldsymbol{r}}\,, \\
v'_{k+G} &= -\frac{1}{v}e_k(\boldsymbol{k}+\boldsymbol{G})V_{k+G}\,, \quad V_k = \int V_A(\boldsymbol{r})e^{i\boldsymbol{k}\cdot\boldsymbol{r}}d\boldsymbol{r}\,, \\
v_k^{(2)} &= -\frac{1}{2}\sum_{G}|e_k\boldsymbol{G}|^2V_{k+G}\,, \quad e_k = \frac{\boldsymbol{u}_k}{u_k}\,, \quad u_k = |\text{Re}\,\boldsymbol{u}_k + i\text{Im}\,\boldsymbol{u}_k|\,.
\end{aligned} \tag{2.5.1}$$

The wave-like variation of the potential energy given by (2.5.1) causes a redistribution of the electron density, that is, generation of a charge density wave associated with the static displacement wave, and a variation of the electron energy $E_e(\boldsymbol{k})$. For displacement waves of small enough amplitudes u_k the electron energy variation can be calculated from the perturbation theory and described by the same equations (2.4.3–10) as in the case of the concentration waves in alloys. It can be seen from (2.4.2) and (2.5.1) that we have only to replace c_k with u_k and v_k with v_k' for this purpose. For instance, if we ignore the excitation processes and find the permittivity in the random phase approximation then we obtain from (2.4.3, 5, 10) for $u_k \to 0$

$$\begin{aligned}
E_e(\boldsymbol{k}) &= E_e^{(2)}(\boldsymbol{k}) = \frac{1}{2}N[\tilde{w}_e(\boldsymbol{k}) + w_{e2}]u_k^2\,, \\
\tilde{w}_e(\boldsymbol{k}) &= \frac{w_e(\boldsymbol{k})}{\varepsilon(\boldsymbol{k})} = -|v_k'|^2\frac{Pk}{1+v_e(\boldsymbol{k})P(\boldsymbol{k})}\,,
\end{aligned} \tag{2.5.2}$$

where w_{e2} is determined by the first-order correction for the electron energy due to the terms $\sim u^2$ in $V(r)$ and does not depend on k (in the limit of almost free electrons we have $w_{e2} = -w_e(k=0)$).

The displacement wave produces also a change in the energy of direct interaction between ions, $E_0(k) = Nw_0(k)u_k$. If the polarization of the displacement wave coincides with the polarization of the normal mode k of the pth branch then $w_0(k) = m_i\omega_{0kp}^2$ (here m_i is the ion mass) is expressed in terms of the frequency

ω_{0kp} of this mode calculated in the absence of the interaction (2.5.1). Below we shall include w_{e2} independent of k in $w_0(k)$.

In the crystals where the electron-phonon interaction is fairly strong and $P(k)$ is sufficiently large the sum $E_0(k)+E_e(k)$ may be negative in a certain k range. Then the generation of a bound charge density wave and a displacement wave is energetically favourable. It happens if the following condition is satisfied [2.86, 87]:

$$w_0(\boldsymbol{k}) + w_e(\boldsymbol{k}) = w_0(\boldsymbol{k}) - |v'_{\boldsymbol{k}}|^2 \frac{P(\boldsymbol{k})}{1 + v_e(\boldsymbol{k})P(\boldsymbol{k})} < 0$$

or

$$\frac{|v'_{\boldsymbol{k}}|^2}{w_0(\boldsymbol{k})} > \frac{1}{P(\boldsymbol{k})} + v_e(\boldsymbol{k}) \,. \tag{2.5.3}$$

In a similar way, when we take into consideration the exchange interaction the singularities of $P(k)$ can produce spin density waves in chromium [2.88, 89] and spiral structures in metallic magnetic materials [2.90, 91].

The instability of the crystal lattice with respect to generation of the charge density waves must occur, for instance, in quasi-one-dimensional systems in which, according to (2.4.8), the flattened parts of the Fermi surface give rise to a sharp peak $-w_e(\boldsymbol{k})$ for $\boldsymbol{k} = 2\boldsymbol{k}_F$ and hence (if $\varepsilon(2\boldsymbol{k}_F)$ is not very large) to a a fairly sharp peak $-\tilde{w}_e(\boldsymbol{k})$ (see Fig. 2.1). This instability of a quasi-one-dimensional system causes the structural Peierls phase transition which produces a gap at $\varkappa = \boldsymbol{k}_F$ in the electron spectrum which decreases the energy of the occupied electron states and makes the dielectric phase stable [2.92, 93]. Such a structural transition was observed, for instance, in the quasi-one-dimensional system consisting of the chains of tetrathiofulvalene and tetracyanquinodimethane molecules (see the reviews [2.94–98]). In strictly one-dimensional systems the fluctuations are known [1.3.1] to disrupt the ordered phase at finite temperatures but the weak coupling between the chains caused both by electron tunneling and the Coulomb interaction significantly suppresses the fluctuations and makes the phase transition possible [2.99–102]. This coupling may produce a regular three-dimensional distribution of the charge.

In quasi-two-dimensional systems the Fermi surface has an almost cylindrical shape and we can also obtain a minimum of $\tilde{w}_e(\boldsymbol{k})$ for $\boldsymbol{k} = 2\boldsymbol{k}_F$ (especially if the appropriate parts of the Fermi surface ar almost flat and $\varepsilon(\boldsymbol{k}) \sim 1$; see Fig. 2.1, curve 2). According to (2.5.3), a strong electron-phonon interaction then will give rise to the charge density waves. Such waves have been observed, for instance, in the laminar dichalcogenides containing transition metals, for instance, in the 2H modification of $TaSe_2$. [2.98, 103-105]. Since the Fermi surface in these compounds has a six-fold symmetry axis the three charge density waves which can be generated correspond to the vectors $\boldsymbol{k}$ in the basal plane rotated by 120°. The process is realized as a combination of these waves producing a two-dimensional regular structure.

When the excitation processes are taken into consideration $w_e(\boldsymbol{k})$ in the one-electron approximation is given by (2.4.5) where $v_{\boldsymbol{k}}$ must be replaced with $v'_{\boldsymbol{k}}$.

If the interaction between electrons (described by the permittivity matrix) does not change the singular features of $w_e(\boldsymbol{k})$ then they can be produced not only for $\boldsymbol{k} = 2\boldsymbol{k}_F$) but also for $\boldsymbol{k} + \boldsymbol{G} = 2\boldsymbol{k}_F$. The equations (2.4.5–8) and (2.5.2) for $w_e(\boldsymbol{k}) = w_e^{(2)}(\boldsymbol{k})$ derived in the second-order approximation of the perturbation theory are valid only in the limit of infinitesimal perturbation amplitudes $u_{\boldsymbol{k}} v'_{\boldsymbol{k}}(\boldsymbol{r})$ or $c_{\boldsymbol{k}} v_{\boldsymbol{k}}(\boldsymbol{r})$. For $\boldsymbol{k} + \boldsymbol{G} \approx 2\boldsymbol{k}_F$ a perturbation of a finite amplitude produces a mixing of the electron states with $\boldsymbol{\varkappa} \approx -\boldsymbol{k}_F$ and $\boldsymbol{\varkappa} + \boldsymbol{k} + \boldsymbol{G} \approx \boldsymbol{k}_F$ with similar energies $E_{\boldsymbol{\varkappa}} \approx E_{\boldsymbol{\varkappa}+\boldsymbol{k}+\boldsymbol{G}} \approx E_F$ and a rearrangement of the electron spectrum in the "singular" range of interest which makes the major contribution to the singular energy component $w'_e(\boldsymbol{k})$. This rearrangement of the electron spectrum must be treated with a generalized perturbation theory which takes into account degeneracy of the electron states [2.106, 107]. For finite amplitudes of the perturbation potential the singular features of $w_e(\boldsymbol{k})$ exhibit a significantly different behaviour (in comparison with $w_e^{(2)}(\boldsymbol{k})$) [2.107, 108].

The variation of the singular component of the electron energy $Nw'_e(\boldsymbol{k})|c_{\boldsymbol{k}}|^2$ caused by the perturbation $v_{\boldsymbol{k}}(\boldsymbol{r})c_{\boldsymbol{k}}\exp(-\boldsymbol{k}\cdot\boldsymbol{r}) + \text{c.c.}$ (see (2.4.2)) is determined by $w'_e(\boldsymbol{k})$ which depends on $U_{\boldsymbol{k}}^2 = v_{\boldsymbol{k}}^2|c_{\boldsymbol{k}}|^2$ where $v_{\boldsymbol{k}}$ equals $|v_{\boldsymbol{k}}(\boldsymbol{\varkappa}, \boldsymbol{\varkappa} + \boldsymbol{k} + \boldsymbol{G})|^2$ averaged over $\boldsymbol{\varkappa}$ and $\boldsymbol{\varkappa} + \boldsymbol{k} + \boldsymbol{G}$ lying in the singular range ($\boldsymbol{k} + \boldsymbol{G} \approx 2\boldsymbol{k}_F$) as it was done for (2.4.6–8). In a similar way, for the perturbation $v_{\boldsymbol{k}}(\boldsymbol{r})u_{\boldsymbol{k}}\exp(-i\boldsymbol{k}\cdot\boldsymbol{r})$ given by (2.5.1) we have $U_{\boldsymbol{k}}^2 = v'^2_{\boldsymbol{k}}u_{\boldsymbol{k}}^2$. For $U_{\boldsymbol{k}} \to 0$ we have $w'_e(\boldsymbol{k})$ tending to the singular component $w'^{(2)}_e(\boldsymbol{k})$ but for finite $U_{\boldsymbol{k}}$ it differs considerably from the latter. In the case of weak interaction between electrons (with $\varepsilon(\boldsymbol{k}) \approx 1$) explicit expressions for $w'_e(\boldsymbol{k})$ have been derived for the Fermi surfaces of different geometries [2.107, 108].

Assume, for instance, that the major contribution to $w'(\boldsymbol{k})$ is made by the flattened parts of the Fermi surface with the areas $S_{\boldsymbol{k}}$ shaped as spherical segments with large curvature radii and small heights $\tilde{k}$. Assume also that the parameters $U_{\boldsymbol{k}}$, $E_m = \hbar u_z k$, and $\delta E = \hbar u_z \delta k_z$ (where u_z is the component of the electron velocity along the normal to the flattened segment k_z and $\delta\boldsymbol{k} = \boldsymbol{k} + \boldsymbol{G} - 2\boldsymbol{k}_F$ are small in comparison with E_F but the ratios between them are arbitrary. Then we obtain [2.108]

$$w'_e(\boldsymbol{k}) = \frac{vS_{\boldsymbol{k}}v_{\boldsymbol{k}}^2}{2\pi^3\hbar|u_z|}\left[\frac{U_{\boldsymbol{k}}}{E_m}g\left(\frac{\delta E}{U_{\boldsymbol{k}}}\right)\right. \\ \left. + \frac{U_{\boldsymbol{k}}}{E_m}g\left(\frac{E_m - \delta E}{U_{\boldsymbol{k}}}\right) - \frac{1}{4} + \frac{1}{2}\ln\frac{U_{\boldsymbol{k}}}{2\hbar|u_z|k_F}\right], \tag{2.5.4}$$

where

$$g(t) = \tfrac{1}{6}t^3 \quad \text{for} \quad |t| \le 1\,;$$

$$g(t) = \tfrac{1}{6}\left[t^3 - (t^2+2)(t^2-1)^{1/2}\operatorname{sign} t + 3t\ln\left(|t| + \sqrt{t^2-1}\right)\right] \tag{2.5.5}$$

$$\text{for} \quad |t| > 1\,.$$

For sufficiently small $U_{\boldsymbol{k}} \ll |\delta E|$ and $|E_m - \delta E|$ the equation (2.5.4) is proportional to $U_{\boldsymbol{k}}^2$ and coincides with the result obtained with the perturbation theory:

$$w_e'^{(2)}(\boldsymbol{k}) = -\frac{vS_{\boldsymbol{k}}v_{\boldsymbol{k}}^2}{4\pi^3\hbar|u_z|} \times \left[\ln\frac{k_\mathrm{F}}{\tilde{k}} - \frac{\delta k_z}{\tilde{k}}\ln\frac{|\delta k_z|}{k_\mathrm{F}} - \frac{\tilde{k}-\delta k_z}{\tilde{k}}\ln\frac{|\tilde{k}-\delta k_z|}{k_\mathrm{F}} - 1\right]. \quad (2.5.6)$$

But for $U_{\boldsymbol{k}} \sim |\delta E|$ (that is, for relatively small $U_{\boldsymbol{k}} \ll E_\mathrm{F}$ if $|\delta k_z| \ll k_\mathrm{F}$) we find that the difference between $w_e'(\boldsymbol{k})$ and $w_e'^{(2)}(\boldsymbol{k})$ is considerable. Equations for the more general case of the flattened and cylindrical Fermi surfaces are given in [2.108].

In the case of the uniform perturbation potential for the perturbation waves with different $\boldsymbol{k}$ the variation of the electron energy E_e can be written as the sum (2.4.3) over $\boldsymbol{k}$ if the differences in k_z are large in comparison with $U_{\boldsymbol{k}}/\hbar|u_z|$. But in this sum $w_e^{(2)}(\boldsymbol{k})$ given by (2.4.5) must be replaced with $w_e(\boldsymbol{k})$ [2.109].

Commensurate, Incommensurate and Soliton Structures. If the charge density wave has a sufficiently small amplitude the most energetically favourable wave has the vector $\boldsymbol{k}_0$ corresponding to the minimum of the energy $E^{(2)}(\boldsymbol{k}) \sim u^2$. If the electron energy component $E_e^{(2)}(\boldsymbol{k})$ has a sharp minimum discussed in Sect. 2.4 the vector $\boldsymbol{k}_0$ is determined by the position of this minimum, that is, by the shape and size of the Fermi surface. Therefore, for a charge density wave with a small amplitude the coefficients H_i in the expansion $\boldsymbol{k}_0 = \sum H_i\boldsymbol{b}_i$ over the basis vectors $\boldsymbol{b}_i$ of the reciprocal lattice of the symmetric phase are generally irrational numbers, that is, the resulting regular structure is incommensurate with the initial phase. The dependence of the minima of $w_e(\boldsymbol{k})$ and $E_e(\boldsymbol{k})$ and hence $\boldsymbol{k}_0$ on the wave amplitude and temperature is relatively weak.

As the amplitude of the incommensurate wave increases, however, its coupling with the periodic potential of the initial phase becomes increasingly significant as it tends to correlate the electron density distribution with the potential and make the wave commensurate with the lattice at the expense of both singular and nonsingular components of the energy. This can be seen from the Landau expansion of the thermodynamic potential in powers of the Fourier component $\varrho_{\boldsymbol{k}}$, for the electron density variation [2.110]. For instance, we have for the solitary modulated wave

$$\Phi = \Phi^0 + \tfrac{1}{2}\textstyle\sum_{\boldsymbol{k}} \Phi^{(2)}(\boldsymbol{k})\varrho_{\boldsymbol{k}}\varrho_{-\boldsymbol{k}} + \tfrac{1}{3!}\sum_{\boldsymbol{k}_1\boldsymbol{k}_2\boldsymbol{k}_3} \Phi^{(3)}(\boldsymbol{k}_1, \boldsymbol{k}_2, \boldsymbol{k}_3)\varrho_{\boldsymbol{k}_1}\varrho_{\boldsymbol{k}_2}\varrho_{\boldsymbol{k}_3} + \dots \quad (2.5.7)$$

Here the coefficients $\Phi^{(3)}$ do not vanish if $\boldsymbol{k}_1 + \boldsymbol{k}_2 + \boldsymbol{k}_3 = \boldsymbol{G}$. Therefore, for an incommensurate wave with the single harmonic $\boldsymbol{k}_0$, ensuring that the second term in the expansion has a minimum, the third term vanishes. If $\boldsymbol{k}_0$ in the incommensurate wave is close to $\boldsymbol{G}/3$ (which is the case in the chalcogenides of transition metals) the second term in (2.5.7) will be only slightly increased by the transition to the comparable wave with $\boldsymbol{k}_0 = \boldsymbol{G}/3$. At the same time the third term of the expansion (with $\boldsymbol{k}_1 = \boldsymbol{k}_2 = \boldsymbol{k}_3 = \boldsymbol{G}/3$) becomes nonvanishing. The third term can be made negative if the appropriate sign is selected for ϱ. This means that for large enough amplitudes the commensurate structure has a lower thermodynamic potential, that is, a phase transition from the incommensurate phase to the commensurate phase must occur when $\boldsymbol{k}_0$ approaches $\boldsymbol{G}/3$, for instance, under variation of the pressure

P. In a similar way the commensurate phase may become energetically favourable at the expense of the fourth term in the expansion of Φ for $\boldsymbol{k}_0 \approx \boldsymbol{G}/4$. (A similar analysis taking into account higher-order terms in the expansion of Φ demonstrates that the coefficients of the expansion of $\boldsymbol{k}_0$ in powers of $\boldsymbol{b}_i$ are, rigorously speaking, always rational numbers though the ranges in the $\boldsymbol{k}_0, P$ diagram where the commensurate phase exists are sufficiently large only for simple fractions and infinitesimal for fractions with large denominators [2.111, 112]. The dependence of $\boldsymbol{k}_0$ on P proves to be a continuous but nonanalytic function in this fascinating situation when an infinite number of rational $\boldsymbol{k}_0$ values exist between any possible irrational $\boldsymbol{k}_0$ (this arrangement is known as a devil's staircase [2.113]). Production of such intermediate commensurate states gives, however, only a negligible energy gain and in practice they are not produced.)

Though the higher-order terms of the expansion (2.5.7) vanish for the incommensurate wave with a single harmonic $\boldsymbol{k}_0$ they can be made nonvanishing by distorting the wave shape with an admixture of higher harmonics [2.114, 115]. For instance, in the wave with the harmonics $(1+\varepsilon)\boldsymbol{G}/3$ and $(1-2\varepsilon)\boldsymbol{G}/3$ the third-order term in (2.5.7) does not vanish (for $\boldsymbol{k}_1 = \boldsymbol{k}_2 = (1+\varepsilon)\boldsymbol{G}/3$ and $\boldsymbol{k}_3 = (1-2\varepsilon)\boldsymbol{G}/3$). The appropriately distorted incommensurate structures give a gain in Φ as the commensurate structures do but they produce a considerably smaller decrease in the harmonic component of Φ than the commensurate waves.

A distortion of the incommensurate wave structure can be produced by modulation of the amplitude or phase of the wave and if the coupling to the lattice is weak the phase modulation produces a greater decrease in Φ [2.115–117] (because no energy gap occurs in the spectrum of the excitations corresponding to the steadily varying phase fluctuations in the incommensurate wave, known as phasons [2.118]). For instance, if $\boldsymbol{k}_0$ is close to $\boldsymbol{G}/3$ then in the wave

$$\varrho(\boldsymbol{r}) = i\varrho_{\boldsymbol{k}_0} \exp\left[-\mathrm{i}\frac{\boldsymbol{G}}{3}\boldsymbol{r} - i\phi(\boldsymbol{r})\right] + \mathrm{c.c.} \tag{2.5.8}$$

in the regions of space where $\phi \approx \pi n/3$ (for even n) we obtain a gain in Φ at the expense of the third-order term in (2.5.7) (for $\Phi^{(3)} < 0$) while in the regions where $\phi \approx \pi n/3$ (for odd n) we obtain a loss in Φ. If we have a single harmonic $\boldsymbol{k}_0$ and $\phi(\boldsymbol{r}) = (\boldsymbol{k}_0 - \boldsymbol{G}/3)\boldsymbol{r}$ linearly depends on $\boldsymbol{r}$ then the above contributions to Φ are cancelled out. If, however, we consider a wave structure with a nonlinearly varying phase in which the regions with unfavourable $\phi \approx \pi n/3$ are considerably compressed and the regions with $\phi = 2\pi n/3$ are stretched out (see Fig. 2.2) then we can obtain an almost full gain in Φ at the expense of the third-order term in (2.5.7) [2.115].

Then the incommensurate phase has a heterogeneous domain structure and consists of domains in which the charge density wave is almost commensurate with the original lattice while the domains are separated from each other with relatively narrow regions of the incommensurate phase. The behaviour of ϕ near the domain boundaries is described by a nonlinear sine Gordon equation whose solution corresponds to static equilibrium solitons so that such domain boundaries are usually

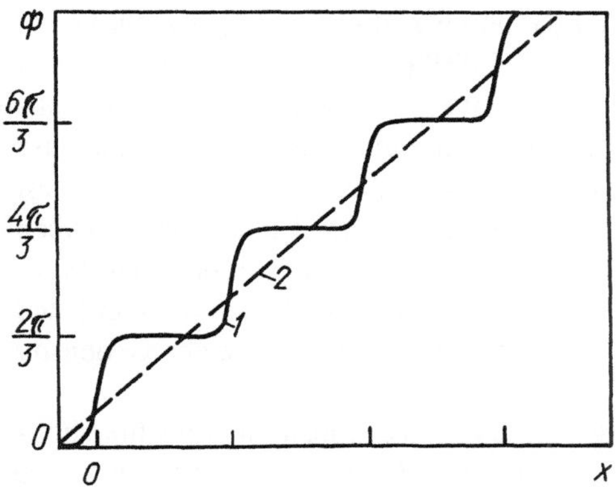

Fig. 2.2. The phase of the distorted incommensurate charge density wave as a function of coordinates. *Curve 1* corresponds to a temperature close to the point of the phase transition to the commensurate state. *Curve 2* corresponds to the distorted wave with a single harmonic

referred to as solitons. (Static soliton solutions are obtained also for some other problems in the theory of ordering [2.111, 119–121].) The distances between solitons become different and their lines are curved owing to the thermal fluctuations. The temperature dependence of the domain structure [2.122–128] demonstrates that the domain size grows as we approach the phase transition to the commensurate phase and a continuous phase transition occurs for a one-dimensional phase structure. (More complicated three-dimensional domain structures of the incommensurate wave regions have been observed in 2H – $TaSe_2$ crystals with three possible directions of the vector $\boldsymbol{k}_0$. [2.129].)

Ordered Alloys with Large Spatial Parameters. The indirect interaction due to the conduction electrons may also give rise to large-parameter structures in ordered alloys. Such structures are typically produced in alloys with the fcc lattice such as CuAu($L1_0$) and Cu_3Au($L1_2$) in certain ranges of composition and temperature and they consist of alternating antiphase ordered regions (domains; see the reviews [2.83, 85, 130]). For instance, in $L1_0$ alloys a one-dimensional large-parameter structure with the axis [010] is produced owing to the displacement of atoms by the vector $(\boldsymbol{a}_1 + \boldsymbol{a}_3/2$ perpendicular to this axis with the spacing of M cell lengths. Such displacement changes the sequence of the atomic (001) planes normally occupied by the atoms A or B in the layered $L1_0$ structure and the sign of the order parameter is reversed. The subsequent displacement of atoms by $(\boldsymbol{a}_1 + \boldsymbol{a}_2/2)$ restores the original sign of the order parameter, that is, we obtain a regular structure with the parameter $D = 2M\boldsymbol{a}_2$ along the axis [010]. In Cu_3Au-type alloys a one-dimensional structure with a large parameter may also be produced owing to the displacement of atoms by the vector $(\boldsymbol{a}_1 + \boldsymbol{a}_3)/2$ which lies in the plane of the domain boundary (010). Such displacements of atoms in the (010) layers containing two types of lattice sites (for the atoms A and B) make them exchange their positions.

The structure parameter $2M$ can be very large in some alloys and it depends on the alloy composition (for instance, in a large-parameter structure of the CuAu alloy (in CuAuII) we have $M \approx 5$, that is, 20 atomic layers are accomodated at a parameter length while for Cu_3AuII we have $M = 6$). Experimental data show a clear correlation between M and the number of electrons per atom [2.83] (this number is varied by adding impurities of different valencies). These results demonstrate convincingly the decisive role played by the conduction electrons in the production of large-parameter structures (the emergence of the antiphase boundaries in an alloy with $|\eta| = 1$ does not alter the number of the nearest neighbours AB, AA and BB, that is, the energy of the direct interaction between them). Such structures are produced by a first-order phase transition when $|\eta| \sim 1$ (for the intermediate values of $|\eta|$ the free energy exhibits a maximum) and typically the $\eta(\boldsymbol{r})$ wave shape is closer to being rectangular than the charge density wave shapes.

The variation of the electron energy E_e under the transition from the short-parameter structure I to the longparameter structure II is caused by the emergence of new Brillouin zones. This variation can be found in the second-order approximation of the perturbation theory from (2.4.3, 5) by numerical integration over k [2.84, 131–133] (for the case of large M we can treat analytically the singular component of the electron energy without using perturbation theory [2.134, 135]).

We must bear in mind here that in the structure I we have for the superlattice site of the reciprocal lattice $c_{\boldsymbol{k}} = c(\boldsymbol{G}') = \xi\eta/2$ where $\xi = 1$ for the $L1_0$ lattice and $\xi = 1/2$ for the $L1_2$ lattice while in the structure II the respective $|c_{\boldsymbol{k}}|^2$ is split into a series of peaks at $\boldsymbol{G}_{\mathrm{II}} = \boldsymbol{G}' + n\boldsymbol{q}_0$. (where $q_0. = (2n/D)e_y$ and n is an integer). For a rectangular wave (when $\eta(r)$ is an odd function) for large M we find that for the alloys of the stoichiometric composition n are odd and $|c(\boldsymbol{G}_{\mathrm{II}})| = \eta\xi/\pi|n|$. If $w_e(\boldsymbol{k})$ has a sufficiently sharp minimum for $\boldsymbol{k} \approx \boldsymbol{G}'$ (for instance, owing to the presence of flattened or cylindrical segments of the Fermi surface; see Fig. 2.1) then the energy E_e can be considerably decreased by the production of a structure with a large spatial parameter and splitting of the $|c_{\boldsymbol{k}}|^2$ peak (see Fig. 2.3).

The main contribution to E_e is made by the terms with $n = \pm 1$ so that for the ordered domains of a low surface energy the parameter $D \approx \sqrt{2}\pi/k^0$ is determined by the distance k^0 between the end of the vector $\boldsymbol{G}'$ and the point corresponding to the minimum of $w_e'(G' + q_0) + w_e'(G - q_0)$. For instance, in the copper-based fcc alloys the vector $\boldsymbol{G}' + \boldsymbol{G}$ where $\boldsymbol{G}' = (001)$ and $\boldsymbol{G} = (1\bar{1}1)$ is close to the diameter $2k_{\mathrm{F}}$ of the Fermi surface in the direction of [110] which connects its segments with a fairly large curvature radius providing for the minimum of $w_e(\boldsymbol{k})$ [2.84]. This fact accounts for the production of the ordered phases of CuAuII and Cu_3AuII.

As it is the case for the charge density waves, the minimum of the singular component of E_e occurs for fractional M but locally the structures I and II must now be commensurate. A considerable part of the charge density $|c(\boldsymbol{G}_{\mathrm{II}}')|^2$ can be displaced to the vicinity of the minimum of $w_e(\boldsymbol{k})$ corresponding to the fractional M if we make the transition to a regular structure with a very large spatial parameter which consists of alternating regions comprising different integral numbers $(M, M \pm 1, \ldots)$ of cells (resulting in further splitting of each of the points $\boldsymbol{G}_{\mathrm{II}}'$) or to a random structure with the fluctuating integral M, that is, the fluctuating distances $D/2$ between the antiphase boundaries (resulting in broadening and dis-

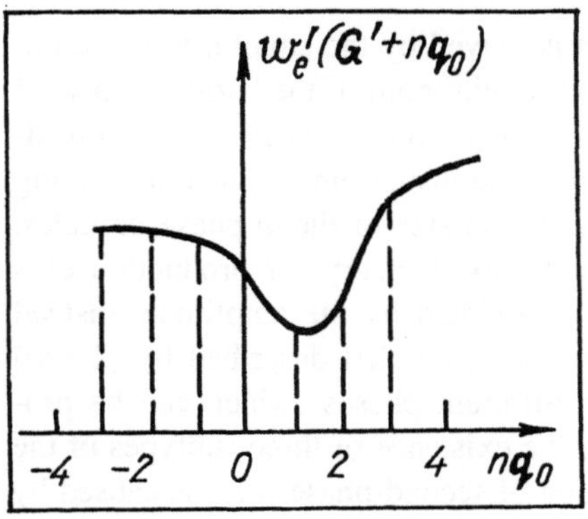

Fig. 2.3. Variation of the singular component of the electron energy caused by the production of the ordered structure with a large spatial parameter

placements of the peaks of $|c(G'_{\text{II}})|^2$ [2.136–139]. The experimental results for various alloys have demonstrated the existence of both types of structures with an effective fractional M. The entropy of the system is increased if the fluctuational changes occur at relatively small parts of the antiphase boundaries. This must make the boundary spiked (especially if the energy of the linear defects at the boundary is small).

In alloys with the nonstoichiometric composition, $c(G'_{\text{II}})$ depends on the modulation of not only the order parameter but also the composition. For instance, the excess Cu atoms in CuAuII alloy can be concentrated at the antiphase boundaries. In the solutions with markedly different atomic radii of the components, for instance, in Cu-Au, the electron energy is reduced not only by the modulation of the order and the composition but also by the static displacements correlated with them. This is why the production of structures with large parameters D is accompanied by a generation of an "optical wave" of static displacements with the period D in which atoms A and B are displaced in the opposite directions and an acoustic wave with the period $D/2$ [2.141, 142].

2.5.2 Heterogeneous States in Metal Systems

Nature of Heterogeneous States in Alloys. In the crystals discussed in Sect. 2.5.1 the long-range order is produced in the entire crystal so that only its sign or phase (and the amplitude within certain limits) vary in space. In some cases, however, the distribution of the order and composition can be essentially inhomogeneous and the regions with a high local order alternate with the disordered (or poorly ordered) regions.

Such inhomogeneously ordered structures have been experimentally observed, for instance, in the studies of the short-range order in some copper-based fcc alloys and in the studies of the displacement ω phase in the bcc alloys, in particular, Zr-Nb (see Sect. 3.5.5). A variety of crystallographic models have been put forward to analyze the diffraction effects in the alloys with inhomogeneous local order [2.143–148] while the systems containing ω phase particles have been described with a number of phenomenological models [2.116, 117, 149–151]. *Cook*

[2.116, 117] suggested that a significant part is played by the small difference between the wave vectors which correspond to the minimum of the harmonic part of the displacement wave and the ω phase. When anharmonicity is taken into consideration this difference makes the modulation of the displacement wave describing the ω phase energetically favourable and limits the size of the ω phase particles. But this factor is not sufficient by itself in the local theory for production of a structure with disordered regions which is modulated by the amplitude, instead of a large-parameter structure modulated by phase (as that described by (2.5.8)) which contains only ordered domains with different phases (which can be produced also under ω transformation owing to the existence of three subtypes of the given type of the ω phase [2.152]). Production of second-phase regions caused by heterophase fluctuations is also practically impossible far from the phase transition point (see Sect. 1.4).

We can see that the heterogeneous structure with disordered regions is more favourable when we take into consideration the non-local effects (the dependence of the thermodynamic potential density $\varphi(r)$ on $\eta(r)$ for $r' \neq r$) caused by the long-range (generally many-body) interaction which produces a non-linear dependence of φ on the ratio p of the ordered regions and limits the growth of p. Such effects can be caused, for instance, by the long-range indirect interaction which are due to the conduction electrons and have a purely quantum character. The simple models treated below demonstrate that the non-local effects due to the long-range interaction indeed must give rise to heterogeneous structures under certain conditions [2.153, 154] (see also [2.155–157]).

The above effects are due to the variation of the electron energy $w_e(\boldsymbol{k})$ caused, as discussed above, by the kth perturbation wave with k close to the diameter $2k_F$ of the Fermi surface connecting flattened or cylindrical parts of this surface. The ordering causes generation of the potential energy waves which correspond to the superlattice vectors $\boldsymbol{G}'$ of the reciprocal lattice and cause a decrease in the electron energy. Note that for $\boldsymbol{G}' \approx 2\boldsymbol{k}_F$ the energy decrease per atom can be considerably higher in the case of small ordered regions produced in the disordered matrix than in the case of homogeneous ordering of the crystal [2.158].

Firstly, this difference is explained by the fact that the conventional perturbation theory is inapplicable in the case of homogeneous ordering because of the transformation of the electron spectrum so that the dependence of $w_e(\boldsymbol{k})$ on the perturbation amplitude is not quadratic in the "singular" range of $\boldsymbol{k} \approx 2\boldsymbol{k}_F$. On the other hand, the effect of the small ordered region on electrons in the disordered crystal is described by second-order perturbation theory where the respective decrease of the energy per atom $|w^{(2)}\boldsymbol{k}|$ can be considerably greater than $|w_e(\boldsymbol{k})|$ (it can be seen for instance from (2.5.4–6)).

Secondly, if we select a suitable long-range or short-range order in the arrangement of the ordered regions and their phases we can "split" the superlattice point $\boldsymbol{G}'$ into a group of points $\boldsymbol{G}' \pm \boldsymbol{q}_n$ and thus decrease the electron energy E_e as in the case of the large-parameter structures discussed in Sect. 2.5.1 if the minimum of $w_e(\boldsymbol{k})$ is somewhat shifted with respect to $\boldsymbol{G}'$ and

$$w_e(\boldsymbol{G}' + \boldsymbol{q}_n) + w_e(\boldsymbol{G}' - \boldsymbol{q}_n) - 2w_e(\boldsymbol{G}') < 0 \,. \tag{2.5.9}$$

The respective decrease in E_e is essentially due to the slowly decreasing oscillating interaction between the ordered regions with the effective length $\sim \pi|\boldsymbol{G}' - 2\boldsymbol{k}_f|^{-1}$ which is produced by superposition of the Friedel oscillations $r^{-n}\cos(2k_F + \phi)$ on the concentration oscillations of the type of $\cos \boldsymbol{G}'\boldsymbol{r}$ in the ordered regions.

Thirdly, considerable local strains are produced as a result of the redistribution of concentration in solutions in which the atomic radii of the components differ significantly and these strains may displace the vectors $\boldsymbol{G}'$ for the small ordered regions with respect to the reciprocal lattice points of the homogeneously ordered phase which may also cause a decrease in E_e and a stabilization of the heterogeneous state.

Thermodynamic Potential of the Heterogeneous Structure. Since the long-range effects are due to the rapidly varying singular component of the energy $w_e(\boldsymbol{k})$ (associated with the Fermi surface or a sharp extremum of v^2 in the equations of Sect. 2.4) it is convenient to divide $w_e(\boldsymbol{k})$) into the smoothly varying component $w_e^0(\boldsymbol{k})$ and the singular component $w_e'(\boldsymbol{k})$. It can be done formally by multiplying $w_e'(\boldsymbol{k})$ by the smoothly varying function $\theta(\boldsymbol{k})$ which equals unity in the singular range and rapidly decreases outside it (the explicit form of $\theta(\boldsymbol{k})$ is not important) and defining w_e^0 as $w - \theta w'$.

The smoothly varying function $w_e^0(\boldsymbol{k})$ is associated only with the short-range indirect interaction. We shall denote by $\varphi^0(c, \eta)$ the respective contribution to the thermodynamic potential density together with the energy of direct interaction between atoms and the entropy term. This contribution depends locally on the order parameter $\eta(\boldsymbol{r})$ and the concentration $c(\boldsymbol{r})$. If we add the singular component of the electron energy we can write

$$\varphi = \int \varphi^0(c(\boldsymbol{r})\,,\ \eta(\boldsymbol{r}))d\boldsymbol{r} + \varphi_e' \,; \quad \varphi_e' = \frac{1}{2v_0} w_e'(\boldsymbol{k})\theta(\boldsymbol{k})|c_{\boldsymbol{k}}|^2 \,, \tag{2.5.10}$$

where the integration over $\boldsymbol{r}$ is performed over the unit volume and v_0 is the atom volume. In the case of the displacement-type ordering $\langle |c_{\boldsymbol{k}}|^2 \rangle$ in (2.5.10) must be replaced with $u_{\boldsymbol{k}}^2$.

In homogeneously ordered solutions the major contribution to the sum for φ_e' is made by μ' reciprocal lattice points $\boldsymbol{G}'$ which appear on ordering and lie in the singular range. For these points $c_{\boldsymbol{k}} = c(\boldsymbol{G}') = \xi\eta/2$ where, for instance, for the $L1_2$ lattice and $\boldsymbol{G}' = \langle 110 \rangle$ we have $\mu' = 12$ and $\xi = 1/2$ while for the $L1_0$ lattice $\mu' = 4$ and $\xi = 1$ (the contribution of the terms with $\boldsymbol{k} \neq \boldsymbol{G}'$ can be ignored since for them $\langle |c_{\boldsymbol{k}}|^2 \rangle \sim 1/N$ while the volume of the singular region in which $\theta(\boldsymbol{k}) = 1$ is small). Therefore, in the case of homogeneous ordering we have

$$\varphi(c, \eta) = \varphi^0(c, \eta) + \varphi_{\text{ehom}}' \,, \quad \varphi_{\text{ehom}}' = \frac{\mu'\xi^2}{8v_0} w_e'(\boldsymbol{G}')\eta^2 \tag{2.5.11}$$

(where $w_e' < 0$). In the disordered alloy (which does not contain ordered regions) we have $\varphi = \varphi^0(c, 0)$.

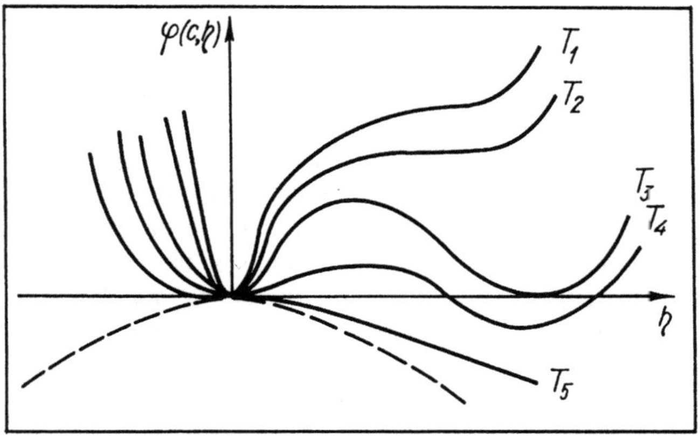

Fig. 2.4. The behavior of $\varphi(c,\eta)$ as a function of η at various temperatures $T_1 > T_2 > T_3 > T_4 > T_5$ for the homogeneous alloys. The plots for heterogeneous alloys may be lower than those shown here owing to a decrease in φ'. The dashed line shows the parabolic dependence of $-K_1 n^2 - K_2 f(x)\eta^2 - K_3\eta^2$ on η

Figure 2.4 shows $\varphi(c,\eta)$ as a function of η for homogeneous alloys with a first-order phase transition. If the disordered phase does not contain ordered regions then the phase transition corresponds to the temperature at which $\varphi(c,\eta) = \varphi(c,0)$ (the point T_3 in the figure). But under the conditions when the singular component φ'_e in (2.5.10) is considerably smaller for the small ordered regions, the heterogeneous structure can be produced in the temperature range in which for the homogeneous ordering we have $\varphi_e > \varphi^0(c,0) - \varphi^0(c,\eta)$ and the reversed condition is satisfied for the heterogeneous ordering. In this case the behaviour of $\varphi(c,\eta)$ as a function of η in first-order phase transitions (see Fig. 2.4) determines the significant part played by the amplitude modulation (in contrast to the phase modulation of the charge density wave discussed in 2.5.1). The second minimum in the curve or a segment with a small slope at $\eta = \eta_{\min} \sim 1$ in addition to the minimum at $\eta = 0$ and the increase in $\varphi(c,\eta)$ for the intermediate values of η favour the production of regions with $\eta(r) \approx \eta_{\min}$ and $\eta \ll 1$.

We shall analyze the physical factors determining the production of an equilibrium heterogeneous state and its structure in the framework of the following simplest model. We shall assume that in the ordered regions $\eta(\boldsymbol{r}) = \eta = \text{const}$ and outside them $\eta = 0$ and that the sizes D_0 and the shapes of the regions are identical (the models specifying a variable dependence $\eta(r)$ give a somewhat better stability of the heterogeneous structure). Since inhomogeneous ordering in alloys must be accompanied by inhomogeneous redistribution of the composition, the concentration c_0 in the ordered regions (whose volume fraction is p) varies by $\delta c = c - c_0$ with respect to the mean concentration c_0 while in the disordered regions it varies by $\delta c' = c' - c_0 = -\delta c p/(1-p)$. In this model the difference $\Delta\varphi$ between the thermodynamic potential densities of the heterogeneous structure φ_{het} and the disordered system $\varphi(c_0,0) = \varphi^0(c_0,0)$ can be written as

$$\Delta\varphi = \varphi_{\text{het}} - \varphi(c_0, 0) = p\psi(c, \eta, p) + \gamma S + \delta\varphi'_e\,, \quad \delta\varphi'_e = \varphi'_{\text{ehet}} - p\varphi'_{\text{ehom}}\,,$$
$$\psi(c, \eta, p) = \varphi(c, \eta) - \varphi(c_0, 0) + \frac{1-p}{p}[\varphi(c', 0) - \varphi(c_0, 0)]$$
$$\approx \varphi(c, \eta) - \varphi(c_0, 0) - \frac{\partial\varphi(c_0, 0)}{\partial c_0}(c - c_0) \tag{2.5.12}$$
$$+ \frac{1}{2}\frac{p}{1-p}\frac{\partial^2\varphi(c_0, 0)}{\partial c_0^2}(c - c_0)^2\,, \quad S = \zeta\frac{p}{D_0}\,.$$

Here γS is the surface energy of the ordered regions per unit volume (with the total surface area of $S, \zeta \sim 1$) and $\delta\varphi'_e$ is the difference between the singular components of φ for the heterogeneous and homogeneous systems with identical parameters η. The higher-order terms of the expansion in powers of $\delta c'$ are eliminated in the approximate equation for Ψ.

According to (2.5.10), the difference $\delta\varphi'_e$ is determined by $c_{\boldsymbol{k}}$ in the singular region for small $\boldsymbol{q} = \boldsymbol{k} - \boldsymbol{G}'$. The latter are related to the qth Fourier components of the order parameter in the model under consideration:

$$c_{\boldsymbol{k}} = \frac{1}{2V}\xi\int e^{i\boldsymbol{q}\cdot\boldsymbol{r}}\eta(\boldsymbol{r})e^{i\phi(\boldsymbol{r})}d\boldsymbol{r} = \frac{1}{2V}\xi\eta s(\boldsymbol{q})p(\boldsymbol{q})\,,$$
$$p(\boldsymbol{q}) = \sum_{\boldsymbol{R}_m}\exp\{i[\boldsymbol{q}\cdot\boldsymbol{R}_m + \boldsymbol{G}'\cdot\boldsymbol{d}(\boldsymbol{R}_m)]\}\,, \tag{2.5.13}$$
$$s(\boldsymbol{q}) = \int s(\boldsymbol{r})\exp(i\boldsymbol{q}\cdot\boldsymbol{r})d\boldsymbol{r}\,.$$

Here $s(\boldsymbol{q})$ is the Fourier component of the shape function $s(\boldsymbol{r})$ of the ordered region, $\boldsymbol{R}_m$ is the radius-vector of the mth ordered region and we have taken into account the fact that when the mth region is displaced by a lattice vector $d(\boldsymbol{R}_m)$ of the disordered crystal the wave phase $\exp(i\boldsymbol{G}'\cdot\boldsymbol{r})$ is varied by $\phi = \boldsymbol{G}'d(\boldsymbol{R}_m)$.

When we identify the constant component $w_e(\boldsymbol{G}')$ in $w'_e(\boldsymbol{k})$ in (2.5.10) and make use of (2.5.11) and the relation

$$\int |s(\boldsymbol{q})|^2|p(\boldsymbol{q})|^2 d\boldsymbol{q} = 8\pi^3 pV$$

(see (1.3.14) in [1.1]) we can write

$$\delta\varphi'_e = \delta\varphi'_1 + \delta\varphi'_2 + \delta\varphi'_3\,, \quad \delta\varphi'_1 = -K_1 p\eta^2\,, \quad K_1 = \frac{\xi^2\mu'}{8v_0}\delta w_1\,,$$
$$\delta\varphi'_2 = -\frac{\xi^2\mu'\eta'}{64\pi^3 v_0 V}\left[\int \delta w_2(\boldsymbol{q})|s(\boldsymbol{q})|^2|p(\boldsymbol{q})|^2\theta(\boldsymbol{k})d\boldsymbol{q}\right.$$
$$\left. - \frac{\tilde{D}_0}{D_0}\int \delta w_2(\boldsymbol{q})\lim_{\tilde{D}_0\to\infty}|s(\boldsymbol{q})|^2|p(\boldsymbol{q})|^2\theta(\boldsymbol{k})d\boldsymbol{q}\right]\,, \tag{2.5.14}$$
$$\delta w_1 = w'_{\text{ehom}}(\boldsymbol{G}') - w'_{\text{ehet}}(\boldsymbol{G}')\,,$$
$$\delta w_2(\boldsymbol{q}) = w'_{\text{ehet}}(\boldsymbol{G}') - w'_{\text{ehet}}(\boldsymbol{G}' + \boldsymbol{q})\,.$$
$$\delta\varphi'_3 = -K_3 p\eta^2\,, \quad K_3 = \frac{\xi^2}{8v_0}\sum_{\boldsymbol{G}'}\frac{\partial w'_e(\boldsymbol{G})}{\partial G'_i}u_{ij}G'_j\,.$$

Here $\delta\varphi_1'$ takes into account the variation of $w_e'(G')$ and the electron energy per unit volume caused by the production of small ordered regions owing to the decrease in the harmonic U_k of the perturbing potential energy, $w_{\text{ehom}}' v_k^{-2}$ corresponds to $U_k = v_k \xi\eta/2$ and $w_{\text{ehet}}' v_k^{-2}$ for small D_0 and p corresponds to the limit of $U_k \to 0$, that is $w_{\text{ehet}}'(G') \approx w_e'^{(2)}(G')$; the variation $\delta\varphi_2$ is caused by the spreading of the peak $|c_{G'}|^2$ and the variation $\delta\varphi_3'$ is caused by the shift of this peak due to the strain u_{ij} in the ordered regions. In the equation (2.5.14) for $\delta\varphi_2'$ the limit in the second term corresponds to a large size $\tilde{D}_0$ of the ordered regions (for a given p and the similarity transformation of the structure). This term is approximately proportional to $1/D_0$ (if we ignore the dependence of $\delta w_2(q)$ on D_0), it takes into account the contribution φ_2' to the surface energy which is included as a component γ_{e2} into the term γS in (2.5.12) and must be subtracted here. (Note that for not too small v_k/E_F a significant part can be played by the singular contributions to $\delta\varphi_e'$ found in the higher-order approximations of the perturbation theory. For instance, the third-order term exhibits a sharper singularity though it includes an additional factor v_k/E_F. As in the problem of the electron-phonon interaction (see, for instance, [2.159]) it manifests itself when the triangle of the superlattice vectors G_1', G_2' and G_3' is almost inscribed into the Fermi surface.) In the equation (2.5.14) for $\delta w_2(q)$ the vector G' corresponds to the strained lattice of the ordered region.

Equations (2.5.12–14) give $\Delta\varphi$ as a function of p, η and c. At equilibrium we have

$$\frac{\partial\Delta\varphi}{\partial c} = 0\,, \quad \frac{\partial\Delta\varphi}{\partial\eta} = 0\,, \quad \frac{\partial\Delta\varphi}{\partial p} = 0\,, \quad \frac{\partial\Delta\varphi}{\partial D_\partial} = 0\,, \quad \frac{\partial\Delta\varphi}{\partial D} = 0\,, \qquad (2.5.15)$$
$$\delta_2 = (\Delta\varphi) > 0\,, \quad \min\Delta\varphi < 0\,,$$

where $\delta_2(\Delta\varphi)$ is the second-order variation of the function $\Delta\varphi(c,\eta,p)$.

Uncorrelated Ordered Regions. Consider first the simplest case when the ordered regions are not distorted and the function $\delta w_2(q) + \delta w_2(-q)$ is sufficiently small. Then we can ignore the terms $\delta\varphi_3'$ and $\delta\varphi_2'$in (2.5.14) and the gain in the electron energy under the transition to the heterogeneous regions depends only on the term $\delta\varphi_1'$ which does not depend on the arrangement of the regions, that is, it can be regarded as being uncorrelated. (The variations δw_1 and $\delta w_2(q)$ and hence $\delta\varphi_1'$ and $\delta\varphi_2'$ for the given function $w(k)$ are related to each other as we can see, for instance, from (2.5.4–6). If one ignores the dependence of v_k^2 on k then in the case described by these equations and considered here, when G' is in the region of the $w_e'(k)$ plot with a large curvature radius, one obtains $\delta w_1 < 0$ and $\delta\varphi_1' > 0$. Typically, however, v_k^2 has a strong dependence on k so that the region where the $w_e'(k)$ plot has a large curvature radius can be shifted to the minimum of $w_e'(k)/v_k^2$ where $\delta w_1 > 0$ and reaches its maximum.)

The positive (for $\gamma > 0$) contribution of γS can be decreased by increasing the size D_0 of the ordered regions. But this size is limited by the fact that the difference between $w_e'(G')$ for the ordered region and for the homogeneously ordered alloy manifests itself only for small $D_0 \lesssim D'_0$ when the additional potential of the ordered regions can be regarded as a small perturbation, If these regions are large then the wave functions inside them are the same as in the bulk of the crystal and we obtain $\delta w_1 = 0$ in the limit of $D_0 \to \infty$ (if we ignore the minor surface effects). Using the well-known estimates for the applicability of the perturbation system (see Sect. 45 in [2.160]) we obtain for the equiaxial ordered regions

$$D'_0 \sim \frac{E_F}{v_k} d \quad \text{for} \quad |\delta k_z| D'_0 \lesssim 1 , \tag{2.5.16}$$

that is, $D_0 \sim (5 \div 10)d$ (where d is the edge length of the cubic cell). For large-size ordered regions δw_1 is considerably decreased so that their production is not favoured.

For sufficiently small D_0 and $p \to 0$ we may assume that when we calculate δw_1 and δw_2 from (2.5.14) we can take $U_{k\mathrm{het}} = 0$ and $U_{k\mathrm{hom}} = v_k \xi \eta / 2$. Under such conditions $\delta\varphi'_e$ and $\Psi(c, \eta, p)$ in (2.5.14) are minimal for $p \to 0$ and $D_0 \to 0$. The function $\Psi(c, \eta, 0)$ decreases with a decrease in temperature and below a certain point T^0 at which

$$\psi(c, \eta, 0) + \frac{\zeta\gamma}{D_0} - K_1 \eta^2 = 0 , \tag{2.5.17}$$

we have, according to (2.5.12), $\Delta\varphi < 0$ and the heterogeneous structure is thermodynamically favourable. The temperature T^0 can be considerably higher than the temperature T_0 of the transition to the homogeneous ordered state (in which $\Delta\varphi = \psi(c_0, \eta_0, 1)$. For instance, for the $L1_0$ alloy which has flattened spherical segments of the Fermi surface and $U_{k\mathrm{hom}} = 2E_m = 4\delta E = 0.7\,\mathrm{eV}$, $\hbar^2 k_F^2/2m_0 = 7\,\mathrm{eV}$, $S_k = \pi k_F^2/4$ and $u_z = \hbar k_F/m_0$. (where m_0 is the mass of the free electron), that is, $\tilde{k} = 0.1 k_F$ and $\delta k_z = k_F/40$ we find from (2.5.4–6, 14) that the parameter $v_0 K_1 \approx 0.06\,\mathrm{eV}$ per atom in the ordered regions. This increase in the effective energy of ordering brings about a difference of approximately 600 K between the temperature T^0 and T_0 even in the absence of other factors stabilizing the heterogeneous structure (see below). A decrease in $\Psi(c, \eta, p)$ when the temperature falls below T^0 gives rise to a continuous growth of p from $p = 0$. But if we take into consideration the term $\delta\varphi'_2$, in addition to $\delta\varphi'_1$, the correlation between the ordered regions becomes energetically favourable for large p and we can obtain a jump-like variation of p at the temperature $T = T^0$. (If the interface energy γ for a single disordered region in the ordered matrix is negative and its absolute value is large enough for making $\Psi + \zeta\gamma D_0^{-1} < 0$ in (2.5.17) then the trend to lower interface energy may give rise to heterogenization. In systems with a short-range interaction this effect would result in fragmentation of the heterogeneous structure down to the scale of $D_0 \sim d$. When G' is at the minimum of the function $w'_e(\boldsymbol{k})$ and $\delta w_2(\boldsymbol{q}) < 0$ (that is, $\delta\varphi'_2 + \gamma_{e2} S > 0$) the singular component of the electron energy density $\delta\varphi_2$ can grow considerably as the small size of the ordered regions cause spreading of is $|s(\boldsymbol{q})|^2$ and $|c_{\boldsymbol{k}}|^2$. When $|\delta\varphi'|$ are small this effect limits the decrease in D_0 and thus stabilizes the heterogeneous structure for large $-\gamma$ (and $\delta w_2(\boldsymbol{q}) < 0$) while the growth of the interface energy can prevent the ordered regions in the disordered matrix from coming closer to each other. Below we shall discuss systems with $\gamma > 0$ or small $|\gamma|$ in which this heterogenization factor does not make any contribution.)

Correlation Between the Ordered Regions. In systems containing a large proportion of ordered regions the correlation in their arrangement becomes significant. The contribution of the correlation is especially significant if $\boldsymbol{G}'$ is in the region where the plot of $w'_e(\boldsymbol{k})$ as a function of $\boldsymbol{k}$ has a negative curvature (point B in Fig. 2.1) and $\delta\varphi'_2\gamma_{e2}S < 0$. A suitable choice of the arrangement and phases of the ordered regions can lead to splitting of the $|c_{G'}|^2$ peak when $|\delta\varphi'_2|$ is increased and thus $\delta\varphi'_e$ is decreased.

To estimate $\delta\varphi'_2$ we shall make use of the model in which the ordered regions are regularly arranged in space (violations of the regularity are discussed below). In such a model $p(q)$ given by (2.5.13) do not vanish only for $q = q_n$, where $\boldsymbol{q}_n$ corresponds to a lattice of the ordered regions, and the integral over $\boldsymbol{q}$ in (2.5.14) reduces to the sum over $\boldsymbol{q}_n$.

The resulting heterogeneous state depends significantly on the type of the ordered structure and the arrangement and geometry of the regions in which the function $\delta w_2(\boldsymbol{q})$ varies sharply. As an illustration we shall consider the fcc alloys with the concentration c about 1/4 which have the structure of Cu_3Au in the ordered state. We shall assume that the vectors $\boldsymbol{G}' + \boldsymbol{G}$ of the (110) type are close to the diameters $2\boldsymbol{k}_F$ of the flattened or cylindrical parts of the Fermi surface (as it is the case in copper-based alloys). Assume for the sake of simplicity that the minima of the functions $w'(\boldsymbol{k})$ and hence $-\delta w_2(\boldsymbol{q})$ are sufficiently sharp and have constant values $\min(-\delta w_2(\boldsymbol{q}))$ at large flattened (cylindrical) Fermi surface segments perpendicular to the vectors $\pm\boldsymbol{k}_F \parallel \boldsymbol{G}$. Then we can retain in the integral (sum) in (2.5.14) only those terms for which $\boldsymbol{q}_n$ is in these segements.

Four types of ordered regions that can be produced differ in the displacements $d_1 = 0$, $\boldsymbol{d}_2 = (\boldsymbol{a}_1+\boldsymbol{a}_2)/2$, $\boldsymbol{d}_3 = (\boldsymbol{a}_1+\boldsymbol{a}_2)/2$, and $\boldsymbol{d}_4 = (\boldsymbol{a}_2+\boldsymbol{a}_3)/2$ and in the respective phases $\phi = \boldsymbol{G}'\cdot\boldsymbol{d}$ (for instance, $0, 2\pi, \pi, \pi$ for the wave with $G' = (110)$). The three-dimensional regular structure of ordered regions typically yields a smaller $\delta\varphi'_2$ than the one-dimensional structure. The former has the cubic symmetry and the same number of ordered regions with different phases in the cell. If a cubic cell with the edge length of D contains ν' ordered regions of the cubic shape with the edge length of D_0 we can find from (2.5.13) for nonvanishing $p(\boldsymbol{q})s(\boldsymbol{q})$ and the respective $\boldsymbol{q} = \boldsymbol{q}_n$:

$$\boldsymbol{q}_n = \frac{2\pi}{D}\sum_{i=1}^{3} n_i \boldsymbol{e}_i\,,$$
$$|p(\boldsymbol{q})s(\boldsymbol{q})|^2 = \frac{8\nu'^2 V}{\pi^3}\delta(\boldsymbol{q}-\boldsymbol{q}_n)\prod_i \frac{\sin^2 \pi x n_i}{n_i^2}\,, \quad x = \frac{D_0}{D}\,. \tag{2.5.18}$$

Here $\boldsymbol{e}_i$ are the unit vectors along the $\langle 100\rangle$ axes and n_i are the integers determined by the structure. For instance, if the ordered regions make up an fcc lattice ($\nu' = 4$)

consisting of four simple cubic sublattices with different displacements d then $\delta\varphi_2'$ is minimal when the sublattices (000), ($\frac{D}{2}$ $\frac{D}{2}$ 0), ($\frac{D}{2}$ 0 $\frac{D}{2}$), and (0 $\frac{D}{2}$ $\frac{D}{2}$) correspond to the displacements $\boldsymbol{d}_1$, $\boldsymbol{d}_4$, $\boldsymbol{d}_2$ and $\boldsymbol{d}_3$. For instance, in the vicinity of the point $\boldsymbol{G}' = (110)$ the numbers n_1 and n_3 must be even for $p(q) \neq 0$ and n_2 must be odd or n_1 and n_3 must be odd and n_2 must be even.

In this model with a sharp peak of the function $\delta w_2(\boldsymbol{q})$ occurring on a flat surface in the k space which is at a small distance $\boldsymbol{k}^0$ from the point $\boldsymbol{G}'$ a significant contribution to the sum in (2.5.14) is made by the terms corresponding to the sets of $(n_1, n_2) = (1,0), (2,-1), \ldots, (0,1), (-1,2), \ldots$ where $n_1 + n_2 = 1$ and any even (or odd) n_3. The minimum of $\delta\varphi_2'$ is reached for $\sqrt{2}\pi/D \approx k^0$. When we find the sum over such $\boldsymbol{q}_n$ in (2.5.14) and make use of (2.5.18) we shall retain only the rapidly changing component of $\delta w_2(\boldsymbol{q}) \approx \delta w_2^{(2)}(\boldsymbol{q})$ whose peak height W is shown in Fig. 2.5 (the contributions of the smoothly varying component of $\delta w_2(\boldsymbol{q})$ to the two integrals in (2.5.14) are cancelled out and the contribution of the rapidly varying component to the second integral can be neglected). We obtain after these transformations

$$\begin{aligned}
&\delta\varphi_2' = -K_2 p \eta^2 f(x)\,, \quad K_2 = \frac{3W}{4\pi^2 v_0}\,, \quad p = \nu' x^3\,,\\
&S = \frac{6\nu' x^2}{D}\,, \quad D = \frac{\sqrt{2}\pi}{k^0}\,,\\
&f(x) = \frac{1}{x}\left(1 - \frac{\sin 2\pi x}{2\pi x}\right) \quad (x < 1)\,.
\end{aligned} \tag{2.5.19}$$

Here the peak height W is the difference between the values of the function $\delta w_2(\boldsymbol{q})$ at the peak and at its smoothly varying part. (In this approach $\delta w_2(\boldsymbol{q})$ is broken down (not precisely) into a smoothly and rapidly varying parts and only those terms in the sum (2.5.14) for which $\boldsymbol{q}_n$ are near the given plane are taken into account and therefore its accuracy is the better the sharper the variation of $\delta w_2(\boldsymbol{q})$ in the singular region.)

When the sublattices (000), ($\frac{D}{2}$ $\frac{D}{2}$ 0), ($\frac{D}{2}$ 0 $\frac{D}{2}$), and (0 $\frac{D}{2}$ $\frac{D}{2}$) correspond to the displacements $\boldsymbol{d}_1$, $\boldsymbol{d}_2$, $\boldsymbol{d}_3$, and $\boldsymbol{d}_4$ we obtain a lower absolute value $|\delta\varphi_2'|$ (then for the point $\boldsymbol{G}' = (110)$ both numbers n_1 and n_2 are even or odd) but D is then doubled, that is the surface energy decreases. Under these conditions we have

$$\begin{aligned}
&D = \frac{2\sqrt{2}\pi}{k^0}\,, \quad f(x) = \frac{1}{4x}\left(1 - \frac{\sin 4\pi x}{4\pi x}\right) \quad \left(x < \frac{1}{2}\right)\,,\\
&f(x) = \frac{1}{4x}\left[1 + \frac{\sin 4\pi x}{4\pi x} + \frac{1-2x}{x}\cos 4\pi x\right] \left(x > \frac{1}{2}\right)\,.
\end{aligned} \tag{2.5.20}$$

If we have a very sharp peak of $\delta w_2(\boldsymbol{q})$ on the cylindrical surface in the $\boldsymbol{k}$-space (with the peak width much smaller than k_{02}/k_{F}) the functions $f(x)$ in (2.5.19, 20) are modified (retaining the form qualitatively) reducing to $2\sin^2 \pi x$ or $(\pi x)^{-2} \sin^4 \pi x$.

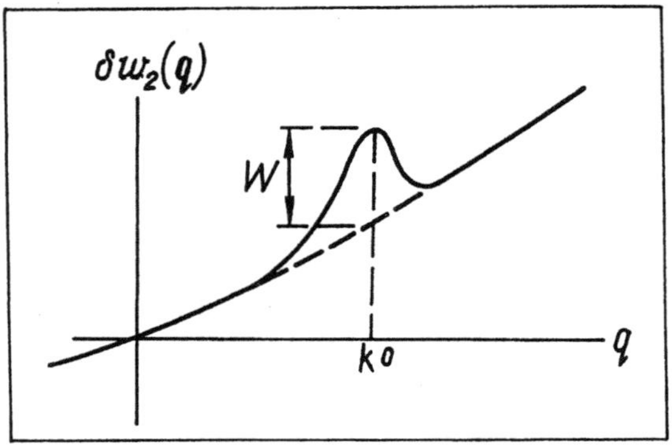

Fig. 2.5. Schematic of the behaviour of $\delta w_2(q)$ as a function of q in the direction of $\boldsymbol{q} \parallel \boldsymbol{G}'$. The dashed line shows the smoothly varying segment of the plot

Equations (2.5.19, 20) describe also the case of a cubic lattice of ordered regions comprising four bcc sublattices with different displacements d when one must bear in mind that $\nu' = 8$.

As it is the case for the systems with uncorrelated ordered regions the equations (2.5.14, 19) for $\delta\varphi_2'$ are valid only for $D_0 \lesssim D_0'$ in the range of applicability of the perturbation theory. When D_0 is considerably larger than D_0' given by (2.5.16) the wave functions of electrons inside the ordered regions (but not in the surface layers) are the same as in a massive crystal. Therefore, for $D_0 \gtrsim D_0'$ the rate of growth of $\delta\varphi_2'$ with an increase in D_0 (or x) must be lower land then the interaction between the ordered regions must become much weaker. This effect is described formally by a significant decrease in the function $f(x)$ for $x = D_0/D > D_0'/D$ in comparison with the values given by (2.5.19, 20).

Though the above equations for $\delta\varphi_2'$ were derived using a model of a regular heterogeneous structure, almost the same gain in the electron energy can be obtained by determining the short-range order in the arrangement of the ordered regions. To do that it is sufficient for the function $|c_{\boldsymbol{k}}|^2$ given by (2.5.13) to have a relatively sharp peak whose width must be smaller than k^0. If, for instance, the short-range order in the arrangement of the ordered regions can be represented by a set of small pseudocrystals whose lattice is formed by the ordered regions and its parameter is $2D$ where D depends on $\boldsymbol{k}^0$ according to (2.5.20) then the $|c_{\boldsymbol{k}}|^2$ peak width is about $\boldsymbol{k}^0/3$ (see (1.3.11) in [1.11). In the paracrystal model the peak is even sharper [2.161]. This means that if the peak of the function $\delta w_2(q)$ is not too sharp the difference in $\delta\varphi_2'$ between systems with short-range and long-range order of the ordered regions is small. Therefore, the heterogeneous state apparently corresponds to the short-range order in their arrangement (for which the entropy is somewhat larger). In the above case of a negative curvature of $w_e'(\boldsymbol{k})$ in the range of $\boldsymbol{k} \approx \boldsymbol{G} + \boldsymbol{G}'$ (that is for $W > 0$) the neighbouring ordered regions must be antiphase.

Heterogeneous Structure with Uncorrelated Antiphase Ordered Regions. The character and structure of the heterogeneous state can be somewhat different if the vector $\boldsymbol{G}'$ is in the range near the minimum of the plot of $w_e'(\boldsymbol{k})$ as a function of $\boldsymbol{k}$

(for instance, at the point A in Fig. 2.1) where the plot has a positive curvature. In contrast to plots of negative curvature discussed above, here $\delta\varphi_2' + \gamma_{e2}S > 0$ and the thermodynamic potential can only grow owing to splitting or spreading out of the reciprocal lattice point under heterogenization. But the heterogeneous state can be thermodynamically favourable even under these conditions at the expense of the negative terms $\delta\varphi_1'$ and $\delta\varphi_3'$ in (2.5.14).

To analyze qualitatively the conditions for production of such structures let us consider first a small fraction p of uncorrelated ordered regions produced at large distances from each other. If these regions are not correlated the point $\boldsymbol{G}'$ is spread out (but not split) and $|c_{\boldsymbol{k}}|^2$ in (2.5.14) is of the order of $p\eta^2|s(\boldsymbol{q})|^2 D_0^{-3}$ where $s(\boldsymbol{q})$ is the Fourier component of the shape function $s(\boldsymbol{r})$ for the region $s(\boldsymbol{r})$ is unity inside the region and vanishes outside it). If the regions are cubic in shape then $|s(\boldsymbol{q})|^2 = D^6$ for $|\boldsymbol{q}| \ll \pi/D_0$, and it rapidly decreases for $|\boldsymbol{q}| \gg \pi/D_0$. For the sake of simplicity we shall limit the discussion to the case when the point $\boldsymbol{G}'$ (displaced owing to the strain of the ordered regions if it is significant) is strictly at a sufficiently sharp minimum of the function $w_e(\boldsymbol{k})$. This minimum is at the point $\boldsymbol{q} = 0$ and we assume that it has the depth of $W > 0$, the width of δk along the axis k_z and a much larger width in the transverse directions while for $q \gg \delta k$ we may assume that $w_e(\boldsymbol{k})$ is constant. Then according to (2.5.14) for a large size of the ordered regions $D_{0z} \equiv D_0$. When $D_0\delta k \gg \pi$ we obtain that $\delta\varphi_2$ is not larger by an order of magnitude than $v^{-1}p\eta^2|W|(D_0\delta k)^{-3}$ and rapidly decreases with an increase in D_0. The main effect produced by the finite size of the ordered regions (spreading of the reciprocal lattice points) in this range of D_0 is described by the term $\gamma_{e2}S = \gamma_{e2}\xi p/D_0$ in the expression for the surface energy where $\gamma_{e2} \sim v^{-1}\eta^2|W|/\delta k$ (this term equals the last term in the equation (2.5.14) for $\delta\varphi'$ taken with the reversed sign). For small D_0 when $D_0\delta k \ll \pi$ we have $\delta\varphi' + \gamma_{e2}S \sim v^{-1}\eta^2 p|W|$.

In contrast to $\delta\varphi' + \gamma_{e2}S$, the term $\delta\varphi'$ in the singular component of the electron energy is negative (in the case treated here when the minimum of $w_e'(\boldsymbol{k})$ is associated with the flattened segment of the Fermi surface). In the limit of small $D_0 \ll D'$ (see (2.5.16)) this term is given by (2.5.14) where $\delta w_1 = \delta w^{(0)} = w_{\text{ehom}}(G') - w'^{(2)}(G')$ and $K_1 = K^{(0)} \sim v^{-1}\delta w^0$. For large $D_0 \gg D'$ it has been shown that δw_1 and K_1 vary as D^{-1} so that $\delta w_1 \sim \delta w^{(0)}D'/D_0$ and $K_1 = \xi_1 K^{(0)}D'/D_0$ where $\xi_1 \sim 1$. In the typical case when $D'\delta k \gg \pi$ we can use these estimates and rewrite the equation (2.5.12) for $\Delta\varphi$ for the range of interest, $D_0\delta k > \pi$, in the following approximate form

$$\begin{aligned} \Delta\varphi &= \varrho\left[\Psi + \frac{\xi\gamma}{D_0} - \xi_1 K_1^{(0)}\eta^2\frac{D_0'}{D_0}\right], \quad D_0 > D_0' > \frac{\pi}{\delta k}, \\ \Delta\varphi &= \varrho\left[\Psi + \frac{\xi\gamma}{D_0} - K_1^{(0)}\eta^2\right], \quad D_0' > D_0 > \frac{\pi}{\delta k} \end{aligned} \tag{2.5.20a}$$

When the parameter $D'\delta k\delta w_1/|W|$ is large the following inequality may be satisfied: $\xi\gamma < \xi_1 K^{(0)}\eta^2 D'$. Under these conditions the thermodynamic potential decreases with a decrease in the size of the ordered regions in the range of $D_0 > D'$

(while their fraction is constant). This decrease stops for $D_0 \sim D'$ and then $\Delta\varphi$ starts to grow for the case of $\gamma > 0$ treated here.

If Ψ is small and the condition $\Psi + \xi\gamma D^{-1} - K^{(0)}\eta^2 < 0$ is satisfied then min$\Delta\varphi < 0$ and the production of the ordered regions in the disordered matrix is thermodynamically favorable. Their size is about D' given by (2.5.16) and it can be much larger than the size of the correlated regions, $D_0 < D/2$ (see (2.5.19, 20)) for $W > 0$.

For temperatures $T < T^0$ for which min$\Delta\varphi < 0$ and the ordered regions are stable, $\Delta\varphi$ is at first proportional to their fraction p as it increases. This is the case, however, only for large distances D between the ordered regions. When the widths of the disordered regions separating the ordered regions become comparable with D', the difference between the wave functions of the electrons in the heterogeneous system and in the homogeneous ordered phase decreases. This is why an increase in p leads to a decrease in the energy gain $-\delta\varphi'/p$ which is under these conditions the main factor stabilizing the ordered regions. At the same time Ψ increases owing to a more difficult redistribution of concentration between the ordered and disordered regions for higher p (see (2.5.12)) while γ grows owing to a more difficult relaxation of the order for shorter distances between the ordered regions. Therefore, p in heterogeneous structures of the above type does not exceed a few tenths and it increases with decreasing temperature. The distances D are, accordingly, of the order of D_0, that is, we have under these conditions

$$D \sim D_q \sim D_0' \sim \frac{2E_{\mathrm{F}}}{\eta v_{\boldsymbol{k}'}} d \tag{2.5.20b}$$

In the case under consideration when G' is at the minimum of the function $w_e(\boldsymbol{k})$ and $W < 0$ a decrease in $\delta\varphi'$ is favoured by a slight spreading out of the $|c_{\boldsymbol{k}}|^2$ peak near a reciprocal lattice point. When the size D_0 of the ordered regions is given this spreading can be decreased somewhat if we correlate the neighbouring regions with the same phases (in contrast to the case of $W > 0$ when the neighbouring regions must be antiphase). When the order distribution inside the ordered region and at its surface is spread out $\delta\varphi'$ may be also accordingly decreased.

Analysis of the Heterogeneous State. Let us study in more detail the heterogeneous state for the case of the negative curvature of the $w'(\boldsymbol{k})$ plot in the range of $\boldsymbol{k} \approx \boldsymbol{G} + \boldsymbol{G}'$ (for $W > 0$). Using (2.5.14, 19, 20) and finding the equilibrium concentration in the ordered regions $c = c\,(\eta, p)$ from the condition $\partial\Delta\varphi/\partial c = 0$ we can write $\Delta\varphi$ as a function of η and $p = \nu' x^3$:

$$\begin{aligned} \Delta\varphi(\eta, p) &= p\left[\psi(\eta, p) + \frac{6\gamma}{xD} - K_1(p)\eta^2 - K_2\eta^2 f(x) - K_3\eta^2\right] , \\ \gamma &= \gamma(\eta, \delta) . \end{aligned} \tag{2.5.21}$$

Here $\Psi(\eta, p) = \Psi[c(\eta, p), \eta, p]$ and γ depends on the distance δ between the edges of the neighbouring ordered regions, that is, on p (when the regions come closer together, the relaxation of the order and composition in space becomes more difficult. The parameter γ rapidly grows for $\delta \sim d$ (typically, $D/2 \sim 5d$, that is for $\nu' = 4$ and $p = 1/16$ we have $\delta \approx 4d$ and for $p = 1/4$ we have $\delta \approx 1.5d$).

The behaviour of $\Psi(\eta, p) + 6\gamma/xD$ as a function of η is described by the plots similar to those in Fig. 2.4 while the negative terms in (2.5.21) exhibit a quadratic dependence on η and produce an additional contribution to the ordering energy. They lead to a decrease in $\Delta\varphi$ as a function of η (for a given p). As in the case of normal homogeneous ordering, for sufficiently high temperatures Ψ is large, $\Delta\varphi > 0$ for all p and η and the heterogeneous state is thermodynamically unfavourable. We have, however, $\min\Delta\varphi < 0$ and the heterogeneous state is in thermodynamic equilibrium starting from a temperature T^0 at which the plot of $\Delta\varphi(\eta, p)$ for the optimal $p = p_0$ contacts the horizontal axis at the point $\eta^0 \sim 1$.

Since for small p we have $K_1 = K_1^{(0)} - K_1^{(2)} p^2$ and $f(x) \sim p^{1/3}$ we obtain for the model with a correlated arrangement of the ordered regions and $K_2 > 0$ (in contrast to the case of $K_2 < 0$ discussed above) a finite fraction $p = p_0 \neq 0$ even for $T = T_0$. Therefore, the transition to the heterogeneous state must appear as a spread-out (owing to the finite size of the ordered regions) first-order phase transition. The fraction p_0 depends on the ratio between $K_1^{(2)}$, K_2, $D^{-1}(\partial\gamma/\partial p)$, and $\partial\Psi/\partial p$. For $T = T_0$ we typically obtain p_0 varying approximately between 0.05 and 0.2. (If K_2 is considerably larger than the other parameters then p_0 corresponds to the maximum of $f(x)$. According to (2.5.19), for $D_0 = Dx_o < D_0'$ we have $x_0 = 1/2$, that is, $p_0 = 1/2$. But when Dx_0 is larger than D' given by (2.5.16) the above-mentioned decrease in $f(x)$ for $x > D_0'/D$ causes a decrease in x_0 (down to about D_0'/D), and accordingly a large decrease in p_0, because the condition for applicability of the perturbation theory is not satisfied. In addition, when γ strongly depends on δ (that is, on p) we can obtain a decrease in $\Delta\varphi$ (at the expense of a decrease in γ/D) if the ordered regions comprise a lattice with a doubled parameter D while $f(x)$ is given by (2.5.20). Under these conditions we have $x_0 = 1/4$ and $p_0 = 1/16$. Typically $K_1^{(2)} \sim K_2$ and the above p_0 must decrease because of that and also because of the increase in $\Psi(\eta, p)$ and γ with an increase in p.) The distance between the ordered regions is the largest (that is, γS has the minimum) in the above fcc lattice of regions with different d (with simple cubic sublattices of regions with different displacements d and $\nu' = 4$) and for T about T^0 this lattice apparently describes the heterogeneous state for $K_2 > 0$.

As the temperature decreases the function Ψ decreases sharply and it can be seen from (2.5.21) that the minimum of $\Delta\varphi$ corresponds to increasing p and η. At temperatures well below T^0 the function Ψ decreases rapidly (and may become negative) and then an increase in p is restricted only by a fast increase in γ for small δ, a drop in $K_1(p)$ and a slower rate of growth for $f(x)$ (especially for $D_0 > D'$) as well as an increase in Ψ with an increase in p. Under these conditions p may grow to about unity.

In the range of p around unity the short-range order may be thermodynamically more favourable in the arrangement of the ordered regions which gives rise not to a lattice of regions with $\nu' = 4$ (and $p \leq 1/2$) but to a more closely packed lattice with $\nu' = 8$. It may be, for instance, a cubic lattice consisting of four bcc sublattices with different displacements d of the ordered regions which can contain a large fraction $p \leq 1$ of these regions or it may be a layered structure accomodating two ordered layers with the thickness D_0 and different displacements (for instance, $d_\perp$ and d_2) at the distance of the lattice parameter D. In the latter case we have in (2.5.19) $f(x) = 2(3x)^{-1} \sin^2 \pi x$, $p = 2x$, $S = 4/D$, that is, $\delta\varphi_2$, is lower (approximately by a factor of 1.5) but γS is also much smaller. The transitions between structures with different short-range order in the arrangement of the ordered regions with different D corresponding to (2.5.19, 20) must be regarded as blurred phase transitions of the first order.

At a certain temperature T_0 we have

$$\Delta\varphi(\eta, p) = \psi(c_0, \eta_0, l) \equiv \varphi(c_0, \eta_0) - \varphi(c_0, 0) \quad (T = T_0) , \tag{2.5.22}$$

and then the alloy undergoes a transition to the homogeneous ordered state with the order parameter η_0. For systems with $p(T_0) \ll 1$ we find similar temperatures T_0 from the conventional condition of phase equilibrium (in the absence of a heterogeneous structure), $\varphi(c_0, \eta_0) = \varphi(c_0, 0)$, and from the condition (2.5.22). If $p(T_0) \sim 1$ the latter condition yields a noticeably lower T^0. For larger K_1, K_2, and K_3 the range $T^0 - T_0$ in which the heterogeneous phase occurs can be fairly wide and comparable with T_0. For instance, if we assume that the Fermi surface for the Cu_3Au alloy is the same as for copper we can obtain the estimates $v_0 K_1 \approx 0.01$ eV for $E_F = 7$ eV, $v_k = 1$ eV, $\delta k_z = 0.05 k_F$ (it is not possible to find a more exact estimate for v_k that is, for K_1, K_2, and K_3 because of the lack of data on the electron wave functions), and $v_0 K_2 f(x) \approx 0.02 \div 0.03$ eV for $x \approx 1/4$. (If we ignore the dependence of v_k on k the numerical results for copper-based alloys [2.84] indicate that the minimum of the electron energy of the concentration wave $w_e(\boldsymbol{k})$ treated as a function of $\boldsymbol{k}$ appears to be not very sharp and depends on both singular and nonsingular components $w_e'(\boldsymbol{k})$ and $w_e^0(\boldsymbol{k})$. Therefore, the equations (2.5.19, 20) for $\delta\varphi_2'$ derived under the assumption of a sharp minimum of $w_e(\boldsymbol{k})$ and the numerical results [2.84] yield only a rough estimate of K_2 (that is, $\delta\varphi_2'$). The sharpness and the position of the $w_e(\boldsymbol{k})$ minimum can be changed by the dependence of v_k on $\boldsymbol{k}$ which is determined by the fine features of the potential difference between atoms in the alloy and the wave functions of electrons which are not known well enough. Numerical calculations with the general equations (2.5.14) and sufficiently accurate v_k can yield more precise estimates for the variation of the electron energy caused by heterogenization. The above order-of-magnitude estimates may, however, still be valid.)

Using (2.5.14) and the estimate for $\partial w_e(G')/\partial G'$ given in [2.84] we obtain an estimate about 0.02 eV for $v_0 K_3$ which depends on the strains in the ordered regions, for instance, for $u_{ij} \approx -(0.03 \div 0.05)$. All these factors act in the same direction and in combination with the effects due to the inhomogeneities of $\eta(\boldsymbol{r})$ and $c(\boldsymbol{r})$ they can give $T^0 - T_0 \sim 1000$ K (we have $T_0 = 667$ K in Cu_3Au).

(In elastically isotropic crystals for the diagonal self-strain tensor L_{ij} the strain $u_{ij} = \Gamma(1/v)(\partial v/\partial c)(c - c_0)$ (here $\Gamma \approx 2/3$; see (18.1) in [1.1]) is determined only by the concentration variation $(c - c_0)$ (in the most ordered region in Cu_3Au a decrease in the Au concentration in the ordered regions brings G' and $2k_F$ together). In real elastically anisotropic crystals or when L_{ij} is nondiagonal the concentration variation in the disordered crystal at the boundary with the ordered region can give rise to considerable strain in the latter. In Cu_3Au a decrease in φ'_e is especially favoured by the production of regions with the $L1_0$ structure at the boundaries of regions with the $L1_2$ structure which gives rise to considerable strains of the required sign. Note that a significant redistribution of the concentration facilitating heterogenization can be especially easily achieved near the two-phase range in the phase diagram. When we take into consideration the elastic energy in elastically anisotropic crystals a nonequiaxial plate-like shape of the ordered regions becomes more favourable (see Sect. 2.1.2). The plate-like or needle-like shapes of the ordered regions can be caused also by the negative surface energy γ as well as by the electron energy for some structures of the ordered regions (for instance, $L1_0$) and (for $\varepsilon(k) - 1 \sim 1$; see below) for structures with significant screening effects.) At temperatures above T_0 the local order is not necessarily of the same type as for $T < T_0$. For example, in the ordered regions of the AB_3 alloy for small p the local order may be the same as that for the CuAu-type structure (with a larger $|\delta\varphi'_2|$ or a mixed system can be produced containing ordered regions of the type of Cu_3Au and CuAu. (In the CuAu alloy $|\delta\varphi'_2|$ grows because the potential energy amplitude for $\boldsymbol{k} = \boldsymbol{G}'$ in this case is twice that for the $L1_2$ structure and thus $|\delta\varphi'_2|$ is increased by a factor of 4/3. In addition, in CuAu where $\boldsymbol{G}'\ (110) < 2\boldsymbol{k}_F$ the superlattice vector $\boldsymbol{G}'$ is close to $2\boldsymbol{k}_F$ that is, to the minimum of $w(\boldsymbol{k})$, since the lattice parameter of CuAu is smaller than that of Cu_3Au (by about 3% for $\eta = 1$). The production of regions with the $L1_0$ structure at lower temperatures and with larger p is not favoured because of the loss of entropy which accompanies a redistribution of concentration.)

When p and η are sufficiently large the third-order terms in the expansion of $\delta\varphi'_2$ in powers of $v_{\boldsymbol{k}}/E_F$ may provide for the stability of the ordered regions with the DO_{22} (structure [2.162] which have the vectors $\boldsymbol{G}'$ of the type of (1 1/2 0) and (010) (a triangle of the vectors (1 1/2 0), (1 1/2 0) and (010) is almost inscribed into the Fermi surface of the copper-based alloys, see the discussion after (2.5.14)).

Small ordered regions in the heterogeneous state must give rise to some peculiar features in the diffuse scattering intensity distribution. They will be discussed in Sects. 3.3.2 and 3.5 in connection with the experimental results on scattering by the fcc alloys.

Heterogeneous States with ω-phase Particles in the bcc Alloys. Another illustration of heterogeneous states is given by bcc alloys undergoing a displacement-type phase transition giving rise to the trigonal ω-phase. Such phases are produced in alloys of the transition metals Ti, Zr, and Hf with the transition metals to the right of titanium in the periodic table, for instance, in Zr-Nb alloys, if the transition to the hexagonal low-temperature phase is suppressed by rapid cooling (see [2.149, 163]). The diffraction results [2.164, 165] demonstrate that the ω-phase is produced owing to the displacement of atomic (111) planes in the bcc lattice with respect to each other by the distance $u_0 \approx \sqrt{3}d/12$ while each third plane is fixed. This can be described by the static displacement waves [2.166]

$$\boldsymbol{u}_s = \boldsymbol{u}_0 \sin(\boldsymbol{G}' \cdot \boldsymbol{R}_s + \phi) , \quad \boldsymbol{u}_0 \parallel \boldsymbol{G}' = \left(\frac{2}{3}\frac{2}{3}\frac{2}{3}\right)\frac{2\pi}{d} . \tag{2.5.23}$$

Clearly we have four types of the ω phase corresponding to different directions $\{111\}$ of the vectors $\boldsymbol{u}_0$ and $\boldsymbol{G}'$ and three subtypes for each type which differ in the numbers of the undisplaced atomic planes $(3n + 0, 1, 2)$ with the phases $\phi = 0, 2\pi/3, 4\pi/3$. The phase transition involving the displacements described by (2.5.23) is a first-order transition [2.149, 167].

In some alloys, for instance, in Zr-Nb, a heterogeneous state comprising alternating regions of the ω- and bcc-phases is produced instead of the homogeneous ω-phase in a certain range of compositions (see Sect. 3.5). This state can be produced owing to the above nonlocal effects which have the quantum-mechanical nature and caused by the sharp dependence of the electron energy in the displacement wave $w_e(\boldsymbol{k})$ for $\boldsymbol{k} \approx \boldsymbol{G}'$ (this dependence was found in the numerical calculations for the system Zr-Nb [2.168, 169]).

We can make use of a simple model for the heterogeneous structure in which the amplitude u_0 is constant in the ω-phase regions in order to calculate the electron energy from (2.5.10, 13) replacing in them in $c_{\boldsymbol{k}}$ with the modulus of the Fourier component of the displacement $u_{\boldsymbol{k}}$, and $\xi\eta$, with u_0. Note that $v_{\boldsymbol{k}}(\boldsymbol{r}) \equiv v'_{\boldsymbol{k}}(\boldsymbol{r})$ here is given by (2.5.1) instead of (2.4.2). Accordingly, $\delta\varphi'_e$ is given by (2.5.14) where $\xi\eta$ is replaced with u_0 and $s(\boldsymbol{q})$ correspond to the nonequiaxial shape of the particles (since the symmetry is lower than cubic here).

We shall represent the heterogeneous structure as a lattice of ω-phase particles in which the unit cells and the particles are rectangular paralellepipeds with the edge lengths D_i and $D_{0i} = x_i D_i$. For $\delta w_2(\boldsymbol{q}) > 0$ (the point $\boldsymbol{G}+\boldsymbol{G}'$ coresponds to the negative curvature range of the function $w_e(\boldsymbol{k})$ we can reduce $\delta\varphi'_2$ in (2.5.14) to a minimum if along the axis $Z \parallel \boldsymbol{u}_0 \parallel [111]$ for a given type of the ω-phase we place alternating particles of different subtypes with the phases $\phi = 0, 2\pi/3, 4\pi/3$ (which are generally separated by regions with the bcc lattice). For such a lattice the only nonvanishing $p(\boldsymbol{q})$ have $\boldsymbol{q} = \boldsymbol{q}_n$ for which $\boldsymbol{q}_{nz} = 2\pi n_z/D_z$ (where $n_z = 3l + 1$ and l is an in integer).

Let us consider, for the sake of definiteness, systems in the bcc regions of which $\delta w'_2(\boldsymbol{q})$ exhibits a sharp peak of almost constant height W for a large part of the (111) plane in the $\boldsymbol{k}$ space which is at a small distance $\boldsymbol{k}^0$ from the point $\boldsymbol{G}'$ (directed away from the origin of coordinates). It can be associated with the flattened or cylindrical segments of the Fermi surface [2.168, 169]. The main contribution to $\delta\varphi'_2$ is made by two points $\boldsymbol{G}'$, (111) and (111) while for other $\{111\}$ $\boldsymbol{G}'$ we have a much smaller $(\boldsymbol{u}_0 \cdot \boldsymbol{G}')^2$ and, hence, $v_{G'}$ (see (2.5.1)). If we assume that a significant contribution to the integral (sum) in (2.5.14) is made only by the terms with n_z for which $\boldsymbol{q}_n$ is in the above plane (111) we find then that the minimum of $\delta\varphi'_2$ is at $D_z = 2\pi/k^0$ and it equals

$$\delta\varphi_2 = -K_2 u_0^2 p \frac{\sin^2 \pi x_z}{x_z}, \quad K_2 = \frac{3W}{4\pi^2 v_0}, \qquad (2.5.24)$$
$$x_z = \frac{D_{0z}}{D_z}, \quad D_z = \frac{2\pi}{k^0}.$$

The equations (2.5.14) for $\delta\varphi_1'$ and $\delta\varphi_3'$ and (2.5.24) for $\delta\varphi_2'$ determine $\delta\varphi_e'$ and thus $\Delta\varphi$, according to (2.5.12). If K_1, K_2, and K_3 are sufficiently large then $\Delta\varphi < 0$ and the heterogeneous state is more stable than the bcc phase. For transition metals $P(\boldsymbol{k})$ in (2.4.10) is typically large and $\varepsilon(\boldsymbol{k}) - 1 \sim 1$. Therefore, $|\tilde{w}'^{(2)}_{\text{ehet}}(\boldsymbol{G}')|$ can be much higher than $|\tilde{w}'^{(2)}_{\text{ehom}}(\boldsymbol{G}')|$ when the ω-phase particles are stretched along the [111] axis so that their thickness $D_{0\perp} \sim d$ is considerably smaller than the length $D_{0z} \approx 2\pi/3k^0$. The points $\boldsymbol{G}'$ (or $\boldsymbol{G}' + \boldsymbol{k}^0$) for such particles spread out to form disks with the width $\sim \pi D_{0\perp}$ so that the mean value of $\boldsymbol{k}^2 \approx G'^2 + k^2$ increases by a factor of 1.5–3 and in (2.4.10) we obtain a considerable decrease in $v_e(k) \sim k^{-2}$ for almost constant $P(\boldsymbol{k})$ and thus a significant increase in $|\tilde{w}'^{(2)}_{\text{ehet}}(\boldsymbol{G}')|$. This increase is even sharper for $|\tilde{w}'_{\text{ehet}}(\boldsymbol{G}')|$ because for small p we have $U_{\boldsymbol{k}\text{het}} < U_{\boldsymbol{k}\text{hom}}$. These factors can make K_1 and $|\delta\varphi_1'|$ in (2.5.14) large and, together with the term $\delta\varphi_2'$ given by (2.5.24), they stabilize the heterogeneous state containing elongated particles. Note that the perturbation theory is applicable for small $|2k_{\text{F}} - G'|$ when a condition similar to (2.5.16) is satisfied:

$$\frac{D_{0z}}{z} < D_0' \sim \frac{E_{\text{F}}}{u_0 v_{\boldsymbol{k}}'} d, \quad \text{if} \quad D_{0\perp}^2 > D_{0z} d \frac{E_m}{E_{\text{F}} \zeta \boldsymbol{k}},$$

or (2.5.24a)

$$D_{0\perp} < \left(\frac{E_m}{u_0 v_{\boldsymbol{k}}' \zeta_{\boldsymbol{k}}} \right)^{1/2} d, \quad \text{if} \quad D_{0\perp}^2 < D_{0z} d \frac{E_m}{E_{F\zeta \boldsymbol{k}}} \quad (\zeta_{\boldsymbol{k}} = S_{\boldsymbol{k}} d^2).$$

According to (2.5.24), $\delta\varphi'$ has a minimum for the largest $x_z = 1/3$, that is, for $D_{0z} = D_z/3$. But the particle dimensions D_{0z}, especially $D_{0\perp}$, and the fraction p of the particles are significantly limited by the dependences of $U_{\boldsymbol{k}\text{het}}$ and w' on p and $D_{0\perp}$ by the growth of Ψ with p (see (2.5.12)) caused by redistribution of the concentration (which is very strong since typically $dT_0/dc \sim 10^4$ K), and by the sharp increase in the surface energy on the sides of the particles caused by a decrease in distance between them. Owing to the combined effects of these factors the fraction p remains small (about 10^{-1}) near the temperature T^0 at which the heterogeneous state is produced. As the temperature reduces, Ψ decreases and therefore p and the "order parameter" u_0 grow until the transition to the homogeneous ω-phase occurs at $T = T_0$. As in the case of the substitutional ordering treated above, the temperature range $T^0 - T_0$ here can be wide (about 10^2–10^3 K) and the particles can be arranged in a short-range, rather than long-range, order. Systems with metastable heterogeneous states can also exist down to $T = 0$ K (that is, $T_0 < 0$) and in some systems the concentration is not redistributed during experimental observation.

As is the case in the systems with substitutional ordering, the heterogeneous state of alloys containing ω-phase particles exhibits a somewhat different structure if the point $\boldsymbol{G}'$ is near the minimum of the function $w_e'(\boldsymbol{k})$ where its plot has a positive curvature ($W < 0$). As noted above, under such conditions the heterogeneous state can be thermodynamically favourable at the expense of a decrease in the electron energy $\delta\varphi'$ of the small ordered regions (particles). In alloys of transition metals with $\varepsilon(\boldsymbol{k}) - 1 \approx 1$ an additional gain in the electron energy is due to the suppression of the Coulomb repulsion between electrons in small rod-shaped particles. For small p and $D_{0\perp}E_F > D_{0z}dE_m$ an expression for $\Delta\varphi$ is given by (2.5.20a) where D_0 is replaced with D_{0z} and γD^{-1} with $\gamma(D^{-1} + D_{0z})$ while the Coulomb interaction gives rise to a strong dependence of $k^{(0)}$ on the smallest transverse dimension of the particle $D_{0\perp}$. As it was found for (2.5.20a) we can show that below a temperature T^0 we obtain a heterogeneous structure containing a fraction p of the particles which is about a few tenths (it increases with a decrease in temperature) whose characteristic lengths are

$$D_z \sim (1-2)D_{0z}\,, \quad D_{0z} \sim 2D' \sim (2E_F/n_0v')d \quad D_{0\perp} < D_{0z} \qquad (2.5.24b)$$

For instance, for $E_F/u_0v')^{-1} \sim 10$ we obtain D_{0z} about 10^2 Å from (2.5.24b). In contrast to the case of $W > 0$, for $W < 0$ and a positive curvature of the $w_e'(\boldsymbol{k})$ plot, $\delta\varphi'$ is smaller the smaller the width of the $|u_{\boldsymbol{k}}|^2$ peak near the point G'. If the phases of the neighbouring particles are identical (that is, they belong to the same subtype of the ω-phase) the peak width decreases and the displacements $u(\boldsymbol{r})$ in the particles are not distributed uniformly but decrease with increasing distance from the centers.

The lack of correlation between particles of different subtypes of the ω-phase and the large particle length in the heterogeneous structure with $W < 0$ must produce considerably smaller shifts of the diffuse scattering intensity peaks from the $\boldsymbol{G}'$ points while the peak widths must be smaller than in the case of $W > 0$. Such narrow diffuse scattering intensity peaks without almost any shifts have been recorded for the hardened Ti-23 at.% V alloy [1.38] and for the Zr-Nb alloy with the Nb content below 17%. These peaks may be attributed to the heterogeneous structure discussed here.

2.5.3 Heterogeneous States in Semiconductor Systems

The heterogeneous states discussed above are produced under the effect of the long-range indirect interaction in metallic systems for $\boldsymbol{G}' \approx 2\boldsymbol{k}_F$. In single-valley semiconductors the Fermi sphere radius $\boldsymbol{k}_F$ is small (or the electron gas is not degenerate at all) and these conditions are not met. But semiconductor systems can give rise to heterogeneous states of a different type which involve redistribution of electrons and ions near the regions where the parameters of order or composition are changed and long-range Coulomb forces are thus generated.

Let us consider, for instance, a fully ionized impurity semiconductor in which a first-order phase transition can occur. Such transitions cause redistribution of impurities (free charges in the case under consideration) between the phases. Under such conditions in massive regions electrons and ions must migrate from one phase to another in identical numbers to prevent a build-up of the electrostatic energy. When thin regions of the second phase are produced, however, the Coulomb energy can be made small, especially in semiconductors with high static permittivities ε. Therefore, if the energy gains U and U' for electrons and ions migrating to the second phase differ considerably, for instance, if $U > 0$ and $U' < 0$ then it can be thermodynamically favourable for charges of the same sign (electrons for $U > U'$) to migrate preferentially to small regions of the phase II so that a heterogeneous state is produced which is not uniform in structure and charge concentration. The resulting decrease in the electron energy can be compensated with increases in the electrostatic energy and the boundary and volume energies of the crystal so that the heterogeneous state becomes stable in the temperature range where the phase II is not produced in the absence of electrons and ions. Therefore, such a heterogeneous phase can occur on the phase diagram between the phases I and II for considerable charge carrier concentrations. Similar heterogeneous states can be produced in solutions or systems with readily variable order parameters η if a considerable decrease in the energy of the system occurs when charges of one sign migrate to the regions with a higher concentration c or a higher order parameter η while charges of the opposite sign migrate to regions with lower c or η.

Such heterogeneous states were treated theoretically in [2.170–172]. Their structures depend significantly on the ratio between U and the Fermi energy E_{F} of the degenerate electron gas before heterogenization. If U is sufficiently large in comparison with E_{F} or $k_{\mathrm{B}}T$ for a nondegenerate electron gas while $U' < 0$ and $|U'|$ is also large then under heterogenization conditions almost all electrons migrate to the second-phase regions (or regions of a changed composition or order) and produce particles with metallic character, with a high electron density, while the ions remain in the first-phase regions forming a shielding atmosphere. In highly doped semiconductors where the charge carrier density is large the reversed condition $E_{\mathrm{F}} \gg U$ can be satisfied. The conditions for the production of heterogeneous states and their parameters are changed in this case while, for instance, the relative changes in the electron density must be small.

We can analyze the heterogeneous state in a system with a first-order phase transition in terms of a model in which the regions with the phase II structure and the surrounding regions with the phase I structure make up a lattice of identical "Wigner–Seitz cells". Then the variation $\Delta\varphi$ of the thermodynamic potential density caused by the transition from the homogeneous phase I to the heterogeneous state where the volume fraction of the phase II regions is p and the interface area is S can be written in the macroscopic approximation as

$$\Delta\varphi = p\psi + \gamma S - U\delta n - U'\delta n' + \delta E_k + k_{\mathrm{B}}T\int n'(\boldsymbol{r})\ln\frac{n'(\boldsymbol{r})}{n_0}d\boldsymbol{r}$$
$$+\frac{1}{2}e\int[n'(\boldsymbol{r}) - n(\boldsymbol{r})]V_p(\boldsymbol{r})d\boldsymbol{r}\ . \qquad (2.5.25)$$

Here Ψ is the difference between the thermodynamic potential densities φ for the phases II and I for the uniform mean density of electrons and ions n_0, δn and $\delta n'$ are the variations of the electron and ion densities in phase II regions, δE_k is the variation of the kinetic, exchange and correlation energies of electrons and their entropy contribution to φ caused by heterogenization, $n(\boldsymbol{r})$ and $n'(\boldsymbol{r})$ are the densities of electrons and ions, respectively, $V_p(\boldsymbol{r})$ is the potential ($e > 0$), and the last term in (2.5.25) gives the electrostatic energy. Since the electron density inside the particles with the phase II structure is large, $n(\boldsymbol{r})$ in these regions and the respective contribution $\delta E_k^{(2)}$ to the kinetic energy can be treated in the quasiclassical approximation including the quantum-mechanical correction terms in the surface energy γ (which is regarded as a parameter of the theory):

$$n(\boldsymbol{r}) = \frac{1}{3\pi^2}\left[\frac{2m}{\hbar^2}\xi(\boldsymbol{r})\right]^{3/2} , \quad \delta E_k^{(2)} = \frac{3}{5}\int_{\mathrm{II}} n(\boldsymbol{r})\xi(\boldsymbol{r})d\boldsymbol{r} , \qquad (2.5.26)$$

$$\xi(\boldsymbol{r}) = \mu - U + eV_p(\boldsymbol{r}) .$$

Here m is the effective electron mass and μ is the chemical potential of electrons taken from the bottom of the conduction band of the initial phase I (as the energy $-U$ is taken from the bottom of the conduction band of the phase II).

The potential distribution $U_p(\boldsymbol{r})$ is described by the Poisson equation which reduces to the Fermi-Thomas equation inside particles of a sufficiently large size. When we find the solutions of these equations we can obtain $V_p(\boldsymbol{r})$, $n(\boldsymbol{r})$, $n'(\boldsymbol{r})$ and then $\Delta\varphi$. Minimization of the expression for $\Delta\varphi$ gives then the equilibrium parameters of the heterogeneous state, that is, the particle thickness D_0, the particle volume fraction p, and the mean density n^0 of electrons in the particles while the condition that $\Delta\varphi < 0$ yields the range of the heterogeneous state in the phase diagram. The thickness D_0 for a given p is determined by the ratio between the last term in (2.5.25) which grows with an increase in D_0 and the term $\gamma S \sim \gamma p/D_0$. A decrease in p reduces the positive term $p\Psi$ but for small p we have an increase in n^0 so that, according to (2.5.26), we obtain an increase in μ and the kinetic electron energy. The balance of these factors determines the equilibrium parameters of the system.

Let us consider first the case of sufficiently large $U \gg E_{\mathrm{F}}$ and $U' < 0$ in which practically all electrons are concentrated in the particles while the ions remain in the ambient medium and the resulting structures exhibit sharp inhomogeneities (in the electron concentration among other parameters) [2.170, 171]. The behavior of such systems depends significantly on whether ions have a mobility sufficiently high for the system to reach full equilibrium or whether they are "frozen" at the given temperature (then we have $n'(\boldsymbol{r}) = n_0$). An important part is also played by the dimensionless parameter γ_0 (see (2.5.29)) which describes the contribution of the surface energy. Within the framework of the above model the plate-like shape of the particles is thermodynamically favorable under the full equilibrium while the parameters of the system in various cases are given by

$$D_0 = 0.84 \left(\frac{\Delta E}{\psi v}\right)^{2/5} \left(\frac{\gamma v^{2/3}}{E_c}\right)^{1/3} v^{1/3} , \quad n^0 = 1.77 \left(\frac{\psi v}{\Delta E}\right)^{3/5} \frac{1}{v} , \tag{2.5.27}$$

$$P = \frac{n_0}{n^0} \quad (\gamma_0 < 1) ,$$

$$D_0 = 2.53 \left[1 - 0.757\gamma_0^{-\frac{1}{10}}\right] \left(\frac{\Delta E}{E_c}\right)^{1/2} \left(\frac{\Delta E}{\psi v}\right)^{1/10} v^{1/3} ,$$

$$n^0 = 0.68 \left[1 - 0.757\gamma_0^{-\frac{1}{10}}\right] \left(\frac{E_c}{\Delta E}\right)^{5/6} \left(\frac{\gamma v^{2/3}}{E_c}\right)^{5/9} \left(\frac{\psi v}{\Delta E}\right)^{1/10} \frac{1}{v} , \tag{2.5.28}$$

$$p = \frac{n_0}{n^0} \quad (\gamma_0 \gg 1) .$$

Here we have

$$\gamma_0 = \left(\frac{E_c}{\Delta E}\right)^{2/3} \left(\frac{\gamma v^{2/3}}{E_c}\right)^{1/9} \frac{\gamma}{\psi v^{1/2}} ;$$
$$\Delta E = \frac{5\hbar^2}{m v^{2/3}} ; \quad E_c = \frac{e^2}{\varepsilon_1 v^{1/3}} , \tag{2.5.29}$$

where ΔE is of the order of the conduction band width and E_c describes the Coulomb energy ($\varepsilon_1$and ε_2 are the permittivities of the phases I and II).

In a system without electrons and ions we have $\psi = q_0(T - T_0)/T$ near the phase transition temperature T_0 where q_0 is the specific heat of the transition, that is, $\psi = 0$ for $T = T_0$. Therefore, for small differences $T - T_0$ the heterogeneous state is described by equations (2.5.28) and at temperatures considerably higher than T_0 by the equations (2.5.27) for small γ. For example, for the typical parameters $v = 10^{-22}\,\mathrm{cm}^{-3}$, $\Delta E = 4\,\mathrm{eV}$, $\varepsilon_1 = 20$, $E_c = 1.5 \times 10^{-2}\,\mathrm{eV}$, we have $q_0 v = 2.5 \times 10^{-2}\,\mathrm{eV}$ for $T - T_0 = T_0/30$ and $\gamma_0 = 3.7$ and we obtain from (2.5.28) $D_0 = 50\,\mathrm{Å}$, $vn^0 = 7.7 \times 10^{-2}$ while for $T - T_0 = T_0$ and $\gamma_0 = 0.12$ we obtain from (2.5.27) the particle width $D_0 = 35\,\mathrm{Å}$ and $vn_0 = 8.6 \times 10^{-2}$. We see that heterogenization in a semiconductor crystal gives rise to particles with a high electron density (which is lower only by one or two orders of magnitude than that in metals with a good conductivity) and the sizes between 10 and 100 Å. For n_0 between about 10^{18} and $10^{20}\,\mathrm{cm}^{-3}$ their volume fraction p varies between 10^{-3} and 10^{-1}. The size and concentration of the particles increase with a decrease in temperature.

At high temperatures or very low electron concentrations $\Delta\varphi > 0$ and no heterogeneous state is produced. This state is thermodynamically favourable at temperatures below T^0 at which the following condition is satisfied

$$\frac{0.43}{\chi[\gamma_0]^{5/2}} \left(\frac{\tilde{U}}{\Delta E}\right)^{3/2} \frac{\tilde{U}}{v q_0} = \frac{\psi(T^0)}{q_0} \approx \frac{t^0 - T_0}{T^0} . \tag{2.5.30}$$

Here the approximate equation is valid in the range where $T^0\psi(T^0) = q_0(T^0 - T_0)$ and

$$\tilde{U} = U + U_1 - 2k_\mathrm{B}T \ln \frac{D_0}{2r_\mathrm{D}p} \sim U\,, \quad r_\mathrm{D}^2 = \frac{\varepsilon_2 k_\mathrm{B}T}{8\pi e^2 n_0}\,,$$

$$U_1 = \frac{3}{5}E_\mathrm{F}(E_\mathrm{F} \gg k_\mathrm{B}T)\,, \quad U_1 = \ln\left[\frac{1}{2}n_0\left(\frac{2\pi\hbar^2}{mk_\mathrm{B}T}\right)^{3/2}\right] \quad (E_\mathrm{F} < k_\mathrm{B}T)\,, \tag{2.5.31}$$

$$\chi(\gamma_0) = 1 + 1.4\gamma_0^{3/5} \quad (\gamma_0 < 1)\,, \quad \chi(\gamma_0) = 1.86\gamma_0^{2/5} \quad (\gamma_0 \gg 1)\,.$$

The temperature range $T^0 - T_0$ of the heterogeneous state for large U may be fairly large; for instance, for $U = 1\,\mathrm{eV}$ and the above parameters we have $T^0 - T_0 = T_0$. The heterogeneous state obviously can exist only for not too high surface energies γ (even in a narrow temperature range $T^0 - T_0$). Using (2.5.30) and the equations (2.5.29, 31) for γ_0 and $\chi(\gamma_0)$, we see that this is the case under the condition $8.6(\Delta E)^{3/4}E^{1/2}\gamma v^{2/3} < U^{9/4}$, that is, for the above parameters $\gamma < 250\,\mathrm{erg/cm^2}$ for $U = 1\,\mathrm{eV}$ or $\gamma < 17\,\mathrm{erg/cm^2}$ for $U = 0.3\,\mathrm{eV}$. At temperatures below T_0 no heterogeneous state is produced for large U. In the framework of the above model the transition to the heterogeneous state occurs continuously (with a logarithmic singularity in $p(T)$). The first to be produced are extremely distant particles with a low electron density (so that the equations (2.5.28, 31) are not applicable in a small temperature range of δT of the order of $(T^0 - T_0)k_\mathrm{B}T/U$). But when we take into consideration some factors acting in real crystals, such as the Van-der-Waals interaction, heterogenization must be regarded as a first-order phase transition.

A similar heterogeneous state can be produced in solutions with a sufficiently large constant A in the equation $V(\boldsymbol{r}) = A[c(\boldsymbol{r}) - c_0]$ describing the electron energy as a function of the concentration or in the ordering systems if the function $V(\boldsymbol{r}) = A\delta\eta(\boldsymbol{r})$ exhibits a strong dependence on $\delta\eta(\boldsymbol{r})$. In such systems we obtain regions with sharply changed composition or order parameters. These regions have sharply defined boundaries where the conduction electrons are concentrated and the parameters of the resulting structure are given by (2.5.27–31) where U must be replaced with Ac_0 (for $A > 0$) or with $-A(1 - c_0)$ (for $A < 0$) and ψ must be replaced with $c_0[\partial\varphi(c_0)/\partial c_0] - \varphi(c_0)$ [2.170, 171]. Heterogenization is favored in the decomposing solutions, especially those that are near the critical point [2.171, 173, 174]. The heterogeneous states in systems with the "frozen" ions are described by similar equations with other numerical coefficients. But in this case, for small n_0 and $p < 10^{-1}$, the particles must apparently have a spherical or rod-like, rather than plate-like, shape (for $p \sim 1$ the metallic particles are coming into contact and make up a unified skeleton which manifests itself as an insulator-metal transition).

In the case of a high charge carrier concentration when we have $E_\mathrm{F} > U$ in a homogeneous semiconductor, production of second-phase regions can clearly cause only a relatively small redistribution of the electron density. It can be, however, sufficient for stabilization of the heterogeneous state which contains alternating plate-like particles of the phases I and II, with thicknesses D_1, and D_2 if the surface energy is small [2.172].

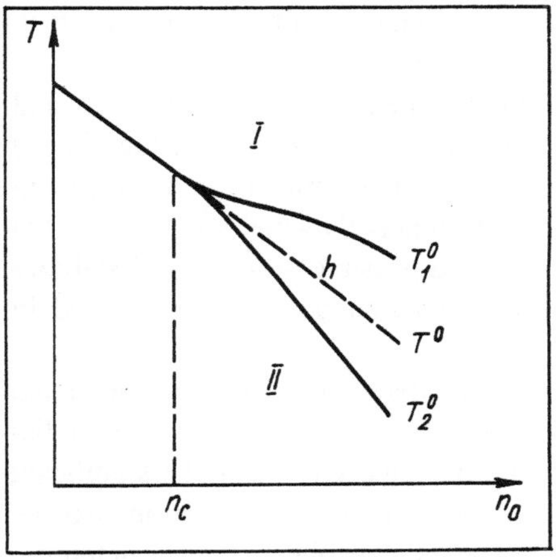

Fig. 2.6. Phase diagram for a semiconductor system with a first-order phase transition in the case of fixed ions and $U \ll E_F$. The dashed line T^0 continues the line T_0 of the equilibrium between the homogeneous phases I and II. The range of heterogeneous states for $n_0 > n_c$ is denoted by h. The temperatures of the transitions between the phases I and II and the heterogeneous phases, respectively, are T_1^0 and T_2^0

Using the quasiclassical approximation we can rewrite the equation (2.5.25) for $\Delta\varphi$ for the system with "frozen" ions in the form

$$\Delta\varphi = \frac{D_2}{D}\psi + \frac{2\gamma}{D} - \frac{U^2}{2\mu_1}\frac{\sinh\varkappa D_1 \sinh\varkappa D_2}{\varkappa D \sinh\varkappa D}$$

$$\mu_1 = \pi\left(\frac{\pi}{3}\right)^{1/3}(a^3 n_0)^{2/3}E_0\,, \qquad (2.5.32)$$

$$E_0 = \frac{me^4}{\hbar^2\varepsilon^2}\,, \quad a = \frac{\hbar^2\varepsilon}{m e^2}\,, \quad \chi^{-1} = \left(\frac{\pi}{3}\right)^{1/6}\frac{a}{(a^3 n_0)^{1/6}}\,, \quad D = D_1 + D_2\,,$$

where a is the Bohr radius and $\varepsilon_1 = \varepsilon_2 = \varepsilon$. According to (2.5.32), we have $\Delta\varphi < 0$ and the heterogeneous state is produced at temperatures both above and below T_0 if the charge carrier concentration n_0 is higher than the critical concentration n_c where

$$n_c = \frac{\pi}{3}\left(\frac{8\pi a^2 E_0}{U^2}\right)^6 a^{-3}\,. \qquad (2.5.33)$$

Figure 2.6 shows the phase diagram for this system. In the quasiclassical approximation for the electron gas we obtain $T^0 - T_0 = T_0 - T^0 \approx 0.03(u^2 n_0/E_0 q_0)(a^3 n_0)^{-2/3}$ for $n_0 \approx 2n_c$ and $D_1 \approx D_2 \approx 1.5a(a^3 n_0)^{-1/6}$ for $T = T_0$ and $n_0 \approx 2n_c$. For instance, for $\varepsilon m_0/m = 30$ (where m_0 is the free electron mass), $T = 300\,\mathrm{K}$, $\gamma = 25\,\mathrm{erg/cm^2}$ and $U = 0.5\,\mathrm{eV}$ we have from (2.5.33) $n_c = 0.5 \times 10^{21}\,\mathrm{cm^{-3}}$ and for $n_0 = 0.7 \times 10^{21}\,\mathrm{cm^{-3}}$ we have $D_1 = 30\,\mathrm{Å}$. If the particle thickness D is small (of the order of k^{-1}) then the quasiclassical approximation is no longer applicable and the heterogeneous structure exhibits characteristic features determined by the sharp variation of the permittivity of the electron gas for $\boldsymbol{k} \approx \boldsymbol{k}_F$ (see (2.4.10) and [2.172]). Note that for large D_1 and D_2 the last term in (2.5.32) can be regarded as a negative electron contribution γ_e to the surface energy determined by redistribution of electrons and giving rise to heterogenization for $\gamma + \gamma_e < 0$. In this connection heterogenization may occur in ferroelectrics [2.175].

In semiconductor solutions, and in ordering systems with "frozen" ions and $U < E_F$, states with relatively small variations $\delta c(\boldsymbol{r})$ or $\delta\eta(\boldsymbol{r})$ (as well as $\delta n(\boldsymbol{r})$) are produced near the heterogenization point T^0. Their structure is discussed in [2.173] (the condition for stability with respect to relatively small variations $\delta n(\boldsymbol{r})$ and $\delta\eta(\boldsymbol{r})$ in such systems is treated in [2.176] and possible violations of stability in the electron-phonon system of conventional semiconductors are discussed in [2.177]).

If the mobility of ions is sufficiently high and heterogenization is accompanied by a redistribution of both electrons and ions then for large $E_F \gg U$, when redistribution of electrons in a highly degenerate electron gas can hardly be significant for sufficiently large $|U'| \gg k_B T$, the main cause for heterogenization may be a gain in the energy of ions, rather than electrons. We obtain then a heterogeneous state with a highly nonuniform distribution of ions and an almost uniform distribution of electrons (details see in [2.172, 173]).

3. Diffuse Scattering of X-Rays and Neutrons in Solid Solutions and One-Component Ordered Crystals

Diffraction studies of diffuse scattering by systems at thermodynamic equilibrium are most widely employed for analyzing the short-range order, the interactions between atoms, and other parameters of solid solutions. These techniques are also increasingly used for analyzing structural details in one-component crystals with orientational disorder or displacement-type disorder.

The intensity distribution for scattering of X-rays or thermal neutrons by a perfect crystal with atoms occupying only lattice sites exhibits sharp peaks of regular Laue-Bragg reflections. This distribution is significantly changed by imperfections which are always found in real crystals. Imperfections of various types give rise to a continuous diffuse scattering intensity distributions corresponding to arbitrary scattering angles (which differ from the Bragg angles) and to attenuation of the regular reflection peak intensities (and to shifts of their maxima) without changing the shapes and widths of these peaks, or to broadening of the peaks.

The nature of the diffraction effects produced in imperfect crystals depends on the asymptotic behaviour of the distortion fields around the defects producing the imperfections [3.1]. For instance, near point defects and other defects bounded in all three dimensions the displacements decrease fairly rapidly as $1/r^2$ and therefore these defects do not produce broadening of regular reflection peaks but only decrease their intensity and give rise to the diffuse scattering intensity distribution (see details in Sect. 1.5 of [1.1].

Both a solid solution and a not-fully-ordered one-component crystal with the orientational disorder or displacement-type disorder can be represented as a perfect crystal made up of "average" atoms. Imperfections, that is, changes in the atomic scattering factors and the static displacements caused by the differences between real atoms in the solution and the "average" atoms, are superimposed on it. The static distortions in such crystals can be regarded as being caused by superposition of the contributions of individual point defects (or molecule-type defects) which describe the local deviations of the real crystal from the "average" one. Therefore, as indicated by the above general results, the scattering intensity distribution for a solid solution or a not-fully-ordered crystal exhibits regular reflection peaks of the same widths. The only diffraction effects that are in evidence are the line shifts in the X-ray or neutron diffraction patterns, attenuation of the line intensities and the diffuse scattering by the static inhomogeneities in the imperfect crystal. The first two effects are discussed in Sects. 3.1.3,4 in [1.1] and the diffuse scattering is treated in this chapter.

As can be seen from Chaps. 1 and 2, the inhomogeneities caused by the differences between atoms of different species occupying lattice sites in a solid solution as well as the static displacements produced by them can be conveniently treated as superpositions of the fluctuation waves of the concentration, the long-range order parameters, the correlation parameters and the respective waves of static displacements. In a similar way, the inhomogeneities in a not-fully-ordered one-component crystal are described by the fluctuation waves of the parameters of long-range order and correlation. As noted above, such a description is particularly convenient for calculating the intensities of diffuse scattering of X-rays or thermal neutrons. Therefore, to analyze diffuse scattering in solid solutions and one-component ordered systems in this chapter we shall employ mostly the method of fluctuation waves and some results of the fluctuation theory given in Chaps. 1 and 2.

To begin with Sects. 3.1,2 give the general results on the theory of diffuse scattering and, in particular, a procedure for separating the effects caused by the difference between the atomic scattering factors and by the static displacements of the solution atoms.

A more detailed treatment of diffuse scattering intensity distribution is given in Sect. 3.3 for undistorted solid solutions and in Sect. 3.4 for solutions with significant static displacements of atoms. The diffuse scattering data yield information on the correlation parameters, ordering energies, Fermi surfaces for metal solutions, static distortions, and short-range order structure which is analyzed in these sections. Experimental diffraction studies of these characteristics for some crystals are described in Sect. 3.5. Diffuse scattering by long-range order fluctuations is treated in Sect. 3.6. Section 3.7 presents the results on the scattering features caused by specific correlations (Debye screening) produced by Coulomb forces in the ionic crystals containing impurities of a different valence.

3.1 Diffuse Scattering Intensity in the Kinematical Theory

The most productive diffraction technique for obtaining information on crystal imperfections is the one using diffuse scattering of monochromatic radiation by single crystals. As shown schematically in Fig. 3.1 the crystal can be rotated around two axes and the scattering angle 2θ can be varied by moving the detector (A) crystal. The directions of the wave vectors $\boldsymbol{k}_1$ and $\boldsymbol{k}_2$ of the incident and scattered waves can be varied with respect to the crystal axes. The diffraction vector

$$\boldsymbol{Q} = \boldsymbol{k}_2 - \boldsymbol{k}_1 \tag{3.1.1}$$

is accordingly arbitrarily aligned with respect to the crystal axes while its length

$$Q = \frac{4\pi}{\lambda} \sin\theta \tag{3.1.2}$$

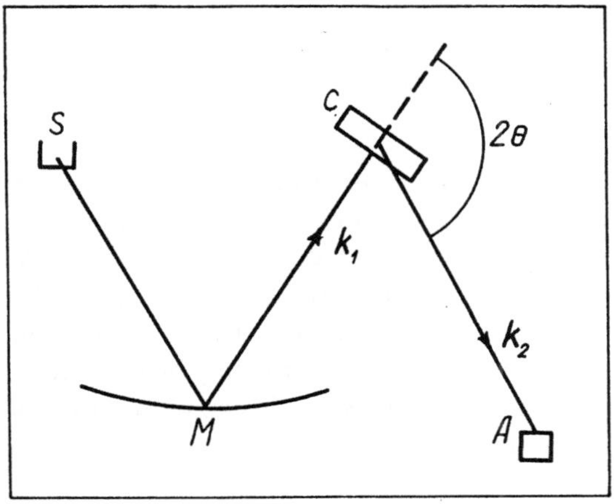

Fig. 3.1. Schematic of the diffraction technique for analyzing diffuse scattering of monochromatic radiation by a single crystal. S is the source of X-rays or neutrons, M is the monochromator, K is the specimen, A is the analyzer (detector). The wave vectors of the incident and scattered waves are $\boldsymbol{k}_1$ and $\boldsymbol{k}_2$ and the scattering angle is 2θ

varies in a wide range which expands with decreasing radiation wavelength $\lambda (k_1 = k_2 = 2\pi/\lambda)$.

A rigourous treatment of diffraction of X-rays or thermal neutrons must make use of the dynamical theory based directly on the Maxwell or Schrödinger equations to describe the general wave field produced by interference of the incident and scattered waves (and multiple scattering processes). But if the size of the crystal or a "region of coherent scattering" in an imperfect crystal is small then we can analyze diffraction within the simpler approximation of the kinematical theory of scattering in which a crystal is regarded as a small perturbation producing scattering of the incident wave (see, for instance, [3.2–7, 29] and Sect. 1.2 in [1.1]). The theory ignores some attenuation of the incident wave intensity as it passes through a small (or imperfect) crystal, as well as the effects of multiple scattering.

The kinematical theory is applicable to all diffraction effects if the crystallite size is small in comparison with the characteristic extinction length so that the intensity of the wave diffracted at the Bragg angle is small in comparison with the intensity of the incident wave. The intensity of diffuse scattering at an arbitrary angle is considerably smaller than the Bragg reflection intensity and thus the applicability conditions for the kinematical theory are almost always satisfied. It is only for some special orientations and highly perfect large crystals that diffuse scattering gives rise to a two-wave field and effects of the dynamical theory become significant. Therefore we shall employ the kinematic theory here for analyzing diffuse scattering.

If the monochromatic X-ray wave with the wave vector $\boldsymbol{k}_1$ is incident on the atom $s\gamma$ which is at the position $\boldsymbol{R}_{s\gamma} + \boldsymbol{u}_{s\gamma}$ (where s is the number of the unit cell, γ is the number of the site in the cell, $\boldsymbol{R}_{s\gamma}$ is the position of the site $s\gamma$ in the perfect lattice, and $\boldsymbol{u}_{s\gamma}$ is the displacement of the atom from this site in the imperfect crystal) then the amplitude of the scattered wave with the wave vector $\boldsymbol{k}_2$ is proportional to the Fourier component $\varrho(\boldsymbol{r} - \boldsymbol{R}_{s\gamma} - \boldsymbol{u}_{s\gamma})$ of the probability density for electrons of this atom:

$$a_{s\gamma} = \int \varrho_{s\gamma}(\boldsymbol{r} - \boldsymbol{R}_{s\gamma} - \boldsymbol{u}_{s\gamma}) \exp[\mathrm{i}(\boldsymbol{k}_2 - \boldsymbol{k}_1)\cdot \boldsymbol{r}]d\boldsymbol{r} = \int \varrho_{s\gamma}(\boldsymbol{r} - \boldsymbol{R}_{s\gamma} - \boldsymbol{u}_{s\gamma}) \times \exp[\mathrm{i}\boldsymbol{Q}\cdot(\boldsymbol{r} - \boldsymbol{R}_{s\gamma} - \boldsymbol{u}_{s\gamma})]d\boldsymbol{r} \exp[\mathrm{i}\boldsymbol{Q}\cdot(\boldsymbol{R}_{s\gamma} + \boldsymbol{u}_{s\gamma})] \,. \tag{3.1.3}$$

We have used here (3.1.1).

The distribution $\varrho_{s\gamma}(\boldsymbol{r} - \boldsymbol{R}_{s\gamma} - \boldsymbol{u}_{s\gamma})$ generally depends on the type of the atom $s\gamma$ and on the type of the neighbouring atoms which distort the wave functions of the outer electrons of the given atom (and on the displacements of atoms, the concentrations and the order parameters). In most cases, however, (with the exception of crystals comprised of light atoms) the resulting corrections are small and the distribution $\varrho_{s\gamma}$ to a good approximation is determined only by the species of the atom (A, B, ...) occupying the site $s\gamma$. Then (3.1.3) is expressed in terms of the atomic factor $f_{s\gamma} = f_A, f_B, \ldots$ of this atom:

$$a_{s\gamma} = f_{s\gamma} \exp[\mathrm{i}\boldsymbol{Q}\cdot(\boldsymbol{R}_{s\gamma} + \boldsymbol{u}_{s\gamma})] \,, \quad f_{s\gamma} = \int \varrho_{s\gamma}(\boldsymbol{r}') \exp(\mathrm{i}\boldsymbol{Q}\cdot\boldsymbol{r}')d\boldsymbol{r}' \,. \tag{3.1.4}$$

For $Q \to 0$ the parameters f are clearly equal to the numbers of electrons in the respective atoms and they have been tabulated for finite Q (see, for example, [1.29]).

In the kinematical theory we can ignore attenuation of the incident wave (by scattering) as it passes through the crystal and therefore at large (in comparison with crystal size) distances the amplitude of scattering of a monochromatic plane wave by all the atoms of the crystal is proportional to the sum of the parameters (3.1.4) for scattering by individual atoms:

$$a = \sum_{s=1}^{N} \sum_{\gamma=1}^{\nu} f_{s\gamma} \exp[\mathrm{i}\boldsymbol{Q}\cdot(\boldsymbol{R}_{s\gamma} + \boldsymbol{u}_{s\gamma})] \,. \tag{3.1.5}$$

It is convenient to express the scattering intensity I for X-rays in terms of electron units. An electron unit is the intensity of scattering by a classical electron and I is found as the ratio of the intensity by a crystal to the intensity of scattering by the classical electron. The parameter (3.1.5) expressed in terms of these units is explicitly the amplitude of scattering of X-rays by the crystal (rather than a function proportional to it) and the scattering intensity $I = |a|^2$ has the following form in the kinematical theory:

$$I = I(\boldsymbol{Q}) = \left\langle \left| \sum_{s=1}^{N} \sum_{\gamma=1}^{\nu} f_{s\gamma} \exp(\mathrm{i}\boldsymbol{Q}\cdot\boldsymbol{R}_{s\gamma}) \exp(\mathrm{i}\boldsymbol{Q}\cdot\boldsymbol{u}_{s\gamma}) \right|^2 \right\rangle \,. \tag{3.1.6}$$

Here the angle brackets denote averaging over the statistical ensemble corresponding to various detailed distributions of atoms (for the given macroscopic parameters describing the composition, order and correlation in the crystal) and to various thermal displacements. We have also taken into account the fact that when small terms (of the order of $1/\sqrt{N}$) are ignored the scattering intensity for the crystal at equilibrium coincides with the intensity averaged over the ensemble (which can be theoretically analyzed in the framework of the statistical physics).

The transition from the conventional scattering intensities recorded in experiments to the intensities expressed in terms of electron units involves taking into consideration some factors which describe the (photo-electric) absorption of X-rays in the crystal, their polarization, the Compton scattering and so on. These are discussed in detail (as well as the details of the specific experimental diffraction techniques) in a number of general textbooks on diffraction (see, for instance [3.4–7; 1.29]).

The scattering intensity (3.1.6) expressed in terms of electron units depends only on the difference $\boldsymbol{Q}$ between the wave vectors $\boldsymbol{k}_2$ and $\boldsymbol{k}_1$ (rather than the wave vectors directly). It can be conveniently treated as a function of the diffraction vector $\boldsymbol{Q}$ in the reciprocal lattice space. A general analysis of scattering by an imperfect crystal containing finite defects (in particular, solid solutions discussed here or not-fully-ordered one-component crystals) demonstrates that in the limit of the infinite crystal the scattering intensity can be divided into the component I_0 including delta functions and describing the Bragg peaks and the component I_1 describing the continuous intensity distribution (see Sect. 1.5 in [1.1]).

The peaks of the function $I(\boldsymbol{Q})$ are at the reciprocal lattice sites of the average perfect crystal, that is, at

$$\boldsymbol{Q} = \boldsymbol{G} \tag{3.1.7}$$

where $\boldsymbol{G}$ are vectors of this reciprocal lattice. The peaks are produced by interference of the radiation scattered by atoms $s\gamma$ and $s'\gamma'$ which are at a large distance of the order of the crystal size. Occupation of such distant sites by atoms of different species and their static displacements are apparently uncorrelated and therefore when we calculate I_0 we can take the product of averaged individual factors instead of the average over the products in (3.1.6). Then we obtain

$$I_0 = \left| \sum_{s=1}^{N} \sum_{\gamma=1}^{\nu} f_\gamma \mathrm{e}^{\mathrm{i}\boldsymbol{Q}\cdot\boldsymbol{R}_\gamma} \mathrm{e}^{\mathrm{i}\boldsymbol{Q}\cdot\boldsymbol{R}_s} \right|^2 = 8\pi^3 \frac{N}{v} |f|^2 \delta(\boldsymbol{Q} - \boldsymbol{G}) \,. \tag{3.1.8}$$

We have used here (1.3.15) in [1.1], $\boldsymbol{R}_{s\gamma} = \boldsymbol{R}_s + \boldsymbol{R}_\gamma$, $\boldsymbol{R}_\gamma$ is the radius vector of the site γ in the first unit cell (for $\boldsymbol{R}_s = 0$), $\boldsymbol{R}_s$ is the radius-vector of the first site (or a conventional center of the sth cell, v is the unit cell volume, and

$$f_\gamma = \langle f_{s\gamma} \exp(\mathrm{i}\boldsymbol{Q}\cdot\boldsymbol{u}_{s\gamma})\rangle_\gamma \,, \quad f = \sum_{\gamma=1}^{\nu} f_\gamma \exp(i\boldsymbol{Q}\cdot\boldsymbol{R}_\gamma) \,. \tag{3.1.9}$$

Averaging above is performed over all atoms in the sublattice γ. In a binary A-B solution, $f_{s\gamma}$ may have two values f_A and f_B (if the dependence on the environment is neglected). Then we can rewrite (3.1.9) as

$$f_\gamma = c_\gamma f_A \mathrm{e}^{-M_{A\gamma}} + (1 - c_\gamma) f_B \mathrm{e}^{-M_{B\gamma}} \,, \tag{3.1.10}$$

where c_γ is the concentration of atoms A in the sublattice, $\exp(-M_{B\gamma})$ and $\exp(-M_{A\gamma})$ are the Debye-Waller factors describing attenuation of the scattering amplitudes for the atoms A and B in the sublattice γ caused by displacement of atoms from the sites of the averaged lattice. These factors are given by

$$\begin{aligned} \exp(-M_{A\gamma}) &= \langle \exp(\mathrm{i}\boldsymbol{Q}\cdot\boldsymbol{u}_{s\gamma})\rangle_{A\gamma}\,, \\ \exp(-M_{B\gamma}) &= \langle(\exp(\mathrm{i}\boldsymbol{Q}\cdot\boldsymbol{u}_{s\gamma})_{B\gamma}\rangle\,, \end{aligned} \tag{3.1.11}$$

where averaging is performed only over atoms A or B which are in the sublattice γ (rather than over all atoms in this sublattice as was done in (3.1.9).

The scattering intensity (3.1.8) vanishes outside small vicinities of the reciprocal lattice sites (in the limit of an infinite crystal and there remains only the diffuse scattering intensity I_1 determined by crystal imperfections. It is determined by the general equation (3.1.6) for $\boldsymbol{Q} \neq \boldsymbol{G}$.

Equations similar to (3.1.6) describe also the cross section for nuclear scattering of thermal neutrons (see, for instance, Sects. 1.2 and 3.2 in [1.13], Sect. 1.2 in [1.1] or [3.8–11]). We can describe scattering of neutrons by the nuclei of each element $\mathfrak{a}$ in the crystal by the amplitude $b_\mathfrak{a}$ averaged over the isotopes and the cross section $\sigma_{\mathfrak{a}c} = 4\pi b^2$ for coherent scattering of neutrons and the cross section σ for incoherent scattering. (These cross sections are tabulated for come elements and isotopes, for instance, in [1.1; 3.10, 11].) Under the conditions of incoherent scattering, due to a completely random arrangement of the isotopes of a given element and possible change in the spin state of the nuclei under scattering, no interference occurs between the waves produced by various nuclei and their contributions to the scattering intensity are additive. Under the conditions of coherent scattering the interference is significant and the amplitudes must be added with the appropriate phase factors as is done in equation (3.1.6) for X-rays.

In contrast to X-ray photons, neutrons have an energy which is comparable to the phonon energy and thus the neutron energy may be considerably changed by scattering. The most comprehensive characteristic of neutron scattering is the differential cross section $\sigma(\boldsymbol{Q}E)$ per unit solid angle (for a given variation of the neutron momentum $\hbar\boldsymbol{Q} = \hbar\boldsymbol{k}_2 - \hbar\boldsymbol{k}_1$) per unit transferred neutron energy E. In some cases, however, it is enough to consider the integrated cross section over energies (integrated over E) and the differential cross section with respect to the scattering angles, $\sigma(\boldsymbol{Q})$. If the energy of the incident neutrons is high enough and exceeds significantly the energy variation caused by scattering then the total cross section of coherent ($\sigma_c(\boldsymbol{Q})$) and incoherent ($\sigma_i(\boldsymbol{Q})$) scattering of neutrons by the system of nuclei in a crystal is given by

$$\begin{aligned} \sigma(\boldsymbol{Q}) &= \sigma_c(\boldsymbol{Q}) + \sigma_i(\boldsymbol{Q})\,, \\ \sigma_c(\boldsymbol{Q}) &= \left\langle \left| \sum_{s=1}^{N}\sum_{\gamma=1}^{\nu} b_{s\gamma}\exp(\mathrm{i}\boldsymbol{Q}\cdot\boldsymbol{R}_{s\gamma})\exp(\mathrm{i}\boldsymbol{Q}\cdot\boldsymbol{u}_{s\gamma}) \right|^2 \right\rangle\,, \\ \sigma_i(\boldsymbol{Q}) &= \frac{N\gamma}{4\pi}\sum_{\mathfrak{a}} c_\mathfrak{a}\sigma_{i\mathfrak{a}}\,. \end{aligned} \tag{3.1.12}$$

Here c_α, is the total concentration of all isotopes of the element and $u_{s\gamma} = u_{s\gamma}^{\text{st}} + u_{s\gamma}^{\text{th}}$ is the sum of the static and thermal displacements of an atom from the site $s\gamma$.

A similar equation can be derived for the differential (with respect to angles and energy) cross section for the purely elastic neutron scattering (see Sect. 1.2 in [1.1]):

$$\sigma_{\text{el}}(\boldsymbol{Q}, E) = \sigma_{c,\text{el}}(\boldsymbol{Q}, E) + \sigma_{i,\text{el}}(\boldsymbol{Q}, E) \; ,$$

$$\sigma_{c,\text{el}}(\boldsymbol{Q}, E) = \left\langle \left| \sum_{s=1}^{N} \sum_{\gamma=1}^{\nu} b_{s\gamma} \exp(\text{i}\boldsymbol{Q} \cdot \boldsymbol{R}_{s\gamma}) \exp(\text{i}\boldsymbol{Q} \cdot \boldsymbol{u}_{s\gamma}^{\text{st}}) \exp(-M_{s\gamma}^{\text{th}}) \right|^2 \right\rangle_c \delta(E) \; , \tag{3.1.13}$$

$$\sigma_{i,\text{el}}(\boldsymbol{Q}, E) = \frac{1}{4\pi} \sum_{s=1}^{N} \sum_{\gamma=1}^{\nu} \sigma_{is\gamma} \exp(-M_{s\gamma}^{\text{th}}) \delta(E) \; ,$$

$$\exp(-M_{s\gamma}^{\text{th}}) = \langle \exp(\text{i}\boldsymbol{Q} \cdot \boldsymbol{u}_{s\gamma}^{\text{th}}) \rangle_T \; .$$

Here $\exp(-M_{s\gamma}^{\text{th}})$ is the Debye-Waller factor which describes attenuation of the scattering amplitude caused only by the thermal displacements of atoms (in imperfect crystals $M_{s\gamma}^{\text{th}}$ are different for different lattice sites), the angle brackets with the subscript T denote averaging over thermal vibrations of atoms for the given configuration of atoms, and the angle brackets with the subscript c denote averaging over configurations of atoms for the given composition and the parameters of order and correlation.

In addition to the above nuclear scattering of neutrons, magnetic scattering of neutrons may occur in crystals containing atoms with unfilled electron shells which have uncompensated magnetic moments. In some experiments magnetic scattering can be identified separately from the nuclear scattering, for instance, when polarized neutrons and external magnetic fields are used (see, for instance [3.12]) and we shall not discuss it here.

It can be seen from (3.1.6, 12) that the cross section $\sigma_c(\boldsymbol{Q})$ of coherent nuclear scattering of neutrons and the intensity $I(\boldsymbol{Q})$ of X-ray scattering are given by essentially identical equations which differ only in the replacement of the atomic factors f_α with the coherent scattering amplitudes b_α (the cross section $\sigma_i(\boldsymbol{Q})$ can be simply calculated from the known concentrations and cross sections of incoherent neutron scattering by individual nuclei). In a similar way, the cross section of elastic coherent neutron scattering by nuclei $\sigma_{c,\text{el}}(\boldsymbol{Q}E)$ given by (3.1.13) differs from (3.1.6) in that $f_{s\gamma}$ is replaced with the parameters $b_{s\gamma} \exp(-M_{s\gamma}^{\text{th}})$ which include the Debye-Waller factors caused by vibrations. This is why the same theoretical analysis is applicable to scattering processes involving X-rays and neutrons. We shall derive below equations for X-ray scattering but they describe just as well neutron scattering (when f_α is replaced with b_α as noted above).

On the other hand, some features of neutron scattering can be conveniently studied in experiments. In contrast to X-ray scattering for which the atomic factors f_α are determined by the number of electrons in the atom and grow steadily with the number of the element in the periodic table, the amplitudes b_α of coherent neutron scattering averaged over isotopes depend nonmonotonically on the element number and can be varied by changing the isotopic composition of the crystal. Therefore, the differences $b_A - b_B$ are generally large even for the solid solutions A-B of the elements which have close positions in the periodic table (for which the differences $f_A - f_B$ are small) and thus high-intensity diffuse scattering of neutrons may occur. The nuclei of light elements typically cause high-intensity scattering of neutrons, in contrast to X-rays. A potentially useful feature of neutron diffraction is that, in contrast to f_α, the amplitudes b_α are practically independent of the diffraction vector $\boldsymbol{Q}$ (because the ratio between the nuclear dimensions and the wavelength is negligibly small), that is, the angular dependence of the diffuse sccattering intensity is fully determined by the correlation of the atoms in the crystal and by their displacements. Since the energy distribution of the scattered neutrons can be analyzed with a sufficiently high resolution (between 10^{-4} and 10^{-7} eV) the experimentally recorded diffuse scattering by the static inhomogeneities of the imperfect crystals can be separated from scattering by thermal vibrations.

Similar separation can be performed in the experiments on diffraction of the Mössbauer photons. In some cases their wavelengths are of the same order as the X-ray wavelengths but the natural widths of their lines are smaller by several orders of magnitude (about 10^{-8} eV for Mössbauer effect with ^{57}Fe nuclei) making possible extremely precise energy analysis.

As noted above, the most complete data can be derived from the results on scattering of monochromatic radiation by a single crystal (see Fig. 3.1) but faster diffraction methods, such as the Debye technique, are also widely used. In this technique monochromatic radiation is scattered by a polycrystal or a powder and the angular distribution of the scattering intensity is analyzed. To find the scattering intensity $I_D(Q)$ in the Debye diffraction pattern (according to (3.1.2), Q corresponds to the given scattering angle 2θ we must average $I(\boldsymbol{Q})$ over all orientations of the crystal for a given $\boldsymbol{Q}$ or, equivalently, over all directions $\boldsymbol{Q}$ for a given orientation of the crystal and the given length of the vector $\boldsymbol{Q}$. The scattering intensity for a specimen containing $\mathfrak{N}^0$ powder grains with arbitrary orientations can be written as

$$I_D = \frac{\mathfrak{N}}{4\pi Q^2} \int_S \langle\langle I(\boldsymbol{Q})\rangle\rangle dS \, . \tag{3.1.14}$$

Here integration is performed over the vectors $\boldsymbol{Q}$ ending on the sphere with the radius Q in the reciprocal lattice space and the angle brackets $\langle\langle\ldots\rangle\rangle$ denote averaging over the sizes and shapes of the crystals and over configurations of atoms for the given macroscopic parameters.

3.2 Diffuse Scattering by Solid Solutions: General Analysis

Diffuse scattering by point defects of an arbitrary type and strength was discussed in Sect. 4.1 of [1.1] for solid solutions with a low defect concentration. These results are also applicable to weak solutions with a low concentration of impurities, $c \ll 1$. Here we shall consider a more general case of high-concentration solutions but we shall assume, however, that the distortions are not very large and for all atoms in the solid solution we have

$$|\boldsymbol{Q} \cdot \boldsymbol{u}_{s\gamma}| < \tfrac{1}{2} . \tag{3.2.1}$$

This condition is typically satisfied in substitutional solutions for not too large Q.

The intensity of diffuse scattering of X-rays by a solid solution is given by the general equation (3.1.6) (for $\boldsymbol{Q} \neq \boldsymbol{G}$). The atomic factor $f_{s\gamma}$ has the values f_A or f_B in a binary solution depending on which atom (A or B) occupies the site γ of the s th crystal cell. The factors $f_{s\gamma} = c_{s\gamma} f_A + (1 - c_{s\gamma}) f_B$ can be expressed in terms of the numbers $c_{s\gamma}$ defined by (1.1). Then we can rewrite the equation (3.1.6) for the intensity of scattering of monochromatic radiation by a single crystal in the form

$$I_1(\boldsymbol{Q}) = \left\langle \left| \sum_{s=1}^{N} \sum_{\gamma=1}^{\nu} [c_{s\gamma} f_A + (1 - c_{s\gamma}) f_B] \exp(\mathrm{i}\boldsymbol{Q} \cdot \boldsymbol{u}_{s\gamma}) \exp(\mathrm{i}\boldsymbol{Q} \cdot \boldsymbol{R}_{s\gamma}) \right|^2 \right\rangle . \tag{3.2.2}$$

For disordered solutions containing one atom per crystal cell we can omit the subscript and introduce the numbers $c_{s\alpha}$ (where $\alpha = A, B$) writing $c_{sA} \equiv c_s$ and $c_{sB} \equiv 1 - c_s$. Then we obtain

$$I_1(\boldsymbol{Q}) = N \sum_{\boldsymbol{\varrho}} I(\boldsymbol{\varrho}, \boldsymbol{Q}) \mathrm{e}^{\mathrm{i}\boldsymbol{q} \cdot \boldsymbol{\varrho}} ,$$
$$I(\boldsymbol{\varrho}, \boldsymbol{Q}) f_\alpha f_\beta^* \langle c_{s\alpha} c_{s'\beta} \mathrm{e}^{\mathrm{i}\boldsymbol{Q} \cdot \boldsymbol{u}_{\varrho}} \rangle , \boldsymbol{q} = \boldsymbol{Q} - \boldsymbol{G} . \tag{3.2.3}$$

Here $\boldsymbol{\varrho} \equiv \boldsymbol{R}_{ss'} = \boldsymbol{R}_s - \boldsymbol{R}_{s'}$ is the vector passing from the site s to the site s', $u_{\varrho} \equiv u_{ss'} = u_s - u_{s'}$ are the differences between the total (thermal and static) displacements of the atoms at these sites, summation over repeated subscripts $\alpha, \beta = A, B$ is implied, $\boldsymbol{G}$ is the reciprocal lattice vector which is the closest to the vector $\boldsymbol{Q}$, and the vector $\boldsymbol{q}$ passes from the respective point of the reciprocal lattice to the end of the vector $\boldsymbol{Q}$ (note that $\exp(\mathrm{i}\boldsymbol{G} \cdot \boldsymbol{\varrho}) = 1$ for any $\boldsymbol{G}$ and $\boldsymbol{\varrho}$).

Separation of the Effects Due to the Differences Between the Atomic Factors and the Crystal Distortions. According to (3.2.2, 3), there are two reasons for violations of the order in solid solutions which cause diffuse scattering of radiation – the differences between the scattering factors f_A and f_B and the displacements of atoms from the crystal lattice sites. The respective contributions to the scattering intensity depend on the diffraction vector $\boldsymbol{Q}$. The effects caused by the concentration inhomogeneities on the atomic scale and the crystal distortions can be analyzed independently with the use of the techniques developed for their separation [3.13–15].

Let us consider, for instance, a disordered solution for which we can ignore the imaginary parts of f_α. Assume that the condition (3.2.1) is satisfied and only the quadratic term can be retained in the expansion of $\exp(i\boldsymbol{Q}\boldsymbol{u}_{ss'})$. Then we obtain from (3.2.3)

$$I((\varrho)Q) = f_\alpha f_\beta p_{\alpha\beta}(\varrho) + i f_\alpha f_\beta p_{\alpha\beta}(\varrho) Q_i \langle u_{\varrho i}\rangle_{\alpha\beta} - \tfrac{1}{2} f_\alpha f_\beta p_{\alpha\beta}(\varrho) Q_i Q_j \times \langle u_{\varrho i} u_{\varrho j}\rangle_{\alpha\beta} \,. \tag{3.2.4}$$

Here $p_{\alpha\beta}(\varrho) = \langle c_{s\alpha} c_{s'\beta}\rangle$ are the probabilities that the atoms α and β occupy a pair of sites connected by the vector ϱ which are expressed in terms of the correlation parameters $\varepsilon(\varrho)$ according to (1.1.4). The averaged components of the displacements, $\langle u_{\varrho i}\rangle_{\alpha\beta}$ and $\langle u_{\varrho i} u_{\varrho j}\rangle_{\alpha\beta}$ correspond to these sites. Since I_1 is not changed when we add a constant to $I(\varrho, Q)$ (because $\boldsymbol{Q} \neq \boldsymbol{G}$) and $p_{\alpha\beta}(\varrho)\langle u_{\varrho i}\rangle_{\alpha\beta} = 0$ we obtain

$$\frac{I_1}{N(f_A - f_B)^2} = H(q) = H_0(q) + k_A Q_i H_i^A(q) - k_B Q_i H_i^B(q) + k_\alpha k_\beta Q_i Q_j H_{ij}^{\alpha\beta}(\varrho) \,, \tag{3.2.5}$$

where

$$\begin{aligned} H_0(q) &= c(1-c) + \sum_{\varrho \neq 0} \varepsilon(\varrho) \mathrm{e}^{\mathrm{i}q\cdot\varrho} \,; \\ H_i^\alpha(q) &= i \sum_{\varrho} p_{\alpha\alpha}(\varrho) \langle u_{\varrho i}\rangle_{\alpha\alpha} \mathrm{e}^{\mathrm{i}q\cdot\varrho} \,; \\ H_{ij}^{\alpha\beta}(q) &= -\frac{1}{2} \sum_{\varrho} p_{\alpha\beta}(\varrho) \langle u_{\varrho i} u_{\varrho j}\rangle_{\alpha\beta} \mathrm{e}^{\mathrm{i}q\cdot\varrho} \,; \quad k_\alpha = \frac{f_\alpha}{f_A - f_B} \,; \\ p_{AA}(\varrho) &= c^2 + \varepsilon(\varrho) \,; \quad p_{AB}(\varrho) = c(1-c) - \varepsilon(\varrho) \,; \\ p_{BB}(\varrho) &= (1-c)^2 + \varepsilon(\varrho) \,. \end{aligned} \tag{3.2.6}$$

The functions H_0, H_i^α and $H_{ij}^{\alpha\beta}$ in (3.2.6) depend only on $\boldsymbol{q} = \boldsymbol{Q} - \boldsymbol{G}$, that is, they are periodic functions of the diffraction vector $\boldsymbol{Q}$ in the reciprocal lattice space. In (3.2.5) these functions are multiplied by the components Q_i raised to various powers. Therefore, we can measure the intensity I_1 at some points of different cells in the reciprocal lattice space corresponding to identical q and different Q and thus find separately the function H_0 determined only by the short-range order (the concentration inhomogeneities in the solution) and the functions H_i^α and $H_{ij}^{\alpha\beta}$ determined by the displacements (thermal and static) of the atoms. Such points can be conveniently selected if the symmetry of the function H determined by the point symmetry of the crystal is taken into account (for instance, in a cubic crystal we have $H_x^\alpha(q_x, q_y, q_z) = H_y^\alpha(q_y, q_x q_z)$). Practical applications of this technique are discussed in detail, for instance, in the reviews [3.16, 17].

The most widely used modification of this technique assumes that k_A and k_B are independent of the scattering angle [3.13]. Then (3.2.5) includes ten independent periodic functions (H_0, three functions $k_A H_i^A - k_B H_i^B$, and six functions $k_\alpha, k_\beta H_{ij}^{\alpha\beta}$) and measurements must be made at ten points $\boldsymbol{q}$ in order to determine them. The assumption that k_A and k_B are constant is rigourougly valid, however, only for neutron scattering while for X-ray scattering it is satisfied only approximately and then (3.2.5) contains, strictly speaking, 25 independent periodic functions (H_0 six functions H_i^α and 18 functions $H_{ij}^{\alpha\beta}$). A more cumbersome procedure of their determination involves making measurements of I_1 at 25 points $\boldsymbol{q}$ [3.14, 15]. In addition to a more accurate identification of the scattering by crystal distortions, this technique makes it possible to determine separately the functions H_i^A and H_i^B, $H_{ij}^{AA}, \ldots$, which describe the averaged relative displacements and the correlation parameters for the displacements $\langle u_{si} u_{s'j} \rangle_{\alpha\beta}$, for the atom pairs AA, AB and BB separately.

Method of Fluctuation Waves. A theoretical analysis of diffuse scattering in solid solutions must yield the intensity $I_1(\boldsymbol{Q})$ (or the functions $H(\boldsymbol{q})$ in (3.2.5)) as a function of the macroscopic or microscopic parameters of the solution, for instance, the ordering energies or the force constants which determine the static displacements. Two approaches may be realized in such analysis. In the first one we shall consider scattering by the inhomogeneities produced by each atom (the variation of the atomic factor and the resulting field of static distortions), determine from (3.2.2) the interference pattern produced by the secondary waves scattered by different atoms for the given atomic configuration, and then perform averaging of the results over the configurations. It was precisely this approach that was employed in most early analyses of X-ray and neutron scattering in solid solutions whether they ignored the static displacements of the atoms in solution [1.28; 3.18–24] or not [3.25–30].

In the framework of the second approach the crystal is treated as a regular structure consisting of effective "average" atoms on which fluctuations of the composition and the order parameters are applied. The regular structure causes regular reflections and the fluctuations produce diffuse scattering. In this approach diffuse scattering is attributed not to individual scattering centers but to the fluctuation waves of concentration and internal parameters and the static displacement waves they generate. As a result, the scattering probability is directly related to the thermodynamic or microscopic parameters of the solution. As noted above, this approach simplifies significantly treatment of some complicated problems in the theory of diffuse scattering, for instance, involving scattering of X-rays near critical points or second-order phase transitions, scattering in the presence of long-range forces and so on. It is especially convenient to use this approach when distortions must be taken into consideration since the Fourier representation considerably simplifies the equations of the elasticity theory and the atomic theory. The results on X-ray and neutron scattering obtained with the method of fluctuation waves wers reported in [1.2–4] (for undistorted crystals) and in [1.8–12, 15, 16, 20; 3.31–33].

We shall consistently apply this method below using the results on the theory of fluctuation waves derived in Chaps. 1 and 2.

When we consider diffuse scattering (that is, for $\boldsymbol{Q} \neq \boldsymbol{G}$) the constant mean component of the term $[c_{s\gamma} f_A + (1 - c_{s\gamma}) f_B] \exp(\mathrm{i}\boldsymbol{Q} \cdot \boldsymbol{u}_{s\gamma})$ in (3.2.2) is eliminated from the equation for the amplitude and a significant contribution is made only by its fluctuating component which is determined by the difference between $c_{s\gamma}$ and their means c_γ and the difference between the displacements $\boldsymbol{u}_{s\gamma}$ and zero. The fluctuations of $c_{s\gamma} - c_\gamma$ and $\boldsymbol{u}_{s\gamma}$ may be described by their Fourier expansions (1.1.8) and (1.5.1) (or (2.2.4) in [1.1] for crystals with several atoms per unit cell).

If the crystal distortions are small and the condition (3.2.1) is satisfied we can expand (3.2.2) in powers of $\boldsymbol{Q} \cdot \boldsymbol{u}_{s\gamma}$ and retain only the first terms. In the expression for the scattering amplitude at first we shall retain only linear terms in $f_A - f_B$ and $\boldsymbol{Q} \cdot \boldsymbol{u}_{s\gamma}$. Replacing $c_{s\gamma} - c_\gamma$ and $\boldsymbol{u}_{s\gamma}$ with the respective Fourier series (1.1.8) and (1.5.1) and using the fact that $\boldsymbol{Q} = \boldsymbol{G} + \boldsymbol{q}$ and $\boldsymbol{R}_{s\gamma} = \boldsymbol{R}_s + \boldsymbol{R}_\gamma$ we can write this expansion as

$$I_1 = \left| \sum_{\gamma=1}^{\nu} \mathrm{e}^{\mathrm{i}\boldsymbol{G}\cdot\boldsymbol{R}_\gamma} \sum_{s=1}^{N} \left\{ f_\gamma + \sum_{\boldsymbol{k}} [(f_A - f_B) c_{\boldsymbol{k}\gamma} - f_\gamma \boldsymbol{Q} \cdot \boldsymbol{u}_{\boldsymbol{k}\gamma}] \mathrm{e}^{-\mathrm{i}\boldsymbol{k}\cdot\boldsymbol{R}_{s\gamma}} \right\} \mathrm{e}^{\mathrm{i}\boldsymbol{q}\cdot\boldsymbol{R}_{s\gamma}} \right|^2 \tag{3.2.7}$$

Here f_γ is given by (3.1.10). In a more rigourous treatment the mean $c_\gamma f_A + (1 - c_\gamma) f_B$ must be used instead of f_γ but for $|\boldsymbol{Q} \cdot \boldsymbol{u}_{s\gamma}| \ll 1$ the difference between these two parameters is small. For small q, however, the factors $\exp(-M_{A\gamma})$ and $\exp(-M_{B\gamma})$ in (3.1.10) improve the accuracy of the equation for I_1 since they correctly take into consideration the terms of a higher order in displacements. This means (see Sect. 3.4) that the term proportional to q^{-2} in the expression for I_1 describing scattering by the distortion wave with a large wavelength is decreased by the same Debye-Waller factor as the expression (3.1.8) (see [1.13; 3.34, 35] and the experimental evidence in [3.36–38]).

Since in the limit of infinite crystal the sum $\sum_s \exp[\mathrm{i}(\boldsymbol{q} - \boldsymbol{k}) \cdot \boldsymbol{R}_{s\gamma}]$ vanishes for $\boldsymbol{q} \neq \boldsymbol{k}$ and equals N for $\boldsymbol{q} = \boldsymbol{k}$ (for the sake of simplicity we assume that k satisfy cyclic conditions) we can write the intensity as a sum of two terms, $I = I_0 + I_1$. The first term is the regular reflection intensity (3.1.8) while the second term describes diffuse scattering:

$$I_1 = N^2 \left| \sum_{\gamma=1}^{\nu} \exp(\mathrm{i}\boldsymbol{G} \cdot \boldsymbol{R}_\gamma) [f_\gamma \boldsymbol{Q} \cdot \boldsymbol{u}_{\boldsymbol{q}\gamma} - (f_A - f_B) c_{\boldsymbol{q}\gamma}] \right|^2 . \tag{3.2.8}$$

The squared first term in (3.2.8) describes scattering caused by distortions, the squared second term describes scattering caused by the difference in the atomic factors and the doubled product of the terms is the interference term dependent both on the distortions and the difference between the atomic factors. Equations (3.2.7, 8) may include averaging over the configurations of atoms corresponding to the given c_γ and the correlation parameters.

When we analyze the higher-order terms in the expansion of I_1 in the displacements we can find the conditions under which we can retain only the quadratic terms and use (3.2.8). According to the results of [1.12], for $|f_A - f_B| \lesssim f_\gamma$ we can write such conditions in the following form (see also Sect. 3.4.1):

$$M_\gamma \frac{q}{k_m} \ll 1\,, \quad \frac{M_\gamma}{c_\gamma(1-c_\gamma)} \left(\frac{q}{k_m}\right)^2 \ll 1\,. \tag{3.2.9}$$

These conditions are satisfied in weakly distorted crystals where $M_\gamma \ll c_\gamma(1-c_\gamma)$ and also in strongly distorted crystals if the vector $\boldsymbol{Q}$ is in the vicinity of a reciprocal lattice point and $\boldsymbol{q}$ is significantly smaller than the largest wave vector in the unit cell, $k_m \sim \pi/d$. When $\boldsymbol{q}$ are large and the conditions (3.2.9) are satisfied for $M_\gamma \ll 1$ the squared first term in (3.2.8) is considerably smaller than the squared second term and has the same order of magnitude as the terms ignored in the derivation of (3.2.8) (which have been analyzed in [1.12]). In this case it would be logical to retain only the terms linear in displacements. But even in weakly distorted crystals ($M_\gamma \ll 1$) for sufficiently small $\boldsymbol{q}$ the squared first term in (3.2.8) makes the main contribution to I_1 (see Sect. 3.3 in [1.1]) and it is considerably larger than the omitted terms. Therefore we must take into consideration also the terms of the second order in displacements in (3.2.8).

According to (3.2.8), the diffuse scattering intensity is expressed in terms of the squares and products of the Fourier components of the concentration fluctuations c_γ and the atom displacements (for the wave vector $\boldsymbol{q}$) which were treated in Chaps. 1 and 2. The equation for the scattering intensity has a much simpler form for the disordered solid solutions containing one atom per unit cell ($\nu = 1$) in which the static displacements are caused only by the concentration fluctuations and are independent of the fluctuations of the correlation parameters. According to (1.5.3), under these conditions $\boldsymbol{u}_{\boldsymbol{k}} \sim c_{\boldsymbol{k}}$ and (3.2.8) is transformed into

$$I_1 = N^2 \langle |c_{\boldsymbol{q}}|^2 \rangle \left[f\boldsymbol{Q} \cdot \boldsymbol{A}_{\boldsymbol{q}} - (f_A - f_B)|^2 \right] \,. \tag{3.2.10}$$

We can analyze diffuse scattering by disordered solutions with the use of the equations for $\langle |c_{\boldsymbol{q}}|^2 \rangle$ given in Sects. 1.1–3 and 2.1 and the equations for the coefficient $\boldsymbol{A}_{\boldsymbol{q}}$ of proportionality between the amplitudes of the fluctuation waves of displacements and concentration given in Sect. 1.5 (and in Sects. 2.2, 3 in [1.1]). We shall apply this analysis to undistorted crystals in Sect. 3.3 and to crystals with static displacements in Sect. 3.4.

3.3 Diffuse Scattering in Undistorted Solid Solutions

The simplest pattern of diffuse scattering intensity distribution is observed for the solid solutions in which the components have similar atomic radii so that diffuse scattering by the static distortions is not strong and the long-range elastic forces produce a comparatively weak effect on the short-range order. Therefore, we shall first consider undistorted solid solutions and the effects caused by distortions will be analyzed in Sect. 3.4. Since experimental techniques can separate directly the scattering effects due to distortions (see Sect. 3.2) many results derived for undistorted crystals are also applicable to the component obtained after such separation from the scattering results for the distorted crystals ($N(f_A - f_B)^2 H_0(\boldsymbol{q})$ in (3.2.5)).

3.3.1 Scattering by Homogeneous Solutions, Determination of the Correlation Parameters, Ordering Energies and the Fermi Surface Diameters

Correlation Parameters of a Solution. Consider first diffuse scattering by single crystals of disordered solutions containing one atom per unit cell. The scattering intensity I_1 here is given by (3.2.10) where $A_{\boldsymbol{q}} = 0$. Writing $\langle |c_{\boldsymbol{q}}|^2 \rangle$ as (1.1.14) we can express the intensity (as it is done in (3.2.5, 6)) in terms of the correlation parameters $\varepsilon(\boldsymbol{\varrho})$:

$$I_1 = N(f_A - f_B)^2 \left[c(1-c) + \sum_{\boldsymbol{\varrho} \neq 0} \varepsilon(\boldsymbol{\varrho}) \cos \boldsymbol{q} \cdot \boldsymbol{\varrho} \right] . \tag{3.3.1}$$

If there is no correlation ($\varepsilon(\boldsymbol{\varrho}) = 0$) the intensity of diffuse scattering by such ideal solution is given by the *Laue* equation [3.18]

$$I_1 = N(f_A - f_B)^2 c(1-c) . \tag{3.3.2}$$

In the ideal solution the ratio $I_1/(f_A - f_B)^2$ is independent of $\boldsymbol{Q}$ and remains the same at all points of the reciprocal lattice space. If the solution is not ideal and the parameters $\varepsilon(\boldsymbol{\varrho})$ do not vanish this ratio depends on $\boldsymbol{Q}$ as can be seen from (3.3.1). Note that, according to (3.3.1), $I_1(f_A - f_B)^{-2}$ is a periodic function of $\boldsymbol{Q}$ with the period equalling the reciprocal lattice parameter (any variation of $\boldsymbol{Q}$ by a reciprocal lattice vector $\boldsymbol{G}$ does not change $\cos \boldsymbol{Q} \cdot \boldsymbol{\varrho} = \cos \boldsymbol{q} \cdot \boldsymbol{\varrho}$ because $\boldsymbol{G} \cdot \boldsymbol{\varrho}$ are multiples of 2π). This periodic variation of the diffuse scattering intensity is observed only for undistorted crystals (see (3.2.5)) and an experimental evidence of it may indicate that crystal distortions are small. The function $I_1(f_A - f_B)^{-2}$ must also have identical values at the points $\boldsymbol{q}$ which coincide under transformations of the point group of crystal symmetry.

Equation (3.3.1) is essentially a Fourier expansion of the function $I_1(f_A - f_B)^{-2}$ where the Fourier coefficients are the correlation parameters $\varepsilon(\boldsymbol{\varrho})$. This expansion can be used for finding the entire set of the parameters $\varepsilon(\boldsymbol{\varrho})$ for different $\boldsymbol{\varrho}$ (or a similar set of the parameters $\alpha(\boldsymbol{\varrho})$ given by (1.1.5)) from the experimental data on the scattering intensity distribution $I_1(\boldsymbol{Q})$ in a reciprocal lattice cell [1.28; 3.39]. indeed, when we perform the Fourier transformation we obtain from (3.3.1)

$$\varepsilon(\boldsymbol{\varrho}) = \frac{v}{8\pi^3} \int_{v_r} d\boldsymbol{Q} \frac{I_1(\boldsymbol{Q})}{N} (f_A - f_B)^{-2} \cos \boldsymbol{Q} \cdot \boldsymbol{\varrho} , \tag{3.3.3}$$

where integration over k is performed over the volume $v_r = 8\pi^3/v$ of the unit cell of the reciprocal lattice. When $\varrho = 0$ we can take in (3.3.3)

$$\varepsilon(0) = c(1-c) \quad \text{or} \quad \alpha(0) = 1 \tag{3.3.4}$$

and use this equation as the normalization condition (or as a criterion of accuracy for the above technique).

We see that we can determine directly the correlation parameters for all coordination spheres by applying the Fourier transformation technique (3.3.3) to the results on diffuse scattering by the concentration inhomogeneities in single crystals. This technique can be also applied to distorted solutions after the effects due to distortions and differences in the atomic scattering factors have been separated with the method described in Sect. 3.2 and the function $I_1/N(f_A - f_B)^2$ in (3.3.3) has been replaced with $H_0(\boldsymbol{q})$ from (3.2.5, 6) (another approximate method for crystals with distortions is discussed in Sect. 3.4). Practical applications of this procedure are treated in detail in [1.27–29; 3.17] and illustrated in Sect. 3.5.

The above results for disordered solutions can be readily extended to ordered solutions in which the unit cell contains a single atom after disordering. If we take then $\boldsymbol{q} = \boldsymbol{Q} - \boldsymbol{G}$ for the vector $\boldsymbol{G}$ corresponding to the "structural" regular reflection for which $\boldsymbol{G}\cdot\boldsymbol{R}_\gamma$ is a multiple of 2π for all $\boldsymbol{R}_\gamma$ then in (3.2.8) we have $\exp(\mathrm{i}\boldsymbol{G}\cdot\boldsymbol{R}_\gamma) = 1$ and I_1 is proportional to $|\sum c_{q\gamma}|^2$. Using (1.1.12, 13) we can then write the scattering intensity in the form similar to (3.3.1):

$$I_1 = N\nu(f_A - f_B)^2 \sum_{\varrho} \varepsilon(\varrho)\cos\boldsymbol{q}\cdot\boldsymbol{\varrho}\,, \quad \varepsilon(0) = \frac{1}{\nu}\sum_{\gamma=1}^{\nu} c_\gamma(1-c_\gamma)\,. \tag{3.3.5}$$

Here the correlation parameters $\varepsilon(\varrho)$ are given by (1.1.6). For instance, in ordered solutions $\varepsilon(\varrho)$ can be found from (3.3.3) where the integration must be performed over the volume $v_r = 8\pi^3\nu/v = 8\pi^3/v_0$ of ν unit cells of the reciprocal lattice for the ordered solid solution (over the volume of a unit cell of the reciprocal lattice for the disordered solution), N is replaced with $N\nu$, v with $v_0 = v/\nu$ and in (3.3.4) we have

$$\varepsilon(0) = \frac{1}{\nu}\sum_{\gamma=1}^{\nu} c_\gamma(1-c_\gamma)\,.$$

Let us consider the diffuse scattering intensity distribution $I_{1\mathrm{D}}(Q)$ in the Debye diffraction pattern recorded for an undistorted crystal. This intensity distribution is obtained by averaging (3.1.1) or (3.3.5) over crystal orientations (or over directions of $\boldsymbol{Q}$) in accordance with (3.1.14). Since the spherical surface S with the radius Q in (3.1.14) passes through several points in the reciprocal lattice space it is more convenient in (3.3.5) to replace $\cos\boldsymbol{q}\cdot\boldsymbol{\varrho}$ with the identical $\cos\boldsymbol{Q}\cdot\boldsymbol{\varrho}$. Since we have

$$\int \cos\boldsymbol{Q}\cdot\boldsymbol{\varrho}\,dS = Q^2 \int\!\!\int \cos(Q\varrho\cos\vartheta)\sin\vartheta\, d\vartheta\, d\varphi = 4\pi Q^2 \frac{\sin Q\varrho}{Q\varrho}$$

where ϑ is the angle between $\boldsymbol{Q}$ and $\boldsymbol{\varrho}$ we obtain after substituting (3.3.5) into (3.1.14)

$$I_{1\mathrm{D}} = N\nu(f_A - f_B)^2 \left[\frac{1}{\nu}\sum_{\gamma=1}^{\nu} c_\gamma(1-c_\gamma) + \sum_{l=1}^{\infty} z_l \frac{\sin Q\varrho_l}{Q\varrho_l}\varepsilon(\varrho_l)\right] . \tag{3.3.6}$$

Here ϱ_l is the radius of the lth coordination sphere with the coordination number z_l. According to (3.1.2), Q varies only in a finite range of $Q < 4\pi/\lambda$ and therefore we cannot transform the expansion (3.3.3) accurately enough and find all parameters $\varepsilon(\varrho_1)$ from the Debye diffraction pattern data. If, however, we assume that only a small number (from one to three) to the correlation parameters for the first coordination spheres do not vanish then we can use (3.3.6) for calculating these parameters from experimental results, for instance, with the least squares technique (see details in the book [1.27]). This simple method gives, of course, less accurate results than the analysis of results obtained with single crystals. (To obtain more accurate results an the correlation parameters than those given by the technique of least squares one can employ sophisticated statistical methods to the experimental data on the diffuse scattering intensity distribution. For instance, the regularization method [3.41] yielded a stable solution of the equations for the short-range order parameters $\varepsilon(\varrho_1)$ allowing for the experimental errors in the results for three-component alloys [3.40]. Note that when we determine the correlation parameters in three-component alloys the most reasonable approach is to analyze neutron scattering in alloys with different isotope compositions, simultaneous neutron and X-ray scattering or X-ray scattering near the absorption edge. The difference between the ratios f_B/f_A and f_C/f_A recorded in these experiments may be used to calculate different combinations of the parameters $\varepsilon_{AB}(\varrho_1)$, $\varepsilon_{AC}(\varrho_1)$ and $\varepsilon_{BC}(\varrho_1)$ and to find them separately.)

Ordering Energies and Thermodynamic Characteristics of Solutions. Equations (3.3.2, 5) can be applied to solutions which are or are not at equilibrium to find, for instance, the parameters $\varepsilon(\boldsymbol{\varrho})$ in nonequilibrium systems. At equilibrium the parameters $\varepsilon(\boldsymbol{\varrho})$ are determined by the thermodynamic and energetic characteristics of the solution and these characteristics may be found from the known parameter $\varepsilon(\boldsymbol{\varrho})$. It is more convenient, however, to express I_1 in terms of these characteristics with the use of equations for $\langle|c_{\boldsymbol{q}}|^2\rangle$ derived in Sects. 1.2, 3 and thus build a framework for their direct determination from the diffraction results.

According to (3.2.10), in the vicinities of the reciprocal lattice points, that is, for small q, the scattering intensity $I_1(Q)$ is determined by the long-wave fluctuations of the concentration for which $\langle|c_{\boldsymbol{q}}|^2\rangle$ is given by the general equations (1.2.10, 11) of the macroscopic theory of fluctuations which is not based on any simplified models. When we substitute these equations into (3.2.10) or (3.2.8) we obtain the following expression for the scattering intensity in the vicinity of reciprocal lattice points (or in the vicinities of the points corresponding to the regular reflections for which $\boldsymbol{G}\cdot\boldsymbol{R}_\gamma$ are multiples of 2π) for crystals without distortions:

$$I_1(\boldsymbol{Q}) = N_0 \frac{k_B T}{v_0} ((f_A - f_B)^2 \frac{1}{\frac{d^2\varphi}{dc^2} + \beta_{ij} q_i q_j} \quad (q \ll k_m) \tag{3.3.7}$$

(here $N_0 = N\nu$, $v_0 = V/N_0$ is the volume of the atom). For sufficiently small q the second term in the denominator in (3.3.7) can be ignored and then $I_1(\boldsymbol{Q})$ depends only on $d^2\varphi/dc^2$ which can be determined from the independent thermodynamic data on the concentration dependence of the chemical potentials or activities (see (1.2.3)). According to (3.3.7), in the absence of distortions the identical intensity distributions $I_1(f_A - f_B)^{-2}$ appear in the vicinities of all the above-mentioned sites of the reciprocal lattice and also for small scattering angles (when q coincides with $\boldsymbol{Q}$). In decomposing solutions the intensity (3.3.7) can be significantly higher than in the ideal solutions while in the ordering solutions it can be lower than that.

In a similar way, in the vicinities of the reciprocal lattice points of the ordered solutions corresponding to the superlattice reflections we can express the intensity I_1 in terms of the derivatives of the thermodynamic potential with respect to the parameters of the long-range order (and concentration). For example, if a unit cell contains two lattice sites and the long-range order is described by a single parameter η then we can take $\exp(\mathrm{i}\boldsymbol{G}\cdot\boldsymbol{R}_\gamma) = (-1)^{\gamma+1}$ for the superlattice reflections in (3.3.8) while the difference $c_{q1} - c_{q2}$ (for $\nu_1 = \nu_2 = 1$) can be replaced with the Fourier component of the long-range order parameter η_q according to (1.1.3). Taking $u_{q\gamma} = 0$ in (3.3.8) for undistorted crystals and describing $\langle|\eta_q|^2\rangle$ by the equation (1.2.26) derived in the macroscopic theory of fluctuations we can write the scattering intensity near the superlattice reflections in the following form (for cubic crystals):

$$\begin{aligned} I_1 &= \frac{N_0^2}{4}(f_A - f_B)^2 \langle|\eta_q|^2\rangle \\ &= \frac{1}{4} N_0 \frac{k_B T}{v_0} (f_A - f_B)^2 \frac{\varphi_{cc} + \beta' q^2}{\varphi_{\eta\eta}\varphi_{cc} + (\alpha\varphi_{cc} + \beta'\varphi_{\eta\eta} - 2\gamma\varphi_{c\eta})q^2} \,. \end{aligned} \tag{3.3.8}$$

Here $q \ll k_m$ give the distance from the superlattice point of the reciprocal lattice. As it is the case for the vicinities of the lattice reflection, the intensity I_1 for small q is expressed only in terms of the thermodynamic parameters. In a disordered solution $\varphi_{c\eta} = 0$ and the last factor in (3.3.8) reduces to a simpler form $(\varphi_{\eta\eta} + \alpha q^2)^{-1}$.

When we determine the intensity I_1, for large q we must make use of the expression for $\langle|c_q|^2\rangle$ derived with the microscopic theory for a specific solution model. For instance, if a disordered solution can be described with the model of pair interaction (see Sect. 1.3) then (1.3.4) can be substituted (it is correct in this model) into (3.2.10) and, since $A_q = 0$ for disordered solutions, the intensity I_1 can be expressed in terms of $x(\boldsymbol{\varrho})$ which depend on the ordering energies $w(\boldsymbol{\varrho})$ for various coordination spheres [1.4–6]:

$$I_1 = \frac{N(f_A - f_B)^2}{x_q} = \frac{N(f_A - f_B)^2}{\sum_{\boldsymbol{\varrho}} x(\boldsymbol{\varrho}) \cos \boldsymbol{q}\cdot\boldsymbol{\varrho}} \,. \tag{3.3.9}$$

The functions $x(\varrho)$ depend on $w(\varrho)$, temperature, and composition. At high temperatures or low-concentration solutions $x(\varrho)$ are expressed in terms of $w(\varrho)$ for the same lattice vectors ϱ with the use of the simple equations (1.3.9, 12, 16) while in the nearest-neighbour approximation for the bcc crystals they are given by (1.3.15).

The ordering energies for various coordination spheres can be found directly from the experimental results on the $I_1(\boldsymbol{Q})$ distribution in the reciprocal lattice space with the use of (3.3.9). This is done by performing the Fourier transformation not for the scattering intensity (divided by $(f_A - f_B)^2$ as it was done in determination of the correlation parameters) but rather for the inverse function [1.4]. Then we obtain from (3.3.9)

$$x(\varrho) = \frac{v}{8\pi^3}\int_{v_r} d\boldsymbol{Q}\frac{N}{I_1(\boldsymbol{Q})}(f_A - f_B)^2 \cos \boldsymbol{Q}\cdot\boldsymbol{\varrho}\,. \tag{3.3.10}$$

Using this equation and (1.3.9, 12, 15, 16) relating $x(\varrho)$ and $w(\varrho)$ we can determine $w(\varrho)$ for all ϱ (the equation for $x(0)$ can be employed for normalization of the distribution I_1). For instance, at high enough temperatures we obtain from (1.3.10) and (3.2.10)

$$\begin{aligned} I_1(\boldsymbol{Q}) &\approx N(f_A - f_B)^2 C\frac{1}{1 + c(1-c)\frac{w(\varrho)}{k_\mathrm{B}T}} \\ &\approx N(f_A - f_B)^2 \frac{c(1-c)}{1 + c(1-c)\frac{w(\varrho)}{k_\mathrm{B}T}} \end{aligned} \tag{3.3.11}$$

and for $w(\varrho) \ll k_\mathrm{B}T$ in (3.3.10) we have $x(\varrho) \approx w(\varrho)/k_\mathrm{B}T$. The function $w(\varrho)$ found from experimental data directly according to (3.3.11) or, more precisely, from (1.3.16) and (3.3.10) approximately (in the self-consistent field approximation) determine the temperature T_0' at which the solution is unstable with respect to decomposition or ordering, as follows from (1.3.13). Therefore, we can calculate approximately the phase diagram for the solution from the diffraction results since T_0' and the phase transition temperature for the undistorted crystal are close or even identical (see illustrations in Sect. 3.5). Finding $x(\varrho)$ from the experimental results with the use of (1.3.19, 20) and (3.3.10) we can determine, in principle, both the derivative $d^2\varphi/dc^2$ and the coefficients β and β_{mij} which determine the gradient (and higher-order) terms in the expansion of the thermodynamic potential in $\delta c(\boldsymbol{r})$ or $\delta\eta_\mu(\boldsymbol{r})$.

According to (1.1.12), (3.2.8) and the expressions for $\langle|c_{\boldsymbol{k}}|^2\rangle$ derived in Sect. 1.3, the equation (3.3.9) is applicable not only to disordered solutions but also to an undistorted ordered solution with the stoichiometric composition AB if $\boldsymbol{q}$ is taken from the reciprocal lattice vector corresponding to the lattice reflection. In this case we have only to replace N with $N_0 = N\nu$ in (3.3.9) and determine $x(\varrho)$ from (1.3.22) for almost fully ordered solutions. Therefore, even for the ordered solutions AB we may use (3.3.10) (where N must be replaced with N_0 in order to find $x(\varrho)$ and hence the ordering energy $w(\varrho)$ from the experimental results on the diffuse scattering intensity distribution.

General Patterns of Scattering Intensity Distribution. We can use (3.3.9) for analyzing the dependence of I_1 on $\boldsymbol{Q}$ for various temperatures and solution concentrations. For instance, in the nearest-neighbour approximation at high temperatures when (1.3.9) does not vanish only for $\boldsymbol{\varrho} = 0$ and for z atoms of the first coordination sphere with $\boldsymbol{\varrho} = \boldsymbol{\varrho}_{\mathfrak{z}}$ (where $\mathfrak{z} = 1, \ldots, z$)) we have the following simple equation:

$$I_1 = Nf(f_A - f_B)^2 \frac{c(1-c)}{1 + c(1-c)\frac{w}{k_B T}\sum_{\mathfrak{z}=1}^{z} \cos \boldsymbol{q} \cdot \boldsymbol{\varrho}_{\mathfrak{z}}} , \tag{3.3.12}$$

Here $w = w(\varrho_1)$. For solid solution of low concentrations at arbitrary temperatures we must replace in (3.3.12) $w/k_B T$ with $1 - \exp(-w/k_B T)$, according to (1.3.12).

It can be seen from (3.3.12) that in the nearest-neigbour approximation the dependence of the intensity I_1 on $\boldsymbol{q}$ is determined only by the sum $\sum \cos \boldsymbol{q} \cdot \boldsymbol{\varrho}_y$ over z points in the first coordination sphere. This sum is the largest for $\boldsymbol{q} = 0$, that is, at the reciprocal lattice points for the disordered solution. For bcc and fcc crystals this sum has the least value at the $\langle 100 \rangle$ points, that is, at the "superlattice" points of the reciprocal lattice which appear when ordered structures of the type of beta brass or Cu_3Au are produced, respectively. It is shown accordingly by (3.3.12) that in the nearest neighbour approximation for the decomposing solutions ($w < 0$) the function $I_1(\boldsymbol{Q})(f_A - f_B)^{-2}$ reaches a maximum at the points $\boldsymbol{Q}$ occupying the reciprocal lattice points for the disordered solution and a minimum at the points corresponding to the above superlattice reflections upon ordering. The pattern is reversed for the ordering solutions ($w > 0$).

In the general case, when the interaction between more distant atoms is taken into consideration the functions x_q in (3.3.9) and $I_1(\boldsymbol{Q})$ may exhibit extrema at other points. As noted in Sect. 1.3, since the function x_q varies regularly, some special high-symmetry points $\boldsymbol{q} = \boldsymbol{k}_m$ must necessarily correspond to its extrema. In bcc and fcc crystals such special points are $\langle 000 \rangle$, $\langle 100 \rangle$, $\langle 1/2\,1/2\,1/2 \rangle$, and $\langle 1\,1/2\,0 \rangle$ [1.26, 69, 70]. The type of the extremum (maximum, minimum or saddle point) depends on the ratio between the parameters $x(\boldsymbol{\varrho})$ (or $w(\boldsymbol{\varrho})$ at high temperatures). Three parameters $w(\boldsymbol{\varrho})$ have been included in such analysis [1.72, 3.42].

According to (1.3.20), near the extremum points $x(\boldsymbol{\varrho})$ exhibits a quadratic dependence on $\boldsymbol{q} - \boldsymbol{k}_m$. Therefore, if $\boldsymbol{k}_m$ corresponds to a minimum of $x_{\boldsymbol{\varrho}}$ then near this point we obtain a bell shaped plot of I_1 as a function of $\boldsymbol{Q}$. The scattering intensity distribution $I_1(\boldsymbol{Q})$ is described by equal intensity surfaces (that is, the surfaces in the reciprocal lattice space at which the scattering intensity $I_1(\boldsymbol{Q})$ is identical) or the equal intensity lines produced by intersections of these surfaces and reciprocal lattice planes. In cubic crystals the equal intensity surfaces are spherical. They are also spherical near the extremum points $\boldsymbol{k}_m \neq 0$ of the cubic symmetry, that is, the points $\langle 100 \rangle$ and $\langle 1/2\,1/2\,1/2 \rangle$ in bcc crystals. But other special points are of a lower symmetry (orthorhombic for the points $\langle 1\,1/2\,0 \rangle$ in bcc crystals, tetragonal for the points $\langle 100 \rangle$ and $\langle 1\,1/2\,0 \rangle$ and trigonal for the points $\langle 1/2\,1/2\,1/2 \rangle$ in fcc crystals) and then the equal intensity surfaces have ellipsoidal shapes.

According to (3.3.9), the equal intensity surfaces have much more complex shapes for large q. The peaks of the intensity distribution lie near the lattice points or at the centers of the cells depending on whether the solution is decomposing or ordering. Figures 3.2, 3 illustrate equal intensity lines found after scattering effects caused by distortions have been eliminated for the fcc ordering alloys Cu_3Au [3.43] and Cu-29.8 at.% Pd [3.44] at temperatures exceeding the ordering temperature ($T_0 = 394°C$ for the first alloy and 330°C for the second alloy).

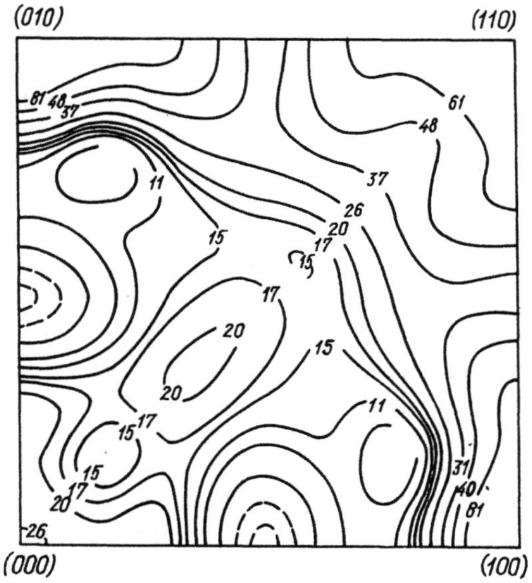

Fig. 3.2. Equal intensity lines in the (001) plane of the reciprocal lattice of the alloy Cu_3Au at 475 K [3.43]

Equation (3.3.12) or more general equations and (1.3.9, 12, 15, 16) and (3.3.9) demonstrate that when the temperature increases or the concentration of a component decreases the ratio $x(\varrho)/x(0)$ decreases sharply, the short range order becomes less significant and the scattering intensity distribution approaches the distribution $I_1 = I_{1L}$ given by (3.3.2) which corresponds to the ideal solution. At high temperatures (3.3.11) indicates that the deviation of $1/I_1$ from $1/I_{1L}$ linearly depends on $1/T$. According to (3.3.12), for small concentrations c the background distribution is more sharply nonuniform for decomposing solutions than for the ordering solutions (for the same $|w|/k_B T$). In the nearest neighbour approximation at temperatures much lower than the ordering temperature $\langle |c_{\boldsymbol{k}}|^2 \rangle$ for AB solutions similar to beta brass is given by (1.3.24). Substituting this equation into (3.2.8) for $\boldsymbol{u}_{q\gamma} = 0$ we obtain

$$I_1 = \frac{1}{2} N (f_A - f_B)^2 \frac{1-\eta^2}{1+\frac{1}{4}(1-\eta^2)\left(\exp\frac{w}{k_B T} - 1\right)\sum_{j=1}^{z} \cos \boldsymbol{q} \cdot \boldsymbol{\varrho}_j} . \quad (3.3.13)$$

As noted in Sect. 1.3, $1-\eta^2$ decreases at a higher rate with a decrease in temperature than $\exp(w/k_BT)$ increases. Therefore, for almost completely ordered solutions at low enough temperatures we can eliminate the second term in the denominator of (3.3.13) since this term is associated with the correlation which becomes insignificant here as is the case at high temperatures, too. Under such conditions the scattering intensity $I_1 \sim (1-\eta^2)$ in a solution at equilibrium decreases as $\exp(-zw/2k_BT)$ with a decrease in temperature. Other equations have been derived for the intensity of scattering by the ordered solutions with different structures and compositions [1.4].

Fig. 3.3. Equal intensity lines in the (001) plane of the reciprocal lattice of the alloy Cu-29.8% Pd quenched from 500°C [3.44]

For weak, ideal and almost completely ordered solutions the general thermodynamic theory describes $\langle|c_k|^2\rangle$ and $\langle|\eta|^2\rangle$ with the equations (1.2.30, 32, 34) which readily demonstrate that in such cases correlations no longer affect diffuse scattering in the framework of this theory as well as in the pair interaction model.

Diffuse Scattering Features Due to the Conduction Electrons in Alloys. The effects of indirect interaction on the probabilities of the fluctuation waves of the concentration in alloys discussed in Sect. 2.4 must give rise to special features of diffuse scattering [1.17]. When we substitute (2.4.11) into (3.2.10) we can write the scattering intensity for disordered undistorted solutions as

$$I_1(\boldsymbol{Q}) = N(f_A - f_B)^2 \frac{c(1-c)}{1 + c(1-c)\frac{w_e(\boldsymbol{q}) + w_0(\boldsymbol{q}) - \bar{w}_e}{k_BT}}, \tag{3.3.14}$$

Here we have explicitly written the contribution $w_e(\boldsymbol{k})$ to the Fourier component of the ordering energies made by the conduction electrons ($\bar{w}_e$ is $w_e(\boldsymbol{k})$ averaged over the reciprocal lattice cell).

As demonstrated by (2.4.6–8), the function $w_e(\boldsymbol{q})$ exhibits singular features for $\boldsymbol{q} = 2\boldsymbol{k}_F - \boldsymbol{G}_0$ where the vector $2\boldsymbol{k}_F$ connects parallel parts of the Fermi surface and $\boldsymbol{G}_0$ is a vector of the reciprocal lattice (which is chosen for $\boldsymbol{q}$ to remain within the Brillouin zone. According to (3.3.14) the singular features of $w_e(\boldsymbol{q})$ must give rise to similar singular features in the behaviour of I_1 as a function of $\boldsymbol{Q}$ for $\boldsymbol{Q} = \boldsymbol{q} + \boldsymbol{G}$ (where $\boldsymbol{G}$ is an arbitrary reciprocal lattice vector). These features are increasingly noticeable for flatter parts of the Fermi surface which are connected by the vector $2\boldsymbol{k}_F = \boldsymbol{q} + \boldsymbol{G}_0$. It can be seen from (2.4.6–8) and (3.3.14) that for the ellipsoidal, cylindrical and flat parts of the Fermi surface the intensity I_1 as a function of $\boldsymbol{Q}$ exhibits a logarithmic singularity of the derivative and a logarithmic or root singularity of the function (which are spread out in the range of δk_z given by (2.4.9).

The dips in the function $\langle |c_{\boldsymbol{k}}|^2 \rangle$ which, as discussed in Sect. 2.4, correspond to small cavities or bridges in the Fermi surfaces, must give rise to dips in the plot of I_1 as a function of $\boldsymbol{Q}$ which lie near $\boldsymbol{q}$ connecting two cavities or bridges.

The equations for $I_1(\boldsymbol{Q})$ change their explicit form for the heterogeneous states of the alloys discussed in Sect. 2.5.2. The dependence of I_1 on $\boldsymbol{Q}$ for such alloys exhibits, however, singular features for $\boldsymbol{q} = 2\boldsymbol{k}_F - \boldsymbol{G}_0$. If we analyze these singularities for homogeneous and inhomogeneous alloys we can, in principle, calculate the diameters of the Fermi surface in alloys and analyze its shape and other characteristics [1.17; 2.61; 3.45]. Diffraction studies of the Fermi surface have an advantage (along with the positron annihilation method) in that they can be applied to the high-concentration solutions (in contrast to most classical techniques involving analysis of the electron properties of metals in high magnetic fields).

The electron energy of an alloy may affect also the behaviour of I_1 as a function of the concentration. For instance, if a variation of the composition changes the number of electrons per atom then near a concentration c_{cr} corresponding to production of new cavities in the Fermi surface the function φ_{cc} exhibits a spread-out root singularity [3.46]. This is why, as shown by (3.3.7), the concentration dependence of $I_1(\boldsymbol{Q})$ for small $\boldsymbol{q}$ includes a term proportional to $|c - c_{cr}|^{1/2}$ at one side of the point c_{cr} (for the concentrations at which new cavities are produced in the Fermi surface) and lacks such a term at the other side of this point.

3.3.2 Scattering by Inhomogeneous Solutions

The equilibrium heterogeneous systems discussed in Sect. 2.5 consisted of small ordered particles in a disordered matrix. Diffuse scattering by such systems must exhibit some singular features in the vicinities of the superlattice points of the reciprocal lattice which have been analyzed separately for systems with randomly distributed particles [2.143] and with correlated particles [2.146].

Scattering by inhomogeneous solid solutions is described formally by the general equation (3.3.1) but the correlation parameters $\varepsilon(\boldsymbol{\varrho})$ in it have components exhibiting characteristic dependence on $\boldsymbol{\varrho}$ (they oscillate and their amplitudes diminish sharply at distances of the order of the particle size and the distance between particles). It is difficult to analyze such complicated functions and the scattering patterns they determine. This is why several simple models of heterogeneous systems may prove convenient for analyzing scattering effects.

Let us consider first a system containing ordered particles with sufficiently sharp boundaries and the order parameter η. The particles belong to several types $a = (\alpha, \mathfrak{a})$ which differ in the orientations α of their axes (for instance, for the ordered structure $L1_0$.) and the vector $d_{\mathfrak{a}}$ describing the displacement with respect to the matrix (see Sect. 2.5.2). We shall assume that only short-range order prevails in the matrix with a characteristic range r_c which is considerably smaller than the particle size D_0 while the particles have both short-range and long-range order. The long-range order can be described by the probabilities $c_\gamma^\alpha(R_s)$ that the lattice sites γ in the particles are occupied by the given atoms.

The diffuse scattering intensity in this model can be conveniently found, both for crystals of a finite size and for the particles in a matrix (see Sects. 1.3, 4 in [1.1]), with the use of the shape functions $s_\alpha(\boldsymbol{r} - \boldsymbol{R}_{ma})$ for individual particles of orientation α. This function equals unity inside the respective particles and vanishes outside them ($\boldsymbol{R}_{ma}$ is the lattice vector giving the position of the mth particle of the type a. The shape function $s_0(\boldsymbol{r})$ for the disordered part of the matrix equals unity inside the matrix and vanishes inside the particles. Then we can rewrite (3.2.2) for undistorted crystals as

$$\begin{aligned} I_1(\boldsymbol{Q}) &= |f_A - f_B|^2 \langle |a_1(\boldsymbol{Q})|^2 \rangle \,, \\ a_1(\boldsymbol{Q})\& &= \int Y_0(\boldsymbol{r}) s_0(\boldsymbol{r}) \mathrm{e}^{\mathrm{i}\boldsymbol{Q}\cdot\boldsymbol{r}} d\boldsymbol{r} \\ &+ \sum_a \sum_m \int [\bar{Y}_a(\boldsymbol{r}) + Y_a'(\boldsymbol{r})] s_\alpha(\boldsymbol{r} - \boldsymbol{R}_{ma}) \mathrm{e}^{\mathrm{i}\boldsymbol{Q}\cdot\boldsymbol{r}} d\boldsymbol{r} \,, \end{aligned} \tag{3.3.15}$$

where

$$\begin{aligned} Y_0(\boldsymbol{r}) &= \sum_s (c_s - c)\delta(\boldsymbol{r} - \boldsymbol{R}_s) \,, \quad \bar{Y}_0(\boldsymbol{r}) = \sum_s [c_\gamma^a(\boldsymbol{R}_s) - c]\delta(\boldsymbol{r} - \boldsymbol{R}_s) \,, \\ Y_a'(\boldsymbol{r}) &= \sum_s [c_s - c_\gamma^a(\boldsymbol{R}_s)]\delta(\boldsymbol{r} - \boldsymbol{R}_s) \,. \end{aligned} \tag{3.3.16}$$

The functions $\langle Y_0(\boldsymbol{r}) Y_0(\boldsymbol{r}')\rangle$ and $\langle Y_a'(\boldsymbol{r}) Y_a'(\boldsymbol{r}')\rangle$ decrease sharply at a distance $r_c \ll D_0$. If we ignore the surface effects of the order of r_c/D_0 we find then that the integral

$$\int\int \langle Y_0(\boldsymbol{r}) Y_0(\boldsymbol{r}')\rangle s_0(\boldsymbol{r}) s_0(\boldsymbol{r}') \exp[\mathrm{i}\boldsymbol{Q}\cdot(\boldsymbol{r} - \boldsymbol{r}')] d\boldsymbol{r} d\boldsymbol{r}'$$

and similar integrals with the functions $Y_a'(\boldsymbol{r}) s_\alpha(\boldsymbol{r} - \boldsymbol{R}_{ma})$ reduce to the term in the square brackets in (3.3.1) multiplied by the number of atoms in the matrix or in the particle of the type a where the correlation parameters $\varepsilon_0(\varrho)$ and $\varepsilon_\alpha(\varrho)$ describe the short-range order in the matrix and particles, respectively. The terms with $s_0(\boldsymbol{r}) s_\alpha(\boldsymbol{r}' - \boldsymbol{R}_{ma})$ and $s_\alpha(\boldsymbol{r} - \boldsymbol{R}_{ma}) s_{\alpha'}(\boldsymbol{r}' - \boldsymbol{R}_{m'a'})$ can be ignored for $am \neq a'm'$

while the terms with $Y_0(\boldsymbol{r})\bar{Y}_a(\boldsymbol{r}')$ and $Y_a'(\boldsymbol{r})\bar{Y}_{a'}(\boldsymbol{r}')$ are eliminated upon averaging. The Fourier component of $\bar{Y}_a(\boldsymbol{r})s_\alpha(\boldsymbol{r}-\boldsymbol{R}_{ma})$ in (3.3.15) is significant only near the superlattice points $\boldsymbol{G}'_\alpha$ of the reciprocal lattice which appear upon ordering of the particles a. This component is calculated as it was for the integral (1.3.2) in [1.1] which yields expression (1.3.7) in [1.1]. Then we obtain the following sum:

$$\begin{aligned} &I_1(\boldsymbol{Q}) = I_1^{(0)}(\boldsymbol{Q}) + I_1^{(1)}(\boldsymbol{Q}) \,, \\ &I_1^{(1)}(\boldsymbol{Q}) = (f_A - f_B)^2 \left(\frac{\xi\eta}{2v}\right)^2 \left|\sum_a s_\alpha(\boldsymbol{q}_\alpha) p_a(\boldsymbol{q}_\alpha)\right|^2 \,, \\ &p_a(\boldsymbol{q}_\alpha) = \sum_{\boldsymbol{R}_{ma}} \exp[\mathrm{i}(\boldsymbol{q}_\alpha \cdot \boldsymbol{R}_{ma} + \boldsymbol{G}'_\alpha \cdot \boldsymbol{d}_a)] \,, \quad \boldsymbol{q}_\alpha = \boldsymbol{Q} - \boldsymbol{G}'_\alpha \,, \\ &s_\alpha(\boldsymbol{q}_\alpha) = \int s_\alpha(\boldsymbol{r}) \exp(\mathrm{i}\boldsymbol{q}_\alpha \cdot \boldsymbol{r}) d\boldsymbol{r} \,. \end{aligned} \tag{3.3.17}$$

Here $I_1^{(0)}(\boldsymbol{Q})$ is given by (3.3.5) (where $c\varrho \ll D_0$) in which $\varepsilon(\varrho)$ equals the mean

$$\varepsilon_0(\varrho) + \sum_a p_a[\varepsilon_a(\varrho) - \varepsilon_0(\varrho)] \,,$$

where p_a is the volume fraction of the particles a in the system. It describes a smoothly varying intensity distribution with the width of the order of π/r_c while $I_1^{(1)}(\boldsymbol{Q})$ describes a considerably more narrow distribution with a width about $2\pi/D_0$ (the constant ξ is of the order of unity in the expression for $I_1^{(1)}$ equals unity for the structure $L1_0$ and $1/2$ for the structure $L1_2$).

When only particles with the same shape functions $s_a(\boldsymbol{q}_\alpha)$ make contributions to a peak near a given point $\boldsymbol{G}'_\alpha$ and only $d_\mathfrak{a}$ and the phases $\boldsymbol{G}'_\alpha d_\mathfrak{a}$ are different for such particles we can write

$$\begin{aligned} I_1^{(1)}(\boldsymbol{Q}) &= N_{p\alpha}(f_A - f_B)^2 \frac{\xi^2\eta^2}{4v^2} |s_\alpha(\boldsymbol{q}_\alpha)|^2 \\ &\times \left\{ 1 + \sum_{\mathfrak{a}\mathfrak{a}'} \sum_{\varrho \neq 0} P_{\alpha\mathfrak{a}\mathfrak{a}'}(\varrho) \cos[\boldsymbol{q}_\alpha \cdot \boldsymbol{\varrho} + \boldsymbol{G}'_\alpha \cdot \boldsymbol{d}_{\mathfrak{a}\mathfrak{a}'}] \right\} \,. \end{aligned} \tag{3.3.18}$$

Here $N_{p\alpha} = \sum_\mathfrak{a} N_{p\alpha\mathfrak{a}}$ is the number of the particles α in the crystal (or the number of all cubic particles when we consider the $L1_2$ structure), $P_{\alpha\mathfrak{a}\mathfrak{a}'}$ is the probability that a particle with the displacement $d_{\mathfrak{a}'}$ is at a distance ϱ from the particle $\mathfrak{a}$, and $d_{\mathfrak{a}\mathfrak{a}'} = d_\mathfrak{a} - d_{\mathfrak{a}'}$.

If the arrangement of the particles is not correlated and/or the same numbers of particles have $\cos \boldsymbol{G}'_\alpha \cdot \boldsymbol{d}_{\mathfrak{a}\mathfrak{a}'} = \pm 1$ then the term in braces in (3.3.18) equals unity (or differs from unity by a factor of the order of $P_{\alpha\mathfrak{a}\mathfrak{a}'}(q \ll 1)$) while the intensity distribution $I_1^{(1)}(\boldsymbol{Q})$ depends only on the shape function and exhibits a peak for $\boldsymbol{Q} = \boldsymbol{G}'_\alpha$. Equations (1.3.11, 14, 15) in [1.11] and (3.3.18) demonstrate that its integrated intensity is equal to the integrated intensity of the superlattice reflection of the homogeneously ordered crystal whose volume equals the volume of particles while the integral widths along the axes q_i are $2\pi/D_{0i}$ for slab shaped particles with edge lengths D_{0i}. If the particles are correlated so that neighbouring particles are more likely to have different phases, then an oscillating factor of a constant sign is superimposed on the function $|s_\alpha(\boldsymbol{q}_\alpha)|^2$ in (3.3.18). This factor in brackets modulates the intensity and can produce a minimum for $\boldsymbol{Q} = \boldsymbol{G}'_\alpha$ and several peaks for $\boldsymbol{Q} \neq \boldsymbol{G}'_\alpha$ (near the peak of $s^2_\alpha(\boldsymbol{q}_\alpha)$).

For instance, for systems of strongly correlated particles producing fcc lattice regions (see Sect. 2.5.2) the term in brackets in in (3.3.18) reduces to $|p(\boldsymbol{q})|^2 N_p$ where $p(\boldsymbol{q})$ is given by (3.5.13) and the peaks are at the points $\boldsymbol{q} = \boldsymbol{q}_n$ (see (2.5.18)). For two types of the fcc lattice with $\boldsymbol{d}_a$ corresponding to the lowest electron energy (2.5.19), $\boldsymbol{q}_n$ the G' (110) point correspond to the sets (n_1, n_2, n_3) for which $n_1 + n_2$ is odd and n_3 is an arbitrary integer. Then, according to (2.5.18), the highest intensity in the (110) plane is exhibited by four peaks for $\boldsymbol{n} = (\pm 100)$ and (0 ± 10) forming a "cross" near the point (110). For another lattice of particles with a lower surface energy corresponding to (2.5.20) n_1 and n_2 are even and n_3 is odd or n_1 and n_2 are odd and n_3 is even and the highest intensity in the plane (110) is exhibited by two peaks for $n = (110)$ and $(\bar{1}\bar{1}0)$ (and two peaks for $n = (00 \pm 1)$ outside this plane). The peak widths are dependent on the length L of the correlated regions of the lattice of particles. They can be estimated from the equation $\delta q = 2\pi/L$ in the pseudocrystal model of this lattice. As noted in Sect. 2.5, in other systems production of neighbouring particles with the same phases may be more probable. For them the intensity peak must be at $\boldsymbol{Q} = \boldsymbol{G}'$ and its width must be smaller than that for the uncorrelated particles.

Equation (3.3.18) may be used also for describing the intensity of scattering by the ordered regions with blurred boundaries, that is, for a steadily varying function $\eta_\alpha(\boldsymbol{r})$. If the distribution of $\eta_\alpha(\boldsymbol{r})$ in different regions do not overlap then it is sufficient to make the following replacement in (3.3.18):

$$\eta s_\alpha(\boldsymbol{q}) \to \int \eta_\alpha(\boldsymbol{r}) \exp(i\boldsymbol{q} \cdot \boldsymbol{r}) d\boldsymbol{r} \,. \tag{3.3.19}$$

Then the intensity in the wings of the distribution $I_1^{(1)}(\boldsymbol{Q})$ decreases at a faster rate.

As shown in Sect. 2.5, variation of the order in the particles is accompanied by a change in the concentration $c - c_0$ with respect to the mean concentration c_0 in the solution. This may give rise to scattering intensity peaks near the lattice points $\boldsymbol{G}$ of the reciprocal lattice and for small scattering angles. In undistorted crystals for $p \ll 1$ we have

$$I_1^{(1)}(\boldsymbol{Q}) = \sum_\alpha N_{p\alpha}(f_A - f_B)^2 \frac{(c - c_0)^2}{v^2} |s_\alpha(\boldsymbol{q})|^2 \times \left[1 + \sum_{aa'} P_{aa'}(\boldsymbol{\varrho}) \cos \boldsymbol{q} \cdot \boldsymbol{\varrho}\right] \tag{3.3.20}$$

and $I_1^{(1)}(\boldsymbol{Q})$ exhibits a maximum for $\boldsymbol{q} = \boldsymbol{Q} - \boldsymbol{G} = 0$ irrespective of the phases of the particles.

The above features of scattering by inhomogeneous solutions are found mainly near the reciprocal lattice points while far from them the only effect of inhomogeneities is the replacement of the $\varepsilon(\varrho)$ in (3.3.1) with their averages. Therefore, for small $p \ll 1$ the Fermi transformation method used above for finding $\varepsilon(\varrho)$ and $w(\varrho)$ can also give these parameters for the first coordination spheres (where the vicinities of the reciprocal lattice points make only small contributions to the Fourier integrals) in the inhomogeneous solutions (with an error of the order of p or pr_c/D_0). But for ϱ of the order of unity the errors in $w(\varrho)$ calculated in this way become significant. Note that the equations (3.3.7, 8) of the thermodynamic theory are rigourously applicable to highly inhomogeneous solutions only for very small q/π which are smaller than the reciprocal of the distance between the particles.

3.4 Diffuse Scattering in Distorted Solid Solutions

3.4.1 Scattering Caused by Concentration Fluctuations

The static displacements caused by the difference between the atomic radii of the solution components produce two kinds of effects on diffuse scattering patterns. Firstly, the imperfections produced by displacements cause additional scattering which changes qualitatively the scattering intensity distribution in the vicinities of the reciprocal lattice points. Secondly, the long-range elastic interaction between atoms caused by displacements gives rise to special features in the behaviour of energies $w(\varrho)$ for large ϱ and changes in the results derived in the conventional theory of fluctuations (see Sect. 2.1). They give rise to additional singular features in the scattering for small q.

It has been shown in Sect. 1.5 that in homogeneous high-concentration solutions the distortions are caused by the fluctuations of both the concentration and the correlation parameters $\varepsilon(\varrho)$ (see (1.5.2)). The part played by these parameters may be insignificant if the Vegard rule is satisfied and the dependence of the elastic moduli on the composition is weak, that is, if the dependence of the lattice parameters on $\varepsilon(\varrho)$ is weak. First we shall discuss solutions of this kind while the effects caused by the fluctuations of the correlation parameters will be treated in Sect. 3.4.2.

Effect Produced by Distortions in Solutions on Diffuse Scattering. If distortions are not very large (the conditions (3.2.9) are satisfied) the intensity of diffuse scattering by disorderred substitutional solution with a single atom per crystal cell is given by (3.2.10). Taking (1.1.14) for $\langle|c_k|^2\rangle$ and using the fact that for small k, according to (1.5.4–7), $A_k \sim k^{-1}$ we obtain for the scattering intensity [1.8]

$$I_1 = N\left[c(1-c) + \sum_{\varrho\neq 0}\varepsilon(\varrho)\cos \boldsymbol{q}\cdot\boldsymbol{\varrho}\right][f\boldsymbol{Q}\cdot\boldsymbol{A}_q - (f_A - f_B)]^2\,, \qquad (3.4.1)$$

$$\boldsymbol{A}_q = \frac{a_q \boldsymbol{e}_q}{q}\,,$$

where $|e_q| = 1$ and $a_q e_q$ tends to a constant limit for $q \to 0$. The squared first term in square brackets in (3.4.1) is determined by scattering caused only by distortions; the squared second term is determined by scattering caused by the difference in the atomic factors; and the doubled product of these terms gives the interference contribution due to both factors. The explicit expressions for the unit vector e_q and for the parameter a_q which describes the distortions are given in Sects. 2.2, 3 of [1.1] (see also the equations for A_k derived in Sect. 1.5). If $f_A - f_B$ is replaced with the variation $\Delta f(Q)$ of the structure amplitude caused by the introduction of an impurity then equation (3.4.1) (and (3.4.3)) is also applicable to the interstitial solutions where impurities occupy interstices of only one type.

If the crystal has no distortions ($a_q = 0$) then (3.4.1) reduces to (3.3.1) and the function $H(Q) = I_1 N^{-1}(f_A - f_B)^{-2}$ depends regularly on Q and has identical values for the points q related to each other by point symmetry group transformations. In order to take into consideration the distortions one just has to make the replacement

$$f_A - f_B \to f_A - f_B - f a_q \frac{Q \cdot e_q}{q} \tag{3.4.2}$$

in the equation for the scattering intensity. Distortions are particularly significant if the lattice parameters depend strongly on the concentration (a_q is large) or if the the structure amplitudes are close. But even for small $|a_q| \ll 1$ and for $|f_A - f_B| \sim f$, the last term in (3.4.2) makes the major contribution in the vicinities of the reciprocal lattice points ($q \ll k_m$), that is, the intensity $I_1 \sim Q^2/q^2$ of scattering by distortions plays the dominant part.

According to (3.4.1), in crystals with distortions $H(Q)$ no longer is a regular function of Q since the intensity of scattering by distortions is proportional to Q^2 while the interference intensity term is proportional to Q. The function $H(Q)$ also no longer has identical values at the points q which coincide under the symmetry transformations applied to the crystal. It can be seen, for instance, from the plot in Fig. 3.4 for the ratio $H(2\pi(1 - \Delta h)/d, 0, 0)/H(2\pi(1 + \Delta h), 0 0)$ between the scattering intensities in the disordered alloy Cu_3Au at the points which are at the same distance from the superlattice point (100) along the [100] axis (which are related by the symmetry transformations). For undistorted crystals this ratio must be unity while in the strongly distorted Cu_3Au crystal it is considerably smaller and the ratio calculated [3.4.7] from (1.5.17, 18) and (3.4.1) is in a good agreement with the experimental results [3.48]. This agreement confirms that the intensity of scattecing by distortions can be described approximately by performing the replacement (3.4.2) and using the equations for A_q derived in Sect. 1.5 (or in Sects. 2.2.2, 3 in [1.1]).

In weak solutions without correlation (3.4.1) reduces to (4.1.4) in [1.1] derived for crystals with low concentrations of point defects while in ideal solutions it differs from this equation in the constant factor $(1 - c)$. Therefore, all the results found from (4.1.4) in [1.1] for the scattering intensity distribution (concerning the shapes of the equal-intensity surfaces, the effects of elastic anisotropy, the dependence of I_1 on q, asymmetry of this dependence and so on) are fully applicable to the ideal substitutional solutions. Using an equation similar to (4.1.3) in [1.1] (where c_α is replaced with $c_\alpha(1 - c_\alpha)$) we can also study scattering effects in ideal interstitial solutions in which atoms occupy interstices of different types α with the probabilities c_α which are generally different. As noted in Sect. 4.1 of [1.1], we can find the positions of impurities in the lattice and some parameters describing the distortion field and the interaction between atoms by comparing such theoretical predictions with the experimental data.

Scattering Intensity and Thermodynamic Parameters of Distorted Solutions. For nonideal solutions the expressions for $\langle|c_q|^2\rangle$ have a different form and (3.4.1) then contains terms with $\varepsilon(\varrho)$ which give rise to a significant additional dependence of I_1 on $\boldsymbol{Q}$. Under certain conditions it is more convenient to express the factor containing $\varepsilon(\varrho)$ in (3.4.1) in terms of thermodynamic parameters of the solution. For instance, in the vicinities of the reciprocal lattice points for small q we can make use of the expression for $\langle|c_q|^2\rangle$ derived in the macroscopic theory of fluctuations as it was done for undistorted solutions. The long-range elastic forces which play a significant part in crystals with distortions will, however, change this expression. For small q we obtain then (2.1.5). Substituting (2.1.5) into (3.2.10) we obtain the following equation for the scattering intensity in the vicinity of the reciprocal lattice point [1.16]:

$$I_1 = \frac{N}{v}\frac{k_B T}{\varphi_{cc}(\boldsymbol{q})}\left[f a_q \frac{\boldsymbol{Q}\cdot\boldsymbol{e}_q}{q} - (f_A - f_B)\right]^2 , \quad \varphi_{cc}(\boldsymbol{q}) = \frac{d^2\varphi}{dc^2} . \tag{3.4.3}$$

It can be seen from (3.4.3) that in nonideal solutions a difference between atomic radii of the components has a direct effect on diffuse scattering causing scattering by distortions and an indirect effect associated with the long-range elastic interaction $\varphi_{cc}^{\mathrm{el}}$ (see (2.1.5)) which affects the probability of fluctuations and diffuse scattering character. According to (1.2.3), $d^2\varphi/dc^2$ is determined by the concentration dependence of the chemical potentials of the solution components. According to (2.1.5) and the equations for $\boldsymbol{A}_k$ derived in Sect. 1.5, $\varphi_{cc}^{\mathrm{el}}$ is determined by the elastic moduli and the concentration dependence of the sizes and shapes of the lattice cells. Therefore, for small q all parameters in (3.4.3) can be found from the independent experimental results and then the scattering intensity distribution in the vicinities of the reciprocal lattice points can be calculated.

If the contribution of the elastic energy $\varphi_{cc}^{\mathrm{el}}$ to $\varphi_{cc}(\boldsymbol{q})$ is much smaller than $d^2\varphi/dc^2$ then $\varphi_{cc}(\boldsymbol{q})$ is almost constant. When we compare (3.4.3) and (3.4.1) for the ideal solution (with $\varepsilon(\varrho) = 0$) we see that the shapes of the equal intensity lines for small q are the same as in the case of the weak and ideal solutions (see Sect. 4.1 in [1.1]). Only the numerical value of I_1 is changed. In ordering solutions $\varphi_{cc}(\boldsymbol{q})$ is typically larger than the result for the ideal solutions, $c(1-c)k_B T/v$, that is, I_1 is lower for small q in the presence of short-range order. In decomposing solutions we typically have $\varphi_{cc}(\boldsymbol{q}) < c(1 - c)k_B T/v$ and for small q the intensity I_1 increases in the presence of short-range order (exceptions from this rule may occur). If the lattice parameter strongly depends on the composition and $\varphi_{cc}^{\mathrm{el}}$ is comparable with $d^2\varphi/dc^2$ then in strongly anisotropic crystals the nonanalytic function $\varphi_{cc}(\boldsymbol{q})$ may significantly depend on the direction of the vector $\boldsymbol{q}$ and therefore the shapes of the equal intensity surfaces become more complicated than in the case of the weak and ideal solutions.

Experimental results on diffuse neutron scattering in Nb-D solutions [3.49, 50] indicate that the interaction between impurities is primarily elastic in character. This observation agrees with the introduction of the term φ_{cc}^{el} dependent on the direction of q in the denominator of (3.4.3) (where $f_A - f_B$ must be replaced with $\Delta f(\boldsymbol{Q})$ for the interstitial solutions). According to (2.1.5) and (3.4.3), for a small q the quantities $\langle |c_{\boldsymbol{k}}|^2 \rangle$ and I_1 are considerably diminished at the expense of the elastic energy and $\varphi_{cc}(\boldsymbol{q})$ found from these experimental results decreases significantly when we go over from the directions $\langle 100 \rangle$ to $\langle 111 \rangle$. Another illustration of systems with primarily elastic interactions between atoms is given by carbon solutions in Fe-C martensites. Numerical calculations made for such interaction [1.18, 19] yielded equal intensity surfaces whose characteristic shapes explain the results on the elastic scattering of electrons in martensite [3.51].

For large q the intensity I_1 can be expressed directly in terms of the ordering energy derived in the framework of the pair interaction in addition to (3.4.1) which gives I_1 in terms of the correlation parameters. As in the case of undistorted solutions, we obtain from (1.3.4) and (3.2.10)

$$I_1 = \frac{N}{x_{\boldsymbol{q}}} \left[f a_{\boldsymbol{q}} \frac{\boldsymbol{Q} \cdot \boldsymbol{e}_{\boldsymbol{q}}}{q} - (f_A - f_B) \right]^2 , \quad x_{\boldsymbol{q}} = \sum_{\boldsymbol{\varrho}} x(\boldsymbol{\varrho}) \cos \boldsymbol{q} \cdot \boldsymbol{\varrho} , \tag{3.4.4}$$

where $x(\boldsymbol{\varrho})$ are related to the ordering energies $w(\boldsymbol{\varrho})$ by the equations (1.3.9) (at high temperatures) or (1.3.12) (for low concentrations). In elastically anisotropic crystals with distortions the energies $w(\boldsymbol{\varrho})$ include, in particular, the elastic interaction energy $\mathscr{E}(\boldsymbol{\varrho})$ given by (2.1.13) which decreases with increasing distance at a relatively slow rate (as $1/\varrho^3$) and, accordingly, $x_{\boldsymbol{q}}$ includes the term $\delta x_{\boldsymbol{q}}$ by (2.1.14) which has a nonanalytic form for small q and makes $\varphi_{cc}(\boldsymbol{q})$ in (3.4.3) nonanalytic.

For large q crystal distortions make apparently a relatively smaller contribution. If the dependence of the lattice parameters on the composition is not very strong, that is $|a_{\boldsymbol{q}}| \ll 1$, and f_A is not too close to f_B then for large q we have $|f a_{\boldsymbol{q}} \boldsymbol{Q} \cdot \boldsymbol{e}_{\boldsymbol{q}} / q| \ll |f_A - f_B|$ and the effects caused by distortions are small. The equations for I_1 then reduce to the formulas derived in Sect. 3.3.

Correlation Parameters, Ordering Energies and Force Constants in Distorted Solutions. We can derive the correlation parameters $\varepsilon(\boldsymbol{\varrho})$ and the ordering energies $w(\boldsymbol{\varrho})$ from the intensity distribution $I_1(\boldsymbol{Q})$ using the Fourier transformations and (3.4.1, 4):

$$\varepsilon(\boldsymbol{\varrho}) = \frac{v}{8\pi^3} \int_{v_r} d\boldsymbol{Q} \frac{I_1(\boldsymbol{Q})}{N} \left[f a_{\boldsymbol{q}} \frac{\boldsymbol{Q} \cdot \boldsymbol{e}_{\boldsymbol{q}}}{q} - (f_A - f_B) \right]^{-2} \cos \boldsymbol{Q} \cdot \boldsymbol{\varrho} , \tag{3.4.5}$$

$$x(\varrho) = \frac{v}{8\pi^3} \int_{v_e} dQ \frac{N}{I_1(Q)} \left[f a_q \frac{Q \cdot e_q}{q} - (f_A - f_B) \right]^2 \cos Q \cdot \varrho \,. \qquad (3.4.6)$$

Equations (3.4.5, 6) extend (3.3.3, 10) to the case of distorted crystals. The contribution of the scattering by thermal vibrations to I_1 has been eliminated here. In order to make use of these equations one must know the factors $a_q e_q$. For small q when the contribution of distortions is especially significant $a_q e_q = q A_q$ can be calculated from the exact equations of the macroscopic theory given in Sect. 1.5 (for instance, from (1.5.7) for cubic crystals (see also Sect. 2.2 in [1.1]). For large q when the contribution of distortions is not very large (if $|f_A - f_B| \sim f$) these factors can be found from the approximate equations such as (1.5.17, 18).

Equations (3.4.5, 6) allow us to find approximately $\varepsilon(\varrho)$ and $w(\varrho)$ from the results on the intensity scattering distribution I_1 in a single reciprocal lattice cell. These approximations can be quite rough, however, because of the inaccuracies in $a_q e_q$ for large q and the simplifications made in the derivations of the equation for I_1 which included, in particular, ignoring scattering by the fluctuations of the correlation parameters (see Sect. 3.4.2). Thus, if the available data describe the scattering intensity distribution for several reciprocal lattice cells it is preferable to separate the intensity of scattering by distortions using the technique [3.13–15] described in Sect. 3.2 and only then to perform the Fourier transformations for the functions $H_0(q)$ given by (3.2.6) or $H_0^{-1}(q)$ given by (3.3.10).

In solutions which satisfy the Vegard rule we can, in principle, use (3.4.1) to find the set of the force constants $W_{ss'}$ which describe the effects produced by atoms with different radii on the averaged crystal. This can be conveniently done by analyzing the ratios between $I_1(Q)$ values for Q corresponding to q which are linked by the symmetry transformations (for instance, at the points with q and $-q$ or at the points symmetrically positioned with respect to the superlattice point as shown in Fig. 3.4) [3.47]. The factors (3.4.1) including $\varepsilon(\varrho)$ are cancelled out in these ratios and thus we can directly calculate A_q for various q. Such calculations can be more precise for solutions with small ratios $|f_A - f_B|/f$. The parameters W_k can then be found by substituting these A_q into (1.5.117) (where $A_k^c = A_k$ for the solutions which satisfy the Vegard rule) and calculating V_{kij} from the results of inelastic neutron scattering. Applying the Fourier transformation to them we can then obtain all the force constants $W_{ss'}$ according to (1.5.15) A method has also been suggested [3.33] for calculating the force constants from the results on $I_1(Q)$ along the high-symmetry axes $\langle 100 \rangle$, $\langle 110 \rangle$ and $\langle 111 \rangle$.

When we perform averaging of (3.4.1) over the crystal orientations, according to (3.1.14), we can find the diffuse scattering intensity distribution $I_{1\mathrm{D}}$ in the Debye powder diffraction pattern. In distorted crystals $I_{1\mathrm{D}}$ linearly depends on the correlation parameters $\varepsilon(\varrho)$ and the coefficients in $\varepsilon(\varrho)$ are complex functions of the scattering angle which have been analyzed in [3.52, 53]. If we know these functions we can find the correlation parameters in a solution with distortions from the diffuse scattering intensity distribution in the Debye diffraction pattern although the accuracy of such results will be considerably poorer than in the case of single crystals (see details in [1.27]).

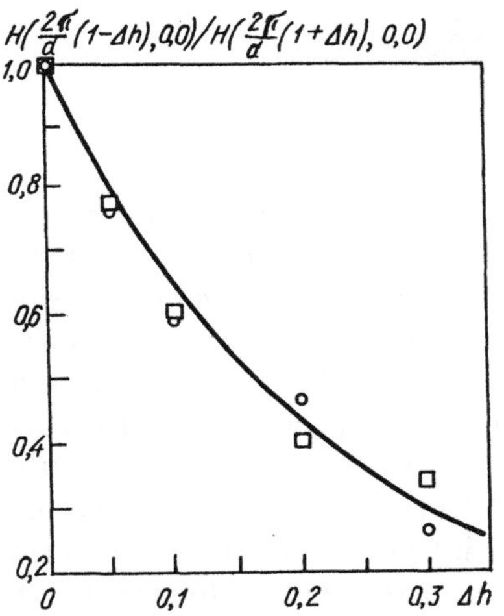

Fig. 3.4. The ratio between the functions $H(Q) = I_1(Q)N^{-1}(f_A - f_B)^{-2}$ at equal distances, Δh, from the superlattice point (100), but in opposite directions (+ or −). Open circles correspond to 405° and open squares to 450°

The expansion (3.2.10) of the scattering intensity in powers of the amplitudes of the concentration waves and the related displacement waves can be employed not only for homogeneous solutions but also for the modulated crystal structures produced in spinodal decomposition.

Higher-Order Terms in the Expansion of Scattering in Powers of Displacements. When we analyzed diffuse scattering above we considered only the first terms in the expansion of the intensity I_1 in powers of the displacements. We considered, in fact, only scattering by a single wave of the concentration fluctuations corresponding to the given $\boldsymbol{q}$. As the distortions and Q grow, an increasing contribution must be apparently made by the higher-order terms in the expansion of I_1 in powers of $\boldsymbol{Q} \cdot \boldsymbol{u}_s$ which describe scattering associated with two or more fluctuation waves and are, in a certain sense, similar to many-phonon scattering by the thermal vibrations. This scattering process is treated in [1.12].

To illustrate the contributions made by the higher-order terms in the above expansion let us discuss the results obtained in the approximation of an elastic isotropic continuum for the simplest case of the vicinities of the reciprocal lattice points in ideal solutions. If we include the terms of the fourth order in $\boldsymbol{u_k}$ and $c_{\boldsymbol{k}}$ in the expansion of I_1 we obtain the additional terms $I_1' + I_1''$:

$$\begin{aligned} I_1' &= 2Nf^2(1-2c)a^0 M\frac{\boldsymbol{Q}\cdot\boldsymbol{q}}{q^2} \\ &\quad + \frac{1}{512}Nf^2c^2(1-c)^2(a^0)^4 v\frac{Q^4}{q}(8\cos^2\varphi + 3\sin^4\varphi)\,, \end{aligned} \quad (3.4.7)$$

$$\begin{aligned} I_1'' &= Nf(f_A - f_B)[2(1-2c)M \\ &\quad + \frac{1}{64}c^2(1-c)^2(a^0)^3 Q^3 v\cos\varphi(1-\cos^2\varphi)\,, \end{aligned} \quad (3.4.8)$$

where φ is the angle between $\boldsymbol{Q}$ and $\boldsymbol{q}$;

$$2M = \langle (\boldsymbol{Q} \cdot \boldsymbol{u})^2 \rangle = \frac{v}{8\pi^3} c(1-c) \int (\boldsymbol{Q} \cdot \boldsymbol{A}_k)^2 d\boldsymbol{k} ,$$

$$a^0 = \frac{1+\sigma}{3(1-\sigma)} \frac{1}{v} \frac{dv}{dc} .$$

It can be seen from (3.4.7) that for $q \to 0$ term I_1' tends to infinity as q^{-1}, rather than q^{-2}, as is the case for I_1. The first term in (3.4.7) reverses its sign when $\boldsymbol{q}$ is replaced with $-\boldsymbol{q}$ and hence we obtain additional asymmetry in the intensity distribution in the vicinity of a reciprocal lattice point. (For small concentrations c this term coincides with the last term in (4.1.12) in [1.1].) The second term in (3.4.7) does not vanish for any angles φ between $\boldsymbol{Q}$ and $\boldsymbol{q}$ and it makes the equal intensity lines differ from the lemniscate-type curves which are typical for crystals where defects occupy lattice sites (see Figs. 4.2, 3 in [1.1]). As a result, the equal intensity lines do not pass through the reciprocal lattice point. These effects are observed though only in cells of the reciprocal lattice well-removed from the origin (for large Q) and for not too small q in strongly distorted solutions. For instance, for $f_A \approx f_B$, $(a^0)^2 = 0.1$ (strongly distorted crystal) and high reflection indices $\sum h_i^2 \approx 60$ (we have here $2M \approx 0.6$ for the fcc lattice according to (3.4.39) in [1.1]), $q = 2\pi/5d$, and $\varphi = 0$ the ratio between the second term in (3.4.7) and (3.4.1) is 0.3 for $c = 1/2$ and $\varepsilon(\boldsymbol{\varrho}) = 0$ (for $\varphi = 90°$ the second term is smaller by a factor of three), while the ratio between the first term in (3.4.7) and (3.4.1) for $c = 1/4$ is 0.1 (or -0.1 for $\varphi = 180°$). Note that a comparison of (3.4.7) and (3.4.1) yields the condition (3.2.9) for disordered solutions.

The contribution of the higher-order terms in the expansion in powers of displacements can be greater in nonideal solutions especially in the presence of the concentration inhomogeneities such as those in ageing alloys (see Sect. 4.2 in [1.1]).

3.4.2 Diffuse Scattering by Fluctuations of the Correlation Parameters

The results derived in Sect. 3.4.1 are strictly applicable only to solutions in which the lattice parameter is not affected by the correlation or the concentration is low (the condition (1.5.24) is satisfied). If considerable deviations from the Vegard rule occur in a concentrated solution and the lattice parameter significantly depends on the order then, according to the results presented in Sect. 1.5, the static distortions are produced not only by the fluctuations of concentration but also by the fluctuations of the correlation parameters. Then, according to (1.5.16), $\boldsymbol{u}_{\boldsymbol{k}}$ linearly depend on $c_{\boldsymbol{k}}$ and on the Fourier $\varepsilon_{\boldsymbol{k}}(\boldsymbol{\varrho})$ of the correlation parameters. This causes additional scattering by the distortions produced by the fluctuations of the correlation parameters which can be significant in such solutions and affect noticeably the intensity distribution [1.20].

Let us consider scattering by the fluctuations of the correlation parameters in ideal disordered solutions with a single atom per unit cell. It is shown by (1.4.3) that under these conditions we have statistically independent fluctuations of the concentration and the correlation parameters as well as the fluctuations of the correlation parameters $\varepsilon(\varrho)$ and $\varepsilon(\varrho')$ with $\varrho' \neq \pm\varrho'$ and the results can be significantly simplified. To find the diffuse scattering intensity let us replace $\boldsymbol{u}_q$ in (3.2.8) (for $\nu = 1$) with (1.5.16) and perform averaging using (1.4.3). Then the diffuse scattering intensity will be written as a sum of two terms $I_1 = I_1^c + I_1^\varepsilon$.

The first term describes scattering by the concentration fluctuations and coincides with (3.4.1) for $\varepsilon(\varrho) = 0$. The second term decribes the fluctuations of the correlation parameters:

$$I_1^\varepsilon = Nc^2(1-c)^2 f^2 \sum_{\varrho} (\boldsymbol{Q} \cdot \boldsymbol{A}_q(\varrho))^2 (1 - \cos \boldsymbol{q} \cdot \varrho) , \tag{3.4.9}$$

Here $\boldsymbol{A}_q(\varrho)$ are the solutions of the equations (1.5.17).

It can be seen from (1.5.15, 17) that for small q the functions $\boldsymbol{A}_q(\varrho)$ are proportional to $1/q^2$. (The expansion of V_{kij} in powers of k begins with the terms proportional to k^2, see, for instance, (2.3.8) in [1.1].) Therefore I_1^ε as well as I_1^c tend to infinity as q^{-2} in the vicinities of the reciprocal lattice points. Using (1.5.21) we can write in these vicinities

$$I_1^\varepsilon = Nc^2(1-c)^2 f^2 \frac{1}{q^2} \sum_{i,j=1}^{2} Q_i Q_j f_{ij}(\boldsymbol{n}) , \tag{3.4.10}$$

where $f_{ij}(\boldsymbol{n})$ depend only on the direction $\boldsymbol{n} = \boldsymbol{q}/q$ but not on the modulus of the vector $\boldsymbol{q}$. In elastically isotropic fcc crystals they are given in the nearest-neighbour approximation by (1.5.22) (the respective expressions for elastically anisotropic crystals ars given in [1.20]).

Though for small q the intensity I_1^ε as well as I_1^c grows as $1/q^2$ these scattering intensities exhibit significantly different types of behaviour as functions of the direction of the vector $\boldsymbol{q}$. In contrast to I_1^c, the intensity I_1^ε does not vanish for any directiong of $\boldsymbol{q}$ and the respective equal intensity lines have oval shapes (in contrast to lemniscate-type shapes for the I_1^c equal intensity lines). Figure 3.5 shows (in arbitrary units) the I_1^c equal intensity lines (for $f_A \approx f_B$) and the I_1^ε lines (the plots 1 and 2) in the (001) plane of the reciprocal lattice space near the point ($h00$) and the equal intensity lines for the total scattering intensities $I_1 = I_1^c + I_1^\varepsilon$ for $c(1-c)(\partial v/\partial \varepsilon)^2(\partial v/\partial c)^{-2} = 0.2$ (curve 3) and for $c(1-c)(\partial v/\partial \varepsilon)^2(\partial v/\partial c)^{-2} = 0.08$ (curve 4). These lines are plotted for elastically isotropic bcc crystals with a Poisson ratio of 0.3. The functions $f_{ij}(\boldsymbol{n})$ in (3.4.10) have been calculated in the nearest neighbour approximation.

In order to estimate the relative contributions of the scattering by the fluctuations of the correlation parameters and the scattering by the concentration fluctuations let us compare for a given q near the point ($h00$) the maximal value (dependent on the angles) of the term proportional to q^{-2} in I_1^c (determined by the distortions) and the maximal value of I_1^ε. According to (1.5.7b, 22) and (3.4.1, 10) for isotropic crystals the latter is considerably smaller than the former if the condition (1.5.23) or the equivalent condition (1.5.24) are satisfied. It can be seen that the intensity of scattering by the fluctuations of the correlation parameters is

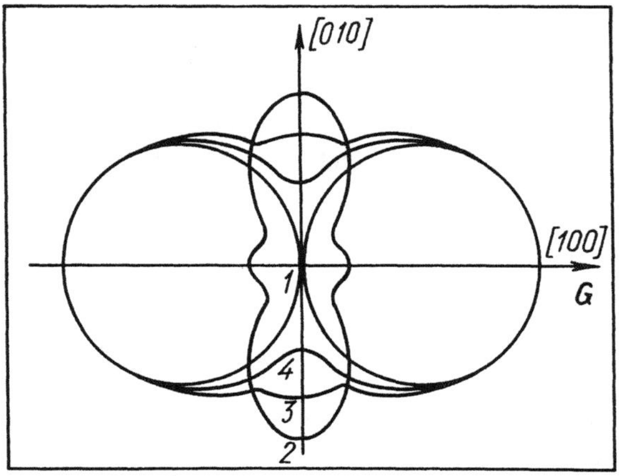

Fig. 3.5. Equal intensity lines in the (001) plane of the reciprocal lattice space near the point ($h00$) with the inclusion of the fluctuations of the correlation parameters

relatively low if the lattice parameter has only a weak dependence on the correlation or (less precisely) if the concentration dependence of the lattice parameter is almost linear (deviations from the Vegard rule are small) and if the concentration of a component in the solution is low. Under certain conditions, however, the inequalities (1.5.23) are not satisfied and I_1^ε becomes comparable with I_1^c. The scattering intensity component I_1^ε can be expected to make a significant contribution for those solutions in which deviations from the Vegard rule are large, in particular, near the peaks and dips of the $v(c)$ plot. Note that even for relatively small parameters $c(1-c)(\partial v/\partial \varepsilon)^2(\partial v/\partial c)^{-2}$, determining whether the condition (1.5.23) is satisfied, the fluctuations of the correlation parameter can considerably alter the character of the angular distribution of the total scattering intensity $I_1(\boldsymbol{q})$ (see Fig. 3.5, curve 4).

As noted in Sect. 4.1 in [1.1] and in Sect. 3.4.1, the equal intensity lines deviate from the lemniscate-type shape also in strongly distorted crystals and interstitial solutions in which interstitials occupy nonsymmetric positions of several kinds. The causes of these deviations can be identified separately in experiments under certain conditions when we take into account the fact that the effects due to the fluctuations of the correlation parameters vanish with an increase in the concentration while the effects due to strong distortions make a relatively insignificant contribution for small q and Q.

In nonideal solutions the equations for I_1^ε have a more complicated form and include the correlation parameters for threes and fours of atoms (see Sect. 1.4). We shall not derive these cumbersome formulas here but note that in nonideal solutions the intensity of scattering by the fluctuations of the correlation parameters can in some cases be considerably higher than in the ideal solutions. For instance, if in low-concentration solutions the impurities form pairs ("diatomic molecules") or groups of several atoms then I_1^ε is proportional to c, rather than to c^2, as is the case in the ideal solutions.

3.5 Analysis of Solid Solutions with Diffraction Techniques Employing Diffuse Scattering

The short-range order and other characteristics of solid solutions are often analyzed with the diffraction techniques making use of diffuse scattering of X-rays and neutrons. The number of experimental studies in this field is growing rapidly (see, for instance, the reviews [1.27, 29; 3.17, 55–58] and the references cited there). In this section we shall only discuss some results of such diffraction studies of solid solutions and the characteristic structural features of some typical systems found in them.

3.5.1 Correlation Parameters

The most direct information found from the experimental data on the diffuse scattering intensity $I_1(\boldsymbol{Q})$ for the solid solutions concerns the pair correlation parameters $\varepsilon(\varrho)$ (or the Cowley-Warren parameters $\alpha(\varrho)$) related to them; see (1.1.5)). According to (3.3.1), in undistorted solid solutions $\varepsilon(\varrho)$ are the Fourier coefficients of the reduced intensity $I_1 QN^{-1}(f_A - f_B)^{-2}$ while in distorted solutions they are the Fourier coefficients of the function $H_0(\boldsymbol{q})$ given by (3.2.6) which is obtained after separating the scattering by distortions with the techniques described in Sect. 3.2. In this procedure suggested by *Borie* and *Sparks* [3.13] (and its modifications [3.14, 15]) the diffuse scattering intensity distribution is analyzed at a large number of points (about 10^3–10^4) in certain regions of several reciprocal lattice cells (chosen according to the symmetry conditions) for a single crystal and subsequently the Fourier transformation is applied to the function $H_0(\boldsymbol{q})$. It is the most precise method among those employed for determining the set of the correlation parameters for a large number of the coordination spheres.

The accuracy of this procedure is characterized by the normalization condition (3.3.4) for $\varrho = 0$. (The errors are analyzed in [3.15, 59]). Improvements in the procedure bring $\alpha(0)$ closer to unity. For instance, in the experimental results [3.60] for Cu-Al the difference between $\alpha(0)$ and unity did not exceed 10%. A fairly high accuracy ($\alpha(0) = 0.95$) in the neutron scattering results [3.61] for the alloy Ni^{62}–23 at.% Fe was achieved owing to the proper selection of the nickel isotope and the separation of thermal and diffuse scattering performed by an energy analyzer. Note that if the error is determined by the component of $I_1(\boldsymbol{Q})$ which varies slowly with variation of $\boldsymbol{Q}$ then it has a considerably weaker effect on the correlation parameters, even if it causes $\alpha(0)$ to differ significantly from unity.

The short-range order in the Cu-Au alloys was analyzed in detail starting with the first diffraction study of polycrystals [3.62] and the first diffraction results for single crystals [1.28; 3.48] in which Fourier transformation was used for finding the correlation parameters $\varepsilon(\varrho) = c(1 - c)\alpha(\varrho)$ for a large number of coordination spheres. Subsequent improvements in the experimental techniques and the procedures for separating scattering by distortions [3.13] made it possible to obtain more precise results for $\alpha(\varrho)$ which differed from the earlier ones (but exhibited similar behaviour). Table 3.1 gives $\alpha(\varrho)$ for the disordered Cu_3Au alloy for various ϱ values (the vector components are expressed in terms of the $d/2$ units) and two temperatures $T > T_0 = 394°C$ calculated from the experimental results [3.4.3] (with the use of more accurate data for f_A and f_B and more accurate extrapolation near the Bragg peaks in [3.17] where the parameters $\alpha(\varrho)$ are given also for other temperatures).

Table 3.1

	$\alpha(\varrho)$					
ϱ	Cu_3Au 420°C [3.17]	Cu_3Au 750°C [3.17]	Cu-29.8 at. % Pd 500°C [3.44]	Cu-14.76 at. % Al 150°C [3.60]	43.5 at. % Cu-Ni 500°C [3.66]	Ni_3Fe 808°C [3.61]
000	1.290	1.706	1.786	1.015		0.9460
110	−0.125	−0.094	−0.105	−0.171	0.107	−0.1100
200	0.154	0.083	0.178	0.125	0.005	0.1380
211	0.023	0.004	−0.028	0.042	0.018	−0.0104
220	0.049	0.008	0.076	−0.004	0.015	0.0550
310	−0.078	−0.041	−0.039	−0.034		−0.0247
222	0.010	0.008	0.046	−0.032		0.0331
321	−0.012	−0.004	−0.007	0.003		−0.0113
400	0.071	0.020	0.040	0.021		0.0328
{411	0.018	−0.009	0.013	0.005		0.004
330	−0.022	−0.013	−0.016	−0.008		−0.0060
420	0.024	0.001	0.009	0.007		0.0216
332	0.000	0.007	−0.007			−0.0067

According to (1.1.4, 5), the parameter $\alpha(\varrho) = 1 - p_B^A(\varrho)/c_B$ describes the difference between the *a posteriori* probability that the lattice site ϱ is occupied by an atom B while the zeroth site is occupied by an atom A and the *a priori* probability c_B. The negative parameters $\alpha(110)$ typical for ordering alloys indicate that the pairs of neighbouring lattice sites are occupied primarily by atoms of different species. In Cu_3Au the values $-\alpha_{max}(110)$ are fairly large and comparable to the maximal value $-\alpha(110) = 0.333$ in the fully ordered $L1_2$ alloy ($\alpha_{max}(200) = 1$). When the distance increases $|\alpha(\varrho)|$ decrease (at a faster rate at high temperatures) but they are still significant even in distant coordination spheres (in the tenth coordination sphere at 420°C).

Similar behaviour of the function $\alpha(\varrho)$ was recorded for other ordering alloys, for instance for the fcc alloy Cu-29.8 at.% Pd annealed at 500°C (the temperature $T_0 = 330°C$) [3.44] (see Table 3.1). Though $|\alpha(\varrho)|$ are small for large ϱ they largely determine the fine structure of the intensity distribution $I_1(\mathbf{Q})$. For instance, the characteristic "crosses" (see Fig. 3.3) could not be obtained in the intensity distribution calculated from (3.3.1) with $\alpha(\varrho)$ taken for 15 coordination spheres so that all coordination spheres with $|\alpha(\varrho)| > 0.005$ had to be taken into consideration.

Highly precise neutron diffraction studies of the disordered alloy $^{62}Ni_3Fe$ quenched from 808 K yielded relatively slowly decreasing alternating parameters $\alpha(\varrho)$ [3.61] (see Table 3.1). In order to plot the experimental results for $I_1(\boldsymbol{Q})$ it was not enough to use $\alpha(\varrho)$ values for 22 coordination spheres and thus the data for all 34 spheres under consideration had to be included.

The correlation parameters were observed to decrease very slowly along the $\langle 100 \rangle$ direction in Au-Mn alloys containing about 20 at.% Mn [3.63, 64]. For $c = 0.203$ at the points (200), (400), (600), (800), and (100) the parameters $\alpha(\varrho)$ have the values 0.222, 0.163, 0.096, 0.042, and 0.061, that is, a chain-type correlation in certain directions acts at distances as large as $5d$. In other directions the correlation weakens at a much faster rate in this alloy. Careful X-ray diffraction study of the Cu-14.76 at.% Al alloy annealed at 150°C (for 1580 hours) yielded short-range order parameters of the same sign as in the Cu-Au alloy [3.60]. The parameter $\alpha(110) = -0.171$ given in Table 3.1 is very close to the parameter -0.173 found under the assumption that the alloy contains no Al-Al pairs of neighbours ($p_{\mathrm{Cu}}^{\mathrm{Al}} = 1$). The short-range order in these alloys is discussed in more detail in Sect. 3.5.5.

In decomposing alloys the maxima of the intensity distribution $I_1(\boldsymbol{Q})$ are near the Bragg peaks and this fact contributes to errors in the analysis of this distribution. But when we analyze neutron scattering we can select an isotopic composition of the alloy, that the scattering amplitudes b_A and b_B will have opposite signs and the mean scattering amplitude will vanish (that is, the Bragg peaks will vanish while the scattering intensity $I_1 \sim (b_A - b_B)^2$ will increase). This "null matrix" technique was applied to the decomposing alloy Cu^{65}-Ni^{62} and sufficiently precise results were in good agreement (for the same equilibrium states) even for Debye scattering by polycrystal specimens (the distortions of the Cu-Ni alloy are small and can be ignored) [3.65–67]. Table 3.1 presents the results for the alloy with the composition corresponding to the "null matrix" (43.5 at.% Cu). Under these conditions $\alpha(110) > 0$ and the correlation parameters are small starting from the second coordination sphere.

The detailed results of the earlier experimental studies of the short-range order in many binary solutions are presented in the book [1.27]. The modified diffuse scattering technique makes it possible also to study more complicated systems containing several atoms per unit cell (see [3.57, 68, 69]) and even solutions of organic compounds [3.70].

Ohshima and Moss have studied diffuse scattering in the sharply anisotropic intercalated laminar compound $Ag_{0.18}TiS_2$ (private communication by S. Moss). The Ag atoms occupy some of the octahedral interstices in the intercalation planes (inserted after each two TiS_2 layers). At room temperature (which is much higher than the ordering temperature) the recorded rod-shaped diffuse scattering intensity distributions are parallel to the axis c^* in the reciprocal space and correspond to a purely two-dimensional short-range order of the Ag atoms and vacancies at the octahedral interstices of each plane. The correlation is effective at a large distance and corresponds to the alternating interaction between atoms which may be associated with the indirect interaction between atoms in the quasi-two-dimensional system mediated by the conduction electrons.

3.5.2 Ordering Energies and Thermodynamic Properties of a Solution

If a solution can be treated in the framework of the pair interaction model then it follows from (3.3.9) that the reduced scattering intensity $I_1(\boldsymbol{Q})(f_A - f_B)^{-2}N^{-1}$ (or the function $H_0(\boldsymbol{q})$ in crystals with distortions) is expressed in terms of $x(\boldsymbol{\varrho})$, that is, in terms of the ordering energies $w(\boldsymbol{\varrho})$ in various coordination spheres (see (1.3.9, 12, 15, 16)). Therefore, $x(\boldsymbol{\varrho})$ and $w(\boldsymbol{\varrho})$ can be found with the method used for calculating the correlation parameters from the diffuse scattering results in which the intensity of scattering by distortions is identified separately. A convenient approach is to use the Fourier transformation (3.3.10) of the function $N(f_A - f_B)^2/I_1$ (or the function $H_0(\boldsymbol{q})$) [1.4]. Since the correlation parameters $\varepsilon(\boldsymbol{\varrho})$ (or $\alpha(\boldsymbol{\varrho})$) are definitely expressed in terms of $x(\boldsymbol{\varrho}')$ by the equations (1.3.4, 17) we can also select $w(\boldsymbol{\varrho}')$ (that is, $x(\boldsymbol{\varrho}')$) in order to obtain the best possible description for the entire set of the parameters $\alpha_{l\,\mathrm{exp}}$ determined experimentally for various coordination spheres [1.57, 73]. To do this we just have to find the minimum of the expression

$$\sum_l W_l(\alpha_{l\,\mathrm{exp}} - \alpha_{l\,\mathrm{th}})^2 \qquad (3.5.1)$$

with respect to the parameters $w(\boldsymbol{\varrho})$ which do not vanish for a certain number of the coordination spheres. Here $\alpha_{l\,\mathrm{th}}$ is given by (1.3.4, 17) and W_l are the weighting factors which are determined by the estimated error in $\alpha_{l\,\mathrm{exp}}$. (This method perhaps produces poorer results than calculations made with (3.3.10) because the expansion (1.3.18) for $\varepsilon(\boldsymbol{\varrho})$ converges at a slower rate than the expansions (1.3.15, 16).)

We see that in homogeneous solid solutions the set of energies $w(\boldsymbol{\varrho})$ for all coordination spheres can, in principle, be found. Rigourously speaking, equations similar to (1.3.9) which relate $x(\boldsymbol{\varrho})$ to $w(\boldsymbol{\varrho})$ are inapplicable to heterogeneous states of alloys described in Sect. 2.5.2 (the interaction between atoms in the ordered regions (particles) is effective at large distances and the expansion of $\varepsilon(\boldsymbol{\varrho})$ for these atoms which is an extension of (1.3.18) converges at a slow rate or even diverges in the case of correlated particles). But if the volume fraction of the particles is $p \ll 1$ then their presence significantly affects only the energies $w(\boldsymbol{\varrho})$ with large $\boldsymbol{\varrho}$ and causes small variations of $w(\boldsymbol{\varrho}) = w_1, w_2, \ldots$ for the first coordination spheres. Therefore, even under such conditions the energies $w_1, w_2, \ldots$ found in this way can serve as useful semiphenomenological parameters which describe approximately the properties of the alloy. In systems with significant many-body interactions the energies $w(\boldsymbol{\varrho})$ calculated with this procedure can also be regarded as semiphenomenological parameters which significantly depend on the alloy composition and the order parameters.

The energies $w(\boldsymbol{\varrho})$ for various alloys were determined by comparing the calculated parameters $\alpha(\boldsymbol{\varrho})$ with the experimentally measured ones and by performing the Fourier transformations (3.3.10) in numerous studies (see the early reports [1.28, 73; 3.71]). The ratios w_2/w_1, $w_3/w_1, \ldots$ can be determined with a better accuracy than the energies $w(\varrho)$ and therefore only the results on these ratios are reported. The functions $x(\varrho) = w(\varrho)/k_BT$ and I^{-1} are usually taken in the form of the simple linear approximation given by (1.3.9) and (3.3.11) which was derived in the self-consistent field approximation. The accuracy of this approximation improves with an increase in temperature (over the ordering temperature T_0). More accurate equations such as (1.3.16) are employed less frequently [3.72].

Improvements in the experimental accuracy make some changes in the recorded values of $w_1, w_2/w_1, \ldots$ which are not usually very significant. We shall give below some parameters w_l only for small l values. In Cu_3Au at 420°C we have $w_1 = 7.3 \times 10^{-2}$ eV, $w_2/w_1 = -0.22$, $w_3/w_1 = 0.06$, and at 750°C we have $w_1 = 6.0 \times 10^{-2}$ eV, $w_2/w_1 = -0.21$, and $w_3/w_1 = 0.03$. The temperature dependence of w_1 is quite noticeable and not monotonic (at 475°C we have $w_1 = 4.8 \times 10^{-2}$ eV). One of the causes of this may be a transformation of the heterogeneous structure (see Sect. 3.5.5). The results [3.73] obtained at 527°C for the CuAu alloy were $w_2/w_1 = -0.40$ and $w_3/w_1 = 0.20$. The Fourier transformation (3.3.10) and the equations (1.3.9) yielded for the Cu -29.3 at% Pd alloy $w_2/w_1 = -1.68$, $w_3/w_1 = 0.15$, and $w_4/w_1 = -0.43$ that is, the nearest-neighbour model is not even approximately correct for this alloy [3.44].

The data [3.66] given in Table 3.1 for the decomposing alloy Cu-43.5 at.% Ni correspond to $w_1 \approx -1.9 \times 10^{-2}$ eV and $w_2 \approx 0.9 \times 10^{-2}$ eV. The parameters $|w_1|$ and w_2 decrease with a decrease in temperature and exhibit a noticeable dependence on the alloy composition indicating a significant contribution made by the many-body interaction between atoms. Other alloy systems, for instance, Cu-Au [1.73; 3.74] and Mg-In [3.75] also exhibited a significant dependence of the behaviour of $w(\varrho)$ on the alloy composition. (For some alloys the many-body interaction was estimated also from the phase diagram [1.43; 3.76] and it was analyzed theoretically in [1.43, 60, 62; 3.77, 78]). These energies $w(\varrho)$ include all types of interaction between atoms including the elastic interaction whose energy $E(\varrho)$ given by (2.1.13) varies as $1/\varrho^3$, and the oscillating indirect interaction mediated by the conduction electrons. Since (3.3.11) derived in the self-consistent field approximation is not accurate at $T \approx T_0$ (see Sect. 1.3) the errors in the calculated energies $w(\varrho)$ may be significant (but decrease with an increase in T or with inclusion of the higher-order terms in the expansions of $x(\varrho)$ in $w(\varrho)/k_BT$ as was done in (1.3.16)).

The energies $w(\varrho)$ found in this way can be used, in principle, for calculating various thermodynamic parameters of the solution in the framework of the pair interaction model (with the inclusion of elastic interaction terms). For instance, for undistorted crystals we may use (1.3.19) to determine $d^2\varphi/dc^2$ and the coefficient β in the gradient expansion of Φ. According to (3.3.7), we can obtain the same results, however, directly from the experimentally measured scattering intensities $I_1(\boldsymbol{Q})$ in the vicinities of the reciprocal lattice points. Since for very small q the Bragg peak gives rise to large errors in the intensity I_1 it is better to make measurements at some distance from the lattice points and determine $d^2\varphi/dc^2$ and β from

$$N_0 \frac{k_B T}{v_0} \frac{(f_a - f_b)^2}{I_1} = \frac{dc^2\varphi}{dc^2} + \beta q^2 = \frac{k_B T}{v_0} x(\boldsymbol{k} = 0)$$
$$+ \beta q^2 \approx \frac{1}{v_0}\left[\frac{k_B T}{c(1-c)} + w(\boldsymbol{k} = 0)\right] + \beta q^2 \qquad (3.5.2)$$

which is a linear function of q^2 (for cubic crystals). The last equation in (3.5.2) corresponds to the self-consistent field approximation. The diffraction data for the relatively weakly distorted alloy Fe_3Al at 570°C yield $w(\boldsymbol{k} = 0)/k_B = 5220\,K$ with this procedure [3.79] which differs by 3% from the result calculated from the activity data (reduced to the same temperature [3.79]). According to (3.4.3), for distorted solutions a similar analysis of the behaviour of $N[fa_{\boldsymbol{q}}\boldsymbol{Q} \cdot \boldsymbol{e}_{\boldsymbol{q}}q^{-1} - (f_A - f_B)]^2/I_1$ (or $1/H_0(\boldsymbol{q})$) as a function of q^2 yields $\varphi_{cc}(\boldsymbol{q})$ and we can find $d^2\varphi/dc^2$ by subtracting the nonanalytic term φ_{cc}^{el} calculated from (2.1.5).

We can also find $d^2\varphi/dc^2$ for undistorted solutions from the experimental data on the correlation parameters $\varepsilon(\boldsymbol{\varrho})$ (or $\alpha(\boldsymbol{\varrho})$) which must be substituted into (3.3.1) for $q \to 0$ and the result must be substituted into (3.3.7). The accuracy of this procedure is improved by a rearrangement of the terms in the equation [3.80, 81]. For distorted solutions, however, before performing summation over a finite number of the vectors $\boldsymbol{\varrho}$ we must subtract the contribution of the terms with large $\boldsymbol{\varrho}$ which vary at a slow rate (as $1/\varrho^3$) (or take into consideration the nonanalytic contribution φ_{cc}^{el} to $\varphi_{cc}(\boldsymbol{q})$ in a different way).

The phase diagram for the solution can be calculated from the set of the parameters $w(\boldsymbol{\varrho})$ found from the diffraction results. The phase diagram can be found aproximately with a simpler procedure, however, for which the intensities $I_1(\boldsymbol{Q})$ must be known only at several high-symmetry reciprocal lattice points. In the self-consistent field approximation we can use (3.3.11) to find the minimal values $w(\boldsymbol{k}_m)$ of the Fourier components $w(\boldsymbol{\varrho})$ directly from the experimental results on $I_1(\boldsymbol{Q})$. Then we can find directly from (1.3.13) the instability temperature T_0 which is close or equal to the phase transition temperature T_0. In this approximation the equation for the thermodynamic potential Φ of the alloy includes only the parameters $w(\boldsymbol{k}_m)$ and $w(\boldsymbol{k} = 0)$ [1.25] so that the entire phase diagram can be found from the condition for the minimum of Φ with these parameters.

Such calculations of the phase diagram with the use of the X-ray diffraction results on $w(\boldsymbol{k}_m)$ were made for the Fe-Al system in the Fe_3Al range [3.79, 82]. For this alloy the minima of $w(\boldsymbol{k})$ are at the points $\boldsymbol{k}_m = \boldsymbol{k}_1 = \{100\}$ which correspond to the ordered structure B2 and the points $\boldsymbol{k}_m = \boldsymbol{k}_2 = \{1/2\,1/2\,1/2\}$ which correspond to the DO_3 structure (see Fig. 1.2). The parameter $w(\boldsymbol{k}_2) \approx -3700\,K$ calculated from the X-ray diffraction results for the alloys with $0.16 < c_{Al} < 0.22$ exhibited only a weak temperature dependence. The strong temperature dependence found for $w(\boldsymbol{k}_1)$ and $w(0)$ was caused by the exchange interaction between the spins of the neighbouring atoms in the ferromagnetic which makes a contribution to $w(\boldsymbol{\varrho})$. This dependence could be described by the approximations

$w(\boldsymbol{k}_1)/k_B = -5400\,K - C\sigma_m^2$ and $w(0)/k_B = 5800\,K + C\sigma_m^2$ where $\sigma_m(T)$ is the relative magnetization, $C = -1710\,K$ for $c_{Al} < 0.3$ and $C = 0$ for $c_{Al} > 0.3$. The phase diagram calculated in the self-consistent field approximation with these parameters is in good agreement with the experimental results. For instance, the temperature of the transition of the disordered bcc alloy Fe_3Al to the ordered B2 structure (1200 K) agrees with the experimental result of 1050 K. The predicted temperature of the B2–DO_3 transition for $c_{Al} = 0.29$ is 920 K, while the experimentally recorded point is 820 K. The agreement between the experimental and predicted results is improved when the correlation terms are taken into consideration. Similar results on the phase diagram for Fe-Si were obtained from the X-ray diffraction data in [3.83]. The diffraction technique was used also for determination of the parameters which must be known for describing the phase diagrams of the more complex semiconductor solutions with two atoms per unit cell [3.84–87] and solutions of organic compounds [3.70–88]. To describe these decomposing solutions in the self-consistent field approximation it is sufficient to know $w(0)$. This parameter directly determines the solution energy in undistorted solutions according to (1.3.1) (where $c_{s\gamma}$ in this approximation must be replaced with c_γ and (1.3.10). As noted in the discussion of (2.1.16), when we consider distorted crystals and the contribution of the elastic interaction to the solution energy we must add to $w(0)$ the term $\bar{\xi} - vc_{ijlm}L_{ij}L_{lm}$ where $\bar{\xi}$ is given by (2.1.13). In the approximation of the elastic isotropic continuum this additional term equals $-\varphi_{cc}^{el}$ and it can be obtained from (2.1.9).

This approach was used [3.86] for plotting the curves of phase equilibria between solid and liquid solutions $w(0)/k_B$ found from the experimental diffraction results (at room temperature) in a practically undistorted HgTe-CdTe solution. In a strongly distorted GaP-GaAs solution $w(0)/k_B \approx -1000\,K$ and $-\varphi_{cc}^{el}$ are almost identical and weakly depend on the concentration [3.85]. The results on $w(0)$ derived from the diffraction data were also used for obtaining the quantitative descriptions of the phase diagrams for this and some other solutions (Si-Ge [3.81] and GaAs-AlAs [3.87]).

The long-range forces were found to cause nonanalytic behaviour of the diffuse scattering intensity $I_1(\boldsymbol{Q})$ for small q in an organic compound tolan diphenyl mercury [3.70]. The specific features in the behaviour of such solutions could be explained with the use of the estimated concentration at the spinodal decomposition (at room temperature) found from the results on the scattering intensity in the directions in which $1/I_1$ has a minimal value. These results on $w(\boldsymbol{\varrho})$ are in a satisfactory agreement with the parameters calculated with the method of the atom-atom potentials [3.89].

3.5.3 Diffraction Studies of the Kinetics of Diffusion Processes

In non-equlibrium alloys the short-range order parameters undergo relaxation which determines the time dependence of the diffuse scattering intensity. The results on this dependence can be useful for analyzing the diffusion processes involving changes in the concentration inhomogeneities at distances of the order of the interatomic spacing.

According to (3.2.10), the kinetics of the scattering intensity variation, determined by the time dependence of the Fourier component $c_q(t)$ which is affected by kinetic and thermodynamic factors. This dependence has the simplest form if the impurities or trace atoms perform a random walk in the perfect lattice making jumps at distances ϱ with the probabilities $W(\varrho)$. If at the moment t a concentration $c_s \equiv c(\boldsymbol{R}_s)$ while other concentrations vanish then at the moment $t + dt$ the mean value of c_s is $1 - dt\sum_\varrho W(\varrho)$ and $c(\boldsymbol{R}_s + \varrho)$ have the values $W(\varrho)dt$. Therefore, according to (1.1.8)

$$\frac{dc_{\boldsymbol{k}}(t)}{dt} = -\tau_{\boldsymbol{k}}^{-1} c_{\boldsymbol{k}}(t)\,, \quad \tau_{\boldsymbol{k}}^{-1} = \sum_{\varrho} W(\varrho)(1 - \mathrm{e}^{\mathrm{i}\boldsymbol{k}\cdot\varrho})\,, \tag{3.5.3}$$

$$c_{\boldsymbol{k}}(t) = c_{\boldsymbol{k}}(0)\exp(-t/\tau_{\boldsymbol{k}})\,. \tag{3.5.4}$$

The correlation function $\langle c_{\boldsymbol{k}}(t)c_{\boldsymbol{k}}^*(0)\rangle$ exhibits the same dependence on t and its spectral representation is described by a Lorentz profile with a finite integrated width $\pi/\tau_{\boldsymbol{k}}$. We obtain as a result diffusion broadening of the spectra recorded for the Mössbauer line, $\sigma((\boldsymbol{q})\omega)$, [3.90, 91] and for quasielastic neutron scattering [3.91, 92] (caused by the phase misalignment generated by diffusion jumps of the atoms). This broadening also amounts to π/τ_q where $\boldsymbol{q} = \boldsymbol{Q} - \boldsymbol{G}$ and $\boldsymbol{Q}$ is the wave vector of the Mössbauer photons or the diffraction vector of the scattered neutrons. In the limit of small k we can express $\tau_{\boldsymbol{k}}^{-1} = Dk^2$ in terms of the diffusion coefficient D for the atoms under consideration so that the coefficient D can be related to $W(\varrho)$ if we have $W(\varrho) \neq 0$ only for the nearest neighbours. The Mössbauer line intensity can also decrease owing to diffusion [3.93, 94]. The characteristic dependence of $\tau_{\boldsymbol{Q}}$ on the orientation of the vector $\boldsymbol{Q}$ with respect to the axes of the single crystal yields extremely valuable information on diffusion processes at distances of the order of d, namely on diffusion mechanisms and elementary events of atom migration (which determine the set of the vectors ϱ with significant $W(\varrho)$) [3.91]. Experimental studies have demonstrated such sharp angular dependences of the Mössbauer and neutron spectra for single crystals [3.95–102].

The second equation (3.5.3) is changed when we take into consideration the correlation between successive diffusion jumps. This correlation is always effective in diffusion via vacancies [3.103] since even if an atom can migrate only to a neighbouring vacancy the latter does not immediately migrate away from the atom and in (3.5.3) $W(\varrho)$ for more distant neighbours will become non-vanishing as a result of several possible successive jumps of atoms to this vacancy. This effect has been found to be small for trace atoms in a perfect crystal [3.91, 104] but it can significantly alter the set of the probabilities $W(\varrho)$ and the relation between τ_k^{-1} and D for impurities [3.102, 105] especially if they attract vacancies.

The Mössbauer results can be useful for diffusion studies only at high temperatures when D varies between 10^{-9} and $10^{-8}\,\mathrm{cm^2/s}$ and $\tau_k^{-1} \sim zD/d^2$ is comparable to the natural width of the Mössbauer line. The results on the variation of the diffuse scattering intensity for quenched alloys at low temperatures may be used for analyzing the diffuse relaxation at interatomic distances [3.106–109]. The range of the coefficients D between 10^{-18} and $10^{-21}\,\mathrm{cm^2/s}$ can then be analyzed if the characteristic time varies between 10^2 and 10^5 s.

If $c_k(t)$ is described by (3.5.4) then according to (3.2.10) (where $\langle|c_k|^2\rangle = \langle|c_k(t)|^2\rangle$) we obtain

$$I_1(\boldsymbol{Q},t) = I_1(\boldsymbol{Q},0)\exp(-2t/\tau_k)\,. \tag{3.5.5}$$

When we determine τ_k in this equation we must bear in mind that the above correlation effects in high-concentration solutions cause changes in $W(\varrho)$ in (3.5.3) and in the simple relationships between $W(\varrho)$ and D. Moreover, in contrast to the Mössbauer studies in which diffusion of the trace or impurity atoms occurs in a homogeneous crystal, in solutions diffusion takes place in the presence of small-scale concentration inhomogeneities and the non-uniformity of the chemical potential does not reduce simply to concentration non-uniformity just as in the case of chemical diffusion. In particular, this must give rise to terms nonlinear in c_k on the right-hand side of (3.5.3) which make an especially significant contribution if the deviations from the equilibrium are large. The experimental data on the behavior of $I_1(\boldsymbol{Q})$ are typically described, however, with a simple equation for τ_k (its accuracy has not been verified clearly) which is an extension of the well-known Darken equation of the continuum theory of diffusion applied to the case of the atomic-scale inhomogeneities [3.108]. In this approximation in the sum over ϱ in (3.5.3) we retain only the terms for the nearest neighbours, $W(\varrho) = W$ is expressed in terms of the coefficient of chemical diffusion,

$$\tilde{D} = [cD_B + (1-c)D_A]vc(1-c)\varphi_{cc}/(k_\mathrm{B}T)$$

where D_A and D_B are the diffusion coefficients for the trace atoms A and B while the thermodynamic factor $\partial^2\Phi/\partial c_k\partial c_k^*$ in the equation for dc_k/dt is taken into consideration in the factor x_k which is given by (1.3.4, 9) (in the self-consistent field approximation). In this approximation we have, according to [1.26, 75; 3.108],

$$\begin{aligned}\tau_k^{-1} &= \frac{\tilde{D}k_\mathrm{B}Tx_q}{v\varphi_{cc}d^2}\sum_{\varrho}(1-\cos\boldsymbol{q}\cdot\boldsymbol{\varrho})\,,\\ x_q &\approx \frac{1}{c(1-c)} + \sum_{\varrho}\frac{w(\varrho)}{k_\mathrm{B}T}\cos\boldsymbol{q}\cdot\boldsymbol{\varrho}\,.\end{aligned} \tag{3.5.6}$$

Experimental diffraction studies of the short-range order behavior in alloys were reported in [3.110–115]. For the alloys Cu-Au and $CoPt_3$ the time dependence of $I_1(\boldsymbol{Q},t)$ for various $\boldsymbol{Q}$ was analyzed for the temperature variation of 15–25°C in the disordered alloys at equilibrium [3.112]. The intensity of scattering by distortions was separated from the total diffuse scattering intensity. The parameters τ_k determined with this method as functions of q in the directions of $\boldsymbol{q}$ parallel to [100] and [110] for the alloy $Cu_{0.815}Al_{0.185}$ can be described reasonably well with (3.5.6) if x_q are determined straight from the experimental results on scattering by an alloy at equilibrium according to (3.3.9) (see Fig. 3.6). When we analyze the temperature dependence of τ_q we can find the activation energy Q_D for the diffusion coefficient at relatively low temperatures. The resulting energy Q_D = 44 kcal/mole is in a good agreement with the results on diffusion of trace atoms at high temperatures (42.6 kcal/mole for Au diffusion in Cu and 40.65 kcal/mole for Cu diffusion in Au). The diffraction results yielded Q_D = 33.5 kcal/mole for an ordered Fe-Al alloy containing 29.2 at.% Al in the 307–416°C range [3.112]. For the alloy containing 16 at.% Al experiments produced a markedly different Q_D = 54 kcal/mole [3.111].

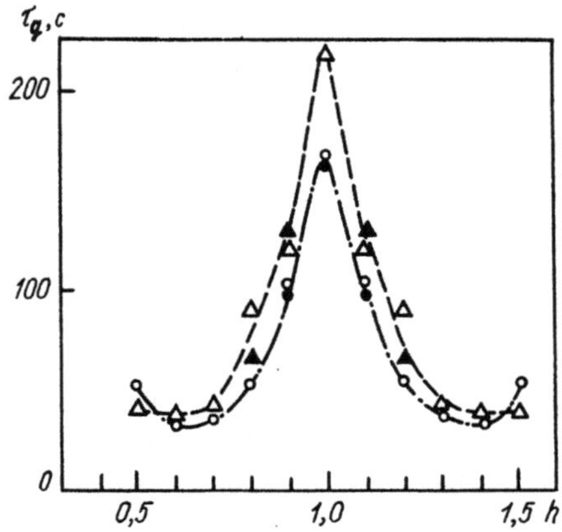

Fig. 3.6. The time τ_q of the relaxation of the short-range order as a function of $q_x = 2\pi h/d$ in the alloy Cu-18.5 at.% Au quenched from the equilibrium state at 355° down to 330°. Open triangles and circles correspond to the calculated results for the directions $\langle hh0 \rangle$ and $\langle h00 \rangle$ and the black triangles and circles are the experimental results

The behavior of the short-range order in a decomposing Cu-Ni alloy was analyzed with the neutron diffraction technique [3.113–115]. Measurements were performed for polycrystalline specimens and the equation (3.5.6) was averaged over angles (while x_q was found from the scattering data for the equilibrium alloy). The equation gave a good description [3.113] of the experimental data for the alloy containing 56.5 at.% Ni in the temperature range between 340 and 700°C for $q < 0.8\,\text{Å}^{-1}$ while for large q the agreement was poorer (perhaps, owing to the above correlation and nonlinear efforts which make a greater contribution for high q. For the decomposing alloy containing 41 at.% Cu bombarded with 3 MeV electrons below the decomposition temperature [3.114, 115] (according to diffuse scattering data [3.115], the coherent decomposition temperature was 527 + 25 K while the incoherent decomposition temperature was 575 + 45 K) small-angle neutron scattering exhibited a heterogeneous concentration distribution with a characteristic wavelength of 45 Å (this distribution is perhaps associated with the electron bombardment).

The short-range order behavior exhibits also some characteristic features for strained or irradiated alloys and also for alloys quenched from a high to a much lower temperature. In the case of plastic strain the dislocations passing through the crystal displace the atoms with respect to each other and partially disrupt the short-range order. Restoration of the short-range order during subsequent annealing of the alloy has been shown to proceed non-monotonically for some alloys, for instance for Ni_3Pt, Ni-Si, Cu-Pt, etc. [1.27; 3.116–119]. In the transition stage a non-homogeneous short-range order is produced in the alloy in which local non-homogeneities are considerably greater than in the equilibrium state. The general variation of the short-range order in such alloys is discussued in [1.27; 3.120, 121].

The short-range order variation may also be non-monotonic during quenching of a disordered solution in the temperature range of $T < T_0$ if the peak of the function τ_k^{-1} corresponds to $k = k_m$ which do not coincide with the high-symmetry points while at the initial stages of the quenching process c_k with $k = k_m$ grow considerably and only later at almost full equilibrium the major contribution is made by the thermodynamic, rather than kinetic, factors which make it favourable to produce structures with k at the high symmetry points [1.74; 3.17].

The thermodynamic factors determine the character of the ordering process somewhat below the first-order phase incoherent transition. If at even lower temperatures we have $T < T_{sp}$ a disordered crystal becomes unstable with respect to production of a long-range order wave and the process of spinodal ordering may become effective. For a discussion of the associated experimental effects see [3.123].

3.5.4 Diffraction Studies of the Fermi Surfaces in Alloys

As can be seen from (3.3.14) and the results discussed in Sect. 2.4. the intensity $I_1(Q)$ of diffuse scattering by homogeneous alloys exhibits characteristic features, for instance, the peaks (corresponding to the flattened parts of the Fermi surfaced at $\boldsymbol{q} = 2\boldsymbol{k}_F - \boldsymbol{G}_0$ dependent on the diameter $2\boldsymbol{k}_F$ of the Fermi surface (here $\boldsymbol{G}_0$ is the reciprocal lattice vector which brings $\boldsymbol{q}$ to the first reciprocal lattice cell). This made it possible to develop a diffraction method for analyzing the Fermi surfaces in alloys [1.17; 2.64; 3.45]. In addition to this characteristic behaviour of the scattering intensity in homogeneous alloys, the indirect interaction mediated by the conduction electrons can also give rise to heterogeneous states at equilibrium (see Sect. 2.5.2). Though the equations for the scattering intensity $I_1(Q)$ have a different explicit form (see Sect. 3.3.2) the positions of the resulting intensity peaks also correspond to flattened parts of the Fermi surface. As in the case of ordered alloys with large periodicity [2.83], the intensity peaks are close to the positions of the scattering intensity peaks for the homogeneous alloy with the same Fermi surface (if the contributions of the surface energy and of the dependence of v_k^2 on k are not very significant). Therefore, the experimental results on the diffuse scattering intensity and its electronic features can be treated in the context of only the geometric aspects, that is, the positions of the intensity peaks and their relation to the Fermi surface and not the magnitude and distribution of the intensity, without taking into account whether the alloy is homogeneous or heterogeneous (the short-range order structure is discussed in Sect. 3.5.5).

The concept of the diffraction analysis of the Fermi surfaces in alloys [1.17; 3.124] was first applied to the Cu-Au alloy system [3.45]. In such alloys quenched from temperatures much higher (from 900°C in [3.126]) than the ordering temperature, T_0, experiments [3.125–128] demonstrated a "cross" of four spots near positions of the type of (110) see a similar pattern for Cu-Pd alloy in Fig. 3.3). An explanation of this effect [3.45] takes into account the fact that according to (3.3.14) the scattering intensity increases at the surface regions in the reciprocal lattice space where $\boldsymbol{q}$ or $\boldsymbol{q} + \boldsymbol{G}_0$ coincides with the Fermi surface diameter $2\boldsymbol{k}_F$ along the [110] direction which is perpendicular to the flattened (cylindrical) parts of the Fermi surface. The scattering intensity growth is especially large at the points of intersection of these surfaces for different $\boldsymbol{G}_0$ giving rise to the spots (peaks) of diffuse scattering intensity. The geometric considerations [3.45] demonstrate that these spots must indeed make up a "cross" near the point (110) of the reciprocal lattice (see Fig. 3.7). For Cu_3Au the distance between the diffraction spots found in experiments was close to the distance calculated from the data on the Fermi surface in pure copper while for $CuAu_3$ a similar agreement was with the results for pure gold. A special analysis of the dependence of this effect on the preliminary annealing parameters demonstrated that the production of the spots is an equilibrium process which has no relation to the processes involved in quenching [3.129].

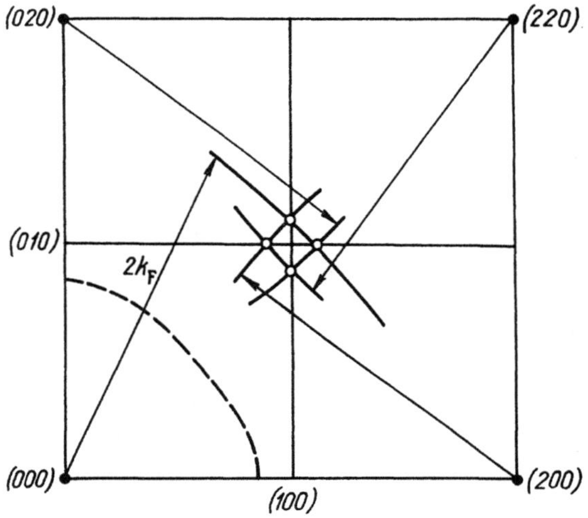

Fig. 3.7. Schematic for the origin of the diffuse scattering peaks determined by the Fermi surface. Solid lines correspond to the Fermi surface on which the vector $\boldsymbol{q}$ passing from different reciprocal lattice points coincides with $2\boldsymbol{k}_F$ and the intensity I_1 increases. At the points of their intersection the growth I_1 is especially significant and four intensity peaks are produced

The observed dependence of the distance m between the spots on the electron concentration e/a is the most convincing proof of the electronic character of the effects under the Fermi surface diameter $2\boldsymbol{k}_F$, that is, m, with the use of the Korringa-Kohn-Rostocker approximation and the coherent potential approximation

[1.46] gave a quantitative account of the experimental results on the dependence of m on c for the Cu-Pd alloy [3.13]. These calculations demonstrated that for significant concentrations of Pd in the alloy the Fermi surfaces exhibit sufficiently flat regions perpendicular to the $\langle 110 \rangle$ directions. It is a common feature for such alloys of noble and transition metals in a certain concentration range (the curvatures of the Fermi surfaces have the opposite signs for $c = 0$ and $c = 1$.

The Fermi surface diameters can also be found with the positron annihilation technique [3.133, 134] which is similar to the diffraction technique in that it is not too sensitive to the fact that the high-concentration solution is strongly non-ideal. The results on $\boldsymbol{k}_F$ derived for Cu-Al with the two techniques are in a good agreement [3.130]. A similar agreement was found for the results derived for Cu-Pd alloy from the scattering data [3.131] and the positron annihilation experiments [3.136].

The scattering data for the copper-based alloys was analyzed in the above studies only from a geometric viewpoint, that is, only the peak positions but not the intensities were taken into consideration. It has been noted [2.64] that the relatively weak Kohn feature of the vibration spectrum for the pure copper [3.137] (the $w_n(\boldsymbol{k})$ plot exhibits a change in slope, rather than a dip) indicates that a mechanism must be suggested for enhancement of the effect in order to account for the prominent $I_1(\boldsymbol{Q})$ peaks observed in experiments. Such enhancement can be determined by a stronger dependence of the difference between pseudopotentials of Cu and Au on k than that for the pseudopotential of copper. It is more probable, however, that it is due to the heterogeneous short-range order consideration. The addition of trivalent indium to Cu_3Au, which increases $e/a = 1 + 2c_{In}$ and $2\boldsymbol{k}_F$, has been shown [3.129] to increase m while the addition of the "zero-valent" palladium $e/a = 1 - c_{Pd}$) decreases m.

Similar effects involving splitting of the diffuse scattering intensity peaks and noticeable correlation between the splitting distance m and the electron concentration have been observed for other copper-based alloys, such as Cu-Al alloys containing 11 and 16 at.% Al [3.130] and for Cu-Pd, Cu-Pt [3.131], Au-Pd, Au-Zn, and Ag-Mg [3.132] alloys. The recorded behaviour of m as a function of e/a agreed with the simple relation derived in the rigid band model,

$$m = \left(\frac{12}{\pi} \frac{e}{a} \right)^{1/3} t - \sqrt{2} \tag{3.5.7}$$

(here m is taken in terms of the distance between the reciprocal lattice points (000) and (200)), the appropriately selected parameter t describes the non-sphericity of the Fermi surface (this parameter equals the ratio between $\boldsymbol{k}_F$ in the [110] direction and the radius of the Fermi sphere corresponding to the given e/a value). This agreement can be seen, for instance, in Fig. 3.8 where the results obtained for Cu-Pd (t = 0.91) and Cu-Pt (t = 0.93) are given [3.131]. Similar results for t have been found for the alloys Au-Pd (0.91) Au-Zn (0.945) and Ag-Mg (0.955) [3.132]. The results for the alloy Cu_3Au doped with In and Pd [3.129] and for Cu-Al [3.140] also give t = 0.95. Though the rigid-band model and equation

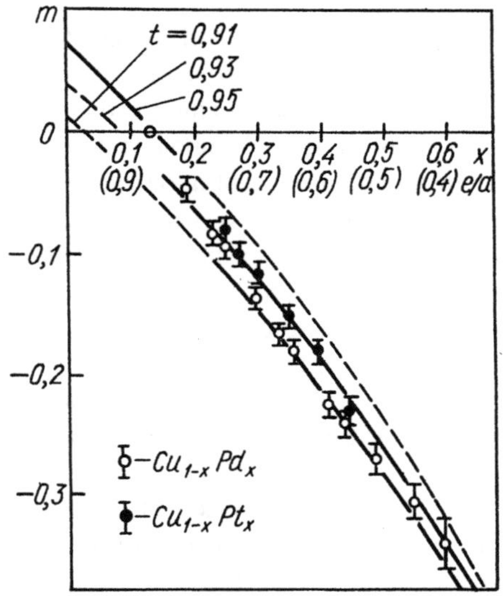

Fig. 3.8. The distance m between the intensity peaks as a function of the number of electrons per atom in Cu-Pd and Cu-Pt alloys [3.131]

(3.5.7) are obviously simplified, while t is apparently an adjustment parameter taking into account not only the non-sphericity of the Fermi surface but also other factors, the correlation between the scattering features and the electron structure characteristics of the alloy found in these studies cannot be questioned.

Detailed calculations made from first principles for structure in these alloys (see Sect. 3.5.5) because of which $I_1(\boldsymbol{Q})$ is described not by (3.3.14) but by the equations derived in Sect. 3.3.2 which give rise to sharper peaks whose positions, however, are almost the same as those given by (3.3.14).

Diffuse scattering results for other metal systems have also been analyzed in terms of the electron structure effects. These effects have been suggested [3.138] to produce a strong dependence of $w(\boldsymbol{k}_1)$ in Fe-Al and Fe-Si alloys with low concentrations of Al or Si (see Sect. 3.5.2). Some features of the diffuse scattering processes in alloys of Ti with Nb, Mo, and Cr and in Nb-Al allloys have been attributed to electron effects [3.139].

The diffuse scattering results for more complicated systems, for instance, oxides of transition metals, also exhibit electron effects. For instance, in the TiO_x oxides with the NaCl structure (which are stable in a fairly wide concentration range of $0.90 < x < 1.25$ at 900°C) both sublattices exhibit high concentrations of structural vacancies ($\sim$ 15% in TiO) which are stabilized owing to filling of the resulting defect states with the conduction electrons and deformation of the electron bands [3.140, 141]. Such high vacancy concentrations cause high-intensity diffuse scattering determined by the short-range order in their arrangement. It has been shown [3.142] that for a crystal with only short-range order in the vacancy distribution the positions of the characteristic scattering intensity peaks are not associated directly with the superlattice peaks which appear at lower temperatures when the long-range order is established. These peaks may, however, be explained by the relation between scattering processes and the Fermi surface and the calculated results for the Fermi surface agree with $2k_F$ values derived from the diffraction data and with their dependence on deviations from the stoichiometric composition of the oxide. Additional evidence of the vacancy nature of the specific behaviour of the intensity $I_1(Q)$ in TiO_x is given in [3.143].

3.5.5 Analysis of Short-Range Order in Alloys

Computer Simulation of Solutions. The diffraction experiments with diffuse scattering yield directly the correlation parameter $\varepsilon(\varrho)$ (or $\alpha(\varrho)$) for pairs of atoms (see Sect. 3.5.1). But these statistical characteristics do not describe fully the short-range order in solution. A more detailed description of the short-range order required for a better understanding of many properties of an alloy must include the degree of its spatial inhomogeneity (for instance, the size, shape, composition, structure, and mutual arrangement of the small partially ordered regions if they exist in the alloy while the correlation functions must be specified for larger groups of lattice sites, rather than only for pairs (see Sect. 1.1).

Such correlation parameters for groups of lattice sites $\varepsilon(\varrho_1, \varrho_2, \ldots)$ cannot be derived directly from the diffraction results and calculated precisely from the given $\varepsilon(\varrho)$. (In principle, the Mössbauer spectra may yield the probabilities for different configurations of atoms on the groups of sites making up the first coordination sphere around the given lattice site or several such spheres but in practice the resulting complicated spectra cannot always be interpreted accurately enough; see [3.144–146].) For instance, if we regard the crystal as a mechanical mixture of sufficiently large regions with significant differences in the local order and composition we can obtain the same mean $\varepsilon(\varrho)$ (determined in the conventional diffraction experiments) for completely different sets of $\varepsilon(\varrho)$ and $\varepsilon(\varrho_1, \varrho_2, \ldots)$, which are not related to each other and exhibit no regular behaviour, for different alloy regions and parts of the regions.

On the other hand, the probabilities that groups of lattice sites are substituted by the atoms A and B are related to the concentrations c and $1 - c$ and the probabilities $p_{AA}(\varrho)$, $p_{AB}(\varrho)$, and $p_{BB}(\varrho)$ (that is, to $\varepsilon(\varrho)$) by the obvious linear equations (such as $p_{AAA}(\varrho, \varrho') + p_{AAB}(\varrho, \varrho') = p_{AA}(\varrho)$). The number of such equations is smaller than the number of the unknown $\varepsilon(\varrho_1, \varrho_2, \ldots)$ and the latter parameters cannot be determined unambiguously but the virtual ranges of their variation prove to be significantly restricted [1.28; 3.147, 148]. Therefore, if we exclude from consideration the special cases of the equilibrium solution for the above case, which are physically hardly realizable, we may hope to derive from the diffraction data on the parameters $\varepsilon(\varrho)$ a "fairly reasonable" and more detailed description of the short-range order which can have a considerably heuristic value while being not entirely unambiguous. The accuracy of the description will clearly improve if the number of sufficiently accurate $\varepsilon(\varrho)$ is increased. For instance, if a disordered solution contains regions of a high local order owing to some physical reasons then a regular oscillation of $\varepsilon(\varrho)$ in these regions must give rise to oscillating components in the net parameters $\varepsilon(\varrho)$ for the entire solution. These components are caused by the presence of the ordered regions and in principle they can be identified and yield information on these regions if their sizes are sufficiently large while the parameters $\varepsilon(\varrho)$ in the "disordered" part of the solution do not exhibit such noticeable regular variations.

Computer simulation is widely used for analyzing the local order in solid solutions starting from the early study described in [3.147]. The three-dimensional models used in computer simulations consist, for instance, of 4000 or 108 000 cells (lattice sites in which atoms A and B are denoted by 1 and 0 and cyclic boundary conditions are applied. The initial arrangement of atoms is specified in the model. Then a random change is produced in the model (for instance, an exchange of atoms) but the simulation program accepts only those changes which make the parameters $\varepsilon(\varrho)$ for several coordination spheres closer to their experimentally determined values in the alloy under consideration. The simulation process is repeated until the difference between two successive sets of $\varepsilon(\varrho)$ vanishes.

The resulting arrangements of atoms over the lattice sites in some disordered alloys exhibit fairly distinct ordered regions of the given structure while in other alloys identification of such regions is largely arbitrary. Such arrangements of atoms can be analyzed with statistical techniques; for instance, they can be described by the numbers of threes or fours of atoms of different species, the numbers of the atoms A and B and their configurations in the first coordination sphere around the given atom and so on [3.148, 149].

When the above random process is repeated for a model of a comparatively small number of atoms different arrangements of atoms are obviously obtained for the given parameters $\varepsilon(\varrho)$. But these arrangements correspond to not very different averaged statistical characteristics; for instance, the dispersion of the configuration numbers of the neighbouring atoms was about 5% in the alloy analyzed in [3.149]. It has also been shown that for strongly different initial states the resulting characteristics are similar [3.149, 150]. The computer simulations in which the parameters $\varepsilon(\varrho)$ were found for a simulated atom arrangement yielded results from which the synthesized atom arrangements were similar to the initial ones [3.17, 148]. These semi-empirical verifications demonstrate the obvious usefulness of computer simulation techniques for describing the short-range order patterns.

Fcc Alloys. A detailed statistical analysis [3.150] was performed by means of computer simulation for Cu-Al alloys for highly accurate parameters $\varepsilon(\varrho)$ [3.60] (see Table 3.1). The number of various atom configurations was determined for 12 lattice sites around a given one and the correlation in the positions of such clusters in the lattice was analyzed. The main configurations of atoms describing the short-range order in this solution is the tetrahedral group of four Al atoms around a Cu atom [3.151] and the ring configuration of four Al atoms around a Cu atom in a (100) plane. In an alloy with 9.13 at.% Al annealed for 1580 hours at 423 K these clusters contain 5.5 and 6.0% Al, respectively, (computer simulation for an alloy of the same composition with $\varepsilon(\varrho) = 0$ did not exhibit such configurations) and all tetrahedral and ring clusters were isolated (that is, there was no contact between them either in the first or second coordination spheres).

As the aluminium concentration increases the content of such clusters increases and the arrangement of them is increasingly correlated. For instance, in the alloy with 14.76 at.% Al annealed under the above conditions the model of 8000 atoms gives the following results. Out of 87 $CuAl_4$ clusters 74 are isolated and 13 are paired with Cu atoms remaining at the distance d from each other so that together with the ring clusters we obtain small fragments of the Cu_3Au-type structure with the antiphase boundaries randomly distributed throughout the alloy. These locally ordered regions account in total for 14% of the crystal volume and 32% of the Al content but their average size is very small (3.4 Å for a spherical approximation of their shape). This is why no concentration inhomogeneities are found in a large volume. This has been directly demonstrated by the small-angle scattering of slow neutrons which exhibited only a relatively weak intensity of scattering by dislocations [3.152]. The character of the short-range order was also analyzed for Cu-Al alloys of various compositions subjected to various types of treatment [1.27; 3.153, 154].

Computer simulation of the atom arrangement in a Au-20.3 at.% Mn alloy demonstrated a qualitatively different structure of the short-range order [3.64]. The simulation results described a three-dimensional irregular structure made up of relatively long linear segments of Mn atom chains aligned along the $\langle 100 \rangle$ directions, that is, which are in contact in the second coordination sphere. In addition, a considerable number of Mn atoms are in contact with the atoms in chains in the first coordination sphere. Assuming that the exchange interaction has a ferromagnetic character in the second and third coordination spheres and an antiferromagnetic character in the first coordination sphere we can use this short-range order structure for explaining the characteristic magnetic properties of the disordered alloy Au-Mn by the formation of ferromagnetic domains of a very small size (owing to the presence of antiferromagnetic bonds).

A computer simulation approach was used for analyzing the short-range order structure at various temperatures in the high-concentration disordered Cu_3Au solution for which the parameters $\alpha(\varrho)$ were calculated in six coordination spheres [3.43]. The short-range order structure was shown to be significantly inhomogeneous in these alloys. At high temperatures the solution consists of alternating ordered and disordered regions. For instance, at 750°C the alloy contains approximately $8 \times 10^{20}\,cm^{-3}$ ordered regions with the structure of Cu_3Al ($L1_2$) each including from 18 to 25 atoms whose total volume fraction is $p \sim 0.2$. The alloy contains also approximately $6 \times 10^{20}\,cm^{-3}$ regions of an altered composition with the structure of CuAu ($L1_0$) which contain on the average nine atoms and typically are at the boundaries of the Cu_3Au regions. The ordered regions are separated by disordered regions of a significantly larger width ($> 3d$). At 930°C the alloy primarily contains ordered regions of the CuAu type at the ends of which regions with the CuPt-type structure are located (with $p \sim 0.1$–0.2).

At lower temperatures (475°C and lower) the character of the local order is changed somewhat. The content of the regions with the $L1_0$ structure and a considerably altered composition is decreased by an order of magnitude but DO_{22}-type regions are produced at the same time. The total fraction of the ordered regions with the $L1_2$ structure increases with a decrease in temperature and at 396°C (two degrees above T_0) a larger part of the crystal is occupied by such regions with the size of about $4d \times 3d \times 3d$ which have different phases. (In the narrow temperature range of a few degrees around T_0 radiation scattering by the alloy Cu_3Au exhibits some characteristic features [3.43, 155, 156]. The intensity of diffuse scattering by crystal distortions increases significantly while the regular reflection intensity decreases. The one-phonon peaks in the spectrum of inelastic neutron scattering have been found to broaden at the same time (for the same vibration frequencies) while the contribution of the third-order terms in $\boldsymbol{u}_s$ to the intensity I_1 is increased. The results obtained at temperatures somewhat exceeding T_0 demonstrate weak superlattice reflections corresponding to scattering by ordered domains with the size about 100 Å whose volume fraction is about 10^{-2}. These observations indicate that larger and/or sharper inhomogeneities exist near the transition than at temperatures far from T_0 and according to [3.156] they are produced by heterophase fluctuations.) The plot of the scattering intensity I_1 as a function of temperature exhibited break points at the point (001) of the reciprocal lattice space in the temperature ranges where the local order is changed. The positions of the break points corresponded to the temperatures (400, 600, and 830°C) at which the specific heat and other characteristics of the alloy Cu_3Au have anomalies. The ordered regions with the size about 15 Å in the disordered matrix were observed also with the high-resolution electron microscope in hardened Cu_3Au alloys [3.158–160] in full agreement with the X-ray scattering results.

The above results on the inhomogeneous local order in Cu_3Au alloys generally agree with the theory of the heterogeneous alloy states at equilibrium discussed in Sect. 2.5.2 and in [2.155]. (The simplified alloy model treated in Sect. 2.5.2 cannot, of course, give a quantitative description of all structural features. In real crystals there must occur sharp fluctuations of the order in small ordered regions with inhomogeneous order as well as fluctuations of their shapes and positions. This can be an intermediate structure between the model treated in Sect. 2.5 and the model of homogeneous short-range order.) The theory accounts for the very fact of the inhomogeneous ordering in a wide temperature range above T_0 estimates correctly the volume fraction of the ordered regions, $\varrho \sim 0.1$, at $T \gg T_0$ and explains the continuous growth of ϱ with a decrease in temperature. According to the discussion of Sect. 2.5.2, the formation of the $L1_0$ regions at 930°C can be caused, firstly, by the two-fold increase at $k = G'$ compared with the $L1_2$ structure (which increases the contribution to the electron energy $|\delta\varphi_2'|$ by a factor of 4/3) and, secondly, by the fact that the superlattice vector G' approaches the minimum of $w_e(\boldsymbol{k})$ (that is, the Fermi surface diameter; see Fig. 3.7) because the lattice parameter is smaller in CuAu than in Cu_3Au (by $\sim$ 3% for $\eta = 1$). As the temperature decreases and p increases the production of $L1_2$ regions with a strongly changed composition in a significant part of the crystal volume is less favorable thermodynamically than the favored production of $L1_2$ regions with slightly altered compositions which are only fringed with $L1_0$ regions (the latter strain the $L1_2$ regions and shift G'

in the desired direction; see the end of Sect. 2.5.2). As the temperature decreases further the increase in ϱ is accompanied with the growth of the contribution to the electron energy made by the third-order terms in the perturbation theory which exhibits a characteristic minimum for the DO_{22} structure with the G' vectors of the type of (1 1/2 0) and (010) in the case of the copper-based alloys (see the end of Sect. 2.5.2 and the discussion of equations (2.5.14)). This is why a significant proportion of the DO_{22} regions is produced for high p values when the energy of the individual DO_{22} regions is higher than the energy of the $L1_2$ regions by not more than $(1 \div 3)\ k_B T$. As discussed in Sect. 2.5.2, the spread-out singular features exhibited by the plots of $I_1(100)$, the specific heat and other properties of Cu_3Au as functions of temperature are explained by possible spread-out phase transitions between crystal states with different structures of the ordered regions or their different arrangements.

Since the distance k^0 between the end of the vector G' and the Fermi surface in Cu_3Au is $0.82/d$ [3.129] the parameter D of the fcc lattice of the $L1_2$ regions is $5.4d \simeq 20$ Å which agrees with the result of 17 Å corresponding to the density of the ordered regions of 8×10^{20} cm−3 at 750°C. The significant magnitude of k^0 corresponding to the negative curvature of the plot of $\tilde{w}'_e(\boldsymbol{k})$ at $\boldsymbol{k} = \boldsymbol{G}'$ accounts for the difference between the phases neighbouring ordered regions found in experiments. If the phases and positions of these regions are correlated at large distances then, according to the discussion of Sects. 2.5.2 and 3.3.2, the intensity distribution near the lattice sites of the (100) type must exhibit a fine structure. It will consist of the cross of four peaks observed in the hardened alloys [3.125–128] for the configuration of ordered regions corresponding to (2.5.19) and the pair of peaks observed at 475°C for the configuration corresponding to (2.5.20) and providing for a lower surface energy [3.43].

Computer simulation analysis of the short-range order was performed also for some other alloys (with the use of the parameters $\varepsilon(\varrho)$ found in earlier studies with less sophisticated techniques for identification of scattering by distortions). These results were used for estimating the numbers of threes, fours, and larger groups of atoms (for instance, the number of Ni atoms in the first coordination sphere for the Cu-Ni alloy which explained the production of giant magnetic clusters in these alloys [3.161]) and for analyzing larger-scale inhomogeneities. For instance, computer simulation was performed for the alloy Ni_3Fe quenched from 808 K (for this alloy T_c = 773 K) with the use of six parameters $\varepsilon(\varrho)$ obtained from the neutron scattering results (see Table 3.1). The simulation results presented a structure with plate-like ordered regions (domains) of 10–20 atoms (with a thickness of two–three atomic layers) whose total content in the disordered matrix was 35%. Computer simulation for CuAu at 700 and 800 K suggested the existence of ordered domains with the size of 15–20 Å in the disordered matrix [3.73]. Similar results were obtained for the alloy Au-40 at.% Pd [3.162] and for some other alloys discussed in the review [3.148]. These results agree with the electron microscopic observations for some hardened alloys (see, for instance, [3.160–163]) and with the discussion below of the direct data on the diffuse scattering intensity peaks (other experimental data are analyzed in [1.27; 3.164]). In some alloys atom positions can be rearranged somewhat in the process of quenching and the resulting short-range order can depend on the kinetic factors and differ from that at equilibrium.) In most cases we may establish a correlation between the structure, phases and arrangement of the ordered regions and the Fermi surface of the alloy (the correlation is similar to that discussed for Cu_3Au).

Computer simulation results for the partially ordered alloys ($T < T_0$), for instance, Cu_3Au [3.147, 148] or $CoPt_3$ [3.165], demonstrated that the atoms occupying foreign sites also make up small groups. In the above alloys these groups produce small antiphase plates with respect to the ordered matrix which lie in the $\{100\}$ planes. (According to the Babinet principle, a variation of the amplitude or the phase of the order parameters in some regions must give rise to spreading, splitting and shifting (owing to the strain) of the $|c_{G'}|^2$ peak, that is, it can make a contribution $\delta\varphi_2'$ stabilizing the heterogeneous state to the electron energy caused by the factors noted in Sect. 2.5.2. These effects must, of course, exhibit different features for $T < T_0$ and $T > T_0$.) The above X-ray scattering results agree with the direct field ion microscopic observations of the atom order in $CoPt_3$ [3.166].

In addition to computer simulation, other techniques can be employed for analyzing the probabilities P_k of the atom configurations on the groups of lattice sites corresponding to the given c and $\varepsilon(\varrho)$. For instance, in the approximate method of probability variations (3.167, 168] applicable to systems with pair interactions it is assumed that the probabilities P_k are determined by the maximum of the entropy function $\sum_k p_k \ln P_k$ calculated under the restrictions corresponding to the given c and $\varepsilon(\varrho)$. This technique was applied to the probabilities P_k for the groups of three and more nearest neighbours in some alloys and the results for the groups of three atoms were in a good agreement with the computer simulation data.

For the systems in which the ordered regions give rise to sufficiently sharp peaks of the diffuse scattering intensity the theory of scattering by inhomogeneous solutions (see Sect. 3.3.2) can be applied to the data on the scattering intensity distribution to find directly the parameters of the ordered regions. The fine structure of the diffuse scattering peak (11)) found for the alloy Cu-29.8 at.% Pd (see Fig. 3.3) reveals a correlation in the arrangement of the antiphase ordered regions. The ordered $L1_2$ regions were found to have the size $D_0 \cong 7$ Å and their volume fraction was about 0.5 according to the results [3.169] of the theoretical analysis [2.146] of the diffuse scattering intensities $I_1(\boldsymbol{Q})$ (see (3.3.17)). The electron microscopic results for the hardened alloy were from 10 to 20 Å for D_0 and from 30 to 40% for the volume fraction ϱ [3.170]. Such analysis [3.171] for CuAu gave $D_0 \cong 14$ Å (the volume of the ordered domain $V_0 = 160v_0$) and $p = 0.4$ which agreed with the computer simulation results [3.73] and demonstrated correlation in the arrangement of the antiphase regions. For a hardened CuPt alloy the analysis of (1/2 1/2 1/2) diffuse scattering intensity peaks [3.171, 172] suggests that they are caused by the uncorrelated $L1_2$ regions. Their sizes found from (3.3.18) where $P_{\alpha aa} = 0$ and from (3.3.19), with a smoothly varying function $\eta_\alpha(r)$, were 17 Å in the $\langle 111 \rangle$ directions and 19 Å in the perpendicular direction ($V_0 = 410v_0$; the electron microscopic results for D_0 varied between 10 and 100 Å [3.173]).

The diffuse X-ray [3.176] and neutron [3.174, 175] scattering results have been analyzed in detail for the alloys Cu-Mn and Ag-Mn of various compositions (from 1.3 to 24 at.% Mn) which are typical spin glasses. The diffuse scattering intensity peaks for Ag-Mn with $c_{\mathrm{Mn}} > 0.08$ have been found near (1± 1/2 0) positions (rather than (100) positions). The models put forward for accounting for these results assumed the existence of ordered regions in the disordered matrix; for instance, the regions with the structure $I4_1/amd$ (corresponding to the alternating pairs of the (021) planes with higher concentrations of Mn and Ag) [3.176]. The dimensions of these plate-like regions are from $3d$ to $4d$ in one direction and from $5d$ to $6d$ in two other directions. This structure (or other structures which give rise to diffuse scattering intensity peaks near the positions of the type of (1± 1/2 0) may be energetically favourable owing to the third-order additional terms in the electron energy see (2.5.14) and the discussion after it and the text following (2.5.223) and magnetic energy.

The Au-Fe alloys containing from 15 to 19 at.% Fe which are also spin glasses exhibited a different diffuse scattering pattern [3.177]. High diffuse scattering intensity was recorded for small angles and the lines connecting the reciprocal lattice points and $(1\,1/2\,0)$ points. The intensity distribution along the lines is asymmetric with respect to the reciprocal lattice points. This scattering pattern is explained if we assume that plate-like ordered regions with a higher Fe concentration are produced along the twelve $\{420\}$ planes (Fe atoms have a smaller radius than Au atoms and the resulting distortion of the crystal lattice causes asymmetry of the scattering intensity distribution). The structure of the intensity distribution parallel and perpendicular to the above lines indicated the existence of thin plates with the thickness of 30 Å consisting of two atomic layers.

The diffuse scattering results were also used for analyzing the short-range order in more complicated systems, for instance, in non-stoichiometric oxides of transition metals TiO_x and VO_x [3.17, 68]. A correlation was found between the short-range order and the character of the conductivity (in metallic VO_x with $x < 1$ and in TiO_x laminar vacancy regions are produced while in the semiconducting VO_x with $x > 1$ the character of the short-range order is changed). This correlation may be attributed to the effects caused by the Coulomb forces as discussed in Sects. 2.5.2, 3 (for instance, the ionic dispersion mentioned at the end of Sect. 2.5). The Coulomb interaction can also be responsible for production of relatively large ordered regions (with D_0 varying between 10 and 100 Å where the composition and order parameters are changed and which were found in superionic conductors [3.178–182]. The diffuse scattering patterns for other compounds (and alloys) have also been described in the framework of other models, for instance, that of a transition state consisting of characteristic polyhedra of atoms [2.145, 147; 3.17, 183–186]. In some compounds of heavy elements, for instance, in bronzes of the type of Nb_2O_5-WO_3 the arrangement of atoms in the disordered solution can be observed directly with a high-resolution electron microscope [3.186].

The following factors introduce some uncertainty in the studies of the short-range order structure and identification of the systems exhibiting non-uniform local order in heterogeneous state. These factors affect the results obtained from computer simulations and from the direct analysis of the sharp scattering intensity peaks and make it difficult to determine directly the heterogeneous character of a structure. In computer simulation we must bear in mind that for any, even sufficiently uniform, statistical arrangement of atoms over the lattice sites a certain (perhaps small) number of small regions with the given local order must exist in the matrix owing to the purely probabilistic factors. Identification of such regions (especially since the local order in them may be partial, rather than completed) is, to some extent, a subjective decision. Identification of a genuinely heterogeneous structure produced owing to some physical factors (such as those discussed in Sects. 2.5.2, 3), rather than a structure with the ordered regions produced by fluctuations, will be the more precise the larger the sizes of these regions and their volume fraction, the greater the correlation in phases for the neighbouring regions and the smaller the concentration of one of the components of the alloy. In some of the above cases these conditions are apparently satisfied (especially if we take into the account the fact that in the AB_3 alloys the concentration $c = 1/4$ is relatively low).

When we analyze directly the diffuse scattering intensity peaks we must bear in mind that this approach often *a priori* assumes the existence of a heterogeneous state. This simplified model assuming a rapidly vanishing correlation of the ordered regions yields only qualitative results. On the other hand, the above model of the heterogeneous structure corresponds to the scattering intensity distribution with several peaks in the vicinities of the superlattice points G' of the reciprocal lattice which can be also produced by scattering in the homogeneous solution if the magnitudes of G' are close to the diameters of the flattened parts of the Fermi surface for the alloy (see Sect. 2.4). The difference between these two cases consists in the fine structures of the scattering intensity distribution. In copper-based homogeneous alloys the fine structure must apparently be considerably less pronounced than that found in experiments (see Sect. 3.5.4) and thus the treatment of the experimental data in terms of heterogeneous alloy structure seems more reasonable.

The short-range order in alloys, and in particular, the conditions for realization of the heterogeneous structure must be studied with specially developed highly sophisticated techniques. They will include precise methods for analyzing the diffuse scattering results, computer simulations with large numbers of the parameters $\varepsilon(\varrho)$ and blocks of lattice cells simulating the crystal as well as objective statistical methods for analyzing the crystal realizations and identifying definitely the ordered regions. In this connection it may prove highly useful to observe directly the mutual arrangement of the solution atoms with the use of a high-resolution electron microscope or a field ion microscope.

Alloys in Which ω Phases are Produced. The diffraction studies can yield important results for bcc alloys in which displacement-type phase transitions can occur giving rise to ω phases. As noted in Sect. 2.5.2, the ω phases are produced in the alloys of Ti, Zr and Hf with a number of transition metals, for instance, with V, Mo, Ni, and Fe [2.163; 3.187]. They can be produced by fast quenching (athermally) or by isothermal annealing. In the case of athermal phase transformation of alloys in a certain concentration range, and in the case of isothermal transformation, the scattering intensity distribution exhibits, apart from the sharp peaks at the points G of the reciprocal lattice for the bcc (β) phase, also sharp peaks at the points $G + G'$ of the ω phase where G' for one of the ω phases is given by (2.5.23). But for hardened alloys of other compositions the peaks of the ω phase prove to be blurred and shifted with respect to the points $G + G'$ (see [3.188–193]).

Some scattering features observed even at much higher temperatures than the phase transition point indicate that a peculiar short-range order is established in such bcc alloys which involves significant correlated displacement of atoms, rather than occupation of lattice sites by atoms of different species (as in the cases treated above). It has been found that it is very convenient to analyze such short-range order using the data on diffuse scattering intensity distribution of X-rays, neutrons and Mössbauer photons in the reciprocal lattice space and the energy spectrum of the scattered radiation. Such studies were performed for Zr-Nb alloys of various compositions [3.194-198] (these alloys are highly suitable for neutron diffraction studies since the atom masses of the components and the scattering amplitudes b_{Zr} and b_{Nb} are similar and the direct effects of atom displacements can be explicitly analyzed).

In hardened Zr-Nb alloys with a considerable Nb concentration the above-mentioned diffuse scattering intensity peaks were observed to be shifted with respect to the $G + G'$ points of the ω phase. Figure 3.9 shows the distribution of the diffuse scattering intensity in the $(1\bar{1}0)$ plane of the reciprocal lattice recorded for the alloy Zr-20% Nb at 300 K [3.194]. The horizontal axis is parallel to the [110] direction and the vertical axis to the [001] direction. A similar pattern was recorded for X-ray scattering [3.197] (in a larger part of the reciprocal lattice cross section). The scattering intensity distributions do not exhibit the symmetry typical of the bcc lattice and, in addition to the sharp Bragg peaks produced by the bcc lattice they include the effects of scattering by the ω phase regions of various kinds aligned along the $\langle 111 \rangle$ axes of the bcc crystal. Broadened diffuse scattering intensity peaks can be seen near the ω phase lattice points (for instance the points $(4/3\,4/3\,4/4)$ in Fig. 3.9 for the alloys containing 20% Nb and the broadening is considerably smaller along the [111] axis than in the perpendicular direction. These results suggest that the scattering ω phase regions are rod-shaped and aligned along the $\langle 111 \rangle$ directions of the bcc crystal. The lengths of these regions have been estimated [3.195–197] (under the assumption that scattering by individual regions occurs independently) to be 20 Å in the alloy with 20% Nb and 60 and 40 Å in the alloys with 8 and 15% Nb (which exhibited fairly sharp reflections of the ω phase). The broadened peaks for the alloys with 20% Nb are noticeably displaced from the reciprocal lattice points of the ω phase to the origin

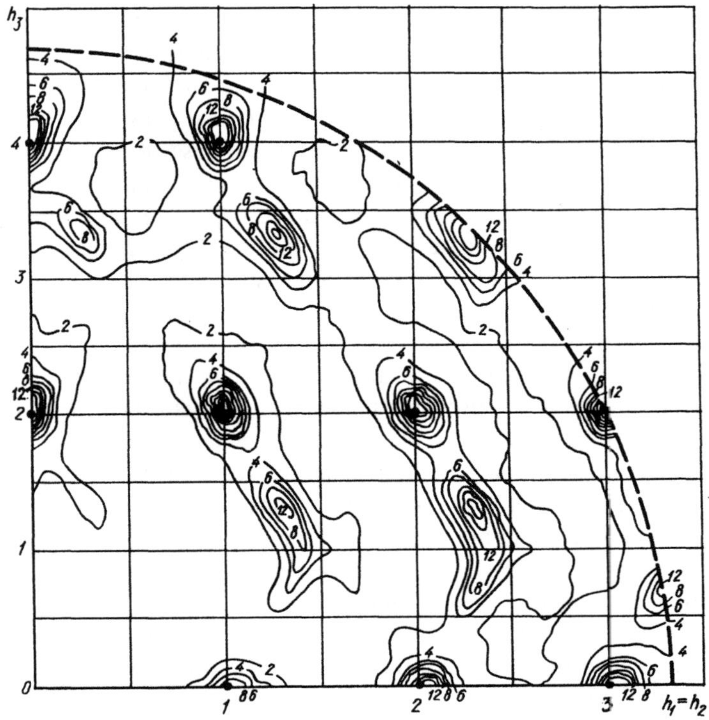

Fig. 3.9. The intensity distribution for scattering by the Zr-Nb alloy [3.194]

of the coordinates (the greater the magnitude of Q the larger the displacement with the exception of the peak at (2/3 2/3 2/3) which is displaced in the opposite direction [3.197].

The blurred diffuse scattering intensity peaks occur also near the reciprocal lattice points of the bcc phase. Since they are independent of the reflection order it can be suggested that these peaks are caused not by the bcc matrix distortions but by the replacement of parts of the bcc matrix with the ω phase regions [3.197] (see Sect. 1.4 in [1.13]. According to (2.5.23), two thirds of the atoms in the ω phase regions are displaced by $\pm u_0$ and therefore the decrease in the Bragg peak intensities for the bcc matrix yields the estimate for the volume fraction of these regions of $p \cong 0.5$ and $u_0 \cong 0.94\sqrt{3}d/2$ for the alloys with 20% Nb at 300 K (while $u_0 = 0.167\sqrt{3}d/2$ for the "perfect" ω phase) [3.195].

The energy distributions of inelastically scattered neutrons [3.194, 196, 198] or Mössbauer photons [3.197] demonstrate that this diffuse scattering is to a considerable extent elastic (or quasielastic) and corresponds to energy transfers of not more than 0.2×10^{-6} eV as shown in [3.198]. This means that scattering is caused by inhomogeneities which are static or very slowly vary with time, rather than by lattice vibrations. It is especially significant that the central peak of quasielastic diffuse scattering in the vicinities of the reciprocal lattice points of the ω phase was recorded in a wide temperature range which is much higher (by about 1000 K) than the temperature T_0 of the $\beta \to \omega$ transition. For instance, for the alloy with 20% Nb where $T_0 \sim 300$ K the elastic scattering intensity $I_{1el}(2l/3\,2l/3\,2l/3)$ exhibited a peak near $l = 2$ corresponding to the point (4/3 4/3 4/3) not only at $T = 300$ K but also at 1300 K [3.194] (see Fig. 3.10) though the area of the peak is much smaller (by a factor of approximately five). This means that even at such high temperatures the bcc matrix contains ω phase regions (though their volume fraction p is lower by an order of magnitude than at 300 K). Such peak was recorded also for pure Zr in the temperature range of the bcc β phase (between 1200 and 1400 K) [3.199].

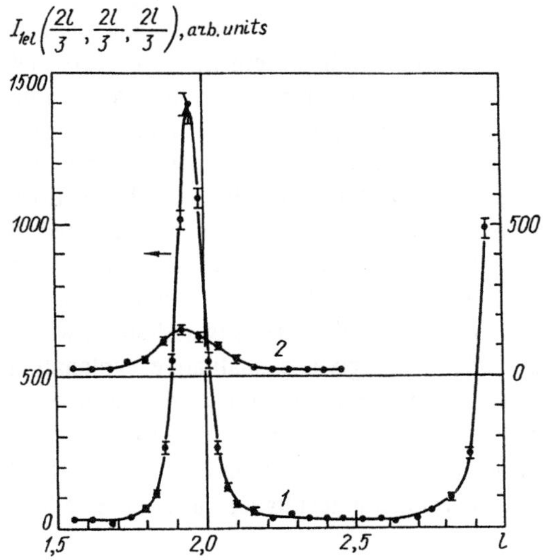

Fig. 3.10. The intensity distribution for the elastic diffuse scattering of neutrons in the [111] direction of the reciprocal lattice space for the Zr-20% Nb alloy at 300 K (*curve 1*) and 1273 K (*curve 2*) [3.194]

The theory presented in Sect. 2.5.2 accounts for these displacement-type features of the short-range order structure in the bcc phases of the alloys of transition metals [2.153, 154]. In these alloys the electron energy $w_e(\boldsymbol{k})$ of the longitudinal displacement wave exhibits a singularity at $k \approx \boldsymbol{G}'$ (this theoretical prediction [2.168, 169] has been confirmed by the neutron diffraction results for the vibration spectrum of Zr showing a sharp drop in the longitudinal vibration frequencies near $\boldsymbol{k} \approx \boldsymbol{G}'$ [3.199]). According to the theory of Sect. 2.5.2, under these circumstances the greatest gain in the electron energy can be obtained if a heterogeneous state is produced. This state consists of the bcc matrix and the correlated rod-shaped particles of various ω phase types (which differ in the phases 0, $2\pi/3$ and $4\pi/3$ of the displacement wave under consideration) which are aligned along the $\langle 111 \rangle$ axes. The non-local effects caused by the interaction between the ω phase particles, a possible decrease in the electron energy of a small particle in comparison with a larger one, and a decrease in the effective permittivity for thin particles make such a state more favourable thermodynamically than either the homogeneous bcc and ω phases or a structure consisting only of the antiphase ω particles without bcc regions (one such structure would be more favourable in a theory that ignores the non-local effects). This is why the estimates of Sect. 2.5.2 show that a heterogeneous state at equilibrium can exist in a wide temperature range of about 1000 K.

This gives a natural explanation to the observations [3.194–199] indicating that quasistatic ω phase regions exist at high temperatures (sometimes exceeding T_0 by 1000 K). These regions are equilibrium elements of the structure rather than fluctuations. The small size of the particles ($D_{0z} < 2\pi/3k^0$) and their rod-like shape ($D_{0\perp} < D_{0z}$) account for the spreading of the diffuse scattering intensity peaks which is especially strong in the directions perpendicular to $\langle 111 \rangle$. As noted in Sect. 2.5.2 (see the text after (2.5.23)), the correlation between the particles of various ω phase types with the phases 0, $2\pi/3$ and $4\pi/3$ produces a decrease in the electron energy since the vector $\boldsymbol{G}' + \boldsymbol{q}_n$ of the displacement wave generated by the correlated particles is effectively increased by such correlation to a value corresponding to a minimum of $w_e(\boldsymbol{k})$. At the same time the addition of $\boldsymbol{q}_n$ displaces the weak diffuse scattering intensity peak near the point (2/3 2/3 2/3) from the origin of the coordinates in agreement with the experimental observations (the effect is similar to shifting of the peaks for the structural models suggested in [2.144, 148] and [2.152] which include alternating layers of various ω phase types as in Fig. 2.2 but not β phase regions). The shifting of all far peaks towards the origin of the coordinates can be attributed [3.197] to the effects of self-strain which are enhanced with an increase in Q. This theory gives a qualitative explanation for the experimentally recorded decrease in the volume fraction ϱ with an increase in temperature. In addition to displacements of atoms the decrease in the electron energy can be attributed to their rearrangement over the lattice sites. This effect was suggested as an explanation for the experimental results for the Ti-based alloys [3.200].

Considerably smaller shifts of the diffuse scattering intensity peaks with respect to the reciprocal lattice of the ω phase were recorded for the hardened alloy Ti-23 at.% V than for Zr-20 at.% Nb. A detailed analysis of the intensities and widths of some ω peaks [3.201] took into account significant anisotropy and anharmonicity of the Debye-Waller factor indicating that fairly large rod-shaped particles of the ω phase were produced in this alloy at room temperature. They had a width of about 45 Å and the length of about 70 Å. The displacements in these particles are essentially non-uniform and the amplitudes u_0 (see (2.5.23)) at the particle centers are twice as large as their average values. Since the boundaries of the particles are blurred they occupy almost the entire crystal volume.

These features of the heterogeneous state of the Ti-V alloy may be attributed to the structure described at the end of Sect. 2.5.2 which is produced if the point $\boldsymbol{G}'$ is at the range where the plot of the function $w_e'(\boldsymbol{k})$ has a positive e curvature. In this structure particles of various ω phase types are not correlated and this fact accounts for the small displacements of the diffuse scattering intensity peaks with respect to the reciprocal lattice points of the ω phase. The particle length of about

10 Å agrees also with (2.5.24b) for this model if we assume that $E_F(u_0v'_{\boldsymbol{k}})^{-1}$ has a typical value of about 10. A similar heterogeneous structure may be produced in the Zr-Nb alloys with a Nb concentration below 17%. These alloys also exhibited rather narrow diffuse scattering intensity peaks which have hardly been displaced from the reciprocal lattice points of the ω phase, that is, which correspond to uncorrelated particles of various ω phase types with longer lengths.

The diffuse scattering peaks corresponding to small ω phase particles were observed not only in high-concentration solutions but also in pure niobium containing small amounts of oxygen and nitrogen in the octahedral interstices of the bcc Nb lattice [3.202]. They can be attributed to the ω phase regions (whose electron energy is significantly lower owing to their small size; see Sect. 2.5.23) produced in strongly distorted crystal regions near the interstitial impurities.

We must obtain more detailed information on the electron structure of the alloys and the electron energy $w_e(\boldsymbol{k})$ of the displacement wave and develop a more sophisticated model in order to understand the nature of the heterogeneous state and to describe quantitatively the features of the scattering intensity distribution. It should be noted that rather different microscopic models (possessing though some important features in common, namely, the rod-like shape of the particles and alternating ω phase types) can yield largely similar scattering intensity distributions and at the same time similar electron energies (which makes it difficult to select a model). For instance, the diffuse scattering results [3.194] and the high-resolution electron microscopic results can be described within the framework of a model postulating the production of ω-like linear defects (regions of two ω phase types separated with a vacancy whose sizes were $25 \times 5 \times 5$ Å for the Zr-20 at.% Nb alloy; for the alloy with 13%Nb content the particle length was 60 Å). Note that the main factors which explain according to [2.154] the origin of the heterogeneous state in the above alloys remain effective (though modifiedd when we make such (or similar, as in the models with inhomogeneous particles [3.205]) alterations in the models under consideration. Other structural models of the alloys containing ω phases were discussed in [3.188, 205–208].

3.6 Diffuse Scattering by Long-Range Order Fluctuations

In Sects. 3.2–5 we treated the diffuse scattering caused by disordered occupation of the lattice sites of a solid solution by atoms of different species. In one-component crystals diffuse scattering is caused by the disorder in the orientations of the molecules in crystals with orientational ordering or by the disorder in the positions of atoms (ions) in crystals with displacement-type ordering. For vectors $\boldsymbol{Q} \approx \boldsymbol{G}$ in the vicinities of the reciprocal lattice points this scattering can be regarded as being caused by the long-wave fluctuations of the long-range order parameters η_μ. A similar treatment can be applied also to the solid solutions in the vicinities of the superlattice points of the reciprocal lattice. In the special case of undistorted solutions with a single long-range order parameter, scattering by long-range order fluctuations was in fact treated in Sect. 3.3 (see (3.3.8)). If the long-range order is described by several parameters and especially if static distortions are taken into account the results can be quite different [1.10, 16]. Therefore, we shall discuss in this section diffuse scattering by one-component crystals and solid solutions under such general conditions. The range of $\boldsymbol{Q} \approx \boldsymbol{G}$ will be analyzed in more detail because in this range macroscopic treatment of the fluctuations of the order parameters can be performed and in some systems the diffuse scattering intensity can be expressed in terms of the properties found from independent experiments, for instance, the dielectric susceptibility in ferroelectrics.

3.6.1 One-Component Crystals

We can use the general equation (3.1.6) for calculating the X-ray scattering intensity in one-component crystals in which the long-range order is described by n parameters η_μ and the static distortions are significant. The structure amplitude of the sth cell in the distorted crystal is

$$f_s = \sum_{\gamma=1}^{\nu} f_{s\gamma} \exp(\mathrm{i}\boldsymbol{Q} \cdot \boldsymbol{R}_\gamma) \exp[\mathrm{i}\boldsymbol{Q} \cdot (\boldsymbol{u}_{s\gamma} - \boldsymbol{u}_s)] \,, \tag{3.6.1}$$

where $\boldsymbol{R}_\gamma$ is the radius-vector passing to the site γ of the given lattice cell from its center or the first site, and $\boldsymbol{u}_s$ is the displacement of the center of the sth cell (or one of its sites, for instance, the first one) in an imperfect crystal. Then the equation (3.1.6) for the scattering intensity can be rewritten as

$$I = \left\langle \left| \sum_{s=1}^{N} f_s \mathrm{e}^{\mathrm{i}\boldsymbol{Q}\cdot\boldsymbol{R}_s} \mathrm{e}^{\mathrm{i}\boldsymbol{Q}\cdot\boldsymbol{u}_s} \right|^2 \right\rangle . \tag{3.6.2}$$

Here $\boldsymbol{R}_s$ is the radius-vector of the center of the first site of the sth lattice ($\boldsymbol{R}_{s\gamma} = \boldsymbol{R}_s + \boldsymbol{R}_\gamma$).

The fluctuations $\delta\eta_\mu(\boldsymbol{r})$ of the long-range order parameters produce changes in the structure amplitudes and the displacements f_s. The variation δf_s of the structure amplitude of the sth cell caused by the long-wave fluctuations under consideration can be written as

$$\delta f_s = \chi_\mu \delta\eta_\mu \tag{3.6.3}$$

(here and below summation is implied over the repeated subscripts μ and μ'. The coefficients χ_μ corresponding to the vicinity of the given reciprocal lattice point are determined from the dependence of the average structure amplitude (3.1.9) (which describes the intensity of the regular Bragg reflections) on the order parameters η_μ (for some reciprocal lattice points $\chi_\mu = 0$).

The coefficients χ_μ in (3.6.3) and in the respective equation for solid solutions are functions of $\boldsymbol{Q}$. In the case of substitution-type ordering in solutions or orientational ordering the dependence of χ_μ on $\boldsymbol{Q}$ is related to the behaviour of the atomic scattering factors and the Debye–Waller factors as functions of Q (in the first case $\chi_\mu \sim [f_A \exp(-M_A/2) - f_B \exp(-M_B/2)]$ and in the second case χ_μ is proportional to the difference between the structure amplitudes for different orientations of the molecules). If the reduction in the symmetry of the crystal and long-range ordering involve displacements of atoms (as in perovskites) then the parameter η_μ may be determined as the ratio between the amplitude $u^\mu_{s\gamma} - u^\mu_s$ of the wave of displacements of the atoms of the sublattice with respect to its geometric center (in the commensurate or incommensurate ordering wave) and its magnitude at $T = 0$. Under these conditions we have

$$\chi_\mu = \mathrm{i}\sum_{\gamma=1}^{\nu} f_\gamma Q \cdot (u^\mu_{s\gamma} - u^\mu_s) \exp \mathrm{i} Q \cdot R_\gamma) \eta_\mu^{-1} \tag{3.6.4}$$

(if $Q \cdot u_{s\gamma} \ll 1$) and the dependence of χ_μ on Q is associated not only with $f_\gamma(Q)$ but also with $Q \cdot (u^\mu_{s\gamma} - u^\mu_s)$. Finding $I_1(Q)$ at several reciprocal lattice points corresponding to the same q we can determine $\chi_\mu(Q)$ for different Q and the character of the displacements $u^\mu_{s\gamma}$ in the fluctuation wave of the long-range order.

If we ignore the distortions caused by the fluctuations of the correlation parameters we can write according to (1.5.8) the static displacement u_s of the center (or the first sited of the sth cell as the following expansion:

$$u_s = \mathrm{i}\sum_k A_{\mu k} \eta_{\mu k} \exp(-\mathrm{i} k \cdot R_s) \,. \tag{3.6.5}$$

(Strictly speaking, (3.6.5) must include also terms second order in the parameters $\eta_{\mu k}$ or equivalent terms to first order in the Fourier components of the fluctuations of the correlation parameters (see (1.5.8)). Under certain conditions the behavior of the displacements as functions of the correlation parameter fluctuations is described by the same small parameters as those describing their behavior as functions of the long-range order fluctuations so that it is not a quite consistent approach to consider only the latter. In some case of interest, however, for instance, near a second-order phase transition, when long-range order fluctuations cause anomalously strong scattering, the scattering by the correlation parameter fluctuations (which is similar to that treated in Sect. 3.4.2) makes a relatively small contribution and may be ignored.)

In (3.6.5) the coefficients $A_{\mu k}$ of proportionality between the Fourier components of the displacements and the Fourier components $\eta_{\mu k}$ of the fluctuations of the parameter η_μ are given by the equations similar to (1.5.9). Since for small k we have $P_k(\hat{L}_\mu) \sim 1/k$ according to (1.5.5) for the long-wave fluctuations we have $A_{\mu k} \sim 1/k$ as in the case of concentration waves. The identification of the cell center and the terms χ_μ and $f Q \cdot A_{\mu k}$ is obviously somewhat arbitrary but the difference $\chi_\mu - f Q \cdot A_{\mu k}$ is determined unambiguously here.

Using (3.6.2, 3, 5) we can write that the contribution to the scattering amplitude made by the sth lattice cell is

$$\left[f + \sum_{k} (\chi_\mu - fQ \cdot A_{\mu k}) \eta_{\mu k} e^{-iQ \cdot R_s}\right] e^{iQ \cdot R_s} , \tag{3.6.6}$$

where f is the average structure amplitude (3.1.9) for the regular reflection whose vicinity is under consideration (including the Debye-Waller factor). It is assumed in (3.6.6) that the distortions are not very large (conditions similar to (3.2.9) are satisfied) and only the linear terms may be retained in the expansion in u_s. Taking the sum in (3.6.6) over the cells (over s) and squaring the result according to (3.6.2) we obtain the following expression for the scattering intensity in the limit of an infinite crystal:

$$I = 8\pi^3 \frac{N}{v} f^2 \delta(q) + N^2 \langle (fQ \cdot A_{\mu k} - \chi_\mu) \eta_{\mu q}|^2 \rangle . \tag{3.6.7}$$

Here the first term gives the intensity of regular reflections and the second term the diffuse scattering intensity I_1.

In the range of small q the mean squares and the products of the Fourier components of the long-wave fluctuations of the long-range order parameters η_μ are given by the equations (1.2.20) of the macroscopic fluctuation theory in undistorted crystals and for distorted crystals by the equations (2.1.20) which take into consideration the long-range elastic forces. Using (2.1.20) we can write the diffuse scattering intensity (the second term in (3.6.7)) in the following form:

$$I_1 = \frac{N k_B T}{v} (fQ \cdot A_{\mu q} - \chi_\mu)(fQ \cdot A_{\mu' q} - \chi_{\mu'}) \varphi^{-1}_{\mu\mu'}(q) \quad (q \ll k_m) . \tag{3.6.8}$$

(The equations (3.6.8, 9, 13–17) derived below take into consideration the scattering by distortions produced by the fluctuations of the parameters η_μ but not the additional scattering by the thermal displacements of atoms (corresponding to the last term in (2.1.21)) which are not related to these fluctuations. If the elastic moduli are not anomalously small, however, then the "thermal scattering" may be relatively weak especially near a second-order phase transition where the intensity (3.6.8) from the fluctuations grows sharply.) The equation (3.6.8) can, in fact, be applied not only to one-component crystals but also to the scattering by the concentration fluctuations and the long-range order parameters in solutions. Then the concentration c must be regarded as one of the parameters η_μ.

In the special case when the long-range order is described by a single parameter η the repeated summation over μ and μ' in (3.6.8) is eliminated and we obtain a simpler equation for the scattering intensity:

$$I_1 = \frac{N k_E T}{v \varphi_{\eta\eta}(q)} (fQ \cdot A_{\eta q} - \chi_\eta)^2 . \tag{3.6.9}$$

Here $\varphi_{\eta\eta}(q) = \partial^2\varphi/\partial\eta^2 + \varphi^{el}_{\eta\eta} + \alpha_{ij}(n) q_i q_j$ is given by (2.1.17). The lattice parameters often have only a weak dependence on the long-range order parameters and the contribution of the elastic energy $\varphi^{el}_{\eta\eta}$ to $\varphi_{\eta\eta}(q)$ is very small (considerably smaller than in the case of the concentration fluctuations) and manifests itself only for very small $\partial^2\varphi/\partial\eta^2$. If in the case under consideration the distortions can be ignored entirely ($\boldsymbol{A}_{\eta q} = 0$) then (3.6.9) reduces to the simple equation

$$I_1 = \frac{Nk_{\mathrm{B}}T}{v} \frac{\chi_\eta^2}{\frac{\partial^2\varphi}{\partial\eta^2} + \alpha_{ij}q_iq_j} . \tag{3.6.10}$$

As in the case of solid solutions (see Sects. 3.3.4), the scattering intensity far from the reciprocal lattice points may be expressed in terms of the correlation parameters for orientations of molecules or arrangement of atoms over lattice cells or in terms of the parameters describing the interaction between molecules. For instance, if orientational ordering means that molecules may have only two orientations (1 and 2) and a lattice cell of the disordered crystal contains only one molecule then the state of the crystal can be described by the numbers $c_s = 1, 0$ for the orientations 1 and 2, as it was done for solutions. Then if we write in (3.6.2) $f_s = f + (c_s - 1/2)\chi$ (where $\chi = \chi(\boldsymbol{Q})$ is the difference between the structure amplitudes for the configurations 1 and 2) and repeat the same transformations as for the solid solutions we can describe the diffuse scattering intensity in undistorted crystals with equations similar to (3.3.1, 9, 11) (with $c = 1/2$):

$$\begin{aligned} I_1 &= N\chi^2 \left[\frac{1}{4} + \sum_{\varrho\neq 0} \varepsilon(\varrho)\cos \boldsymbol{q}\cdot\boldsymbol{\varrho}\right] \\ &= \frac{N\chi^2}{\sum_{\varrho} x(\varrho)\cos \boldsymbol{q}\cdot\boldsymbol{\varrho}} \approx \frac{N\chi^2}{4 + \sum_{\varrho} \frac{w(\varrho)}{k_{\mathrm{B}}T}\cos \boldsymbol{q}\cdot\boldsymbol{\varrho}} . \end{aligned} \tag{3.6.11}$$

Here the parameters $\varepsilon(\varrho)\langle c_s c_{s'}\rangle - 1/4$ describe the correlation in the orientations of the molecules, $w(\varrho) = V_{12}+V_{21}-V_{11}-V_{22}$ is the difference between energies for pairs of molecules with different orientations, and the last approximate equation in (3.6.11) is valid for $|w(\varrho)| \ll k_{\mathrm{B}}T$.

It has been shown [3.209] that (3.6.11) is in good agreement with the experimental results on neutron scattering intensities as functions of $\boldsymbol{Q}$ and T recorded for the crystals of ND_4Br thus indicating that the orientational ordering in these crystals can be described with the Ising model.

Similar expressions may be derived for crystals in which molecules can have several orientations but the energies $w(\varrho)$ are still small (an equivalent case is that of scattering by a multicomponent alloy). It is purely geometric considerations, however, which often cause a sharp increase in the energy for certain mutual orientations of molecules so that the respective configurations cannot be realized in practice. Under such circumstances the approximations similar to the self-consistent field approximation (as in the last equation in (3.6.11)) are inapplicable. Special techniques (for instance, the graph method) which give a better convergence have been employed for calculating scattering characteristics in molecular crystals [3.210] and ice [3.211–213] where some configurations exhibit strong repulsion.

3.6.2 Solid Solutions

The general equation (3.6.8) for the diffuse scattering intensity in the vicinities of the reciprocal lattice points can be written in a more explicit form also for the solid solutions whose structure is described by a single long-range order parameter and where the lattice sites are centers of symmetry (a solid solution with two sublattices similar to beta brass, NaCl, etc.). We shall assume that the condition (1.5.24) is satisfied, that is, the lattice parameter has a much stronger dependence on the concentration than on the long-range order parameter. Then we can ignore the distortions caused by the fluctuations of the long-range order parameters and the correlation parameters and take into consideration only the concentration distortions. In the above case the parameters η_μ in (3.6.8) are the long-range order parameter $\eta \equiv \eta_1$ and the concentration $c \equiv \eta_2$.

Since the distortions caused by the fluctuations of the parameter η are not taken into account, we must take in (3.6.8) $\boldsymbol{A}_{1q} = 0$ while the quantities $\boldsymbol{A}_{2q} = \boldsymbol{A}_q$ are given by the general equations derived in Sect. 1.5 for the amplitudes of the concentration distortion waves. In the vicinities of the Bragg reflections (when $\boldsymbol{Q} \approx \boldsymbol{G}$ and $\exp(\mathrm{i}\boldsymbol{G}\cdot\boldsymbol{R}_\gamma) = 1$) for $M \ll 1$ the equations (1.1.3) and (3.1.9, 10) show that the structure amplitude of the lattice cell, $f = 2[cf_A + (1-c)f_B]$, is not changed by the fluctuations of the long-range order parameter, that is, $\chi_1 = 0$. The variation of f under the concentration fluctuations corresponds to $\chi_2 = 2(f_A - f_B)$. Therefore the sum over μ, μ' in (3.6.8) retains only the term with $\mu = \mu' = 2$ and in cubic crystals we have

$$I_1 = \frac{N}{v}\frac{k_{\mathrm{B}}T(\varphi_{\eta\eta} + \alpha(\boldsymbol{n})q^2)}{\varphi_{\eta\eta}\varphi_{cc}(\boldsymbol{q}) - \varphi_{c\eta}^2 + \alpha(\boldsymbol{n})\varphi_{cc}(\boldsymbol{q})q^2}/f\boldsymbol{Q}\cdot\boldsymbol{A}_q - 2f_A + 2f_B)^2 \,. \quad (3.6.12)$$

Here the partial derivatives with respect to η and c in $\varphi_{\eta\eta}$ and $\varphi_{cc}(\boldsymbol{q})$ are found for constant c and η. Since $d^2\varphi/dc^2 = \varphi_{cc} - \varphi_{c\eta}^2/\varphi_{\eta\eta}$ the expression for the intensity I_1 in the range of small q can be easily seen to be the same as that for disordered solutions (see (3.4.3)).

In the case of superlattice reflections (when $\boldsymbol{Q} \approx \boldsymbol{G}'$ and $\exp(\mathrm{i}\boldsymbol{G}'\boldsymbol{R}_\gamma) = (-1)^{\gamma+1}$) for $M \ll 1$ we obtain $f = \eta(f_A - f_B)$ from (1.1.3) and (3.9, 10). The parameters χ_1 and χ_2 which determine the variations of the structure amplitudes of the lattice cells subject to fluctuations of the order and the concentration reduce then to $\chi_1 = f_A - f_B$ and $\chi_2 = 0$. Then we obtain from (3.6.8) for the vicinities of the "superlattice" points of the reciprocal lattice:

$$\begin{aligned} I_1 &= \frac{N}{v}(f_A - f_B)^2 \frac{k_{\mathrm{B}}T}{\varphi_{\eta\eta}\varphi_{cc}(\boldsymbol{q}) - \varphi_{c\eta}^2 + \alpha\varphi_{cc}(\boldsymbol{q})q^2} \\ &\quad \times [\varphi_{cc}(\boldsymbol{q}) + 2(\varphi_{c\eta} + \gamma q^2)\eta\boldsymbol{Q}\cdot\boldsymbol{A}_q + (\varphi_{\eta\eta} + \alpha q^2)\eta^2(\boldsymbol{Q}\cdot\boldsymbol{A}_q)^2] \,. \quad (3.6.13) \end{aligned}$$

In the equations (3.6.12, 13) the terms proportional to q^2 are retained only near the derivatives $\varphi_{\eta\eta}$ and $\varphi_{c\eta}$ which are small near second-order phase transitions. In the absence of distortions ($\boldsymbol{A}_q = 0$) the equation (3.6.13) reduces to (3.3.8).

For arbitrary vectors $\boldsymbol{Q}$ it is easy to describe the diffuse scattering intensity for distorted ordered solid solutions in terms of the correlation parameters $\varepsilon_{\gamma\gamma'}(\boldsymbol{\varrho})$ and the concentration c_γ of the atoms A in various sublattices. This can be conveniently done by taking the general expression (3.2.8) for I_1 and expressing $\boldsymbol{u}_{q\gamma}$ in terms of $c_{q\gamma'}$ with the use of (1.5.10). If we take into consideration (1.1.11) we obtain then

$$
\begin{aligned}
I_1 &= N \sum_{\gamma,\gamma'=1}^{\nu} \left[c_\gamma(1-c_\gamma)\delta_{\gamma\gamma'} + \sum_{\varrho\neq 0} \varepsilon_{\gamma\gamma'}(\boldsymbol{\varrho}) \cos \boldsymbol{q}\cdot\boldsymbol{\varrho} \right] \\
&\times \sum_{\gamma_1,\gamma_2=1}^{\nu} \exp[\mathrm{i}\boldsymbol{G}\cdot(\boldsymbol{R}_{\gamma_1}-\boldsymbol{R}_{\gamma_2})][(f_A-f_B)]\delta_{\gamma\gamma_1} - f_{\gamma_1}\boldsymbol{Q}\cdot\boldsymbol{A}_{q\gamma_1\gamma}] \\
&\times [(f_A-f_B)\delta_{\gamma'\gamma_2} - f_{\gamma_2}\boldsymbol{Q}\cdot\boldsymbol{A}_{q\gamma_2\gamma'}] \,. \qquad (3.6.14)
\end{aligned}
$$

For instance, in stoichiometric solid solutions with two sublattices we have $c_\gamma(1-c_\gamma) = (1-\eta^2)/4$ and if the correlation can be ignored then the dependence of the scattering intensity on the long-range order is described by $I_1 \sim (1-\eta^2)$ as in undistorted solid solutions.

3.6.3 Intrinsic Ferroelectrics

Most parameters in equation (3.6.8) giving the diffuse scattering intensity can be found from independent experiments. For instance, the parameters f and χ_μ for the given reciprocal lattice point can be easily calculated if the atomic scattering factors and the crystal structure are known. The parameters $\boldsymbol{A}_{\mu q}$ are determined from the data on the elastic moduli and the dependence of the size and shape of the lattice cell on η_μ. It is only the derivatives $\varphi_{\mu\mu'} \equiv \partial^2\varphi/\partial\eta_\mu\partial\eta_{\mu'}$ that are difficult to find from independent experimental data in the general case. But in intrinsic ferroelectrics, where the long-range order is unambiguously described by the components P_i of the polarization vector, the derivatives $\varphi_{\mu\mu'}$ and $\varphi^{-1}_{\mu\mu'}$ can also be determined from the independent experimental results [1.10, 16]. If P_i are taken as the long-range order parameters then according to (1.2.23) the respective derivatives $\varphi_{ij} \equiv \partial^2\varphi/\partial P_i \partial P_j$ are determined by the components $\varkappa_{ij}$ of the dielectric susceptibility tensor of the ferroelectric.

The equations for the mean squares and products of the Fourier components of the polarization fluctuations, that is, for the functions $\varphi^{-1}_{\mu\mu'}(\boldsymbol{q})$ in (3.6.8), are affected significantly by the contributions of the long-range dipole-dipole and elastic forces. If we can neglect the effect produced by the elastic interaction on the fluctuation probability then for sufficiently small q when the wavelength of the fluctuation wave is considerably smaller than the Debye radius r_D (that is, $qr_\mathrm{D} \ll 1$ and the dipole-dipole interaction is shielded, the polarization fluctuations can be calculated from (1.2.24) derived in the absence of long-range forces (see also (2.2.11) and the discussion after (1.2.24)). Then we can take the second term in (3.6.7) for diffuse scattering by the polarization fluctuations in intrinsic ferroelectrics ($\eta_{\mu q} \to P_{qi}$) and write

$$I_1 = \frac{Nk_BT}{v}[\varkappa_{ij} - \varkappa_{\infty ij}(q)](fQ \cdot A_q^i - \chi_i)(fQ \cdot A_q^j - \chi_j) . \quad (3.6.15)$$

Here $\chi_1 = \partial f/\partial P_i$ and A_q^i determine the variation of the structure amplitude and the distortions produced by the fluctuations of the component P_i of the polarization vector. All parameters in (3.6.15) can be found from independent experimental data and then the calculated intensity I_1 can be compared with the experimental limit for $q \to 0$.

For large q when $qr_D) \gg \sqrt{\varepsilon}$ (but $qd \ll 1$) the Debye screening does not have time to occur at the distance of the wavelength and the long-range dipole-dipole forces become effective. Then the polarization fluctuations must be found from (2.2.8) which takes into consideration these forces and the diffuse scattering intensity [1.16] is given

$$I_1 = \frac{Nk_BT}{v}[\varkappa_{ij}(q) - \varkappa_{\infty ij}(q)](fQ \cdot A_q^i - \chi_i)(fQ \cdot A_q^j - \chi_j) . \quad (3.6.16)$$

According to (2.2.5) and (3.6.15, 16) the dipole-dipole forces affect significantly the intensity I_1 and its behavior as a function of q. Then this function must exhibit characteristic features in the range of $q \sim 1/r_D$ where the dipole-dipole interaction becomes effective and the scattering intensity can vary when the concentration of the screening charges (that is, r_D) is varied.

In uniaxial ferroelectrics the tensor $\varkappa_{ij}(q)$ has a strong and non-analytic dependence on q which must give rise to a characteristic dependence of I_1 on $n = q/q$ (for small q) even for non-distorted crystals. For instance, if in strongly anisotropic uniaxial ferroelectrics only the scattering by the fluctuations of P_z is effective and $\varkappa_{zz} \gg 1$, $\varkappa_{xx} = \varkappa_{yy} \sim 1$ and $\varkappa_{ij} = 0$ for $i \neq j$, then we obtain from (2.2.5, 8) and (3.6.16)

$$I_1 = \frac{Nk_BT}{v}\left[\frac{\tilde{\varepsilon}_\perp \varkappa_{zz}}{\tilde{\varepsilon}_\perp + 4\pi\varkappa_{zz}n_z^2} - \frac{\tilde{\varepsilon}_{\infty\perp}\varkappa_{\infty zz}}{\tilde{\varepsilon}_{\perp\infty} + 4\pi\varkappa_{\infty zz}n_z^2}\right](fQ \cdot A_q^z - \chi_z)^2 , \quad (3.6.17)$$

where $\tilde{\varepsilon}_\perp = 1 + 4\pi\varkappa_{xx}(1 - n_z^2)$, that is, I_1 exhibits a strong dependence on n_z for $n_z^2 \ll 1$.

We can make use of microscopic models of ferroelectrics and the self-consistent field approximation and derive equations similar to (3.6.11) for $\langle P_{qi}P_{qi}^*\rangle$ and for I_1 (which must explicitly take into consideration the dipole-dipole interaction) applicable for large q [3.214, 215]. The more rigorous cluster approximation was also used for analyzing the scattering intensity for large q [3.216, 217]. It has been demonstrated that in KH_2PO_4-type ferroelectrics the crystal structure and the character of bonding in the PO_4 groups (hydrogen bonds) give rise to a strong anisotropy of the scattering intensity distribution for large q even if the dipole-dipole interaction is not taken into consideration. This anisotropy is largely similar to the anisotropy which is produced by the dipole-dipole interaction for small q. The special features of scattering by ferroelectrics with two-dimensional properties have been treated in [3.216, 218].

If the distortions caused by the polarization fluctuations are relatively large and the effects of the elastic forces on the fluctuation probability cannot be ignored then in (3.6.15) the tensor $\varkappa_{ij}$ must be replaced with $\varkappa'_{ij}$ given by the equation (2.2.14) where these forces are taken into account. In addition, according to (2.2.14), $\varkappa_{ij}(\boldsymbol{q})$ in (3.6.16) must be replaced with $\varkappa'_{ij}(\boldsymbol{q})$ and then the scattering intensity will decrease. In these cases all the parameters in the equation for the intensity I_1 can also be determined from independent experimental results.

Scattering by the fluctuations of the long-range order parameters must occur both in the ordered and disordered phases (in the latter case fluctuations occur for the vanishing parameters η_μ). The intensity distribution for the scattering by one-component crystals can be essentially different depending on whether the fluctuations produce distortions, that is, for $\boldsymbol{A}_{\mu q} = 0$ and $\boldsymbol{A}_{\mu q} \neq 0$. According to the considerations of symmetry, in some structures, for instance, in crystals with a center of symmetry, in the disordered phase we have the identity $\boldsymbol{A}_{\mu q} = 0$. In other cases, however, for instance, if the non-ferroelectric phase exhibits a piezoelectric effect the considerations of symmetry do not prohibit non-vanishing values of $\boldsymbol{A}_{\mu q}$ in the disordered phase. In the ordered phase $\boldsymbol{A}_{\mu q}$ are generally always non-vanishing (though they may be quite small).

If $\boldsymbol{A}_{\mu q}$ are non-vanishing then the expression for I_1 in the vicinities of the reciprocal lattice points includes a term proportional to q^{-2} for the reflections with $f \neq 0$. The dependence of I_1 on q (for a given $\boldsymbol{n}$) is described by a hyperbola. If only a single order parameter is fluctuating, a comparison of (3.6.9) and (3.4.3) shows that the equal-intensity surfaces are qualitatively the same as in the case of scattering by the concentration fluctuations in disordered solutions and pass through the reciprocal lattice point for small q. But if several order parameters undergo significant fluctuations, $\boldsymbol{A}_{\mu q}$ become perpendicular to $\boldsymbol{Q}$ generally for different $\boldsymbol{n}$ and the general equation (3.6.8) demonstrates that the intensity I_1 does not vanish for any $\boldsymbol{n}$ (for $q \to 0$), that is, the equal-intensity surfaces pass through the reciprocal lattice point only in the case of a special symmetry (see Sect. 4.1.2 in [1.1]).

In weakly distorted crystals the first terms in parentheses in (3.6.8, 9) can, on the contrary, be ignored in comparison with χ_μ for the reflections for which $\chi_\mu \neq 0$ for a q which is not too small. The scattering intensity distributions near the reciprocal lattice points where $f = 0$ has the same character (as in the disordered phase for points corresponding to the superlattice reflections) and also in undistorted crystals with vanishing $\boldsymbol{A}_{\mu q}$ as determined by symmetry considerations (if we can ignore the distortions caused by the correlation parameters). As shown by (3.6.9, 10), under such conditions the dependence of I_1 on q for a given direction of the vector $\boldsymbol{q}$ is described by bell-shaped plots and the equal-intensity surfaces go around the reciprocal lattice points but not through them.

If the condition (1.5.24) is satisfied in solid solutions the distortions are mainly caused by the fluctuations of the concentration rather than the long-range order. It can be seen from (3.5.12, 13) that the term proportional to q^{-2} in the expression for I_1 here also includes the squared structure amplitude as a factor and for the disordered phase this term makes a contribution only in the vicinities of the reciprocal lattice points.

The behavior of the diffuse scattering intensity as a function of temperature must exhibit characteristic features near second-order phase transitions in ferroelectrics and other ordering crystals. They are discussed in Sect. 4.2.

3.6.4 Incommensurate Structures

In some crystals a phase transition gives rise to a wave-like variation of the long-range order parameter $\eta(\boldsymbol{r})$ whose period is incommensurate with the original phase. Charge density waves (see Sect. 2.5.1) or oystems in which the minimum of $w(\boldsymbol{k})$ does not coincide with a high-symmetry point are examples of such structures (which do not have translational symmetry). Depending on temperature, the order in such a structure is described by a single harmonic with the wave vector $\boldsymbol{k}_0$ or (near the transition to the commensurate phase) by a set of harmonics corresponding to the production of regions of the commensurate phase divided by the soliton domain boundaries (see Sect. 2.5.1). In the first case the variation of the order parameter $\eta(\boldsymbol{r})$ is accompanied with a wave of displacements $\boldsymbol{u}_{s\gamma}$ of the atoms in each sublattice whose amplitude is $\boldsymbol{u}_\gamma$ and whose phase is ψ_γ:

$$\boldsymbol{u}_{s\gamma} = \boldsymbol{u}_\gamma \sin(\boldsymbol{k}_0 \cdot \boldsymbol{R}_s + \psi_\gamma) \, . \tag{3.6.18}$$

The amplitude of scattering by the sth cell, $f_s \exp(\mathrm{i}\boldsymbol{Q} \cdot \boldsymbol{u}_s)$ (see (3.6.1, 2)) is related to the factors $\exp(\mathrm{i}\boldsymbol{Q} \cdot \boldsymbol{u}_{s\gamma})$. Using the identity

$$\exp(\mathrm{i}z \sin \alpha) = \sum_{n=-\infty}^{\infty} J_n(z) \exp(\mathrm{i}n\alpha) \tag{3.6.19}$$

where $J_n(z)$ is the Bessel function we can rewrite these factors n in the following convenient form:

$$\exp(\mathrm{i}\boldsymbol{Q} \cdot \boldsymbol{u}_{s\gamma}) = \sum_{n=-\infty}^{\infty} J_n(\boldsymbol{Q} \cdot \boldsymbol{u}_\gamma) \exp[\mathrm{i}n(\boldsymbol{k}_0 \cdot \boldsymbol{R}_s + \psi_\gamma)] \, . \tag{3.6.20}$$

Since (3.6.20) includes harmonics with the wave vectors $n\boldsymbol{k}_0$ a crystal with the displacement wave must give rise to the regular reflections for $\boldsymbol{Q} = \boldsymbol{G} - n\boldsymbol{k}_0$, in addition to the conventional Bragg reflections at $\boldsymbol{Q} = \boldsymbol{G}$. If we can ignore the variations of $f_{s\gamma}$ in the wave and the fluctuations of the amplitude and the phase of the wave and assume that $\boldsymbol{u}_\gamma$ and ψ_γ are constants then we obtain from (3.6.1, 2, 20) the following regular reflection intensity [2.118] for $\boldsymbol{G} - n\boldsymbol{k}_0$:

$$\begin{aligned} I_0(\boldsymbol{Q}) &= 8\pi^3 \frac{N}{v} |f_n|^2 \delta(\boldsymbol{Q} - \boldsymbol{G} + n\boldsymbol{k}_0) \, , \\ f_n &= \sum_{\gamma=1}^{\nu} f_\gamma \mathrm{e}^{\mathrm{i}\boldsymbol{Q} \cdot \boldsymbol{R}_\gamma} J_n(\boldsymbol{Q} \cdot \boldsymbol{u}_\gamma) \mathrm{e}^{\mathrm{i}n\psi_\gamma} \, . \end{aligned} \tag{3.6.21}$$

For structures with a single atom per unit cell we have $f_n = fJ_n(\boldsymbol{Q}\cdot\boldsymbol{u})$ and if $|\boldsymbol{Q}\cdot\boldsymbol{u}|$ is small the intensity $I_0 \sim |\boldsymbol{Q}\cdot\boldsymbol{u}|^n$ rapidly decreases with decreasing number n. In some circumstances what is produced is not a displacement wave described by (3.6.18) but a more complicated incommensurate structure corresponding to a combination of waves with different directions of the vectors $\boldsymbol{k}_0$. The system of the I_0 peaks becomes more complicated accordingly [2.103; 3.219].

The parameters f_γ in (3.6.21) include the conventional Debye-Waller factors associated with the vibrations. The fluctuations of the amplitude and the phase of the wave give rise to fluctuating terms $\delta\boldsymbol{u}_\gamma$ and $\delta\psi_\gamma$ in addition to the constant components in the functions $\boldsymbol{u}_\gamma$ and ψ_γ in (3.6.18). The fluctuating terms produce an additional variation by the factor $\exp[-W_\gamma(n)]$ of the amplitude f_γ. According to (3.6.20) we have

$$\exp[-W_\gamma(n)] = \left\langle e^{in\delta\psi_\gamma}\frac{J_n[\boldsymbol{Q}\cdot(\boldsymbol{u}_\gamma+\delta\boldsymbol{u}_\gamma)]}{J_n(\boldsymbol{Q}\cdot\boldsymbol{u}_\gamma)}\right\rangle . \tag{3.6.22}$$

If the fluctuations of the amplitude and the phase are sufficiently small (if $n^2\langle(\delta\psi_\gamma^2)\rangle \ll 1$ and $n^2\langle(\boldsymbol{Q}\cdot\delta\boldsymbol{u}_\gamma)^2\rangle \ll 1$) then they are statistically independent. If we expand (3.6.22) in powers of $\delta\psi_\gamma$ and $\boldsymbol{Q}\cdot\delta\boldsymbol{u}_\gamma$ we obtain [2.118; 3.220, 221]

$$W_\gamma(n) = \frac{1}{2}n^2\langle(\delta\psi_\gamma)^2\rangle - \frac{1}{2}|n|(|n|-1)\frac{\langle(\boldsymbol{Q}\cdot\delta\boldsymbol{u}_\gamma)^2\rangle}{(\boldsymbol{Q}\cdot\boldsymbol{u}_\gamma)^2} . \tag{3.6.23}$$

The factors $\exp[-W_\gamma(n)]$ affect only the satellite intensities and are independent of Q, in contrast to the conventional Debye-Waller factors. In a simplified model $W_\gamma(n)$ is proportional to u_γ^{-2}, always positive and can be either small or greater or approximately equal to unity. In the latter case the fluctuation probabilities are, apparently, no longer described by a Gaussian distribution and (3.6.23) can be modified (see Sect. 3.2.4 in [1.1]). When $W_\gamma(n) \gg 1$ the Bragg satellite lines may be expected to transform into broadened "quasilines" as in the case of strongly distorted crystals (see Sect. 4.3 in [1.1]).

The fluctuations of the amplitude and the phase of the displacement wave give rise to diffuse scattering. The highest scattering intensity occurs in the vicinities of $\boldsymbol{Q} = \boldsymbol{G} - n\boldsymbol{k}_0$. This scattering is caused by the long-wave fluctuations, in particular, the fluctuations of the phase. Since the thermodynamic potential density $\varphi(\boldsymbol{r})$ remains constant when the phase is shifted by a constant the potential variation $\delta\varphi(\boldsymbol{r})$ caused by the long-wave phase fluctuations is given by an equation similar to (1.2.15) [3.222–224],

$$\delta\varphi = \frac{1}{2}u_0^2\gamma_{ij}\frac{\partial\psi}{\partial x_i}\frac{\partial\psi}{\partial x_j} , \tag{3.6.24}$$

where $\boldsymbol{u}_\gamma = u_0\boldsymbol{e}_\gamma$ and $|\boldsymbol{e}_\gamma| \sim 1$.

Using the $\boldsymbol{k}$ representation as it was done in (1.2.8, 10, 16) we obtain

$$\langle|\psi_{\boldsymbol{k}}|^2\rangle = \frac{k_{\mathrm{B}}T}{V}\frac{1}{u_0^2\gamma_{ij}k_ik_j} , \tag{3.6.25}$$

that is, in the incommensurate phase we have $\langle \psi_{\boldsymbol{k}}|^2\rangle \sim k^{-2}$ in a wide temperature range. This is explained by the possibility of a phase shift for the wave which gives rise to the (always soft when the coupling of the wave to the lattice can be ignored) Goldstone mode of phase oscillations (the phasons) in addition to the mode of amplitude oscillations ("amplitudons") which features a gap [2.118; 3.224, 225].

When we expand (3.6.20) in powers of the long-wave phase fluctuations $\delta\psi_\gamma \cong \delta\psi$ and make use of (3.6.1, 21) we obtain in the approximation linear in $\delta\psi$

$$\delta[f_s \exp(\mathrm{i}\boldsymbol{Q}\cdot\boldsymbol{u}_s)] = in\delta\psi(\boldsymbol{R}_s)f_n \ .$$

Therefore, according to (3.6.2, 25) the intensity I_1^ψ of diffuse scattering by these fluctuations for small $\boldsymbol{q}_n = \boldsymbol{Q} - \boldsymbol{G} + n\boldsymbol{k}_0$ is given by

$$I_1^\psi(\boldsymbol{Q}) = N^2 n^2 |f_n|^2 \langle|\psi_{q_n}|^2\rangle = N\frac{k_\mathrm{B}Tn^2}{vu_0^2}\frac{1}{\gamma_{ij}q_{ni}q_{nj}} \ . \tag{3.6.26}$$

A diffuse scattering intensity peak $I_1 \sim q_n^{-2}$ can be thus observed near each Bragg peak at $\boldsymbol{G} - n\boldsymbol{k}_0$ with $n \neq 0$. The phason mode must also give rise to a low-frequency peak in the spectrum of inelastic neutron scattering [3.223].

As noted in Sect. 2.5.1. with a decrease in temperature and a growth of $\boldsymbol{u}_\gamma$ a growing contribution is made by the higher harmonics of the incommensurate wave and a soliton structure is produced in the case when $\boldsymbol{k}_0$ is close to $\boldsymbol{G}_p = \boldsymbol{G}_0/p$ where $\boldsymbol{G}_0$ is a reciprocal lattice vector and p is a small integer (for instance, 3 or 4). If the thicknesses L_c of the regions of the commensurate phase are large in comparison with the thicknesses of the soliton boundaries then the phase of the wave

$$\phi(z) = (\boldsymbol{k}_0 - \boldsymbol{G}_p)\cdot\boldsymbol{R}_s + \psi(z) = \phi_0 + \frac{2\pi m}{p} \tag{3.6.27}$$

is almost constant in these regions, as in Fig. 2.2, and sharply varies by $2\pi/p$ at their boundaries (here $Z \parallel \boldsymbol{G}_0 \parallel \boldsymbol{k}_0$ and $m > z/L_c$ is an integer closest to z/L_c). It follows then from (3.6.1, 2, 18, 27) that if the soliton boundaries are regularly arranged and $|\boldsymbol{Q}\cdot\boldsymbol{u}_\gamma| = u_0|\boldsymbol{Q}\cdot\boldsymbol{e}_\gamma| \ll 1$ the X-ray scattering amplitude for the crystal is

$$\begin{aligned} -\frac{if_c u_0}{2v}\sum_{m=1}^{L_z/L_c}\exp(\mathrm{i}Q_zL_cm)\Big[s^0(\boldsymbol{Q}-\boldsymbol{G}+\boldsymbol{G}_p)\exp\mathrm{i}\left(\frac{2\pi}{p}+G_{pz}L_c\right)m \\ + s^0(\boldsymbol{Q}-\boldsymbol{G}-\boldsymbol{G}_p)\exp\left(-\mathrm{i}\left(\frac{2\pi}{p}+G_{pz}L_c\right)m\right)\Big] \ , \end{aligned} \tag{3.6.28}$$

$$f_c = \sum_{\gamma=1}^{\nu} f_\gamma \exp(\mathrm{i}\boldsymbol{Q}\cdot\boldsymbol{R}_\gamma)\boldsymbol{Q}\cdot\boldsymbol{e}_\gamma \ ,$$

Here L_z is the crystal size along the axis Z and

$$s^0(q) = \int s^0(r)\exp(iq\cdot r)dr$$

the Fourier component of the shape function $s^0(r)$ for the commensurate region. It can be then seen that [3.223] a series of Bragg peaks is produced at

$$Q = G_l^{\pm} = G \pm G_p + \frac{2\pi}{L_c}\left(l \pm \frac{1}{p}\right) e_z \tag{3.6.29}$$

near each $G \pm G_p$ reflection (here l is an integer and $e_z = G_p/G_p$) whose intensities

$$I_0(Q) = 2\pi\frac{N}{v}u_0^2|f_c|^2\frac{\sin^2\pi/p}{(l \pm 1/p)^2}\delta(Q - G_l^{\pm}), \tag{3.6.30}$$

rapidly decrease with increasing l. The spacing of these equidistant peaks decreases with an increase in L_c but their total intensity remains unchanged (the same as in the commensurate phase with the same u_0). If the solitons are arranged randomly, rather than regularly, then the reflections $G \pm G_p$ must be broadened rather than split (see Sect. 5.1.4 in [1.1] and [2.152]).

3.7 Diffuse Scattering by Ionic Crystals with Charged Defects or Impurities

The characteristic behavior of the scattering intensity $I_1 \sim q^{-2}$ near the reciprocal lattice points is determined by the behavior of the displacement fields near point defects described by $u(r) \sim r^{-2}$. When it was derived in Sect. 3.4 the defects were assumed to be electrically neutral. If impurities (or other point defects) in ionic crystals are charged then the diffuse scattering intensity must exhibit some new features [1.15]. These features are associated with the long-range Coulomb forces which cause polarization of the lattice in addition to its strain (that is, the atoms in a cell are displaced with respect to each other). Owing to this reason crystals possessing a symmetry center may have significantly different relationships between the coefficients at $1/q^2$ in the equations for I_1 in the vicinities of the "superlattice points" and "lattice" points of the reciprocal lattice for the neutral and charged defects. In addition, the slowly decreasing Coulomb force gives rise to a sharp increase in the energy and a characteristic behaviour of the long-wave fluctuations for the charged defects (see Sect. 2.3), that is, the diffuse scattering intensity exhibits characteristic features for $q \gtrsim 1/r_D$.

In piezoelectric crystals the difference between scattering by neutral and charged defects must be especially significant. In these crystals the Coulomb field of a charged defect produces a strain proportional to the field intensity, that is, to $1/r^2$ rather than $1/r^3$ as in other types of crystals (or for neutral defects in piezoelectric crystals). This is why the intensity of diffuse scattering by charged defects must grow as $1/q^4$ when we approach a reciprocal lattice point for $q > 1/r_D$ rather than $1/q^2$ as in other cases.

Nonpiezoelectric Crystals. Since the above effects manifest themselves most strongly for small q we shall consider only this q range which corresponds to scattering by the long-wave fluctuations. We shall assume that the impurity concentration is rather low ($c \ll 1$). Let us consider, as it was done in Sect. 2.3, an ionic crystal AB with two ions per lattice cell (NaCl- or CsCl-type crystal lattices). Assume that the crystal contains positive and negative defects of the same valency (vacancies, interstitial ions, impurity ions of different valencies and so on). The positions which can be occupied by the positive and negative defects will be denoted by s_1 and s_2, respectively and the distribution of the defects over these positions will be described by the numbers $c_{s\gamma'}$, equal to 1 or 0 (here $\gamma' = 1, 2$, see Sect. 2.3). We shall limit the discussion to the cases when the unit cell contains one position s_1 and one position s_2.

The diffuse scattering intensity will be found from the general equation (3.1.6) (for $\boldsymbol{q} \neq 0$). The variation of the scattering amplitude caused by the production of a positive defect in the sth lattice cell will be $f_1' \exp[\mathrm{i}\boldsymbol{Q} \cdot (\boldsymbol{R}_s + \boldsymbol{R}_1')]$ and that caused by the negative defect will be $f_2' \exp[\mathrm{i}\boldsymbol{Q} \cdot (\boldsymbol{R}_s + \boldsymbol{R}_2')]$ (here $\boldsymbol{R}_{\gamma'}' = \boldsymbol{R}_{1,2}'$ are the vectors passing from the first site of the sth cell to the positions $s_{1,2}$). As it was done above, let us expand (3.1.6) in powers of $\boldsymbol{Q} \cdot \boldsymbol{u}_{s\gamma}$ and retain only the linear terms. For $c \ll 1$ we can ignore the displacements of the defects and perform summation in the linear terms only over the lattice atoms. Then in (3.1.6) we have $\gamma = 1, 2$ and $f_{s\gamma} = f_{1,2}$ and we obtain the following equation for the diffuse scattering intensity:

$$I_1 = \left\langle \left| \sum_s \left[\sum_{\gamma'=1,2} f_{\gamma'}' c_{s\gamma'} \mathrm{e}^{\mathrm{i}\boldsymbol{Q}\cdot\boldsymbol{R}_{\gamma'}'} + \sum_{\gamma=1,2} \mathrm{i}\boldsymbol{Q} \cdot \boldsymbol{u}_{s\gamma} f_\gamma \mathrm{e}^{\mathrm{i}\boldsymbol{Q}\cdot\boldsymbol{R}_\gamma} \right] \mathrm{e}^{\mathrm{i}\boldsymbol{q}\cdot\boldsymbol{R}_s} \right|^2 \right\rangle . \tag{3.7.1}$$

It is convenient here to make the transition from $c_{s\gamma'}$ to their Fourier components given by (1.1.8) and then to their sum $c_{\boldsymbol{k}}^+$ and difference $c_{\boldsymbol{k}}^-$ given by (2.3.1). In a similar way, the Fourier components $\boldsymbol{u}_{\boldsymbol{k}\gamma}$ of the displacements $\boldsymbol{u}_{s\gamma}$ can be conveniently replaced with their sum and difference:

$$\begin{aligned} &2\boldsymbol{u}_{\boldsymbol{k}}^+ = \boldsymbol{u}_{\boldsymbol{k}1} + \boldsymbol{u}_{\boldsymbol{k}2} , \quad 2\boldsymbol{u}_{\boldsymbol{k}}^- = \boldsymbol{u}_{\boldsymbol{k}1} - \boldsymbol{u}_{\boldsymbol{k}2} , \\ &\boldsymbol{u}_{s\gamma} = \mathrm{i} \sum_{\boldsymbol{k}} \boldsymbol{u}_{\boldsymbol{k}\gamma} \exp(-\mathrm{i}\boldsymbol{k} \cdot \boldsymbol{R}_s) . \end{aligned} \tag{3.7.2}$$

Here $\boldsymbol{u}_{\boldsymbol{k}}^+$ and $\boldsymbol{u}_{\boldsymbol{k}}^-$ describe the strain and polarization displacements of the crystal atoms, respectively.

The Fourier components introduced by (3.7.2),

$$\boldsymbol{u}_{\boldsymbol{k}}^+ = \boldsymbol{A}_{\boldsymbol{k}}^+ c_{\boldsymbol{k}}^+ + \boldsymbol{A}_{\boldsymbol{k}}^- c_{\boldsymbol{k}}^- , \quad \boldsymbol{u}_{\boldsymbol{k}}^- = \boldsymbol{B}_{\boldsymbol{k}}^+ c_{\boldsymbol{k}}^+ + \boldsymbol{B}_{\boldsymbol{k}}^- c_{\boldsymbol{k}}^- , \tag{3.7.3}$$

can be seen to be linear functions of the Fourier components of the defect concentration fluctuations. In the long-wave approximation for small k the coefficients $\boldsymbol{A}_{\boldsymbol{k}}$ and $\boldsymbol{B}_{\boldsymbol{k}}$ can be expressed in terms of the macroscopic constants found from independent experiments. For instance, the coefficients $\boldsymbol{A}_{\boldsymbol{k}}$ which determine the strain distortions can be derived from the equations of elasticity theory which describe the stresses under the smooth variations $\delta c^+ = (\delta c_1 + \delta c_2)/2$ or $\delta c^- = (\delta c_1 - \delta c_2)/2$. Then we obtain equations for $\boldsymbol{A}_{\boldsymbol{k}}^{+,-}$ similar to (1.5.4–6) in which we must only replace L_{lm} with $\delta u_{lm}/\delta c^{+,-}$ (in cubic crystals we can make use of (1.5.7) replacing in it dv/dc with $dv/dc^{+,-}$). The Fourier components $\boldsymbol{u}_{\boldsymbol{k}}^-$ of the polarization displacements and $c_{\boldsymbol{k}}^-$ for small k may be related to the Fourier components (2.3.2, 3) of the waves of the electric charge and electric displacement. Then we obtain the following simple expressions [1.15]:

$$\boldsymbol{B}_{\boldsymbol{k}}^+ = 0\,, \quad \boldsymbol{B}_{\boldsymbol{k}}^- = \frac{3(\varepsilon - \varepsilon_\infty)}{\varepsilon(2+\varepsilon_\infty)} \frac{z'}{z} \frac{\boldsymbol{k}}{k^2}\,, \tag{3.7.4}$$

Here z is the effective charge of the ions in the crystal and z' is the charge of the defects. The equation $B_{\boldsymbol{k}}^+ = 0$ in (3.6.4) takes into account the fact that in the macroscopic approximation the $c_{\boldsymbol{k}}^+$ fluctuations do not produce an electric field and polarization displacements of ions. As for $\boldsymbol{A}_{\boldsymbol{k}}$, we have $\boldsymbol{B}_{\boldsymbol{k}}^- \sim 1/k$.

When we substitute (2.3.1) and (3.7.2, 3) into (3.7.1) we obtain the diffuse scattering intensity

$$I_1 = N^2 \langle |f'_+ - \boldsymbol{Q} \cdot \boldsymbol{A}_{\boldsymbol{q}}^+ f_+) c_{\boldsymbol{q}}^+ + (f'_- - \boldsymbol{Q} \cdot \boldsymbol{A}_{\boldsymbol{q}}^- f_+ - \boldsymbol{Q} \cdot \boldsymbol{B}_{\boldsymbol{q}}^- f_-) c_{\boldsymbol{q}}^- |^2 \rangle\,, \tag{3.7.5}$$

where

$$f'_\pm = f'_1 \mathrm{e}^{\mathrm{i}\boldsymbol{G}\cdot\boldsymbol{R}'_1} \pm f'_2 \mathrm{e}^{\mathrm{i}\boldsymbol{G}\cdot\boldsymbol{R}'_2}\,, \quad f_\pm = f_1 \pm f_2 \mathrm{e}^{\mathrm{i}\boldsymbol{G}\cdot\boldsymbol{R}_2}\,. \tag{3.7.6}$$

When we substitute into (3.7.5) the expressions (2.3.7) for the mean squares and products of the Fourier components we obtain

$$I_1 = \frac{1}{2} cN |f'_+ - \boldsymbol{Q} \cdot \boldsymbol{A}_{\boldsymbol{q}}^+ f_+|^2 + \frac{1}{2} cN |f'_- - \boldsymbol{Q} \cdot \boldsymbol{A}_{\boldsymbol{q}}^- f_+ - \boldsymbol{Q} \cdot \boldsymbol{B}_{\boldsymbol{q}}^- f_-|^2 \frac{r_{\mathrm{D}}^2 q^2}{1 + r_{\mathrm{D}}^2 q^2}. \tag{3.7.7}$$

For scattering by neutral defects of two types the intensity I_1 is given by a similar equation where the last factor in the second term describing the Debye screening must be omitted and $\boldsymbol{B}_{\boldsymbol{q}}^- = 0$ (see Sect. 4.1 in [1.1]). This is why the behavior of the intensity I_1 as a function of $\boldsymbol{q}$ can be significantly different for scattering by neutral and charged defects giving thus an opportunity for determining whether the given defects are charged or not.

If we can ignore distortions in the range of $\boldsymbol{q}$ under consideration, that is, take $\boldsymbol{A}_{\boldsymbol{q}} = \boldsymbol{B}_{\boldsymbol{q}} = 0$, then, according to (3.7.7), for $r_{\mathrm{D}} q \gg 1$ we have $I_1 = cN(|f'_+|^2 + |f'_-|^2)/2$ and $I_1 = cN|f'_+|^2/2$ for $r_{\mathrm{D}} q \ll 1$. The scattering intensity can thus be changed considerably for $q \sim 1/rt_{\mathrm{D}}$. Since $\boldsymbol{A}_{\boldsymbol{q}}$ and $\boldsymbol{B}_{\boldsymbol{q}}$ are proportional to $1/q$ for small q in most cases, on the contrary, we can ignore the terms with f'_+ and f'_- which are not associated with distortions. Then $I_1 \sim 1/q^2$ both for

$r_D q \gg 1$ and for $r_D \ll 1$ but the proportionality coefficients are different in these two ranges.

The effects caused by the charges of defects must manifest themselves especially sharply in the vicinities of the "superlattice" points of the reciprocal lattice ($\exp(\mathrm{i}\boldsymbol{G} \cdot \boldsymbol{R}_\gamma) = (-1)^{\gamma+1}$) if f_1 and f_2 are similar (as is the case for instance in KCl). Then we obtain from (3.7.6) $f_+ \ll f_-$. Taking $f_+ = 0$ we obtain then $I_1 \sim 1/q^2$ for $r_D q \gg 1$ and $I_1 = \text{const}$ for $r_D q \ll 1$ (since $f_+ \neq 0$ for very small q the intensity I_1 also varies as $1/q^2$ but with a much smaller proportionality coefficient). In the case of neutral defects no scattering intensity component proportional to $1/q^2$ is recorded near the "superlattice" points. The intensity I_1 as a function of q must therefore exhibit significant features for $q \sim 1/r_D$ giving thus an opportunity, in principle, to estimate the Debye length from the diffraction data. For instance, if $\varepsilon = 10$, $z' = 1$, $T = 1000\,\mathrm{K}$, $c = 10^{-4}$, and $v = 60\,\text{Å}$ then we obtain from (2.3.8) $r_D = 4 \times 10^{-7}\,\mathrm{cm}$.

Piezoelectric Crystals. The electric fields of charged defects produce elastic stresses in piezoelectric crystals. Therefore, we must add to the expression for the stress tensor σ_{ij} the following term proportional to the electric displacement

$$\sigma_{ij} = c_{ijlm} u_{lm} - c_{ijlm}(L^+_{lm}\delta c^+ L^-_{lm}\delta c^-) - h_{ijl} D_l \,. \tag{3.7.8}$$

in addition to the terms proportional to the strain u_{ij} and the concentration variations δc^+ and δc^- (see Sect. 2.2.10 in [1.1]). Here h_{ijl} are the components of the third-rank tensor, and the term $-c_{ijlm}L^-_{lm}\delta c^-$ describes the stresses caused by the variation δc^- which are not related to the electric displacement $\boldsymbol{D}$ (that is, the stresses that are preserved when the displacement is compensated by an oppositely directed field). When we replace $\delta c^{+,-}$ and u_{lm} with their values corresponding to the $\boldsymbol{k}$th wave and make use of (2.3.3) we obtain from the equilibrium conditions $\partial\sigma_{ij}/\partial x_j = 0$

$$k c_{ijlm} n_j n_l A^+_{km} = c_{ijlm} L^+_{lm} n_j \,, \tag{3.7.9}$$

$$k c_{ijlm} n_j n_l A^-_{km} = c_{ijlm} L^-_{lm} n_j + \frac{8\pi\sqrt{-1}ez'}{vk} h_{ijl} n_j n_l \quad (i = 1,2,3)\,. \tag{3.7.10}$$

For sufficiently small k we can ignore the first term on the right-hand side of (3.7.10]. Then $\boldsymbol{A}^-_k$ increases as $1/k^2$ rather than $1/k$ with a decrease in k as does $\boldsymbol{A}^+_k$ (or $\boldsymbol{A}^-_k$ in a non-piezoelectric crystal), that is,

$$\boldsymbol{A}^+_k = \frac{a^+_k \boldsymbol{e}^+_k}{k}\,, \quad \boldsymbol{A}^-_k = \frac{a^-_k \boldsymbol{e}^-_k}{k^2}\,, \tag{3.7.11}$$

where $\boldsymbol{e}^+_k$ and $\boldsymbol{e}^-_k$ are unit vectors and for small k a^+_k and a^-_k depend only on the direction (but not on the magnitude) of the vector $\boldsymbol{k}$. Note that all the coefficients in (3.7.9, 10) are determined from independent experimnental results.

For sufficiently small k, that is, near a reciprocal lattice point only the terms with $\boldsymbol{A}^{+,-}_q$ can be retained in (3.7.7). Using (3.7.11) we obtain

$$I_1 = \frac{1}{2} c N f_+^2 \left[\frac{a_q^{+2} (\boldsymbol{Q} \cdot \boldsymbol{e}_q^+)^2}{q^2} + \frac{|a_q^-|^2 (\boldsymbol{Q} \cdot \boldsymbol{e}_q^-)^2}{q^4} \frac{r_D^2 q^2}{1 + r_D^2 q^2} \right] . \qquad (3.7.12)$$

Note that (3.7.12) is applicable not only to the crystals with two ions per lattice cell but also to the general case (then f_+ is given by the equation (3.1.9) for the structure amplitude f). If the Debye length is sufficiently large while a_q^+ is small then the main contribution to (3.7.12) is made by the second term. In the range of $r_D q \gg 1$ the intensity I_1 increases as $1/q^4$ with decreasing q while for $r_D q \ll 1$ the intensity varies as $1/q^2$, that is, we obtain an additional factor q^{-2} in comparison with the similar term in the equation for non-piezoelectric crystals.

Charged defects in ionic crystals must give rise also to sharp characteristic features in the diffuse scattering of electrons in electron-diffraction studies [1.15].

4. Anomalous Scattering Near Second-Order Phase Transitions and Critical Points

The fluctuations of the internal parameters or the concentration become anomalously large near the lines of phase equilibrium on the phase diagrams (or some points on these lines) if the difference between the phases separated by them is small. For instance, near the critical point on the decomposition plot for a solution where the phases at equilibrium have very similar compositions (see Fig. 4.1) the concentration fluctuations grow sharply in undistorted crystals. It can be seen from (1.2.5, 10) when we remember that at the critical point $d^2\varphi/dc^2 = 0$. In a similar way, at a line of second-order phase transitions (which separates states with vanishing and/or extremely small order parameters η) we have $\varphi_{\eta\eta} \equiv \partial^2\varphi/\partial\eta^2 = 0$ and the fluctuations of η grow sharply according to (1.2.16).

Since $d^2\varphi/dc^2$ is small in the vicinity of a critical point the minimal work $\mathfrak{R}$ of production of the fluctuations $\delta c(\boldsymbol{r})$ and the probability of the fluctuations significantly depend on the degree of inhomogeneity of $\delta c(\boldsymbol{r})$, according to (1.2.1, 6, 7), while the averages $\langle |c_{\boldsymbol{k}}|^2\rangle$ in (1.2.10) strongly depend on $\boldsymbol{k}$ in the range of small $k \ll \pi/d$. This means that even in systems with short-range forces, the fluctuations $\delta c(\boldsymbol{r})$ are related to the fluctuations $\delta c(\boldsymbol{r}')$ in rather distant regions, that is, a spatial correlation of the fluctuations [4.1] is effective at large distances $|\boldsymbol{r} - \boldsymbol{r}'| \gg d$. Occurrence of large smoothly varying fluctuations, in turn, affects significantly the behavior of the thermodynamic parameters in the critical range. As a result, the latter acquire a universal character, that is, they are independent of the details of the interaction between atoms and the crystal structure at small distances and are determined only by such general properties as the dimensionality of the system, the number of the long-range order parameters, and the nature of the long-range forces (if they are effective).

The long-range order established below the phase transition affects the intensity of the regular reflections and gives rise to the new "superlattice" Bragg reflections if the transition changes the translational symmetry. A sharp increase in the long-wave fluctuations of the long-range order parameters or the composition in the critical range causes an anomalously strong diffuse scattering of X-rays and neutrons in the vicinities of the reciprocal lattice points which exhibits a characteristic dependence on $\boldsymbol{Q}$ and T.

X-ray and neutron scattering results yield extensive information on the phase transitions. The intensities of the superlattice reflections describe the structure of the long-range order and the temperature dependence of the order parameters.

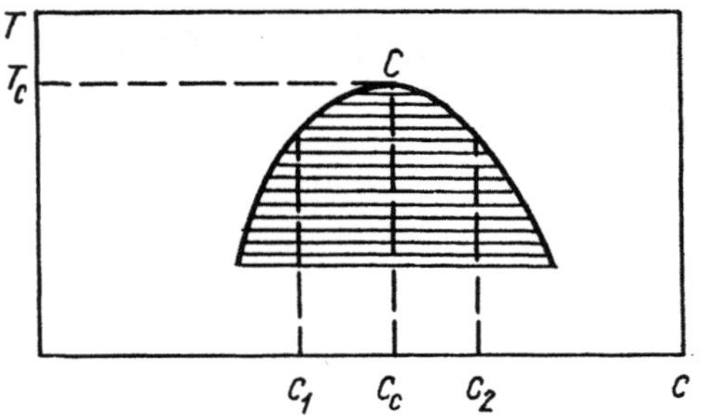

Fig. 4.1. Phase diagram in the vicinity of the critical point on the decomposition curve

Diffuse scattering data give valuable information on the behavior of some thermodynamic parameters (the generalized susceptibilities) which depend not only on the temperature but also on the wave vector and thus this information is more complete than that obtained with other techniques. The energy spectrum of the scattered neutrons yields information on the fluctuation dynamics in the critical range.

In Sect. 4.1 we shall briefly present the theory of second-order phase transitions and critical points and, in particular, the dermination of the fluctuation wave probabilities. Special attention is paid to the effects of the long-range forces which significantly affect the behavior of the thermodynamic parameters.

The general approach presented in Chap. 3 is used in Sect. 4.2 to apply this theory to analyzing the anomalous diffuse scattering in perfect crystals. Crystals containing defects and, in particular, dislocations, which alter significantly the character of the phase transition and critical scattering, are discussed in Sect. 4.3.

4.1 Thermodynamic Parameters and Fluctuation Waves Near Second-Order Phase Transitions and Critical Points

4.1.1 Landau Theory

The anomalously large fluctuations of the order parameters must be analyzed in the framework of a thermodynamic or statistical theory of second-order phase transitions. It can be done in a very simple form with the thermodynamic *Landau* theory [1.31, 67, 68]. The theory assumes that at the second-order phase transition point T_0 the order parameter η continuously vanishes with disordering. The thermodynamic potential density φ is expanded in powers of the order parameter and the expansion coefficients are assumed to depend linearly on the temperature T and the pressure P. Though a consistent analysis of the fluctuations yields a more complicated behavior of the potential φ at the transition point (see Sect. 4.1.2) this theory gives a qualitatively correct description of the phase diagram and the temperature dependences of the thermodynamic properties. What is most important is that this theory graphically reveals the profound relationship between the phase transitions and the crystal symmetry [1.31, 67, 70]. For many crystal systems the

Landau theory yields quantitative results in a wide temperature range and the expansion in η and $T-T_0$ introduced in this theory is used as the basis for the theory of fluctuations. It is a reasonable approach, therefore, to analyze the behavior of the anomalous fluctuations first in the framework of the Landau theory.

One-Component Crystals with a Single Order Parameter. Let us consider at first the simplest case of one-component ordered phases (not solutions) with a center of inversion which are described by a single (one-component) long-range order parameter η. Then the expansion of φ in powers of η includes only the even-order terms and for smoothly varying functions $\eta(\boldsymbol{r})$ we have

$$\varphi = \varphi_0 + \frac{1}{2}A_2\eta^2 + \frac{1}{4}A_4\eta^4 + \frac{1}{6}A_6\eta^6 + \frac{1}{2}\alpha_{ij}\frac{\partial\eta}{\partial x_i}\frac{\partial\eta}{\partial x_j}\,, \qquad A_2 = a(T-T_0)\,. \tag{4.1.1}$$

Here φ_0, a, A_4, A_6 and α_{ij} are constants (for the given pressure P) and the higher-order terms are omitted. For $A_4 \neq 0$ (and $A_4 > 0$) we can also ignore the term with η^6. Then at $T > T_0$ the minimum of φ corresponds to $\eta = 0$ (for the sake of definiteness we consider the conventional case of $a > 0$) and at $T \leq T_0$ the conditions $\varphi_\eta = 0$ and $\varphi_{\eta\eta} > 0$ yield the following temperature dependence of the order parameter:

$$\eta = \sqrt{\frac{a}{A_4}}(T_0 - T)^{1/2}\,. \tag{4.1.2}$$

According to (4.1.1, 2) we obtain

$$\varphi_{\eta\eta} = (T-T_0) \quad \text{for} \quad T \geq t_0\,, \qquad \varphi_{\eta\eta} = 2a(T_0 - T) \quad \text{for} \quad T < T_0\,. \tag{4.1.3}$$

Substitution of (4.1.3) into (1.2.16) leads to the following expressions for the mean-squared Fourier component of the order parameter [1.2]:

$$\langle|\eta_{\boldsymbol{k}}|^2\rangle = \frac{k_\mathrm{B}T}{V}\frac{1}{a(T-T_0)+\alpha_{ij}k_ik_j} \quad (T \geq T_0)\,, \tag{4.1.4}$$

$$\langle|\eta_{\boldsymbol{k}}|^2\rangle = \frac{k_\mathrm{B}T}{V}\frac{1}{2a(T_0-T)+\alpha_{ij}k_ik_j} \quad (T < T_0)\,. \tag{4.1.5}$$

We see that the variation of $\langle|\eta_{\boldsymbol{k}}|^2\rangle$ as a function of k (for the given $\boldsymbol{n} = \boldsymbol{k}/k$) is described by the Lorentzian curves which are sharper the smaller the difference $|T - T_0|$. For a given $\boldsymbol{k}$ the function $\langle|\eta_{\boldsymbol{k}}|^2\rangle$ reaches a maximum for $T = T_0$ and then it linearly decreases with increasing or decreasing temperature (the rate of decrease is twice as large for $T < T_0$).

The sharp dependence of $\langle|\eta_{\boldsymbol{k}}|^2\rangle$ on k near the transition point in the range of relatively small $k \sim k_c \sim (a|T-T_0|/\alpha)^{1/2}$ gives rise to a significant spatial correlation between the fluctuations which is effective at large distances $r \sim r_c \sim k_c^{-1}$. In order to determine explicitly the variation of the macroscopic parameter $\xi(\boldsymbol{r}) = \langle\delta\eta(\boldsymbol{r}')\delta\eta(\boldsymbol{r}'+\boldsymbol{r})\rangle$ of the correlation between the fluctuations $\delta\eta$ of the order parameter at distances $r \gg d$ we must take into account the fact that the parameters $\xi(\boldsymbol{r})$ are related to $\langle|\eta_{\boldsymbol{k}}|^2\rangle$ by the same relationship as that between the Fourier components of the correlation parameters $\varepsilon(\boldsymbol{r})$ and $\langle|c_{\boldsymbol{k}}|^2\rangle$ (see (1.1.14)). Using (4.1.4) for this relationship we obtain for cubic crystals at $T > T_0$

$$\xi(\boldsymbol{r}) = \frac{k_{\mathrm{B}}T}{4\pi\alpha r}\exp(-r/r_c)\,, \quad r_c = \sqrt{\frac{\alpha}{a(T-T_0)}}\,. \tag{4.1.6}$$

Though correlation typically disappears at distances of the order of d when we approach the phase transition point the correlation length r_c tends to infinity while at the very transition point the exponential factor disappears and the parameter $\xi(\boldsymbol{r})$ decreases as a power function, that is, at a much slower rate.

The function $\varphi(\eta)$ has a minimum at the transition point for $\eta = 0$ if the coefficient A_4 is positive. If $A_4 < 0$ then a first-order phase transition occurs at the point $T = \tilde{T}_0 > T_0$. The function $A_4(P)$ may vanish at some point when we vary the pressure P (or another external parameter). The critical point T_c, P_c on the phase transition line on which $A_4(P) = 0$ and the line of second-order phase transitions $T_0(P)$ is transformed into a first-order phase transition line $\tilde{T}_0(P)$ is known as the tricritical point. In the vicinity of this point A_4 is small and we must take into consideration the term proportional to η^6 in (4.1.1) when we analyze the critical fluctuations [4.2]. Under such conditions $\varphi_{\eta\eta}$ and $\langle|\eta_{\boldsymbol{k}}|^2\rangle$ in the disordered phase are still given by (4.1.3, 4) while in the ordered phase at $T < T_0$ or $T < \tilde{T}_0$ we have

$$\varphi_{\eta\eta} = \frac{1}{A_6}(\Delta - A_4)\sqrt{\Delta}\,, \quad \eta^2 = \frac{\sqrt{\Delta} - A_4}{2A_6}\,, \quad \Delta = A_4^2 - 4A_2A_6\,. \tag{4.1.7}$$

Near the line of the second-order phase transitions $T_0(P)$, when the condition $A_4^2 \gg 4|A_2|A_6$ is satisfied, (4.1.7) reduces to (4.1.3) and $\langle|\eta_{\boldsymbol{k}}|^2\rangle$ is given by the same equation (4.1.5) as it is far from the tricritical point. But in the immediate vicinity of this point the reversed condition, $A_4^2 \ll 4|A_2|A_6$, is satisfied if the temperature is brought even slightly below the transition point. According to (1.2.16) and (4.1.7) we have then

$$\begin{aligned} \langle|\eta_{\boldsymbol{k}}|^2\rangle &= \frac{k_{\mathrm{B}}T}{V}\frac{1}{4a(T_0-T) + \alpha_{ij}k_ik_j} \quad (A_4^2 \ll 4|A_2|A_6\,; \quad T < T_0)\,, \\ \eta^2 &= \left[\frac{a(T-T_0)}{A_6}\right]^{1/2}\,. \end{aligned} \tag{4.1.8}$$

Under such conditions $\langle|\eta_{\boldsymbol{k}}|^2\rangle$ for a given $\boldsymbol{k}$ decreases with decreasing temperature at a rate twice as large as in the immediate vicinity of the transition point. The first-order transition line (for $A_4 < 0$) is determined by the conditions $\varphi(0) = \varphi(\eta_0)$ and $\varphi_\eta(\eta_0) = 0$ and corresponds to $a(\tilde{T}_0 - T_0) = A_2(\tilde{T}_0) = 3A_4^2/16A_6$. At this line we have $\varphi_{\eta\eta} = 3A_4^2/4A_6$, that is, the maximum of $\langle|\eta_{\boldsymbol{k}}|^2\rangle$ (for $\boldsymbol{k} = 0$) is inversely proportional to the distance to the tricritical point.

Solutions with a One-Component Order Parameter. In solid solutions the fluctuations of the long-range order parameters and the composition are not, generally, statistically independent and must be treated simultaneously (see Sect. 1.2.3). The respective expression for $\langle|\eta_{\boldsymbol{k}}|^2\rangle$ is given by (1.2.26). According to (4.1.1), at $T \geq T_0$ the derivative $\varphi_{c\eta} \equiv \partial^2\varphi/\partial c\partial\eta = 0$, the expression (1.2.26) coincides with (1.2.16) and (4.1.4) is applicable to such solid solutions.

According to (4.1.1, 2) we obtain however for $T \leq T_0$,

$$\varphi_{c\eta}^2 = a^2\left(\frac{dT_0}{dc}\right)^2\eta^2 = \frac{a^3(T_0-T)}{A_4}\left(\frac{dT_0}{dc}\right)^2 , \quad \varphi_{cc} \approx \frac{\partial^2\varphi_0}{\partial c^2} \neq 0 \tag{4.1.9}$$

and (1.2.26) yields

$$\langle|\eta_{\boldsymbol{k}}|^2\rangle = \frac{k_\mathrm{B}T}{V}\frac{1}{2aM(T_0-T)+\alpha_{ij}k_ik_j} \quad (T \leq T_0) ,$$
$$M = 1 - \frac{\varphi_{c\eta}^2}{\varphi_{\eta\eta}\varphi_{cc}} \approx 1 - \frac{a^2}{2A_4\varphi_{cc}^0}\left(\frac{dT_0}{dc}\right)^2 . \tag{4.1.10}$$

Since $M < 1$ in this case $\langle|\eta_{\boldsymbol{k}}|^2\rangle$ decreases at a slower rate than in one-component crystals.

When $T \to 0$ only completely ordered solid solutions or one-component crystals can be at equilibrium, according to the Nernst theorem. As shown in Fig. 4.2, the second-order phase transition line AC for solutions ends therefore with a critical point where AC meets two decomposition lines, CB and CD, so that the range I corresponds to the disordered solution, the range II to the ordered solution, and the range III to the mixture of these two phases. We can demonstrate with the use of the expansion (4.1.1) that the line AC is continuously transformed into the line CB (as in Fig. 4.2) and that the condition $M = 0$ is satisfied at the critical point C [1.67]. In the Landau theory in the vicinity of the critical point we can expand M in powers of $T - T_0$ and $c - c_0$ (according to [1.67], the difference $c - c_0$ is proportional to $T_0(c) - T_c$) and retain only the linear terms:

$$M = g_1(T_0(c)-T) + g_2(T_0(c)-T_c) . \tag{4.1.11}$$

Equations (4.1.10, 11) demonstrate that when we move away from the temperature T_0 in the vicinity of the critical point $\langle|\eta_{\boldsymbol{k}}|^2\rangle$ decreases at a much slower rate in the ordered phase range than in the disordered phase range.

It can be seen from (1.2.27) and (4.1.3, 9) that the fluctuations of the concentration near the conventional second-order phase. transition are not anomalously large. But near the critical point C under consideration here, the concentration fluctuations grow anomalously as do the fluctuations of the long-range order parameter [1.3]. When we substitute (4.1.3, 9) into (1.2.27) and make use of the expressions (4.1.10, 11) for M we obtain

$$\langle|c_{\boldsymbol{k}}|^2\rangle = \frac{k_\mathrm{B}T}{V}\frac{2a(T_0-T)+\alpha k^2}{2\alpha\varphi_{cc}^0[g_1(T_0-T)+g_2(T_0-T_c)](T_0-T)+\alpha\varphi_{cc}^0k^2}$$
$$(T < T_0) . \tag{4.1.12}$$

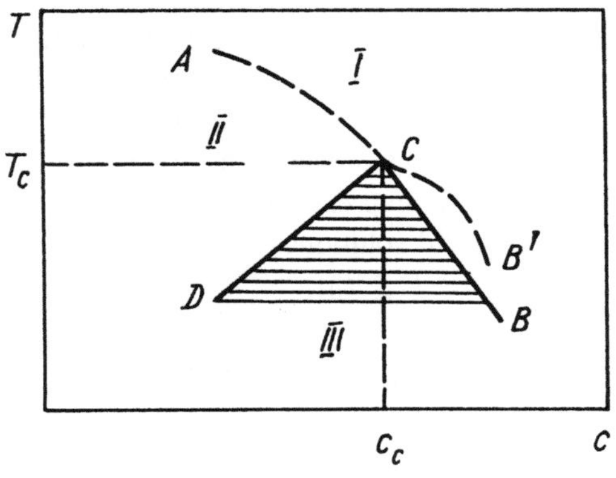

Fig. 4.2. Phase diagram near the tricritical point on the line of second-order phase transitions in solutions. Curve CB' shows the bend in the line ACB caused by the fluctuations

We see that for sufficiently small k the squared Fourier components of the concentration fluctuations near the critical point in the ordered phase become anomalously large and reach the same order of magnitude as $\langle|\eta_k|^2\rangle$. For a given k the sharp peak of $\langle|c_k|^2\rangle$ is somewhat below T_0 (at $T_0 - T \sim \alpha/ag)^{1/2}k$ if $T_0 = T_c$). In the disordered phase the mean squares $\langle|c_k|^2\rangle = k_B T V^{-1}(\varphi_{cc}+\beta' k^2)^{-1}$ are small.

Vicinity of the Critical Point on the Decomposition Line. When we consider the critical concentration fluctuations in the decomposing solutions we may use the expansion of φ in powers of the deviation $c - c_c$ of the concentration from the critical concentration c_c (see Fig. 4.1) which is similar to the expansion (4.1.1) (and also has a restricted applicability) [1.3]. This expansion includes also the odd-order terms and can be written as

$$\varphi = \varphi_0 + A_1(c-c_c) + \frac{A_2}{2}(c-c_c)^2 + \frac{A_3}{3}(c-c_c)^3 + \frac{A_4}{4}(c-c_c)^4 + \dots \tag{4.1.13}$$

At the critical point C (see Fig. 4.1) we have $d^2\varphi/dc^2 = 0$ and $d^3\varphi/dc^3 = 0$ and in the spirit of the Landau theory we can take $A_2 = a(T - T_c)$, $A_3 = b(T - T_c)$, and $A_4 > 0$. Substituting (4.1.13) into the equation

$$\varphi(c_1) - \varphi(c_2) = \varphi_c(c_1)(c_1 - c_2)\,, \quad \varphi_c(c_1) = \varphi_c(c_2)\,, \tag{4.1.14}$$

describing the equilibrium of phases with the concentrations c_1 and c_2, we obtain the equation for the decomposition line $T_0(c)$ for small $c - c_c$:

$$T_c - T_0(c) = \frac{A_4}{a}(c-c_c)^2\,, \quad \frac{d^2\varphi}{dc^2} = a[T + 2T_c - 3T_0(c)]\,. \tag{4.1.15}$$

When we substitute the equation (4.1.15) for $d^2\varphi/dc^2$ into (1.2.10) we obtain for the disordered solutions

$$\langle|c_k|^2\rangle = \frac{k_B T}{V}\frac{1}{a[T + 2T_c - 3T_0(c)] + \beta_{ij}k_ik_j}\,, \tag{4.1.16}$$

that is, in the critical range (4.1.16) grows anomalously reaching a maximum at the critical point at $T = T_c$ where $c = c_c$ (we have $T_0(c_0) = T_c$).

Second-Order Phase Transitions and Crystal Symmetry. Considerations of crystal symmetry must be taken into detailed account in the general analysis of second-order phase transitions [1.34, 67–70; 2.11, 131]. The crystal symmetry can be described by the symmetry of the probability density function $\varrho(\boldsymbol{r})$ (or the electron density) for atoms of a given species. In a high-symmetry phase $\varrho(\boldsymbol{r}) = \varrho_0(\boldsymbol{r})$ remains invariant under all transformations of its symmetry group $\mathfrak{G}_0$. Under the phase transition the symmetry of $\varrho(\boldsymbol{r}) = \varrho_0(\boldsymbol{r}) + \delta\varrho(\boldsymbol{r})$ is decreased and, in addition to the invariant component $\varrho_0(\boldsymbol{r})$, this function includes the term $\delta\varrho(\boldsymbol{r})$ which is altered under the transformations $\mathfrak{G}_0$. This term can be expanded in the functions $\varphi_\mu^p(\boldsymbol{r})$ giving the non-trivial irreducible representations of the group $\mathfrak{G}_0$, that is, the functions which are transformed into each other under the symmetry transformations $\mathfrak{G}_0$ and do not break down into smaller groups (ϱ is the number of the representation and μ is the number of the function in the basis):

$$\varrho(\boldsymbol{r}) = \varrho_0(\boldsymbol{r}) + \delta\varrho(\boldsymbol{r}) \,, \quad \delta\varrho(\boldsymbol{r}) = \sum_p \sum_\mu \eta_\mu^{(p)} \varphi_\mu^{(p)}(\boldsymbol{r}) \,. \tag{4.1.17}$$

The coefficients $\eta_\mu^{(p)}$in this expansion describe the long-range order in the low-symmetry phase. In a second-order phase transition they vary continuously at the transition point so that at $T < T_0$ the parameters $\eta_\mu^{(p)}$ for one basis do not vanish. This causes a jump-like variation of the symmetry which is now determined by the group $\mathfrak{G}$ which is a subgroup of the group $\mathfrak{G}_0$ (since $\varrho_0(\boldsymbol{r})$ must also remain invariant under all $\mathfrak{G}$ transformations). In the special case of one-dimensional irreducible representations a one-component long-range order parameter describes the low-symmetry phase but in the general case the number n of the functions in the $\varphi_\mu^{(p)}$ basis is not unity and the long-range order is described by a multicomponent parameter.

It is convenient to assume that under the $\mathfrak{G}_0$ transformations the functions $\varphi_\mu^{(p)}$ remain unchanged while $\eta_\mu^{(p)}$ are transformed through each other. The density φ of the thermodynamic potential can apparently depend only on various $\eta_\mu^{(p)}$ invariants which are not changed under any symmetry transformations. In the Landau theory φ is written as an expansion in powers of $\eta_\mu \equiv \eta_\mu^{(p)}$ of the order parameters which appear below the transition. There is no invariant linear in η_μ while the second-order one can always be reduced to a sum of squares. Therefore, we obtain the following expansion:

$$\begin{aligned} \varphi &= \varphi_0 + \frac{1}{2} A_2 \sum_\mu \eta_\mu^2 + \frac{1}{3} A_{3\mu\mu'\mu''} \eta_\mu \eta_{\mu'} \eta_{\mu''} + \frac{1}{4} A_{\mu\mu'\mu''\mu'''} \eta_\mu \eta_{\mu'} \eta_{\mu''} \eta_{\mu'''} \\ &\quad + \frac{1}{2} \alpha_{\mu\mu' ij} \frac{\partial \eta_\mu}{\partial x_i} \frac{\partial \eta_{\mu'}}{\partial x_j} + \frac{1}{2} \delta_{\mu\mu' i} \left(\eta_\mu \frac{\partial \eta_{\mu'}}{\partial x_i} - \eta_{\mu'} \frac{\partial \eta_\mu}{\partial x_i} \right) , \\ A_2 &= a(T - T_0) \,. \end{aligned} \tag{4.1.18}$$

Here summation from 1 to n is implied over the repeated subscripts μ (and over i and j from 1 to 3). The coefficients $\hat{A}_3$, $\hat{A}_4$, $\hat{\alpha}$ and $\hat{\delta}$ are such that the respective terms are invariant under η_μ transformations. The terms with the derivatives

$$\partial\eta_\mu/\partial x_i \quad \text{and} \quad \partial(\eta_\mu\eta_{\mu'})/\partial x_i = \eta_\mu\partial\eta_{\mu'}/\partial x_i + \eta_{\mu'}\partial\eta_\mu/\partial x_i \ ,$$

are not written in (4.1.18) since they vanish upon integration of $\varphi(\boldsymbol{r})$.

The transition to a commensurate phase of a lower symmetry discussed here is a phase transition of the second order (that is, the parameters η_μ continuously increase from zero if the following conditions are satisfied:

$$A_{3\mu\mu'\mu''} = 0 \ , \delta_{\mu\mu' i} = 0 \ . \tag{4.1.19}$$

As in the case of one-component order parameter, under this condition for $A_2 > 0$ the minimum of $\varphi(\eta_\mu)$ is at the point $\eta_\mu = 0$ and corresponds to the high-symmetry phase while for $A_2 < 0$ it is the point of the maximum of $\varphi(\eta_\mu)$ and the minimum corresponds to the low-symmetry phase ($\eta_\mu \neq 0$).

If $\hat{A}_3 \neq 0$ then for temperatures slightly exceeding T_0 (that is, for small A_2) and an appropriately selected sign of η_μ the third-order term makes $\varphi(\eta_\mu \neq 0)$ lower than $\varphi(\eta_\mu = 0)$ in the range of small (but not infinitesimal) η_μ where the higher-order terms of the expansion can be ignored. This means that at a temperature $\tilde{T}_0 > T_0$ when the point $\eta_\mu = 0$ still corresponds to a minimum of $\varphi(\eta_\mu)$ the second minimum of $\varphi(\eta)_\mu$ for small η_μ becomes lower than the first one, that is, a first-order phase transition to the low-symmetry phase takes place. Thus, the equation $\hat{A}_3 = 0$ is indeed a necessary condition for second-order phase transitions.

The condition $\hat{\delta} = 0$ must be satisfied if the resulting commensurate phase is to be stable with respect to continuous inhomogeneous variations of $\eta_\mu(\boldsymbol{r})$. If $\hat{\delta} \neq 0$ then we may go over to the regularly varying order parameters $\eta_\mu \sim \sin(\boldsymbol{k}_0\boldsymbol{r}+\phi_\mu)$, that is, to an incommensurate phase with the parameter $2\pi/k_0$ dependent on T and decrease the thermodynamic potential for the appropriate ϕ_μ at the expense of the last term in (4.1.18) which is proportional to $\sin(\phi_\mu - \phi_{\mu'})$.

For some space groups the considerations of symmetry do not allow the existence of the cubic η_μ invariants and the Lifshitz invariants $L_{\mu\mu'} = \eta_\mu\partial\eta_{\mu'}/\partial x_i - \eta_{\mu'}\partial\eta_\mu/\partial x_i$, that is, the necessary conditions (4.1.19) for the second-order phase transitions are satisfied. Then the T, P or T, c phase diagrams may exhibit continuous lines of second-order phase transitions such as the AC line in Fig. 4.2 (the first-order phase transition to the ordered phase is, of course, always possible). If the symmetry conditions do not require a vanishing of $\hat{A}_3$ then a continuous transition is possible only at an isolated point on the line of the first-order phase transitions (at which $\hat{A}_3 = 0$) while in the structures with $\hat{\delta} \neq 0$ the line of the second-order phase transitions can be only a boundary between the disordered phase and the incommensurate phase.

The occurrence of structures for which $\hat{A}_3 \equiv 0$ and $\hat{\delta} \equiv 0$ can be analyzed with the methods of the group theory [1.25, 26, 34, 69, 70; 2.11]. Such analysis is simplified if we take explicitly into account the translational symmetry of the crystal, that is, the Bloch-function form of the functions $\varphi_\mu^{(p)}(\boldsymbol{r})$ realizing the irreducible $\mathfrak{G}_0$ representation. The long-range order occurring by substitutional ordering in a binary AB solution with a single atom per lattice cell may be described as the regular variation of the mean probability $\langle c_s(R_s)\rangle$ that atoms A occupy lattice sites:

$$\langle c_s\rangle = c + \sum_{G'_\mu} [\eta(G'_\mu)\exp(-\mathrm{i}G'_\mu \cdot R_s) + c.c.] \tag{4.1.20}$$

Here G'_μ are the superlattice vectors of the reciprocal lattice making up a star or several stars, that is, which are transformed into each other under group $\mathfrak{G}_0$ transformations. The parameters $\eta(G'_\mu)$ unambiguously describe the long-range order and can be regarded as the order parameters η_μ in (4.1.18). In a similar way, under displacement-type ordering an expansion similar to (4.1.20) is valid for the mean static displacements of atoms and the amplitudes of the displacement waves of certain polarizations may be chosen as the order parameters $\eta_\mu \equiv \eta(G'_\mu)$.

Owing to the translational symmetry, the expansion of φ in powers of $\langle c_s\rangle - c$ includes the quadratic terms $(\langle c_s\rangle - c)(\langle c_{s'}\rangle - c)f(R_s - R_{s'})$, and the third-order terms $(\langle c_s\rangle - c)(\langle c_{s'}\rangle - c)(\langle c_{s''}\rangle - c)f(R_s - R_{s'}, R_s - R_{s''})$, that is, the coefficients depend only on the differences $R_s - R_{s'}$. When we go over to the k representation (4.1.20) we have accordingly in the expansion (4.1.18) in $\eta(G'_\mu) \equiv \eta_\mu$ the term with $\eta(G'_\mu)\eta(G'_\mu)$ containing the factor $\delta(G'_\mu + G'_{\mu'})$, the term with $\eta(G'_\mu)\eta(G'_{\mu'})\eta(G'_{\mu''})$ containing $\delta(G'_\mu + G'_{\bar{\mu}'} + G'_{\mu''})$ and so on. Here $\delta(k) \neq 0$ only for $k = G$ where G is a reciprocal lattice vector of the disordered phase, including the reciprocal lattice origin. Therefore the condition that the cubic invariants vanish (or $\hat{A}_3$ vanishes) reduces to the condition that the equation $G'_\mu + G'_{\mu'} + G'_{\mu''}$ is prohibited for the symmetry under consideration, that is, the line of the second-order phase transitions exists when

$$G'_\mu + G'_{\mu'} + G'_{\mu''} \neq G\,. \tag{4.1.21}$$

The condition (4.1.21) is satisfied, for instance, when a $L2_0$ structure (such as β brass) with $G' = \langle 100\rangle$ is formed in a bcc lattice (for which $\langle 111\rangle$ are not reciprocal lattice vectors). But when the structures $L1_0$ (such as CuAu) and $L1_2$ (such as Cu_3Au) are formed and we also have $G'_\mu = \langle 100\rangle$, the sum $\langle 111\rangle$ of the three vectors G'_μ equals the vector G for the fcc lattice, that is, only a first-order phase transition can take place.

When (4.1.20) has been substituted into $\Phi = \int \varphi(r)dr$ the second-order term $\mathcal{A}_2(G'_\mu)|\eta(G_\mu)|^2$ includes the coefficient $\mathcal{A}_2(G_\mu)$ which depends on G'_μ and includes the contribution of the last term in (4.1.18). The minimum of the function $\mathcal{A}_2(G'_\mu)$ is zero at the second-order phase transition (otherwise, at a higher temperature the stability of the disordered phase would be disturbed by the production of a wave with a different G'_μ). The extremum condition $\partial\mathcal{A}_2(k)/\partial k = 0$ for $k = 0$ can be identically satisfied if the point group of the point G'_μ includes the axes or planes of symmetry intersecting at one point (in particular, a center of inversion). Under these circumstances the vector $\partial\mathcal{A}_2(k)/\partial k$ (which is invariant by definition) cannot remain unchanged under all symmetry transformations, that is, it must vanish for any T, c, P (while the invariants $L_{\mu\mu'} = \eta_\mu \partial\eta_{\mu'}/\partial x_i - \eta_{\mu'}\partial\eta_\mu/\partial x_i$ which give rise to a contribution linear in $k_i - G'_{\mu i}$ to $\mathcal{A}_2(\boldsymbol{k})$ cannot be produced [1.69, 70]). Therefore, the line of second-order phase transitions to the commensurate phase can exist only when structures with such $\boldsymbol{G}'_\mu$ are produced. It has been shown that $G_{\mu i}$ can only equal the basic parameters of the reciprocal lattice of the disordered phase divided by two, three or four [1.69, 70].

If the minima of $\mathcal{A}_2(\boldsymbol{k})$ are not at high-symmetry points (because the structure allows the existence of the invariants $L_{\mu\mu'}$ or owing to arbitrary factors) then the transition to an incommensurate phase takes place. For such $\boldsymbol{G}'_\mu$ the condition (4.1.21) is satisfied, that is, the transition can be a second-order phase transition. The resulting vectors $\boldsymbol{G}'_\mu$ depend on P and c (and correspond to $L_{\mu\mu'}$ vanishing for the given P, c).

Fluctuations of the Multicomponent Order Parameters. In the general case of systems with a multicomponent long-range order parameter, the mean squares and the products of the Fourier components of the fluctuations $\delta\eta_\mu(\boldsymbol{r})$ are given by [1.2.19, 20). Near a second-order phase transition, $\varphi(\eta_\mu)$ in these equations must be replaced with (4.1.18) with $\hat{A}_3 = 0$ and $\hat{\delta} = 0$. For instance, for a three-component parameter $\eta_\mu = \eta_{1,2,3}$ we have only two independent fourth-order invariants and, according to (4.1.18, 19) we obtain

$$\varphi = \varphi_0 + \frac{1}{2}a(T - T_0)\sum_{\mu=1}^{3}\eta_\mu^2 + \frac{1}{4}A_4'\left(\sum_{\mu=1}^{3}\eta_\mu^2\right)^2 + \frac{1}{4}A_4''\sum_{\mu=1}^{3}\eta_\mu^4 \qquad (4.1.22)$$

(see [4.4, 5]). The structure of the long-range order (the ratios η_2/η_1 and η_3/η_1) is determined by the condition that the fourth-order terms reach a minimum for the given $\sum\eta_\mu^2$ (or the condition that the sixth-order terms reach a minimum for $A_4'' = 0$).

If $A_4'' < 0$ and $A_4' + A_4'' > 0$ then the second-order phase transition gives rise to domains of a tetragonal ordered phase with the parameters $\eta_1 \pm \eta$ and $\eta_2 = \eta_3 = 0$ (as in $BaTiO_3$). At temperatures above T_0 the mean squares $\langle|\eta_{\mu\boldsymbol{k}}|^2\rangle$ are given by (4.1.4) and at $T < T_0$ the longitudinal order fluctuations ($\langle|\eta_{1\boldsymbol{k}}|^2\rangle$) for $\eta_1 \neq 0$) are described by (4.1.5). According to (1.2.19, 20), however, the transverse order fluctuations ($\delta\eta_{2,3}$ for $\eta_1 \neq 0$) at $T < T_0$ are described by the equation [3.223]

$$\langle|\eta_{\perp\boldsymbol{k}}|^2\rangle\langle|\eta_{2\boldsymbol{k}}|^2\rangle = \langle|\eta_{3\boldsymbol{k}}|^2\rangle = \frac{k_B T}{V}\left(\frac{A_4'}{|A_4''|} - 1\right)\frac{1}{a(T_0 - T)} \quad (k \to 0) \quad (4.1.23)$$

and may differ significantly from the longitudinal fluctuations.

For $A_4'' > 0$ and $3A_4' + A_4'' > 0$ the phase transition gives rise to domains of a trigonal phase with the order parameters $\pm\eta_1 = \pm\eta_2 = \pm\eta_3$. The longitudinal order fluctuations in these domains and the fluctuations at $T > T_0$ are also described by (4.1.4, 5) while the transverse order fluctuations at $T < T_0$ are given by

$$\langle|\eta_{\perp\boldsymbol{k}}|^2\rangle = \frac{k_B T}{V}\left(\frac{3A_4'}{A_4''} + 1\right)\frac{1}{2a(T_0 - T)} \quad (k \to 0)\,. \qquad (4.1.24)$$

When we consider the dependence of $\langle \eta_{\mu k}|^2 \rangle$ on $\boldsymbol{k}$ we must bear in mind that even in cubic crystals the terms $\alpha_{\mu\mu' ij} k_i k_j$ in (1.2.19) can strongly depend on the direction of $\boldsymbol{k}$ if the points G'_μ do not possess cubic symmetry.

Long-Range Force Effects. As discussed in Chap. 2, the equations for the fluctuation probabilities are changed in the presence of long-range dipole-dipole or elastic forces whose effects can significantly suppress the fluctuations [1.16; 2.14]. These effects are especially strong in the vicinities of the critical points and second-order phase transitions.

The dipole forces produce a sharp effect on the long-range order fluctuations in ferroelectrics. In intrinsic uniaxial ferroelectrics with an axis of easy polarization Z, the respective component of the polarization vector may be selected for the long-range order parameter, $\eta \equiv P_z$. According to (1.2.23), $\varphi_{\eta\eta} \equiv \varkappa_{zz}^{-1}$ is expressed in terms of the susceptibility $\hat{\varkappa}$ and we obtain an equation similar to (4.1.3) in the vicinity of the second-order phase transition:

$$\varkappa_{zz}^{-1} = a(T - T_0) \quad (T \geq T_0)\,, \quad \varkappa_{zz}^{-1} = 2a(T_0 - T) \quad (T \leq T_0)\,. \tag{4.1.25}$$

If the dipole interaction is screened then, according to (1.2.24) and (4.1.25), $\langle |P_{kz}|^2 \rangle$ is determined by the general equations (4.1.4,5) and increases anomalously for all $\boldsymbol{k}$ directions. But if the dipole interaction is effective then (1.2.24) must be replaced with (2.2.10) and using (4.1.25) we obtain for instance at $T > T_0$

$$\langle |P_{kz}|^2 \rangle = \frac{k_B T}{V} \frac{1}{a(T - T_0) + 4\pi\varepsilon_{xx}^{-1} n_z^2 + \alpha_{zzzz} k_z^2 + \alpha_{zzxx} k_\perp^2}$$

$$(n_z^2 \ll 1) \tag{4.1.26}$$

(here $\varepsilon_{xx}/4\pi = \varkappa_{xx} + (4\pi)^{-1} \sim 1$), that is $\langle |P_{kz}|^2 \rangle$ exhibits anomalous growth only near the k_x, k_y plane.

For intrinsic ferroelectrics whose disordered phase possesses cubic symmetry we can take the three components of the polarization vector $\boldsymbol{P}$ as the three-component parameter of the long-range order. At $T > T_0$ their fluctuations are described by (2.2.9) where $\varepsilon - 1 = 4\pi a^{-1}(T - T_0)^{-1}$ (the axis Z is taken parallel to $\boldsymbol{k}$. Therefore, we obtain for $k \to 0$ the following expressions accurate within small terms of the order of $\varepsilon_\infty/\varepsilon$:

$$\langle |P_{kx}|^2 = \langle |P_{ky}|^2 = \frac{k_B T}{V a(T - T_0)}\,, \quad \langle |P_{kz}|^2 = \frac{k_B T}{4\pi V \varepsilon_\infty}\,. \tag{4.1.27}$$

In these circumstances the fluctuations of the transverse polarization components (which are perpendicular to $\boldsymbol{k}$) grow anomalously for any $\boldsymbol{k}$ but the longitudinal order fluctuations are suppressed by the dipole interaction forces. For temperatures $T < T_0$ according to (2.2.5,8) we must invert the matrix $\varkappa_{ij}(\boldsymbol{k})$ in order to calculate $\langle (P_{ki} P_{kj}^* + P_{ki}^* P_{kj}) \rangle$ for $k \to 0$ and according to (4.1.22) in (2.2.5) we have

$$\varkappa_{ij}^{-1} = a(T - T_0)\delta_i + \left(\frac{A_4'}{2}p^2 + 3A_4''P_i^2\right)\delta_{ij}A_4'P_iP_j \,.$$

The long-range elastic forces significantly suppress the concentration fluctuations near the critical points on the decomposition curves (see Fig. 4.1). When we take into consideration the elastic forces (2.1.52 and (4.1.15) show that

$$\langle |c_{\boldsymbol{k}}|^2 \rangle = \frac{k_B T}{V} \frac{1}{a[T + 2T_c - 3T_0(c)] + \varphi_{cc}^{\text{el}} + \beta_{ij}k_ik_j} \tag{4.1.28}$$

remain finite at the critical point ($T = T_c$, $T_0(c = c_c) = T_c$) in the case of $k \to 0$ (but $k \gg L^{-1}$) for all $\boldsymbol{k}$ directions at the expense of the contribution φ_{cc}^{el} made by the elastic forces (see (2.1.5–9)). If the contribution φ_{cc}^{el} is small then $\langle |c_{\boldsymbol{k}}|^2 \rangle$ can grow significantly near the critical point but they do not tend to infinity as they do in undistorted crystals. The denominator in (4.1.28) vanishes and the fluctuations become anomalously large only for some $\boldsymbol{k}$ directions at considerably lower temperatures in supercooled nonequilibrium solutions near the point of spinodal decomposition (see Sect. 2.1). In a similar way, near the critical point on the line of second-order phase transitions in solid solutions (see Fig. 4.2) the mean squares $\langle |c_{\boldsymbol{k}}|^2 \rangle$ for $k \to 0$ are described by (2.1.5) in which $d^2\varphi/dc^2 = \varphi_{cc}^0\{g_1(T_0 - T) + g_2[T_0(c) - T_c]\}$ tends to zero at the critical point according to (4.1.10, 11). The additional term φ_{cc}^{el} suppresses the anomalous concentration fluctuations under these circumstances, too.

The fluctuations of the long-range order parameters near a second-order phase transition are suppressed by the elastic forces in cases where in the disordered phase the stresses u_{ij} vary linearly with the variations $\delta\eta$ of the order parameter and $L_{\eta ij} = \delta u_{ij}/\delta\eta \neq 0$ (as in ferroelectrics exhibiting a piezoelectric effect). Then in crystals described by a one-component long-range order parameter the mean squares $\langle |\eta_{\boldsymbol{k}}|^2 \rangle$ are given by (2.1.17) where $\varphi_{\eta\eta}$ tends to zero for $T \to T_0$ according to (4.1.3) but $\varphi_{\eta\eta}^{\text{el}}$ is generally finite. In some crystals $\varphi_{\eta\eta}^{\text{el}}$ may also vanish for some directions of $\boldsymbol{k}$ so that for such directions $\langle |\eta_{\boldsymbol{k}}|^2 \rangle$ (in contrast to the concentration fluctuations) exhibits anomalous growth near T_c. For instance, if the tensor $\hat{L}_\eta$ has only one non-vanishing component $L_{\eta xy}$ then (2.1.18, 19) show that $\varphi_{\eta\eta}^{\text{el}}$ vanishes for $n_x = 1$ or $n_y = 1$. For instance, for $T > T_0$ and $n_x \cong 1$ (2.1.17, 18) and (4.1.3) yield

$$\langle |\eta_{\boldsymbol{k}}|^2 \rangle = \frac{k_B T}{V} \frac{1}{a(T - T_0) 4c_{66}L_{\eta xy}^2(\zeta_y n_y^2 + \zeta_z n_z^2) + \alpha_{xx}k^2} \,, \tag{4.1.29}$$

where $\zeta_y \sim \zeta_z \sim 1$. While the dipole interaction in uniaxial ferroelectrics allows anomalous growth of $\langle |\eta_{\boldsymbol{k}}|^2 \rangle$ near certain planes the elastic interaction in these crystals allows such growth only near certain lines in the $\boldsymbol{k}$ space. Some ferroelastic crystals exhibit similar features in their fluctuation characteristics [2.19–21; 3.223].

Applicability of the Landau Theory. The large-scale fluctuations of the order parameter $\eta(\boldsymbol{r})$ cannot be properly taken into account with the expansions similar to (4.1.1) with the self-consistent field approximation and other approximations of the microscopic theory discussed in Sect. 1.3. The fluctuations and the effects caused by them grow rapidly near the transition point and the expansion (4.1.1) must be expected to become inapplicable. The conditions for applicability of the Landau theory can be determined by analyzing the additional fluctuation terms in various parameters [4.6–8].

A rough (but qualitatively accurate) estimate for the range of applicability of the expansion (4.1.1) can be obtained from the condition that the fluctuation term $\sim A_4\langle(\delta\eta)^2\rangle$ in the equation (4.1.3) for $\varphi_{\eta\eta}$ be small in comparison with $\varphi_{\eta\eta} \sim |A_2|$. Here $\langle(\delta\eta)^2\rangle$ takes into consideration only rather smoothly varying fluctuations for which the inhomogeneity scale is of the order of the correlation length r_c given by (4.1.6). We obtain for the isotropic fluctuations described by (4.1.4,5)

$$\langle(\delta\eta)^2\rangle = \sum_{\substack{k \\ (k<k_c)}} |\eta_k|^2 \sim k_{\mathrm{B}}T \int_0^{k_c} \frac{k^2dk}{\varphi_{\eta\eta}+\alpha k^2} \sim \frac{k_{\mathrm{B}}Tk_c}{\alpha}\,, \qquad k_c = \frac{1}{r_c} \sim \sqrt{\frac{\varphi_{\eta\eta}}{\alpha}} \tag{4.1.30}$$

and then the condition of smallness for the fluctuations terms $A\langle(\delta\eta)^2\rangle \ll \varphi_{\eta\eta}$ can be reduced with the use of (4.1.3) to the Levanyuk-Ginzburg condition [4.6, 7]

$$|T - T_0| \gg \frac{(k_{\mathrm{B}}T_0A_4)^2}{a\alpha^3}\,. \tag{4.1.31}$$

In a similar way, when we determine $\langle(\delta c)^2\rangle$ near the critical point on the decomposition curve with the use of (4.1.16) we see that the range of $T + 2T_c - 3T_0(c)$ where (4.1.16) is applicable can also be found from the condition (4.1.31) with α replaced with β. The parameters a, A_4 and α or β can be estimated in the microscopic theory of ordering or decomposition, for instance, with equations (1.3.19). We obtain $a \sim k_{\mathrm{B}}/v$, $A_4 \sim k_{\mathrm{B}}T_0/v$ (far from the tricritical point), and $\alpha \sim \beta \sim k_{\mathrm{B}}T_0r_0^2/v$ where r_0 is the radius of the region with the significant energies $w(\varrho)$ of constant sign. Therefore, the right-hand side in (4.1.31) is of the order of $T_0(d/r_0)^6$ where $d^3 \sim v$, that is, in the nearest-neighbour model the expansion (4.1.1) is inapplicable but for a large interaction length the applicability range includes a wide temperature range, $T_0 \gg |T - T_0| \gg T_0(d/r_0)^6$ with the exception of a narrow segment at the transition point. The width of this segment agrees with the estimate (1.3.14) for the range of applicability of the self-consistent field approximation.

In the above case of isotropic fluctuations in a three-dimensional system we always have a temperature range in which the condition (4.1.31) is not satisfied and the expansion (4.1.1) is invalid. On the other hand, in a d-dimensional problem the integral over $dk \sim k^{d-1}dk$ in (4.1.30) yields the following estimate:

$$\langle(\delta\eta)^2\rangle \sim k_B T_0 k_c^{d-2}/\alpha \sim k_B T_0 \varphi_{\eta\eta}^{(d-2)/2}\alpha^{-d/2} ,$$

This means that for $d > 4$ (which is not realized physically) the condition $A_4\langle(\delta\eta)^2\rangle \ll \varphi_{\eta\eta}$ that the fluctuation terms be small is satisfied and the Landau theory is applicable. In the boundary case of $d = 4$ the additional fluctuation terms in $\varphi_{\eta\eta}$ also do not alter qualitatively the dependence of $\varphi_{\eta\eta} \sim |T-T_0|$ which follows from (4.1.1) though they give rise to additional factors of a logarithmic form [2.58].

In systems where fluctuations are suppressed by the long-range forces $\langle(\delta\eta)^2\rangle$ is considerably smaller and varies with $T-T_0$ as a function differing from (4.1.30) and corresponding to the effective space dimensionality $d > 3$. This result justifies application of the Landau theory to some systems of this type [1.51; 2.17–19, 58]. For instance, (4.1.26, 30) show that in uniaxial intrinsic ferroelectrics for $\alpha \sim T_0^{-1}$ the smoothly varying fluctuation waves with $k_\perp < k_c = \sqrt{a|T-T_0|/\alpha}$, $k_z^2/k^2 < \alpha k_c^2$ make a contribution to $\langle(\delta P_z)^2\rangle$ which is of the order of $k_B T_0 \alpha^{-3/2} a|T-T_0|$ and linearly depends on $|T-T_0|$ as in the case of isotropic fluctuations in the four-dimensional space. This is why the general form of the equation (4.126) for $\langle|P_z|^2\rangle$ in this case is not changed by the inclusion of fluctuations which give rise only to a factor $[\ln(|T-T_0|/T_0)]^{1/3}$ which relatively weakly depends on the temperature [2.58].

The applicability conditions become even more favorable for the Landau theory in systems where elastic forces make a significant contribution. For instance, consider systems with the piezoelectric effect where fluctuations are described by (4.1.29) and for $|T-T_0| < c_{66}L_{\eta xy}^2/a$ exhibit anomalous growth only along preferred axes. In such systems the additional fluctuation term in $\varphi_{\eta\eta}$ is proportional to $|T-T_0|^{3/2}$ and becomes insignificant for $T \to T_0$ [2.17]. A similar additional fluctuation term is obtained for ferroelastic phase transitions if the fluctuations grow only along preferred lines (if the fluctuations grow along preferred planes they give rise to logarithmic additional terms) [2.18–21]. The additional fluctuation terms for φ_{cc} may also be ignored in the case of the solid solutions with a considerable elastic energy $\varphi_{cc}^{\rm el} \gtrsim (k_B T_0 A_4)^2/\beta^3 \sim (k_B T/v)(d/r_0)^6$ and then (4.1.28) may be used in the entire temperature range (for considerably smaller $\varphi_{cc}^{\rm el}$ and $r_0 \gg d$ the expansion (4.1.13) has different coefficients far from and near to the critical point).

When we approach the tricritical point $A_4 \to 0$ and we obtain a decrease in the width of the temperature range in which the condition (4.1.31) for applicability of the Landau theory is not satisfied. According to [4.9], this theory gives for the tricritical point itself (see (4.1.8)) $\varphi_{\eta\eta} \sim |T-T_0|$ in the entire range of $|T-T_0| \ll T_0$ while the order parameter $\eta \sim [|T-T_0|\ln(|T-T_0|/T_0)]^{1/4}$.

4.1.2 Fluctuation Range

Systems with Short-Range Forces. In the absence of long-range forces and in the immediate vicinity of the point T_0 the condition (4.1.31) is not satisfied and the fluctuations play a significant part since they determine the peculiar features of the thermodynamic parameters which differ from such features in the Landau theory. The range where the reversed condition (4.1.31) is satisfied is referred to as the fluctuation range. It is discussed in detail in a variety of review papers [4.10–15] and we shall note here only the main results, particularly on the correlation functions.

As is the case in the Landau theory, such characteristics as the order parameter η, the susceptibility $\varkappa = \varphi_{\eta\eta}^{-1}$, the correlation length r_c, and the singular component δC_p of the specific heat exhibit a power-law dependence on the reduced temperature $\tau = (T - T_c)/T_0$:

$$\begin{aligned} &\eta = \eta^0(-\tau)^\beta \,, \quad \varphi_{\eta\eta} = \varkappa^{-1} = \frac{k_B T}{v}\Gamma^{\pm}|\tau|^\gamma \,, \\ &r_c = r_{c0}^{\pm}|\tau|^{-\nu} \,, \quad \delta C_p \sim |\tau|^{-\alpha} \end{aligned} \tag{4.1.32}$$

Here η^0, $\Gamma^{\pm}$, $r_{c0}^{\pm}$ are constants taken at $T = T_0 \pm 0$ and the exponents (the critical indices) differ from the parameters $\beta = 1/2$, $\gamma = 1$, $\nu = 1/2$, and $\alpha = 0$ found in the Landau theory (see (4.1.2, 3, 6)). The exponents in the dependences of $\langle|\eta_k|^2\rangle$ on k and the correlation function $\xi(r)$ on r at $T = T_0$ are also altered in comparison with those in (4.1.4, 6). We have

$$\langle|\eta_k|^2\rangle = \frac{\text{const}}{k^{2-\eta_0}} \,, \quad \xi(r) = \frac{\text{const}}{r^{d-2+\eta_0}} \tag{4.1.33}$$

(here $\eta_0 \neq 0$ in contrast to (4.1.4–6)).

Since the special features exhibited in the fluctuation range are caused by the long-wave fluctuations they, as noted above, are universal in character, that is, the critical indices depend on such general properties as the dimensionality d of the system, the number n of the order parameters, the variation of the interaction as a function of the distance and the angle, and the number and types of the invariants of various orders that can be made up of η_μ. The magnitude of the interaction, the number of the coordination spheres taken into account, and so on, affect only the ordering temperature but not the critical indices. The indices are related to each other and these relations in the most general form can be derived from the principle of scale invariance [4.16, 17] according to which there are only two independent indices. For instance, the following relations can be written for the indices in (4.1.32, 33) (see, for example, [1.31]):

$$\alpha + 2\beta + \gamma = 2 \,, \quad \nu(2 - \eta_0) = \gamma \,, \quad \tau d = 2 - \alpha \,. \tag{4.1.34}$$

The expression (4.1.32) for $\varphi_{\eta\eta}$ and (4.1.33) yield $\langle|\eta_k|^2\rangle$ in the limiting case of $k \to 0$ (then $\langle|\eta_k|^2\rangle = k_B T/V\varphi_{\eta\eta}$) or $\tau \to 0$. In the fluctuation range, according to the principle of scale invariance, the only distance parameter describing the fluctuations is the correlation length r_c. Therefore, in the general case of one-component crystals with a single order parameter the dependence of $\langle|\eta_{\boldsymbol{k}}|^2\rangle$ on k in cubic crystals is determined by a function $g(x)$ of a variable kr_c so that we obtain the following expressions for $\langle|\eta_{\boldsymbol{k}}|^2\rangle$ and its Fourier transform $\xi(\boldsymbol{r})$:

$$\langle|\eta_{\boldsymbol{k}}|^2\rangle = \frac{1}{N\Gamma^{\pm}|\tau|^{\nu}} g(kr_c) , \quad \xi(\boldsymbol{r}) = \frac{1}{r^{d-2+\eta_0}} \tilde{g}\left(\frac{r}{r_c}\right) . \tag{4.1.35}$$

Here r_c is the correlation length which describes the second moment of the function $\xi(\boldsymbol{r})$, the function $g(x)$ for $x \to 0$ and $x \to \infty$ has the following values: $g(0) = 1$, $dg/dx^2|_{x\to 0} = -1$, and $g(x \to \infty) = \text{const}/x^{2-\eta_0}$. These values lead (with the use of (4.1.35)) to the correct limiting expressions for $\langle|\eta_{\boldsymbol{k}}|^2\rangle$ for $k \to 0$ and $\tau \to 0$ while the function $\tilde{g}(x) \to \text{const}$ for $x \to 0$ (for $r \ll r_c$ the function $\xi(\boldsymbol{r})$ must be the same as at the transition point) and $\tilde{g}(x) \to \text{const}\, x^{-(3-\alpha)/2+\eta_0} \exp(-\xi x)$ for $x \to \infty$ (we have $\zeta \cong 1$ and the difference between ζ and unity is related to the above definition of r_c in (4.1.35)) and then for $r \gg r_c$ the function $\xi(\boldsymbol{r})$ is given by (4.1.6).

Various approximate equations for $g(x)$ have been suggested which would make it possible to describe the behavior of $\langle|\eta_{\boldsymbol{k}}|^2\rangle$ for all k and τ (see, for example, [4.18–24]). According to [4.18], in the Ising model we can write

$$g(x) \approx \frac{1}{\left[1 + \left(1 - \frac{1}{2}\eta_0\right)^{-1} x^2\right]^{1-\eta_0/2}} \approx \frac{(1 + \Phi^2 x^2)\eta_0/2}{1 + \left(1 + \frac{1}{2}\eta_0\Phi^2\right) x^2} , \tag{4.1.36}$$

where Φ varies between 0.06 and 0.09 for $d = 3$. For small x these equations accurately describe the terms proportional to x^2 (which correspond to the function $\xi(\boldsymbol{r})$ varying as $r^{-(d-1)/2} \exp(-\zeta r/r_c)$). For large x the first equation (4.1.36) which is simpler but less accurate describes correctly the dependence on x but for $d = 3$ in the Ising model it overestimates the coefficient at $x^{-(2-\eta_0)}$ by about 10–15%.

The second equation (4.1.36) and (4.1.35) indicate that $\langle|\eta_{\boldsymbol{k}}|^2\rangle$ for the given $\boldsymbol{k}$ reach the maximum at a temperature somewhat higher than T_0 for $\tau \sim k^{1/\nu}$. A more exact expression for $g(x)$ for large x includes, in addition to the term proportional to $x^{-(2-\eta_0)}$, a term [4.20) proportional to $x^{-[2-\eta_0+(1-\alpha)/\nu]}$ which gives rise, as shown by (4.1.34, 35), to a $\tau^{-\alpha}$-type singularity in the derivative $\partial\langle|\eta_{\boldsymbol{k}}|^2\rangle/\partial\tau$ [4.15].

The critical indices for the two-dimensional Ising model were found analytically by *Onsager* (see [1.31]). In the three-dimensional case they were calculated numerically with a sufficiently high accuracy from the high-temperature or low-temperature expansions for various characteristics which included large numbers of terms with known coefficients. Such analysis of the dependences of the coefficients in an expansion on their numbers allows us to determine the nature of sufficiently simple features (for instance, if the function has a simple pole as is the case with the logarithmic derivative of $\varphi_{\eta\eta}$) and to calculate the critical indices. Wilson's method of renormalization group was employed for developing analytical techniques for approximately calculating the critical indices (see [4.11, 14, 15, 25–28]). In the more completely developed method of the ε-expansion [4.26] the ordering problem for four-dimensional space is solved [2.58] and the expansion in powers of the parameter $\varepsilon = 4 - d$ is used (it equals unity in three-dimensional systems with short-range forces. Numerical techniques have also been developed for calculating the critical indices with the use of the method of renormalization group (see reviews [4.11, 14, 15]). Another approach to calculating critical indices is based on the technique of collective variables [4.29].

Table 4.1 presents calculated results for some critical indices in (4.1.32, 33). The results for three-dimensional crystals ($d = 3$) have been obtained from the analyses of the high-temperature expansions with the exception of the studies [4.31, 32] where the expansions in powers of the coupling constant (derived with the renormalization group method) were analyzed with the Borel summation technique and an asymptotic formula [4.36] for the highest power terms in the perturbation theory series; another exception was the study [4.33] where such an analysis was performed for the ε-expansion (where the terms of the order of ε^4 were retained).

Table 4.1

d n	α	β	γ	ν	η_0	Refs.
2 1	$O(\ln)$	1/8	7/4	1	1/4	
3 1	0.125 ± 0.02	0.312 ± 0.0005	$1.25^{+0.003}_{-0.007}$	$0.638^{+0.002}_{-0.008}$	0.05 ± 0.01	4.30
	0.110 ± 0.05	0.325 ± 0.0015	1.241 ± 0.0020	0.630 ± 0.0015	0.031 ± 0.004	4.31
			1.2412 ± 0.0011	0.6302 ± 0.0003	0.040 ± 0.001	4.32
				0.628 ± 0.002	0.0333 ± 0.0001	4.33
3 2	-0.02 ± 0.05	0.35 ± 0.02	1.33 ± 0.02	0.670 ± 0.006	0	4.34
	-0.007 ± 0.006	0.3455 ± 0.0020	1.316 ± 0.0025	0.669 ± 0.0020	0.033 ± 0.004	4.31
			1.3172 ± 0.0007	0.672 ± 0.002	0.041 ± 0.0025	4.32
				0.666 ± 0.004	0.0352 ± 0.0001	4.33
3 3	-0.14 ± 0.06	0.38 ± 0.03	1.38 ± 0.02	0.715 ± 0.01	0.04 ± 0.01	4.34
	-0.115 ± 0.009		$1.42^{+0.02}_{-0.01}$	0.725 ± 0.015		4.35
		0.3645 ± 0.0025	1.386 ± 0.0040	0.705 ± 0.030	0.033 ± 0.004	4.31
			1.390 ± 0.005	0.708 ± 0.003	0.042 ± 0.0020	4.32
				0.700 ± 0.007	0.0354 ± 0.0001	4.33
Landau theory	0	1/2	1	1/2	0	
Tricritical point	1/2	1/4	1	1/2	0	

It can be seen from the data in Table 4.1 that the critical indices for the two-dimensional system differ significantly from the results derived with the Landau theory. For three-dimensional systems the difference is less striking but also considerable. The indices α and η_0 which vanish in the Landau theory remain small in the fluctuation region and we have $\alpha(n = 1) > 0$ and $\alpha(n \geq 2) < 0$. According to (4.1.35, 36), smallness of η_0 means that there is little difference between the dependence of $\langle|\eta_{\boldsymbol{k}}|^2\rangle$ on k and the Lorentz curve derived in the Landau theory (see (4.1.4.5)). But there are significant differences between the fluctuation range and the range of applicability of the Landau theory in the temperature dependence of $\varphi_{\eta\eta}$ and the coefficient of k^2. Significant differences are observed also between the ratios of the critical coefficients Γ^-/Γ^+ for $T < T$ and $T > T_0$ (which also are universal characteristics). In the Landau theory (4.1.4,5) yield $\Gamma^-/\Gamma^+ = 2$ while in the fluctuation range $\Gamma^-/\Gamma^+ = 3.7$ for $d = 2$ and $n = 1$ and $\Gamma^-/\Gamma^+ = 5.1$ for $d = 3$ and $n = 1$, that is, $\langle|\eta_{\boldsymbol{k}}|^2\rangle$ for a given $\boldsymbol{k}$ depends much sharper on T at $T < T_0$ than at $T > T_0$ [4.34].

The results derived for $\varphi_{\eta\eta}$, and r_c, and $\langle|\eta_{\boldsymbol{k}}|^2\rangle$ in the Ising model for one-component crystals with a one-component order parameter can also be used for determining $d^2\varphi/dc^2$, r_c and $\langle|c_{\boldsymbol{k}}|^2\rangle$ in undistorted solutions in the vicinity of the critical point on the decomposition curve for the critical concentration $c = c_c$. These results can also be employed for finding $\langle|\eta_{\boldsymbol{k}}|^2\rangle$ near the second-order phase transition point in solid solutions. Note that for the concentration $c = c_m$ for which the transition temperature $T_0(c)$ is maximal and $\varphi_{c\eta} = 0$, according to (1.2.26) the mean square $\langle|\eta_{\boldsymbol{k}}|^2\rangle$ is given by the same expression as in the case of one-component crystals and described by the same critical indices. For concentrations differing from c_m, however, according to [4.37], for $\alpha > 0$ the critical indices are changed somewhat so that $\varphi_{\eta\eta}$ and $\varphi_{\eta\eta} - \varphi^2_{c\eta}/\varphi_{cc}$ are described by the index $\tilde{\gamma} = \gamma(1-\alpha)^{-1}$. This change occurs, however, only in a narrow range of $|\tau| \lesssim [d\ln T_0/d\ln\mu_B]^{1/\alpha}$ (here μ_B is the chemical potential of the atoms B) [4.15]. This range is narrow because α is much smaller than unity ($1/\alpha \cong 9$ for $d = 3$ and $n = 1$) while the expression in square brackets is usually smaller than unity.

The indices for $n > 1$ in Table 4.1 correspond to the isotropic Heisenberg model, that is, they were derived without taking into consideration the invariant $\sum_\mu \eta_\mu^4$ in (4.1.22) and with inclusion of the isotropic gradient terms of the type of $(\nabla\eta_\mu)^2$. In the case of $n = 3$ the cubic anisotropy does not affect the critical indices for $|\tau| \to 0$ but it can "delay" greatly the approach to the limit of $|\tau| \to 0$ (and sometimes it can cause a phase transition of the first order) [4.14, 38, 39].

Examples of ordered structures described in the Heisenberg model with $n = 2$ are the incommensurate structures in low-symmetry crystals described by only two wave vectors $\boldsymbol{k}_0$ and $-\boldsymbol{k}_0$ (the order parameters η_μ are two complex amplitudes) [4.40].

In systems with multicomponent long-range order parameters in the fluctuation range there can occur characteristic effects which produce qualitative differences between theoretical predictions based on the expansion (4.1.1) and the actual behavior of these systems. One such difference is the change in the order of the phase transition caused by the fluctuations. In systems with large n there can be produced numerous ordered domains with different orientations and this makes possible intense development of fluctuations at $T > T_0$ and a strong coupling of the fluctuations of the different η_μ components. These fluctuations cause an effective decrease (and reversal of the sign) of the coefficients A_4 in the expansions of the type of (4.1.22) when the temperature approaches the point T_0, that is, the fluctuations may give rise to phase transitions of the first order. In some systems such first-order phase transitions occur for given ranges of the coupling parameters (which differ from the prediction of the theory using the expansion (4.1.22) with constant coefficients) while second-order phase transitions occur for parameters varying in other ranges [4.26, 41, 42]. It is significant, however, that for particular symmetry types of the long-range order with $n \geq 4$ the fluctuations always (irrespective of the interaction parameters) cause a first-order phase transition though second-order phase transitions are possible according to the results of the symmetry analysis performed in the framework of the Landau theory [4.43–50]. (In some circumstances the critical fluctuations, especially in systems of low dimensionality, may give rise to the reversed effect, that is, the transformation of the first-order phase transitions into continuous transitions [4.51]. For instance, such a transformation of the transition nature occurs near the tricritical point, that is, this point is shifted owing to the fluctuations. The first-order phase transitions caused by the critical fluctuations can be transformed into continuous transitions if we take into account the interactions which break down the symmetry; see, for instance, [4.47, 52, 53].)

Another qualitatively different effect which occurs in the fluctuation range of the systems with $n \geq 2$ is the possibility of enhancement of the crystal symmetry at the exact point of the second-order phase transition in comparison with the high-symmetry phase (in the Landau theory these symmetries are always identical). For instance, in a system with $n = 2$ which has a fourth-order symmetry axis C_4 this enhancement of symmetry occurs if for $T \to T_0$ strongly developed fluctuations cause vanishing of the coefficient A_4 in the expansion (4.1.22) and then the system is invariant with respect to rotation by any angle (not only by 90°) in the η_1, η_2 plane [4.11, 26, 28, 41, 42].

Systems with Long-Range Forces. The critical indices are changed significantly if the interaction between atoms varies with distance as r^{-3}. It has been noted in Sect. 4.1.1 that the suppression of the critical fluctuations by the dipole forces in uniaxial ferroelectrics restores for $|\tau| \to 0$ the validity of the results of the Landau theory with small additional logarithmic terms. In the typical case for ferroelectrics, when the dipole forces are strong (we have $a \sim T_0^{-1}$ in (4.1.26)), an analysis [2.58] of the main diagrams of the expansion in A_4 in the entire range of $|\tau| \ll 1$ has shown that

$$
\begin{aligned}
&\left(\frac{\partial^2 \varphi}{\partial P_z^2}\right)^{-1} = \varkappa_{zz} = \frac{v}{k_B T \Gamma_P^{\pm}} \frac{\psi(\tau)}{|\tau|}, \quad P_z^2 \sim (-\tau)[\psi(\tau)]^2, \\
&\delta C_p \sim [\psi(\tau)]^{-1}, \\
&\psi(\tau) = \left[1 + \zeta \ln \frac{\zeta'}{|\tau|}\right]^{1/3},
\end{aligned}
\tag{4.1.37}
$$

Here, as in the Landau theory, $\Gamma_p^- / \Gamma_p^+ = 2$ and ζ and ζ' are dimensionless constants while $\Gamma^\pm$ are dimensional constants. The same result was obtained with the method of renormalization group [4.54,55]. It can be seen from (2.2.10) and (4.1.37) that the correlation function $\langle |P_{kz}|^2\rangle$ for $\varkappa_{zz}\alpha k^2 \ll 1$ is determined by the equation (4.1.26) (where $a = k_B \Gamma_P^{\pm}/v$) with the additional factor $\psi(\tau) \approx [\zeta \ln(\zeta'|\tau|)]^{1/3}$ at $(T - T_0)$ while for $\varkappa_{zz}\alpha k^2 \gg 1$ this function is given by (4.1.26) with an accuracy to within small terms of the order of $[\ln(\text{const}/k)]^{-1}$ [4.54].

If the coefficient $g = 4\pi(\varepsilon_{xx} a T_0)^{-1}$ which describes the dipole forces according to (4.1.26) is small then the equations (4.1.37) which depend on the dipole interaction are valid only for very small $|\tau| \ll \tau_0$ and for $1 \gg \tau \gg \tau_0$ this interaction can be ignored. Then the critical indices correspond to a system with short-range forces (equations of the Landau theory similar to (4.1.26) but with different coefficients and without logarithmic additional terms are also valid in the range of $1 \gg \tau \gg (d_0/r_0)^6$ for the case when the range of the "short-range" forces is $r_0 \gg d_0$). In this and other cases when additional weak coupling (described by the coupling constant g) causes changes in the universality class and in the critical indices, there is an intermediate range of τ in which the system is described by an additional critical index ψ [4.56,57], apart from the conventional indices $\gamma, \beta, \ldots$. For instance, the susceptibility in this range can be written as

$$\varkappa_{zz} = \frac{1}{\tilde{\tau}^{\gamma}} \tilde{\varkappa}\left(\frac{g}{\tilde{\tau}^{\psi}}\right), \quad \tilde{\tau} = \frac{T - T_0(g=0)}{T_0(g=0)}, \tag{4.1.38}$$

where the function $\tilde{\varkappa}(x) \to \text{const}$ for $x \to 0$ and $\tilde{x}(x) \to (x_0 - x)^{-\gamma'}$ for $x \to x_0 - g/\tau_0^{\psi}$ (here $\tau_0 = \tilde{\tau}(T = T(g))$ and γ and γ' are the critical indices for $g \to 0$ and $\tau \to 0$)). According to (4.1.38), the characteristic temperature $\tau_0 \sim g^{1/\psi}$. The behavior of $\varkappa_{zz}$ as a function of temperature in the intermediate range can be conveniently described with the effective parameter $\gamma_{\text{eff}} = -\partial(\ln \varkappa_{zz})/\partial(\ln \tau)$. For instance, for the uniaxial ferroelectrics we have [4.58] $\psi = \gamma$ and the intermediate range corresponds to the variation of τ by several orders of magnitude.

In cubic ferroelectrics the dipole forces suppress only the longitudinal fluctuations and therefore the critical indices differ from the results obtained with the Landau theory. In the isotropic dipole model for large g these critical indices are close to those in the Heisenberg model (see Table 4.1) [4.54] while for small g they exhibit a characteristic dependence on τ in the intermediate range [4.14]. The cubic anisotropy makes the behaviour of these systems in the critical range more complicated [4.14,60].

In crystals with a linear piezoelectric effect (and in ferroelastic crystals) in which the relationship between the one-component order parameter η and the strains u_{ij} is described by the term $L_{hij}\eta u_{ij}$ in the equation for the thermodynamic potential density φ, as discussed in Sect. 4.1.1, the suppression of the order parameter fluctuations by the elastic forces makes insignificant the additional terms for the results obtained with the Landau theory for $\tau \to 0$. For small $|L_{\eta ij}|$ and significant $|\tau|$ we may, however, obtain the fluctuation range of critical indices as well as the intermediate range). The fluctuation range may be observed in the presence of the piezoelectric effect in the paraelectric phase in crystals described by a multicomponent order parameter [4.61].

When the term $L\eta^2 u_{ij}$ in the expression for φ which describes the relationship between η and the strain is of the second order in η then the resulting striction effects in systems with an infinite specific heat can change the character of the phase transitions. The modulus of dilatation K has been found [4.62] to decrease and vanish for $\delta C_v \to \infty$ so that the resulting elastic instability causes a first-order phase transition at $\tau = \tilde{\tau} > 0$. A more detailed analysis [4.63, 64] has shown that though δC_v remains finite, a first-order phase transition occurs in systems where the singular part of the specific heat δC_p (determined by the fluctuations of the order parameter) tends to infinity when we take into account the coupling of fluctuations to the acoustic phonons and the non-vanishing shear modulus μ. Such a transition occurs in the range where

$$\begin{aligned} &\frac{4\mu}{K(3K+4\mu)}\varkappa_0^2 T_0 \delta C_p(\tau) = -\frac{4\mu}{K(3K+4\mu)}\delta K(\tau) \sim 1 , \\ &\varkappa_0 = -K\frac{1}{T_0}\frac{\partial T_0}{\partial P} , \end{aligned} \tag{4.1.39}$$

that is, the singular part $\delta K(\tau)$ of the dilatation modulus is comparable with its non-singular part.

The magnitude of these striction effects depends on the parameter $\varkappa_0$ which is determined by the behavior of the transition point as a function of the pressure. The smallness of the critical index α and the parameter $k_B T_0/vK$ (which is of the order of 10^{-1} to 10^{-2}) determine the smallness of the shift of the point $\tilde{T}_0$ of the first-order phase transition with respect to T_0 in the typical case of $\varkappa_0 \lesssim 1$. Thus, outside a narrow range with the width of the order of $|\tilde{T}_0 - T_0|$ the additional striction terms (in particular, the additional terms for the correlation functions [4.65]) are small so that the resulting first-order phase transition is close to the second-order transition. These additional striction terms may, however, be large in systems with large $\varkappa_0$. The cubic anisotropy [4.66] and the anisotropic coupling of the strains and the order [4.67, 68] in systems with multicomponent long-range order parameters give rise to these striction effects and to first-order phase transitions.

4.2 Diffuse Scattering by Critical Fluctuations in Perfect Crystals

The sharp increase in the probability of the fluctuation waves of the long-range order parameters near the second-order phase transition point discussed in Sect. 4.1 must give rise to unusually significant large-scale inhomogeneities of the structure amplitude and to anomalous growth of the diffuse scattering intensity in the vicinities of the superlattice points of the reciprocal lattice [1.2]. In a similar way, the growth of the concentration fluctuations near the critical point on the decomposition curve or on the curve of the second-order phase transitions must cause anomalous diffuse scattering in the vicinities of the reciprocal lattice points and at small scattering angles [1.3; 4.69, 70]. When we take into account the strains, the high-intensity scattering must correspond both to the superlattice and regular lattice reflections near the conventional second-order phase transition in some crystal systems [1–10, 16].

The anomalous diffuse scattering by critical fluctuations of the long-range order parameters and the concentration can be conveniently analyzed with the method of the fluctuation waves with the use of the general equations such as (3.2.10) and (3.6.7) which express the intensity I_1 in terms of $\langle|\eta_{\boldsymbol{k}}|^2\rangle$ or $\langle|c_{\boldsymbol{k}}|^2\rangle$ found from the results of Sect. 4.1. We shall apply below this approach to anomalous diffuse scattering in a simpler case first, when the striction strains and the long-range forces can be ignored in the analysis of the fluctuations of the order parameters and later we shall discuss the above respective effects.

Anomalous Scattering in Systems Without Striction Effects. In the simplest case of the one-component crystals with a single long-range order parameter $\eta_\mu \equiv \eta$ (for $n = 1$) which reduces the translational symmetry, the scattering intensity at small distances $\boldsymbol{q}$ from the superlattice points is given by (3.6.7) in which only one term is retained in the sum over μ. If the striction effects are insignificant then in (3.6.7) we can take $A_{\mu\boldsymbol{q}} = 0$ and find $\langle|\eta_{\boldsymbol{q}}|^2\rangle$ without taking into account the long-range elastic forces, that is, from (4.1.4, 5, 8) in the range of validity of the Landau theory and from (4.1.35, 36) in the fluctuation range. The structure amplitude $f = \chi_\eta$ for the superlattice reflections is proportional to the order parameter η (see (3.6.3)). Therefore, if we take into account the expression (4.1.32) for η we can identify in (3.6.7) the regular reflection intensity

$$I_0 = 8\pi^3\frac{N}{v}\chi^2\eta_0^2|\tau|^{2\beta}\delta(\boldsymbol{q}) \quad (T < T_0)\,, \quad I_0 = 0 \quad (T > T_0) \tag{4.2.1}$$

and the diffuse scattering intensity I_1. According to (3.6.10) and (4.1.3, 7), we obtain for the latter in the range of validity of the Landau theory

$$I_1 = \frac{N|\chi|^2}{\Gamma^\pm + \tilde{\alpha}_{ij}q_iq_j}\,, \quad \Gamma^+ = \frac{va}{k_B}\,, \quad \tilde{\alpha}_{ij} = \frac{v\alpha_{ij}}{k_BT_0}\,. \tag{4.2.2}$$

In the Landau theory $\tilde{\alpha}_{ij} \sim r_0^2$ are independent of $\tau = (T - T_0)/T_0$ (where r_0 is the range of the interaction between atoms) and the coefficients Γ^+ and Γ^- differ below and above the transition point. Near a conventional second-order phase transition we have $\Gamma^- = 2\Gamma^+ = 2av/k_B$ while at the tricritical point (for $r = A_4^2(4|A_2|A_6)^{-1} \ll 1$) we have $\Gamma^- = 4\Gamma^+$. In the vicinity of this point at $T < T_0$ we obtain from (4.1.7) for (4.2.2) $\Gamma^- = (4 + 4r - 4\sqrt{r^2 + r})\Gamma^+$ which significantly depends on T and P. (Though the Landau theory is always applicable at the tricritical point T_c, P_0 [4.9] owing to the fluctuation effects it becomes invalid in a certain range $|\tau|$ around the phase transition line whose width grows with $|T_0 - T_c|$. The intensity I_1 is described by more general equations than (4.2.2) in this range and the intermediate range around it [4.71]. The fluctuation effects also cause bending of the ACB line in Fig. 4.2 at the tricritical point C for the solutions indicated by the curve CB'.)

Equation (4.2.2) gives also the diffuse scattering intensity for solid solutions with a one-component order parameter in the range of applicability of the Landau theory. It must only be born in mind that according to (4.1.10) the coefficient $\Gamma^- = 2M\Gamma^+$ where $M < 1$ and it can be seen from (4.1.11) to be very small in the vicinity of the tricritical point C (see Fig. 4.2). In $A - B$ binary solutions $\chi = \zeta(f_A - f_B)$ where $\zeta \sim 1$ (in alloys of the type of β brass $\zeta = 1$).

In the fluctuation range (3.6.7) and (4.1.35, 36) yield the following equations for the intensity of diffuse scattering by cubic crystals:

$$\begin{aligned} I_1 &= \frac{N|\chi|^2}{\Gamma^\pm|\tau|^\gamma} g(qr_c) \approx \frac{N|\chi|^2}{\Gamma^\pm|\tau|^\gamma\left[1+\frac{q^2r_c^2}{1-\eta_0/2}\right]^{1-\eta_0/2}} \\ &\approx \frac{N|\chi|^2}{\Gamma^\pm|\tau|^\gamma}\frac{(1+\Phi^2q^2r_c^2)^{\eta_0/2}}{1+(1+\eta_0\Phi^2/2)q^2r_c^2}\,, \quad r_c = r_{c0}^\pm/|\tau|^{-\nu}\,. \end{aligned} \tag{4.2.3}$$

Equation (4.2.2) can be regarded as a special case of (4.2.3) for $\gamma = 1$, $\nu = 1/2$, $h_0 = 0$, and $\alpha = \Gamma^\pm(r_{c0}^\pm)^2$. The correlation function $g(qr_c)$ in it has been approximated with (4.1.36) [4.8].

As noted in the discussion of (4.1.35, 36), this approximation is accurate for small x while for large x it describes accurately enough (an error between 10 and 20%) the main term $g_0x^{-(2-\eta_0)}$ in $g(x)$ but the additional terms $g_1^{\pm 2-\eta_0+(1-\alpha)/\nu}/x$ and $g_2^{\pm 2-\eta_0+1/\nu}/x$ differ from the appropriate terms in the expansion (4.136) in powers of $1/x$ [4.20]. Accordingly, for large qr_c we obtain

$$I_1 = \frac{N|\chi|^2}{\Gamma^\pm(qr_{c0}^\pm)^{2-\eta_0}}\left[g_0 + \frac{g_1^\pm|\tau|^{1-\alpha}}{(qr_{c0}^\pm)^{\frac{1-\alpha}{\nu}}} + \frac{g_2^\pm|\tau|}{(qr_{c0}^\pm)^{1/\nu}} + \dots\right] \tag{4.2.4}$$

$(qr_c \gg 1)$.

The relationship (4.1.34) between the critical indices has been taken into account here.

According to (4.2.2, 3), near a phase transition of the second order (for $|\tau| \ll 1$) we obtain symmetric high-intensity peaks of diffuse scattering intensity $I_1(\boldsymbol{q})$ whose maxima correspond to the superlattice points of the reciprocal lattice (at $\boldsymbol{q} = 0$). The respective equal intensity surfaces are ellipsoids (or spheres in cubic crystals for $n = 1$) around the superlattice points. It can be seen from (4.2.2) that in the range of applicability of the Landau theory, the behaviour of I_1 as a function of q for $n = 1$ is described by Lorentzian curves. As we approach the transition point the curves become narrower and the intensity I_{1m} at the peak maximum grows as $1/|\tau|$ while the peak width decreases as $\sqrt{|\tau|}$ (see Fig. 4.3). According to (4.2.2), $I(\tau)$ must exhibit a maximum (for a given $\boldsymbol{q}$) at the transition point where its plot must have a break. In one-component crystals the intensity I_1 decreases at a faster rate for $T < T_0$ than for $T > T_0$ (the rate of decrease is twice as great for a normal second-order phase transition and four times greater for a tricritical point). In solid solutions the ratio $2M$ between these rates is lower (according to (4.1.10) $M < 1$) and it decreases with increasing distance from the maximum of the function $T_0(c)$ so that near the critical point C in Fig. 4.2 the intensity I_1 decreases with growing $|\tau|$ at a much slower rate in the ordered phase than in the disordered phase.

In the fluctuation range the dependence of I_1 on q can be seen from (4.2.3, 4) to differ somewhat from a Lorentzian shape and for large q (and for all q at the transition point) I_1 varies not as $1/q^2$ but as $1/q^{2-\eta_0}$, that is, at a slower rate. These differences are small, however, owing to the smallness of η_0 (see Table 4.1). The temperature dependence of I_1 varies considerably greater. The intensity I_{1m} at the maximum is now proportional to $|\tau|^{-\gamma}$ where $\gamma = 5/4$ for $d = 3$ and $n = 1$ and I_{1m} decreases at lower temperatures at a much faster rate than at $T > T_0$ (for the one-component crystals [4.34] with $d = 3$ and $n = 1$ the ratio $\Gamma^-/\Gamma^+ \approx 5.1$ instead of 2 as in the Landau theory). The same ratio is obtained between the rates of decrease of the intensity I_1 with increasing $|\tau|$ for $\boldsymbol{q} \neq 0$ and $|\tau|^{\nu} \gg q_{c0}^{\pm}$. In the immediate vicinity of the transition point for $|\tau|^{\nu} \ll q r_{c0}^{\pm}$ the intensity I_1 is demonstrated by (4.2.4) to have the same dependence on τ as does the energy, that is, $dI_1/d\tau$ is proportional to $|\tau|^{-\alpha}$ and for $\alpha > 0$ it tends to infinity at the transition point [4.15]. The last equation (4.2.3) indicates that the maximum of the plot of I_1 as a function of τ for a given $\boldsymbol{q}$ is above the transition point for $\tau \sim q^{1/\nu}$ (that is, for $r_c(\tau)q \approx 6$ for the bcc lattice in the nearest-neighbour approximation of the Ising model [4.18]).

For crystals with a multicomponent order parameter the equation for the intensity I_1 is more complicated and has the general form of (3.6.8) where in the absence of strain effects $\boldsymbol{A}_{\mu q} = 0$ and according to (1.2.19) for $q \to 0$ we have $\varphi_{\mu\mu'}^{-1} = \varphi_{q\mu\mu'}^{-1}$, tending to $\varphi_{\mu\mu'}^{-1}$, (which is the inverted matrix of the derivatives $\varphi_{\mu\mu'} = \partial^2\varphi/\partial\eta_\mu\partial\eta_{\mu'} \sim \tau^{\nu}$). Therefore, the intensity I_{1m} at the maximum grows as $|\tau|^{-\gamma}$ for $T \to T_0$ as for $n = 1$ but the equal-intensity lines have a more complicated shape and are not spherical even in cubic crystals. The ratios between the intensities I_1 for $\tau < 0$ and $\tau > 0$ are also changed. In the range of applicability of the Landau theory $\varphi_{q\mu\mu'}$ are given by (1.2.19) and the relationships between $\varphi_{\mu\mu'}$ and $\alpha_{\mu\mu' ij}$ with different μ, μ' and, hence, the dependence of I_1 on $\boldsymbol{q}$ and τ, is determined by the long-range order symmetry. For instance, for a three-component order parameter when we have $A_4'' < 0$ and $A_4' + A_4'' > 0$ in the expansion (4.1.22) and equal numbers of domains are produced under ordering; $\eta_1 = \pm\eta$, $\eta_2 = \eta_3 = 0$; $\eta_2 = \pm\eta$, $\eta_1 = \eta_3 = 0$; $\eta_3 = \pm\eta$, $\eta_1 = \eta_2 = 0$ we obtain from (3.6.7) and (4.1.22, 23) for $q \to 0$

$$I_1 = \frac{N|\chi|^2}{\Gamma^{\pm}|\tau|}, \quad \Gamma^+ = \frac{av}{k_B}, \quad \Gamma^- = \frac{av}{k_B}\left(\frac{2A_4'}{3|A_4''|} - \frac{1}{2}\right), \tag{4.2.5}$$
$$|\chi|^2 = |\chi_1|^2 + |\chi_2|^2 + |\chi_3|^2 \quad (q \to 0)$$

(for $\tau > 0$ the equation (4.2.5) is valid also for other relations between A_4' and A_4''). The numerical calculations with (3.6.8) and (1.2.19) can yield the distributions I_1

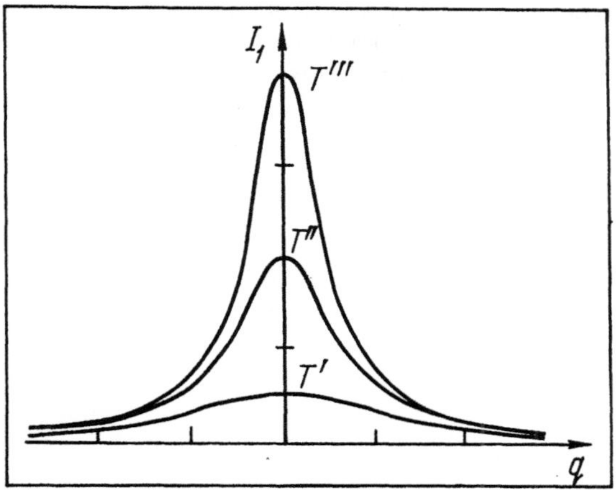

Fig. 4.3. Intensity distribution for the anomalous diffuse scattering near second-order phase transition ($T' > T'' > T''' > T_0$)

for different τ in the range of applicability of the Landau theory for specific crystal systems with specific types of order.

Experimental Results on Anomalous Scattering. As can be seen from (4.2.1–4) the experimental data on the heights and widths of the $I_1(q)$ peaks for various τ enable us to calculate the critical indices γ and ν and the results on $I_0(\tau)$ give the index β. Only the diffraction results can yield directly information on the correlation length r_c and the index ν. The diffraction techniques in most cases are the most direct (though not always the most precise) tools for analyzing the susceptibility $\varkappa$, that is, the index γ, and the order parameter η, that is, the index β (with the exception of such crystal systems as ferroelectrics or ferromagnetics where η is linearly related to a real external field). The data on the behavior of $\partial I_1/\partial\tau$ at the transition point enables us in principle to find the index α (and the transition point).

The structure of the ordered state produced after the transition can be found from the positions and intensities of various peaks of the superlattice reflections and the anomalous diffuse scattering intensity. Figures 1.1a and 1.2 illustrate such structures produced by second-order phase transitions in ordering alloys. They have lower translational symmetry and give rise to the peaks of $I_0(\boldsymbol{Q})$ and $I_1(\boldsymbol{Q})$ in the vicinities of the new superlattice points $\boldsymbol{G}'$. We obtain the same scattering pattern in the case of orientational ordering if in the disordered phase the molecules have the same probabilities of two (or several) possible orientations while in the ordered phase half of the molecules are largely aligned along one direction and the other half along another direction (antiferro-orientational ordering). Systems with such phase transitions are illustrated by $NaNO_3$ (see Fig. 4.4) in the ordered state of which the rhombohedral unit cell contains two molecules while the unit cell of the disordered phase (shown by bold lines in the figure) contains only one molecule.

Under different conditions ferro-orientational ordering takes place and in the ordered phase all molecules have the greatest probability to have the same orienta-

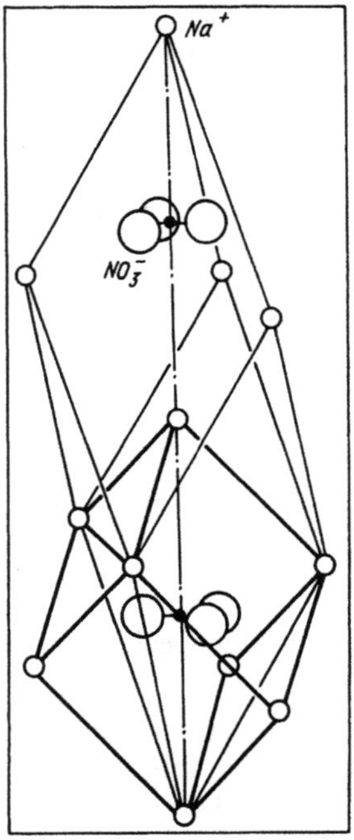

Fig. 4.4. Arrangement of points in the unit cell of the ordered phase of $NaNO_3$. Both orientations of the NO_3^- ion have equal probabilities in the disordered phase

tion. Then ordering involves a decrease in the point (rather than the translational) symmetry, there are no superlattice points (and thus no superlattice reflections) of the reciprocal lattice and the peaks of the anomalous diffuse scattering intensity $I_1(\boldsymbol{Q})$ correspond to the lattice points. Since the variation of the structure amplitude produced by the fluctuations of the order parameter η is still described by (3.6.3) the above expressions for I_1 are also applicable in this case. Ordering of this type occurs, for instance, in ND_4Cl where N and Cl make up a lattice similar to that of CsCl while the atoms D are arranged around an atom N in a tetrahedral structure. In the disordered phase such tetrahedra are randomly aligned along either of the directions $\langle 111\rangle$ and $\langle\bar{1}\bar{1}\bar{1}\rangle$ and in the ordered phase they are preferentially aligned along one of these directions (see Fig. 4.5).

In some systems the second-order phase transitions involve small displacements of the equilibrium positions for the atomic vibrations rather than an ordering process (which requires large displacements of atoms). Such displacement-type transitions typically occur in perovskites. These ABO_3 crystals (where A is a univalent or bivalent metal and B is a tetravalent or pentavalent metal) have a structure illustrated in Fig. 4.6 where the atoms A are at the vertices of the cubes, B are at their centers and the oxygen atoms O are at the centers of the cube faces. In some perovskites the averaged positions of the B and O ions are displaced in opposite directions in all cells by the same amount at temperatures below the transition from the cubic phase. For instance, in $BaTiO_3$, $PbTiO_3$ and $KNbO_3$ such displacements of the ions Ti and O occur in one of the $\langle 100\rangle$ directions (shown by arrows in Fig. 4.6). This gives rise to a spontaneous dipole moment $\boldsymbol{P}$ and the crystal

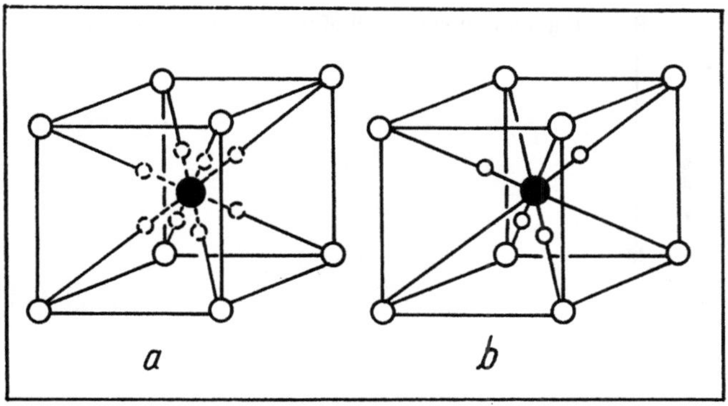

Fig. 4.5a,b. Unit lattice cell of ND_4Cl. (**a**) disordered phase, (**b**) one of two types of the ordered phase, large open circles denote Cl atoms, black circles denote N atoms

becomes ferroelectric. We may choose as the components of the order parameter the polarization components P_i or parameters proportional to the displacements of the ions. The transition produces a decrease in the point symmetry (the cubic crystal is transformed to a tetragonal phase) but the number of the cells and the translational symmetry remain unchanged. This is why the anomalous scattering intensity peaks occur around the points of the reciprocal lattice as in the case of the ferro-orientational transition.

In other perovskites, ions in alternating lattice cells are displaced in different directions, the translational symmetry of the lattice is decreased and we obtain superlattice points of the reciprocal lattice with anomalous scattering intensity peaks in their vicinities. For instance, in $SrTiO_3$, $KMnF_3$ and $LaAlO_3$ antiferrodistortional ordering consists in rotations by small angles of the octahedral structures of TiO_6, MnF_6 or AlO_6 around a [100]-type axis. This gives rise to distortions in the rotation planes shown in Fig. 4.7 (in the plane above the one shown in this figure rotation is in the opposite direction) which correspond to the superlattice point G' (1/2 1/2 1/2) [4.72, 73]. A series of phase transitions resulting in a decrease in the symmetry with decreasing T often occur in the same system.

Displacement-type phase transitions are physically due to the fact that the frequencies of some vibration modes with $\boldsymbol{k} = \boldsymbol{G}'$ tend to zero (in many systems they are transformed into dissipative modes at $T \to T_0$). The resulting instability at $T < T_0$ results in a condensation of the vibrations which gives rise to a static ordering wave. For instance, in $BaTiO_3$ the frequency of one of the transverse modes of the optical vibrations tends to zero for $\boldsymbol{k} = 0$ while in $SrTiO_3$ the frequency of the triply degenerate vibration Γ_{25} of the acoustic mode at the boundary of the Brillouin zone at $\boldsymbol{k} = (111)\pi/d$ tends to zero. These effects were analyzed in detail using inelastic neutron scattering [4.74–77].

Alloys with the structure of $L2_0$ (β brass) or DO_3 (Fe_3Al (in the ordered state) are classical cases of systems with second-order phase transitions of the order-disorder type which are characterized by a single order parameter and can be satisfactorily described in the framework of the Ising model with short-range in-

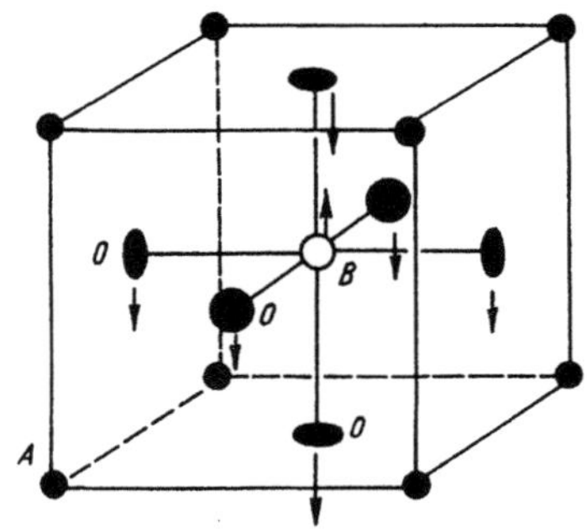

Fig. 4.6 Unit lattice cell of the cubic perovskite ABO_3

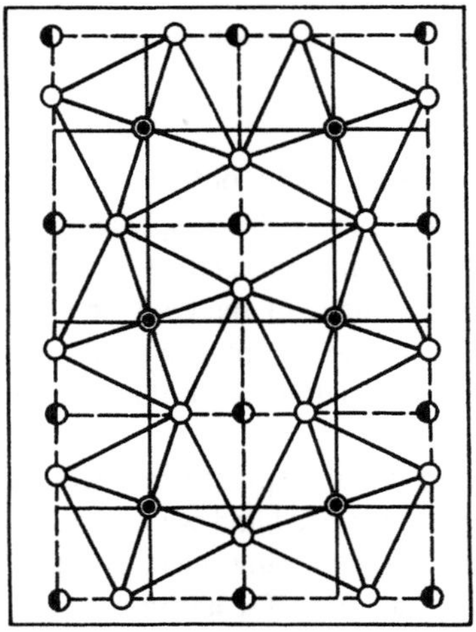

Fig. 4.7. Schematic of the ordered structure in $SrTiO_3$. Open circles denote oxygen atoms, half-black circles denote Ti Atoms, and circles with a black center dot denote Sr atoms

teraction. Anomalous scattering in such systems was analyzed in detail in neutron-scattering experiments with the alloy CuZn with a nearly stoichiometric composition near the order-disorder transition [4.78–81] and in X-ray scattering experiments with the Fe-29.2 at.% Al (at the peak of the $T_0(c)$ plot) near the transition $L2_0 \rightarrow DO_3$ [4.82]. The experimental results could be described by (4.2.3) (see, for instance, Fig. 4.8). The critical indices γ, ν and η_0 were found from the temperature dependences of the scattering intensities (in the ranges of $3 \cdot 10^{-3} < \tau < 3 \cdot 10^{-2}$ for CuZn [4.80] and $2 \cdot 10^{-4} < \tau < 2.5 \cdot 10^{-2}$ for Fe-Al [4.82]. The index β was found from the data on the intensities I_0 of the superlattice reflections for CuZn [4.83–85] and Fe-Al [4.86]. (The experimental separation of I_0 and I_1 must be analyzed carefully in view of the finite width of the $I_0(\boldsymbol{q})$ peak determined by the instrumental and dislocation-produced broadening (see Sect. 4.3.2), the smallness of the intensity I_0 and the considerable anomalous scattering intensity $I_1(\boldsymbol{q})$ (see, for instance, [4.80]). Determination of the indices β and γ from the overlapping distributions $I_0(\boldsymbol{q})$ and $I_1(\boldsymbol{q})$ is treated theoretically in detail in [4.87].)

The experimental results given in Table 4.2 agree approximately with the theoretical predictions for $d = 3$ and $n = 1$ (see Table 4.1) though β for CuZn proved to be somewhat smaller than the calculated result (the agreement between

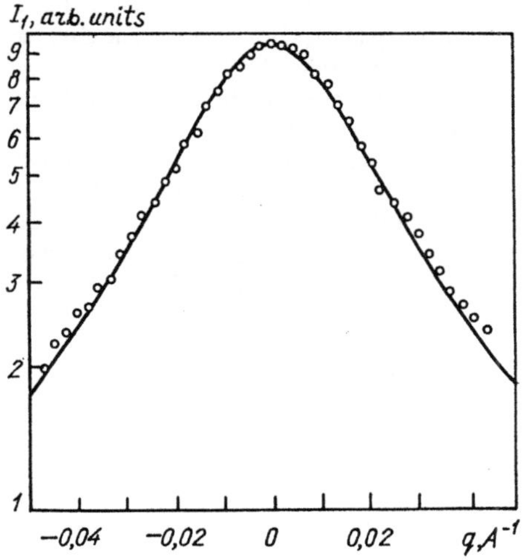

Fig. 4.8. Intensity distribution for diffuse scattering by the alloy Fe-29.2 at.% Al at $T = T_0 + 1.6\,\mathrm{K}$. The curve has been calculated from the last equation (4.2.3) [4.82]. (The intensity I_1 is given in arbitrary units.)

these β values is reached with the simplest approach to taking into account the compressibility of the crystal [4.88] in which $T_0 = T_0(V(\tau))$ is regarded as being dependent on the temperature through the variation of the volume owing to thermal expansion). According to the theoretical predictions for the fluctuation range the ratio $\Gamma^-/\Gamma^+ = 5.46 \pm 0.05$ for CuZn [4.81] is large (the analysis [4.90] of the series yields the ratio of 5.1 while the equation of state [4.91] gives $\Gamma^-/\Gamma^+ = [(1-2\beta)\gamma/2(\gamma-1)\beta^{\gamma-1}\gamma/\beta = 5.3$). The experimental results also agree with the theoretical predictions [4.18] in that the peak of I_1 as a function of τ is at $\tau = \tau_m > 0$ and τ_m grows with an increase in q [4.81]). In contrast to the critical indices and the ratios between the coefficients in (4.1.32), the coefficients $\Gamma^\pm$ ano $r_{c0}^\pm$ are not universal and depend on the chosen model. The experimentally calculated values $r_{c0}^\pm = (0.33 \pm 0.03)d$ for CuZn [4.80] and $r_{c0}^\pm = (3.6 \pm 0.01)d$ for Fe-Al [4.82] (where d is the edge length of the cubic cell) are smaller by 10–20% than the theoretical prediction for the nearest-neighbour approximation in the Ising model [4.18]. This approximation yields considerably smaller values of $\eta(T)$ than the experimental results in the range of $0.4T_c < T < 0.9T_0$ (see Fig. 4.9).

In the above systems (cubic crystals with $n = 1$) the equal-intensity surfaces around the superlattice points are spherical. The equal-intensity surfaces for non-cubic and especially highly anisotropic crystals are essentially nonequiaxial. Such highly anisotropic quasi-two-dimensional crystals are illustrated by the intercalated graphite compounds, such as $C_{12n}M$ (where $n = 2, 3, \ldots$) in which the layers of an alkali metal (M = K, Rb, Cs) are repeated after n hexagonal graphite planes [4.92, 93]. For instance, a phase transition of the second order was found [4.94, 95] to involve ordering of the K atoms in $C_{24}K$. The diffuse scattering intensity $I_1(\boldsymbol{Q})$ in this crystal was found both for $T > T_0$ and for $T < T_0$ to correspond to the two-dimensional system and the equal-intensity surfaces are shaped as cylinders perpendicular to the $(h_1\, h_2\, 0)$ plane [4.96] (a similar two-dimensional pattern of diffuse scattering intensity distribution was also observed in $C_{36}K$ and $C_{48}K$ at high temperatures [4.97]). At the same time the peaks of the superlattice reflec-

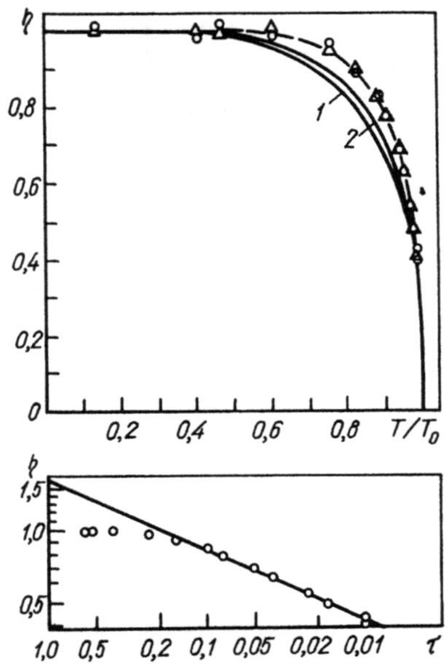

Fig. 4.9. The order parameter η as a function of temperature in CuZn [4.84, 85]. The dashed line gives the X-ray diffraction results [4.84]. Open triangles and circles present the neutron diffraction results [4.85] for the reflections (100) and (111). The solid lines are the theoretical plots derived in the Ising approximation for the rigid (*curve 1*) and compressible (*curve 2*) lattices. In the lower part of the figure the experimental data are approximated with the function $n = 1.64\tau^{0.293}$ (log-log scale) [4.85]

tions observed at $T < T_0$ are at the lattice points (and not on the lines) and correspond to three-dimensional long-range order both within the K planes and between them. The interaction between the atoms K in different planes is significantly weaker than the interaction between the atomic planes (it may be associated with the elastic forces [2.50; 4.98, 99]; note that in order to take them into account in a consistent way we must consider the strong elastic anisotropy of graphite and the effects of spatial dispersion discussed in Sect. 2.5 in [1.1]). Therefore, the critical index $\beta = 0.18 \pm 0.01$ found in [4.96] lies between the values of 0.312 and 0.125 calculated for the three-dimensional and two-dimensional Ising models (see Table 4.1). This temperature range of $10^{-3} < -\tau < 2 \cdot 10^{-1}$ corresponds probably to the intermediate range and the three-dimensional function $\eta(\tau)$ is realized at lower $-\tau$. Experiments with the similar systems $C_{24}Cs$ and $C_{24}Rb$ and a more detailed study of $C_{24}K$ demonstrated orientational ordering of the layers of the alkali metal with respect to graphite and production of a modulated structure in the metal layers which is not quite commensurate with graphite; the latter fact was found to cause displacements of carbon atoms in the ordered phase [4.100–102] (see also [4.103–107]).

Anomalous scattering under orientational ordering exhibits many features similar to those observed for ordering by alloys. In the absence of long-range elastic and dipole forces the equal-intensity surfaces are shaped as ellipsoids around the superlattice points. This is what was observed for anomalous scattering near the second-order phase transition point $T_0 = 548$ K upon ordering of the NO_3^- radicals in $NaNO_3$ (see Fig. 4.4). It can be seen from Fig. 4.10 that these ellipsoids [4.108]

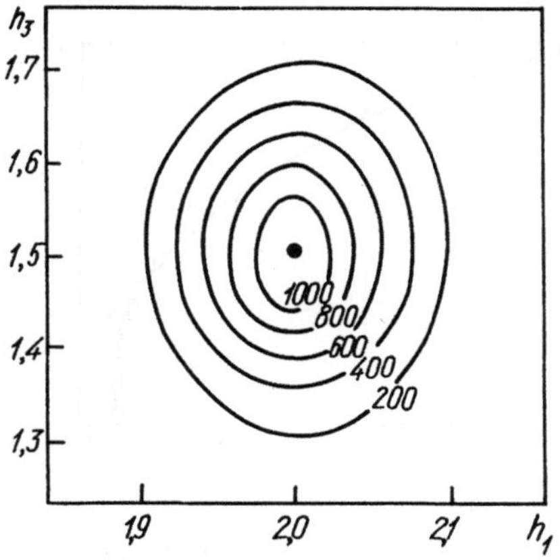

Fig. 4.10. Equal intensity lines in the plane (h_1, h_3) around the point (2 0 1.5) for X-ray scattering in the $NaNO_3$ crystal at $T = T_0 + 5$ K [4.108]

have a small eccentricity and the behaviour of I_1 as a function of q and $T - T_0$ is described by the plots similar to those in Figs. 4.3, 8.

On the other hand, the equal-intensity surfaces in some instances are significantly anisotropic owing to the interaction between the neighbouring molecules which has a complicated and strong dependence on their mutual orientations (in particular, caused by the purely geometric, that is, steric, factors). For instance, the intensity distributions of the critical neutron scattering near the point $T_0 = 27.0$ K of the orientational ordering of CD_4 were considerably extended along the $\langle 111 \rangle$ axes owing to the strong quasi-two-dimensional correlation between molecules in the $\{111\}$ planes (with $r_{c\perp} \approx 20$ Å and $r_{c\parallel} \approx 5$ Å at 27.6 K [4.109].

The orientational phase transition in ND_4Cl is an illustration of a transition close to the tricritical point. As noted above, the ferro-orientational ordering of the tetrahedral ND_4 structures (see Fig. 4.5) does not give rise to new superlattice points of the reciprocal lattice but causes critical scattering near the lattice points and changes the intensities of all regular reflections. Here the structure amplitude $f = f_0 + \eta\chi$ linearly depends on the order parameter $\eta = (N_{[111]} - N_{[\bar{1}\bar{1}\bar{1}]}/(N_{[111]} + N_{[\bar{1}\bar{1}\bar{1}]})$ (where the parameters N are the numbers of the tetrahedral structures with the orientations along [111] and $[\bar{1}\bar{1}\bar{1}]$, respectively). If we select the (221) reflection with a small (owing to accidental reasons) amplitude $f = f_0$ in the disordered phase and with a large χ we can analyze the pure effect of ordering on $I_0 \approx \text{const} \cdot \eta^2$ and I_1.

Such detailed analysis of neutron scattering by crystals of ND_4Cl (containing 93% D and 7% H) at various temperatures and pressures is reported in [4.110]. A first-order phase transition occurs under the atmospheric pressure ($T_0 = 242$ K) and a second-order phase transition occurs under high pressures. The tricritical point is at $P = P_c = (128 \pm 10)$ bars and $T = T_c = (250.2 \pm 0.1)$ K. In a significant range of $|\tau|$ around the transition far from the tricritical point the critical indices correspond to the values for the fluctuation range (see Table 4.1). For instance, for $P = 4140$ bars in the range of $7 \cdot 10^{-4} < -\tau < 5 \cdot 10^{-2}$ the intensity I_1 varies as $\tau^{2\beta}$ where $2\beta = 0.61 \pm 0.04$ (see Fig. 4.11a).

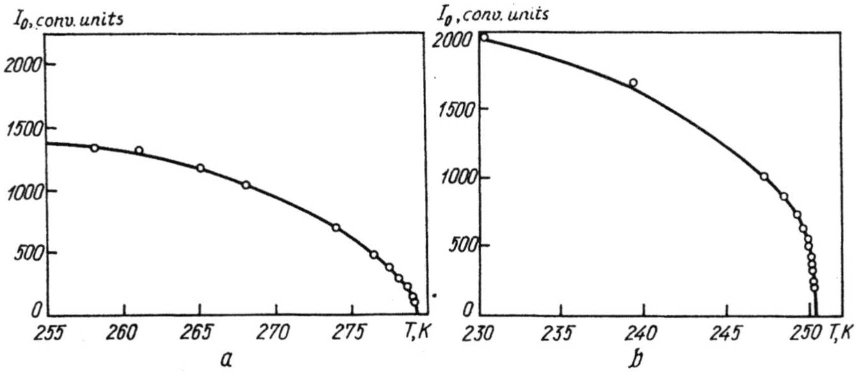

Fig. 4.11a,b. Behavior of the (221) reflection intensity (in arbitrary units) as a function of temperature, $I_0 = \text{const}\,\eta^2(\tau)$, in ND_4Cl [4.110]. (**a**) far from the tricritical point at $P = 4140$ bar, (**b**) near the tricritical point at $P = 157$ bar ($P_c = 128$ bar)

As we approach the tricritical point the fluctuation range around $T_0(P)$ becomes smaller and at the tricritical point the order parameter $\eta \sim (\tau \ln|\tau|)^{1/4}$ is described by the equation derived in the Landau theory with the additional logarithmic term. Therefore, the range of $|\tau|$ under investigation proves to belong to the intermediate range while the indices 2β are effective parameters in the power-law approximation for the function $I_0(\tau)$ which decrease when we approach T_c. Near the tricritical point the function $I_0(\tau)$ exhibits a much sharper variation in the vicinity of the transition point than at a large distance from it (see Fig. 4.11b). The respective $\beta \approx 0.18$ is somewhat smaller than $\beta = 0.25$ in (4.1.7) derived with the Landau theory and this difference may be attributed to the effect of the logarithmic factor. The critical scattering intensity for $P = P_c$ was described by (4.2.3) (with $\Phi = 0$) and the indices corresponding to the Landau theory (see Table 4.2).

In contrast to ND_4Cl (and NH_4Cl) in NH_4Br or ND_4Br the orientational ordering ($T_0 = 215$ K) gives rise to antiparallel arrangement of the tetrahedral NH_4 structures. This means that we obtain the anomalous scattering intensity peak around the superlattice point of the reciprocal lattice at $T \approx T_0$ (for neutron scattering in ND_4Br the peak is caused by the disorder of the tetrahedral structures and for X-ray scattering it is caused by the antiparallel displacements of the Br ions) [3.209; 4.111–113].

Perovskites are the best-known systems demonstrating displacement-type phase transitions. Anharmonicity makes the frequencies of some modes tend to zero in a number of perovskites and this involves the frequencies of other vibration branches, that is, for $T \to T_c$ not only do we obtain anomalous scattering, but the thermal scattering by these vibrations is also changed in character. Therefore, it is convenient to analyze the scattering mechanism near transitions in such systems from the neutron scattering results which allow us to separate the contributions of different vibrational and dissipative modes. The antiferro-distortion phase transition of the second order in $SrTiO_3$ was analyzed in detail with this method (see Fig. 4.7) [4.73, 74, 114–118]. The behavior of the cross section $\sigma(\boldsymbol{Q}, E)$ of inelas-

Table 4.2

Crystal	T_0, K	β	γ	ν	η_0
CuZn	741	0.293 [4.85]	1.25 ± 0.020 [4.80]	0.65 ± 0.02 [4.80]	0.071 ± 0.067 [4.80]
Fe-Al	823	0.307 ± 0.003 [4.86]	1.246 ± 0.010 [4.82]	0.649 ± 0.005 [4.82]	0.080 ± 0.005 [4.82]
FeCo	997	0.308 ± 0.014 [4.89]			
$C_{24}K$	122.9	0.18 ± 0.01 [4.96]			
ND_4Cl (P = 4140 bar)	279	0.305 ± 0.02 [4.110]			
ND_4Cl (P = 157 bar $\approx P_c$)	250.4	0.18 ± 0.005 [4.110]	1.05 ± 0.20 [4.110]	0.52 ± 0.08 [4.110]	
$SrTiO_3$	106	0.34 ± 0.02 [4.117]			
$RbCaF_3$	198	0.25 ± 0.01 [4.122]			
$PrAlO_3$	151	0.50 ± 0.02 [4.161]		0.5 [4.161]	
Al-Zn	T_c = 625		0.95 ± 0.12 [4.182]	0.5 ± 0.03 [4.182]	

tic neutron scattering has demonstrated that as the temperature is decreased the frequency $\omega_0 \sim (T - T_0)^{1/2}$ of the acoustic oscillations at the points $\{1/2\,1/2\,1/2\}$ at the Brillouin zone boundary decreases at first [4.73]. Then a central peak of a very small width (smaller than $8 \cdot 10^{-8}$ eV [4.117]) corresponding to the dissipative mode [4.115] appears in the $\sigma(Q, E)$ distribution near the transition point, in addition to the satellites with $E = \pm\hbar\omega$. (The central quasielastic peak in the spectrum of scattered neutrons can be produced owing to the variation of the character of the motion of atoms near the transition. The vibrations of atoms around the symmetric equilibrium positions are replaced with vibrations with appreciable frequencies about new equilibrium positions of the lower-symmetry phase. The dissipative mode and the central peak-correspond to slow fluctuations of these clusters with the altered lower order parameter $\eta \sim \tau^\beta$; see [4.76, 118]. High-precision measurements of the structure amplitudes and the regular reflection intensities must exhibit the production of new shifted peaks of mean probability density for the atoms. Such data for crystals of the type of $SrTiO_3$ (and anharmonic contributions to the Debye-Waller factors) suggest that the observed effects may be accounted for in the framework of a model with locally ordered clusters occurring at $T > T_0$ [4.118, 120]. In some cases, especially at temperatures significantly higher than T_0 scattering by the ordered clusters produced near defects makes an important contribution to the central intensity peak; see, for instance, Sect. 4.3 and [4.121].)

The scattering cross section $\sigma(\boldsymbol{Q})$ integrated over energies at $T > T_0 + 10\,\mathrm{K}$ is proportional to $(T - T_0)^{-1}$ in accordance with the Landau theory [4.73, 116] but at $T < T_0 + 10\,\mathrm{K}$ the fluctuation effects come into play and the cross section $\sigma(\boldsymbol{Q}) \sim (T - T_0)^{-\gamma}$ with $\gamma > 1$. (The experimentally determined γ [4.116, 117] are not precise enough and they depend on the history of the specimen, that is, on the imperfections in the crystal; the experimental results of $\gamma = 1.53$ and 1.21 were obtained for two specimens in [4.117]; see also Sect. 4.3.) The index $\beta = 0.34 \pm 0.02$ was found from the regular reflection intensities $I_0 \sim \eta^2$ in the range of $T_0 - 10\,\mathrm{K} < T < T_0 - 1\,\mathrm{K}$ (in this system the order parameter η is proportional to the angles of rotations of the octahedral structures TiO_6 in Fig. 4.7 or to the displacements of the oxygen atoms). This result for β is smaller than the calculated value (see Table 4.1) for a system with the three-component order parameter ($n = 3$) corresponding to three possible directions of the $\langle 100 \rangle$ axes of rotation for the octahedral structures (or the triply degenerate Γ_{25} vibrations). This difference may be due to the contribution made by anisotropy in the range of $|\tau|$ under consideration [4.14] and the contribution of the critical scattering to the recorded scattering intensity [4.87]. A similar index β was found for $LaAlO_3$. On the other hand, various diffraction techniques have yielded $\beta = 0.25$ corresponding to the tricritical point for a similar phase transition in $RbCaF_3$ [4.122] (see also [4.123]).

In the systems described by a multicomponent order parameter the symmetry of the points $\boldsymbol{k} = \boldsymbol{G}'$ is lower than cubic and the distribution of the critical scattering intensity around the points $\boldsymbol{G}'$ must be anisotropic. Diffuse X-ray scattering in some perovskites exhibited a sharply anisotropic intensity distribution concentrated along some axes and planes in the reciprocal lattice [4.124–128]. For instance, for $BaTiO_3$ and $KNbO_3$ high intensity scattering was observed in the $\{100\}$ planes [4.125] and for $KMnF_3$ [4.128] and $NaNbO_3$ [4.127] in the $\{100\}$ planes and along the $\langle 100 \rangle$ axes. This strong anisotropy was explained by an assumption about quasi-one-dimensional ordering of the atoms in weakly coupled chains in various equilibrium positions [4.124]. The experimental data on inelastic neutron scattering, however, suggest that these results are determined by the peculiar features of the vibration spectra in perovskites and can be accounted for by the conventional model of the displacement-type phase transitions. It has been shown that the quasi-one-dimensional or quasi-two-dimensional correlation is dynamic in character and involves the thermal vibrations of atoms, rather than their equilibrium positions although some systems can be described within the framework of a more general intermediate model which includes strong anharmonicity of the vibrations or their "multi-well" structure). For instance, the occurrence of bridges in the intensity distributions along the $\langle 100 \rangle$ directions in $KMnF_3$ and $NaNbO_3$ can be explained by the fact that these crystals exhibit low-frequency acoustic vibrations at two points (1/2 1/2 1/2) and (1/2 1/2 0), in at the boundary of Brillouin zone and along the segment connecting them (that is, the rotations in different $\{100\}$ planes shown in Fig. 4.7 are weakly coupled while the rotations of the ions in the given plane which belong to different octahedral structures are strongly correlated). Naturally, we obtain a high intensity thermal scattering by these acoustic vibrations along the $\langle 100 \rangle$ axes [4.129-131]. In the "latent ferroelectric" $KTaO_3$ (with $T_0 < 0$), in $KNbO_3$ and in $BaTiO_3$, where we obtain a soft mode of the transverse optical vibrations, the "quasi-one-dimensional" scattering along the $\{100\}$ planes can be described as thermal scattering by transverse acoustic vibrations [4.132, 135]. For $T \to T_0$ the frequencies of such vibrations polarized along the $\langle 100 \rangle$ axes are decreased considerably in a narrow range of $\boldsymbol{k}$ directions almost perpendicular to the $\langle 100 \rangle$ axes owing to the coupling with the transverse optical modes (but not for small $\boldsymbol{k}$ for which there is no coupling of modes). This results in the quasi-one-dimensional correlation and high-intensity thermal scattering of X-rays. The dipole forces give rise to a strong angular dependence of $\omega_{\perp}(\boldsymbol{k})$ [4.134, 135]. (The dipole interaction is effectively screened by the conduction electrons and makes a much weaker contribution to the scattering processes in the perovskite Na_xWO_3 where $0.4 < x < 1.0$ which has a metallic conductivity. Therefore, the energies of anisotropy and the striction coupling of the order parameter to the strains are small. The X-ray [4.136] and neutron [4.137] diffraction results and the inelastic neutron scattering data [4.137] suggest that as the temperature decreases Na_xWO_3 undergoes a series of phase transitions from the cubic phase 0_h^1 to the tetragonal phase D_{4h}^5, then to the tetragonal phase D_{4h}^{17}, and finally to the cubic phase T_h^5. These transitions correspond to successive softening and freezing of one, two and three modes of the acoustic vibrations at the (110)-type points while the rotations of the WO_6 octahedral structures around the $\langle 100 \rangle$ axes, such as those shown in Fig. 4.7, in the neighbouring planes have the same directions, rather than the opposite directions as in $SrTiO_3$. All transitions are continuous for the superlattice reflection intensities averaged over the domains, in contrast to the series of the phase transitions involving successive condensation of the degenerate modes in non-metallic perovskites, for instance in $BaTiO_3$ or $PrAlO_3$.)

The contributions made by the dipole forces explain also the complicated character of ordering of the ions Na^+ and NO_2^- in $NaNO_2$ which is ferroelectric at $T < T_c = 162°C$ and sinusoidally antiferroelectric in the narrow temperature range of $T_c < T < T_N = 164°C$. In this structure (which is orthorhombic in the disordered phase) the pure dipole-dipole interaction included in the model of the molecular field makes the modulated structure with $\boldsymbol{k}_0 = 2\pi \boldsymbol{e}_x/5d$ energetically favourable. When we take additionally the weak short-range forces into account, the ferroelectric phase becomes more energetically favorable at lower temperatures [4.138] (a more detailed microscopic theory [4.138–141] takes into consideration the contribution made by the coupling of the orientational order parameter and the acoustic vibrations and the macroscopic theory describing the ordering of $NaNO_3$ is presented in [4.142, 143]; an approach using symmetry considerations takes into account the fact that the coupling of two order parameters of different symmetries manifests itself in the incommensurate phase and stabilizes it).

The satellite superlattice intensity peaks are found at $\boldsymbol{G}' = \boldsymbol{G} + n\boldsymbol{k}_0$ (see Sect. 3.6.4) in the modulated phase while the fluctuations of the order parameter at temperatures above the point of transition to this phase give rise to anomalous diffuse scattering with the peaks of the highest intensity at $\boldsymbol{q} = \pm\boldsymbol{k}_0$. Such anomalous scattering was found in $NaNO_3$ [4.138, 145–147] and in other modulated structures, for instance, in the $TaSe_2$ $2H$ modification where charge density waves occur [3.219]. Since in incommensurate structures the order parameter is a complex number, they usually are systems described by a two-component ($n = 2$) order parameter (if the structure is characterized only by two vectors $\pm\boldsymbol{k}_0$). The critical index β was determined from the measured intensities of the superlattice peaks in K_2SeO_4 [4.148] and Rb_2ZnCl_4 [4.149] where such phases are produced (see also [4.150, 151]). The experimental results $2\beta = 0.75 \pm 0.05$ and $2\beta = 0.69 \pm 0.01$ agree with the predicted values (see Table 4.1).

The phase transitions for the case of $n = 4$ have been studied with NbO_2 where $\boldsymbol{G}'$ includes four non-equivalent vectors $\pm(1/2, 1/4, \pm 1/4)$. In the immediate vicinity of the transition (for $1.5 \cdot 10^{-3} < |\tau| < 8 \cdot 10^{-3}$) the index was found [4.152] to be $\beta = 0.40^{+0.04}_{-0.07}$ which agrees with the predicted $\beta \approx 0.39$ calculated with the renormalization group method [4.153]. In a higher temperature range of $6 \cdot 10^{-3} < |\tau| < 6 \cdot 10^{-1}$ the result of $\beta = 0.19 \pm 0.02$ proved to be much lower, which was explained by the transition to the two-dimensional case when the system with $n = 4$ broke down into two two-dimensional systems with $d = 2$ and $n = 2$ which corresponded to the planes (110) and $(1\bar{1}0)$ (for $|\tau| \approx 0.05$ it was found that $r_{c\perp}/r_{c\parallel} \lesssim 0.1$).

Another example of the ordered structure [4.154] with a four-component order parameter is the incommensurate phase in $Ba_2NaNb_5O_{15}$ which (in contrast to most other incommensurate phases) is described by four vectors $\pm\boldsymbol{k}_0$ and $\pm\boldsymbol{k}_0'$. The diffuse neutron scattering intensity distribution for this system exhibited a rod-like shape similar to that observed for NbO_2 indicating that outside the range of very small $|\tau|$ its critical behavior apparently corresponds to $d = 2$ and $n = 2$.

Systems with Significant Striction Effects. When we take into consideration the strains produced by the fluctuations of the order parameters, the intensity I_1 of diffuse scattering by the fluctuations of η in one-component crystals is described by the general equations (3.6.8, 9) instead of (3.6.10). According to (3.6.9), the respective long-wave distortions give rise, firstly, to a growth of $I_1 \sim q^2$ caused by approach to the reciprocal lattice points with $f \neq 0$ (that is, $\boldsymbol{A}_{\eta q} \sim q^{-1}$) and, secondly, to a decrease in the fluctuations and the scattering intensity I_1 caused by the contribution $\varphi^{\mathrm{el}}_{\eta\eta}$ of the elastic energy to $\varphi_{\eta\eta}(\boldsymbol{q})$ (see (2.1.17)). As noted in Sect. 4.1.1, the latter effect can also change the behavior of the thermodynamic properties at the phase transition.

Near a second-order phase transition in systems described by a one-component long-range order parameter we can rewrite (3.6.9) in the range of applicability of the Landau theory with the use of (2.1.17) and (4.1.3) in the following form [1.16]:

$$
\begin{aligned}
I_1 &= N\frac{1}{\Gamma^{\pm}|\tau| + \tilde{\varphi}^{\mathrm{el}}_{\eta\eta} + \tilde{\alpha}_{ij}q_iq_j}(f\boldsymbol{Q}\cdot\boldsymbol{A}_{\eta\boldsymbol{\varrho}} - \chi_\eta)^2\,,\\
\tilde{\varphi}^{\mathrm{el}}_{\eta\eta} &= \frac{v}{k_{\mathrm{B}}T_0}\varphi^{\mathrm{el}}_{\eta\eta}\,,
\end{aligned}
\tag{4.2.6}
$$

where $\Gamma^{\pm}$ and $\tilde{\alpha}_{ij}$ are given in (4.2.2) and its explanation.

The influence of the striction effects on the behavior of I_1 as a function of $\boldsymbol{q}$ and τ depends on whether the striction interaction energy is of the first or second order in the order parameter η. In the former case the strain u_{ij} is linearly related to $\delta\eta$ and, it can be seen from (1.5.5, 9) that $L_{\mu ij} \equiv L_{\eta ij}$ and, hence, $\boldsymbol{A}_{\eta q}$ do not vanish in both phases and can be regarded as being independent of τ at the transition point. Therefore, we have $I_1 \sim q^{-2}$ for $q \to 0$ for the reflections with $f \neq 0$ above and below the transition and for the given $\boldsymbol{n} = \boldsymbol{q}/q$ the dependence of the intensity I_1 on q is described by a hyperbola (for $f = 0$ this dependence is described by a Lorentzian curve and for small $|f\chi_{\eta}^{-1}L_{\eta ij}|$ the plot which is a hyperbola for $q \to 0$ is transformed into a Lorentzian curve with an increase in q). The strong angular dependence of I_1 is determined not only by the term $\boldsymbol{Q} \cdot \boldsymbol{A}_{\eta q}$ in (4.2.6) but also by the dependence of $\varphi_{\eta\eta}^{\mathrm{el}}$ on the direction of $\boldsymbol{q}$. For instance, according to (2.1.18, 19) in crystals for which the only non-vanishing component of the tensor $\hat{L}_{\eta}$ is $L_{\eta xy}$ the parameter $\varphi_{\eta\eta}^{\mathrm{el}}$ vanishes in the directions of $\boldsymbol{q} = \boldsymbol{n}q$ parallel to the axes X or Y and does not vanish in other directions $\boldsymbol{n}$. This is why the intensity (4.2.6) is very high for small $|\tau|$ (it is proportional to $|\tau|^{-1}$ for $q \to 0$) when $\boldsymbol{q} \parallel X$ and $\boldsymbol{q} \parallel Y$ and decreases rapidly in other directions of $\boldsymbol{q}$ (for not too small $|L_{\eta xy}|$). Under these conditions the Landau theory and, hence, the equation (4.2.6) are always applicable near the transition point (see Sect. 4.1.1).

In many structures the symmetry conditions do not allow linear striction in the disordered phase. In such case of quadratic striction we have $u_{ij} \sim \delta\eta^2 \sim \eta\delta\eta$, that is, $L_{\eta ij} = 0$ and $\boldsymbol{A}_{\eta q} = 0$ in the disordered phase, while in the ordered phase we obtain, in the range of applicability of the Landau theory, $L_{\eta ij} \sim \eta(\partial v/\partial\eta^2) \sim \eta$ and $\boldsymbol{A}_{\eta q} \sim \eta$ (see (1.5.5, 9)). According to (4.2.6), this means that for $T > T_0$ the striction effects do not change the scattering by the fluctuations of the long-range order parameters discussed above (see (4.2.2)) while at $T < T_0$ the term $\tilde{\varphi}_{\eta\eta}^{\mathrm{el}} \sim \eta^2 \sim |\tau|$ causes only renormalization of the coefficient Γ^{-1} at $|\tau|$ and $\boldsymbol{A}_{\eta q} \sim \eta \sim |\tau|^{1/2}$ are small.

Therefore, the diffuse scattering intensity (4.2.6) must exhibit anomalous growth and its distribution must have a bell-shaped plot such as that given by (4.2.2) for all q with the exception of a narrow range of $q < \mathrm{const}\sqrt{|\tau|}$ where we have scattering by the distortions produced by the fluctuations of the order parameter η.

In this case of quadratic striction a much greater contribution for small q may be made by the scattering by the distortions produced by the fluctuations of the correlation parameters $\varepsilon(\varrho)$ (see Sects. 1.4; 1.5.2; and 3.4.2). This scattering occurs both below and above the transition point and its intensity is proportional to q^{-2} and $|\tau|^{-1/2}$ (this can be shown for light scattering results [2.2.3]). This scattering intensity may have a complicated angular dependence determined by the model used for its treatment. The intensities of scattering by the distortions caused by the fluctuations of the parameters η and $\varepsilon(\varrho)$ are of the same order of magnitude only near the tricritical point where $\eta^2/\varphi_{\eta\eta} \sim |\tau|^{-1/2}$ (see (4.1.8)) is much higher than near the conventional transition point (for the same $|\tau|$).

In the fluctuation range for a conventional second-order phase transition we must replace $\Gamma^{\pm}|\tau|$ with $\Gamma^{\pm}|\tau|^{\gamma}$ in (4.2.6) while $L_{\eta ij} \sim \partial v/\partial\eta \sim \tau^{1-\alpha-\beta}$ (since $\partial v/\partial\tau \sim \tau^{-\alpha}$). Then using (4.1.34) we find that the intensity of scattering by the distortions produced by the fluctuations of η is proportional to $|\tau|^{-\alpha}$ for $q \to 0$. The intensity of scattering by the distortions produced by the fluctuations of $\varepsilon(\varrho)$ is proportional to $|\tau|^{-\nu(1-2\eta_0)}$ (see [2.23]). The width of the q-range in which scattering by distortions manifests itself is also small under these conditions for $\chi_\eta \neq 0$ and second-order striction (it is proportional to $\tau^{1-\alpha-\beta}$).

In systems with a multicomponent long-range order parameter the expression for the intensity I_1 has a more complicated form. In the vicinity of the transition point, I_1 is given by the general equations (2.1.20) and (3.6.8) in which $\varphi_{\mu\mu'} \sim |\tau|$ in the range of validity of the Landau theory or $\varphi_{\mu\mu'} \sim |\tau|^{\gamma}$ in the fluctuation range. For instance, for the high-symmetry cubic phase we have in the range of validity of the Landau theory

$$
\begin{aligned}
I_1 &= \frac{Nk_BT}{v}(f\boldsymbol{Q}\cdot\boldsymbol{A}_{\mu q} - \chi_\mu)(f\boldsymbol{Q}\cdot\boldsymbol{A}_{\mu' q} - \chi_{\mu'})\varphi^{-1}_{\mu\mu'}(\boldsymbol{q}) ,\\
\varphi_{\mu\mu'}(\boldsymbol{q}) &= aT_0\tau\delta_{\mu\mu'} + \varphi^{\mathrm{el}}_{\mu\mu'} + \alpha_{\mu\mu' ij}q_iq_j ,
\end{aligned}
\tag{4.2.7}
$$

where $\varphi_{\mu\mu'}$ is given by (2.1.20) in which $\hat{c} = \hat{c}^\infty$ are determined for constant η_μ and $\varphi_{\mu\mu'}$ (and (T_0)) for constant stresses. For small q when $aq^2 \ll \varphi_{\mu\mu'} + \varphi^{\mathrm{el}}_{\mu\mu'}$ and $f|\boldsymbol{Q}\cdot\boldsymbol{A}_{\eta q}| \gg |\chi_\mu|$, the total intensity $I_1 = I_{1H} \sim q^{-2}$ of scattering by the distortions (including the thermal scattering by the "rigid" vibration modes) can be expressed in terms of the low-frequency elastic moduli $\hat{c}^0$ which take into account relaxation of η_μ with deformation,

$$
I_1 = \frac{Nk_BT}{vq^2}f^2Q_iQ_jC^{0-1}_{qij} , \quad C^0_{qij} = c^0_{ilm}n_ln_m . \tag{4.2.8}
$$

Here equations (2.1.21, 21a) have been used. The equation (4.2.8) is, of course, applicable to crystals of any symmetry. It may be useful, for instance, for describing scattering in intrinsic ferroelastic crystals in which some elastic moduli vanish at the transition point (but for $n = 2$ and 3 the purely ferroelastic transition which does not involve long-range order parameters of a different symmetry is typically a first-order phase transition, for instance, the transition from the cubic to the tetragonal phase [2.19, 20; 4.155]).

The characteristic behavior of I_1 as a function of $\boldsymbol{q}$ described by (4.2.6–8) is especially typical in systems with linear striction. They include, for instance, crystals with the cooperative Jahn-Teller phase transition which simultaneously cause splitting of some electron levels which are degenerate in the disordered phase and deformation of the crystal which eliminates the degeneracy (see [4.156–159]). The anomalous scattering caused by such a first-order transition ($T_0 \approx 300$ K), similar to a second-order transition, was analyzed in detail for the $NiCr_2O_4$ crystals [4.160]. At temperatures below the transition point this cubic crystal with the spinel-type structure and Ni^{2+} ions occupying the tetrahedral positions becomes tetragonal so that the triple splitting of the $3d$ states of Ni^{2+} is eliminated and the resulting three-component order parameter depends on this splitting. Scattering by distortions was analyzed especially thoroughly for the vicinities of the (0.0.16) and (12.12.0) reflections. Figure 4.12a shows the equal intensity lines [4.160] for anomalous scattering by the tetragonal distortions produced by X-, Y-, and Z-type fluctuations which are similar in shape to the lines obtained for scattering by tetragonal defects in cubic crystals (see Fig. 4.12 in [1.1]). The equal intensity lines near the point $(hh0)$ proved to be strongly contracted in the direction of [110] (see Fig. 4.12b). This effect is explained by strongly anisotropic suppression of the critical fluctuations by the long-range elastic forces corresponding to the term $\varphi^{\mathrm{el}}_{\mu\mu'}$ in (4.2.7). Under these conditions (4.2.7) or (4.2.8) suggests that in the vicinity of the $(hh0)$ reflection for q parallel to the high-symmetry directions [4.160] we have

$$I_1 \sim \{T - T_0[1-\xi(\boldsymbol{q})]\}^{-1}\,, \quad \text{where} \quad \xi(\boldsymbol{q}) = 0 \quad (\boldsymbol{q} \parallel [1\bar{1}0])\,;$$

$$\xi(\boldsymbol{q}) = \frac{2c_{11}+c_{12}}{3c_{11}} \quad (\boldsymbol{q} \parallel [100])\,,$$

$$\xi(\boldsymbol{q}) = \frac{2}{3}\,\frac{c_{11}+2c_{12}+3c_{44}}{c_{11}+c_{12}+2c_{44}} \quad (\boldsymbol{q} \parallel [110])\,.$$

Therefore, the scattering intensity in the $[1\bar{1}0]$ directions increases at a much faster rate for $T \to T_0$. The behavior of I_1 as a function of temperature follows a similar pattern, that is, I_1 grows critically for $\boldsymbol{q} \parallel [1\bar{1}0]$ and is almost constant for $\boldsymbol{q} \parallel [0\bar{1}0]]$ (see Fig. 4.13).

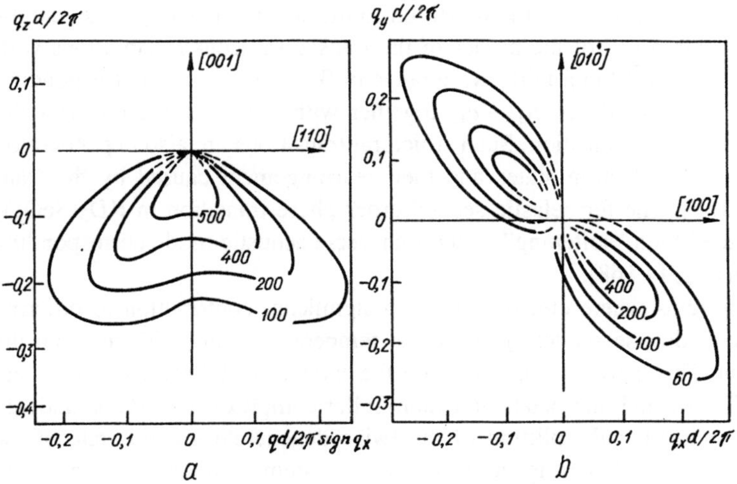

Fig. 4.12a,b. Equal intensity lines for anomalous scattering in $NiCr_2O_4$. **(a)** in the $(1\bar{1}0)$ plane near the point (0, 0, 16), **(b)** in the (001) plane near the point (12, 12, 0) [4.160]

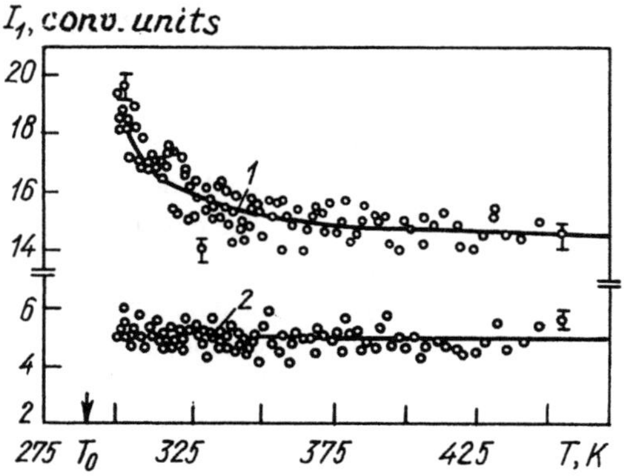

Fig. 4.13. Diffuse scattering intensity I_1 (in arbitrary units) as a function of temperature in $NiCr_2O_4$ near the point (12, 12, 0). *Curve 1* for q = (−0.105, 0.105, 0) $2\pi/d$, *curve 2* for q = (0, −0.148, 0) $2\pi/d$ [4.160]

The Jahn-Teller second-order phase transition between the orthorhombic phase and the monoclinic phase in the perovskite $PrAlO_3$ was studied in detail with a neutron diffraction technique [4.161]. In this crystal system the continuous shifting from the direction [101] to [001] of the axis around which the AlO_6 complexes execute antiferrodistortional rotation corresponds to the long-range ordering. The elastic strain energy corresponding to the linear striction suppresses the critical fluctuations with the exception of some directions of $\boldsymbol{k}$ so that when the temperature approached T_0 a decline occurred only for the phonon frequencies of the acoustic branch with $\boldsymbol{k} \parallel [101]$ and $\boldsymbol{e}_k \parallel [10\bar{1}]$. Since the fluctuations are suppressed, the Landau theory is applicable near the transition (see Table 4.2) and the results can be described in the approximation of the self-consistent field for Jahn-Teller transitions [4.162].

The crystals V_3Si [4.163] and Nb_3Sn [4.164, 165], which have the β-W structure and a high temperature for the transition to the superconducting state, exhibited a similar "softening" of the modes of the transverse acoustic vibrations with $\boldsymbol{k} \parallel [110]$ and $\boldsymbol{e}_k \parallel [1\bar{1}0]$ in a fairly wide range of $0 < k <\approx k_m/2$ corresponding to a decrease in the modulus $c_{11} - c_{12}$ together with critical scattering near the structural "martensite" transition. Such structural transitions are also dependent on the behavior of the electron states and their rearrangement caused by the transition [4.166, 167]. The ferroelastic second-order phase transition in $KD_3(SeO_3)_2$ was accompanied by a "softening" of the transverse acoustic mode corresponding to the modulus c_{44} [4.168].

In some systems where electrons make a significant contribution to ordering processes strong diffuse scattering occurs at temperatures much higher than the transition point. This is true, for instance, for the crystals of $R_2Pb(CuNO_2)_6$ (where R = K, Rb, Tl) in which the octahedral Jahn-Teller complexes $Cu(NO_2)_6$ undergo distortions associated with splitting of the twice degenerate ground state of the Cu^{+2} ion [4.169–171]. Ordering in these crystal systems produces incommensurate structures with $\boldsymbol{k}_0 \parallel [110]$. The characteristic anisotropic diffuse scattering with the intensity $I_1 \sim q^{-2}$ in the vicinity of the lattice points occurs, however, at temperatures exceeding by tens of degrees the transition temperature and its temperature dependence is weak for $T > T_0$ (while the intensity declines sharply at $T < T_0$). To explain this type of scattering it was assumed that at temperatures much higher than the transition point the $Cu(NO_2)_3$ complexes had already been subjected to strong tetragonal distortion (randomly in the three $\langle 100 \rangle$ directions) and thus the tetragonal quasidefects caused diffuse scattering (similar to that illustrated in Fig. 4.5 in [1.1], see Sect. 4.1.2 in [1.1]) [4.170]. A significant contribution to the production of the incommensurate structure is made by the elastic interaction between the defects. Since a first-order phase transition occurs in these systems the critical scattering for $\boldsymbol{k} \approx \boldsymbol{k}_0$ is not strong. The nature of the incommensurate phase and the distorted structure in such Jahn-Teller systems have been treated theoretically in [4.144].

In Fe_3O_4 crystals similar diffuse scattering in the vicinity of the reciprocal lattice points was observed at temperatures 80 K higher than the point of electronic ordering of the Fe^{2+} and Fe^{3+} ions and it was explained by the production of molecular polarons which locally distort the crystal [4.172–174] (see Sect. 4.1.6 in [1.1]). In this model the electronic phase transition is a result of the interaction between such polarons. Since ordering in magnetite is a second-order phase transition the conventional bell-shaped intensity peaks of anomalous scattering caused by the critical fluctuations of the orientations of such polarons were found in the vicinities of the superlattice points several degrees above the transition point [4.175–177].

Critical Point on the Decomposition Curve. In the vicinity of the critical point on the decomposition curve the concentration fluctuations (see Fig. 4.1) give rise to high intensity peaks of diffuse scattering in the vicinities of the reciprocal lattice points. As noted in Sect. 4.1, a suppression of the concentration fluctuations which is typically produced to some extent by the long-range forces caused the Landau theory to be applicable in the immediate vicinity of the critical point on the curve of incoherent decomposition so that the equation (4.1.28) describes the fluctuations. (If a two-phase system partially preserves lattice coherence of the phases then the phase diagram differs from that illustrated in Fig. 4.1 – there is no critical point and the curves of phase equilibrium stretch upwards without touching [4.178, 179]. The explanation is that the stresses in the inhomogeneous, partially incoherent, two-phase system and the dislocations which accommodate the incommensurateness prevent homogenization and do not allow a homogeneous phase to be restored. Such phase diagrams have been analyzed for H-Nb interstitial solutions [4.178] and H-$Pd_{1-x}Ag_x$ (here $x = 0.1$ and 0.2) [4.179] under the conditions of partial coherence and were found to include two-phase ranges well above T_c.) The intensity of diffuse scattering in the vicinity of the critical point is given by (3.2.10) and (4.1.28) or (3.4.3) and (4.1.15)) [1.3, 16]:

$$I_1 = \frac{Nk_\mathrm{B}T}{v} \frac{\left[f a_q \frac{\boldsymbol{Q} \cdot \boldsymbol{e}_q}{q} - (f_A - f_B) \right]^2}{a(T + 2T_c - 3T_0(c)) + \varphi_{cc}^{\mathrm{el}} + \beta_{ij} q_i q_j}, \qquad (4.2.9)$$

Here $\varphi_{cc}^{\mathrm{el}}$ is given by (2.1.5–9).

At the critical point we have $c = c_c$ and $T = T_0(c_c) = T_c$ and in the absence of distortions the intensity $I_1 \to \infty$ for $q \to 0$. But when we take into account the distortions, and the elastic energy associated with them, we always have $\varphi_{cc}^{el} > 0$, that is, for $q \to 0$ the denominator in (4.2.9) determining the temperature dependence of the scattering intensity does not vanish and I_1 remains finite at the transition point though it becomes fairly large if the distortions are small.

As noted in Sect. 2.1, the parameter $\varphi_{cc}(\boldsymbol{q})$ (that is, the denominator in (4.2.9) for $q \to 0$) vanishes and the scattering intensity for the solution which did not have time to decompose in the nonequilibrium range becomes anomalously high at lower temperatures close to the spinodal temperature $T_0'(c)$ for which $\varphi_{cc}(\boldsymbol{k}_{min}) = 0$. In elastically anisotropic crystals $\varphi_{cc}(\boldsymbol{q})$ vanishes only for several directions of $\boldsymbol{q}$ but has finite values for other directions. The critical scattering intensity distribution must therefore be sharply anisotropic near the spinodal line. If the kinetic factors allow coherent spinodal decomposition of the solution then it exhibits characteristic behavior caused by the preferential development of macroscopic concentration modes which are related to the elastic energy and depend on the crystal shape [2.4] (additional complications are produced by critical wetting of the surface of one phase with the second phase causing a decrease in the surface energy [4.180, 181]).

The suppression of the critical fluctuations by the elastic forces which gives rise to the term φ_{cc}^{el} in the denominator in (3.4.3) or (4.2.9) is clearly revealed in the experimental results on the diffuse scattering by a solution near the critical point of decomposition. The critical small-angle scattering ($\boldsymbol{q} = \boldsymbol{Q}$) was studied in detail for the Al-Zn substitutional solution with the concentration $c = 40.1 \pm 0.8$ at.% Zn which is close to the critical one [4.182]. The results presented in Figs. 4.14, 15a show that the scattering cross section $d\sigma/d\Omega$ is still not anomalously large above

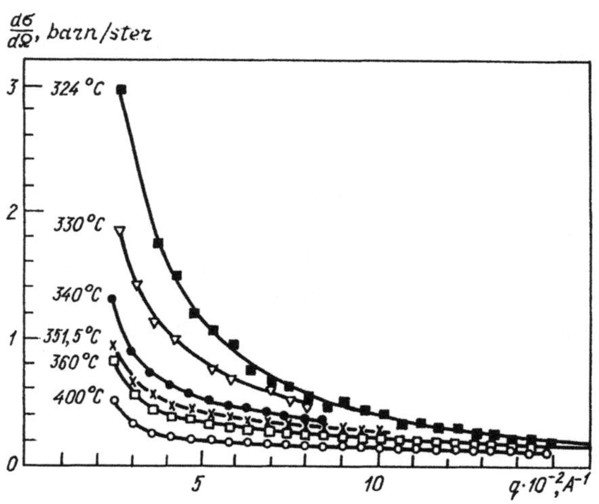

Fig. 4.14. Cross section $d\sigma/d\Omega$ (barns/steradian) of small-angle elastic diffuse scattering of neutrons in the Al-Zn alloy with the critical composition as a function of the length (in 10^{-2} Å^{-1} units) of the diffraction vector $\boldsymbol{q}$ at various temperatures [4.182]

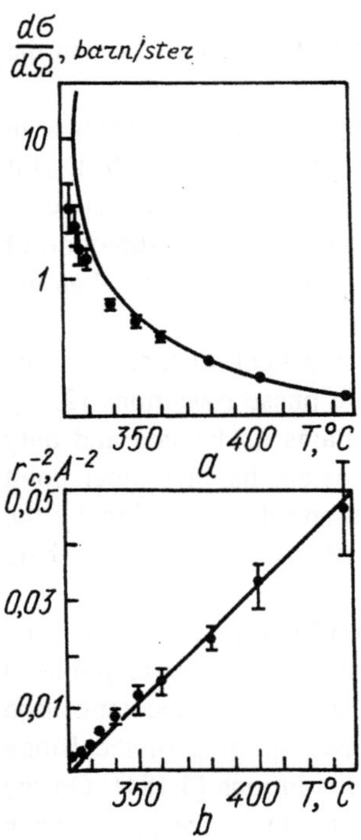

Fig. 4.15. The limiting cross section $d\sigma/d\Omega$ (barns/steradian) (a) for $q \to 0$ and (b) the squared correlation length r_c^{-2} as functions of temperature for diffuse neutron scattering in a Al-Zn alloy of the critical composition. The solid lines give the results calculated from (4.2.9) [4.182]

the critical temperature $T_c = 351.5°C$ (the maximum point on the equilibrium decomposition curve). The scattering intensity grows rapidly at lower temperatures approaching the spinodal point T_0'. In this temperature range at $T > 324°C$ measurements could be made in sufficiently short periods which were shorter than the incubation time for incoherent decomposition and thus the solution was in "quasiequilibrium". For $T < 324°C$ no equilibrium could be obtained and thus it was suggested that $T_0'(c_0') \equiv T_c' \approx 324°C$. The behavior of $d\sigma/d\Omega$ as a function of q could be described by (3.4.3) or (4.2.9). Since the elastic behavior of the aluminium-based alloys is not very anisotropic the parameter φ_{cc}^{el} was calculated from the equation (2.1.9) derived in the approximation of isotropic continuum ($v^{-1}dv/dc = 0.088$). The predicted $T_c - T_c' = 28°C$ (found from the calculated φ_{cc}^{el} and the experimentally determined thermodynamic potential $\varphi(T)$) agrees with the experimental data ($T_c = 351.5°C$ (and $T_c' \approx 324°C$). The temperature dependences of the diffuse scattering intensity peak (that is, $d\sigma/d\Omega$ for $q \to 0$) and the peak half-width (that is, r_c^{-1}) are described by the indices γ and ν which agree with (4.2.9), that is, the approximation of the Landau theory (see Table 4.2). In addition to the contribution made by the elastic forces, another explanation for the above result may be the fact that mean length of interaction is considerably larger than the distance between atoms as can be seen from the slope of the plot in Fig. 4.12b. (Otherwise, in elastically anisotropic crystals we may expect that for very small φ_{cc}^{el}, in addition to the range of applicability of the Landau theory at T_0', the fluctuation range can be realized at $T_c \gg T - T_0' \gg T_c - T_0'$ and then for $T_0(c) = T_{c_\gamma}$ we must replace $a(T - T_c)$ with a parameter $\sim (T - T_c)^\gamma$ in the denominator of (4.2.9)).

The critical fluctuations have been found to be suppressed even stronger in the studies of small-angle neutron scattering by the interstitial solution Nb-D (for which $T_c = 440$ K and $c_c = 0.33$H/Nb) in which the elastic energy of the concentration wave is high ($v^{-1}dv/dc = 0.15$) and strongly depends on the direction of $\boldsymbol{k}$ [3.49, 50]. The spinodal temperatures found in these studies are lower than the temperature of coherent decomposition by a factor of two or more.

In one-component crystals similar suppression of the critical fluctuations must occur at the critical point on the line of the isomorphic phase transitions [2.19]. When such transitions occur the crystal symmetry remains unchanged and only the lattice cell parameters are altered (for instance, cerium has a critical point for isomorphic transitions [4.183]). Under such conditions the scattering by the critical fluctuations of the order parameter can also be described by (4.2.9) where we must take $f_A = f_B$.

As noted in Sect. 2.1, the long-wave fluctuation modes with the characteristic inhomogeneity length of the order of the crystal size are less susceptible to suppression by the elastic forces. In the case of coherent spinodal decomposition ($T' < T < T_c$) such modes must cause broadening and splitting of the Bragg peaks. These effects have been analyzed for the Nb-H solutions [4.184]. On the other hand, the ordering of atoms on the lattice sites or the interstices, even when it is caused by their elastic interaction, involves rearrangement of atoms at small distances of the order of d and therefore the long-wave fluctuations apparently must not be suppressed. This conclusion has been confirmed in experiments with the ordering of the interstitial deuterium atoms in [4.185].

The scattering intensity exhibits the same features near the lattice points at the tricritical point on the line of phase transitions of the second order (see Fig. 4.2) as in the vicinity of the maximum of the decomposition curve but only in the ordered phase. Under these conditions we can use (1.2.28) and (4.1.10, 11) and find I_1 from (3.4.3) replacing in it $d^2\varphi/dc^2$ with $\varphi^0_{cc}\{g_1[T_0(c)-T]+g_2[T_0(c)-T_c]\}$ which vanishes at the tricritical point. But since the difference $\varphi_{cc}(\boldsymbol{q}) - \varphi^2_{c\eta}/\varphi_{\eta\eta}$ remains finite (and positive) at this point, in contrast to $\varphi_{cc} - \varphi^2_{c\eta}/\varphi_{\eta\eta} = d^2\varphi/dc^2$, the intensity I_1 tends to infinity at the tricritical point but at the point where $\varphi_{cc}(\boldsymbol{q}) - \varphi^2_{c\eta}/\varphi_{\eta\eta} = 0$. The latter point corresponds to an incomplete equilibrium state in the two-phase range.

The intensity of anomalous scattering in the vicinities of the superlattice points of the reciprocal lattice in solutions without linear striction is given by the same equations (4.2.2, 3) as in one-component undistorted crystals (as shown by (3.6.13); the distortions produced at $T < T_0$ are now effective for small $T_0 - T$ and η only for very small q because $a_{q\eta} = 0$ and $\varphi_{\eta\eta}(\boldsymbol{q}) = \varphi_{\eta\eta}$ for $\eta = 0$). One consequence of that is the anomalous growth of the intensity I_1 at the tricritical point near both the lattice points and superlattice points. As noted above in this section, near the superlattice points I_1 decreases relatively slowly with decreasing temperature at the tricritical point.

Intrinsic Ferroelectric Crystals Near Second-Order Phase Transitions. As discussed in Sect. 3.6.3, in intrinsic ferroelectrics, in which the long-range order is fully described by the polarization vector components P_i, the diffuse scattering intensity I_1 for small q is given by an expression in which all the parameters are found from the experimental results. For instance, the derivatives $\varphi_{\mu\mu'} = \varphi_{ij} = \partial^2\varphi/\partial P_i \partial P_j$ are fully determined by the data on the dielectric susceptibility $\varkappa_{ij} = \varphi_{ij}^{-1}$. This makes it possible to calculate $I_1(q \to 0)$ and to make comparison with independent experimental data (in particular, for $T \to T_0$ when some components of the susceptibility tensor $\hat{\varkappa}$ tend to infinity and the intensity I_1 is anomalously high). This comparison can be performed without making assumptions about the behavior of φ_{ij} as a function of temperature near the transition point (which can have a complicated form in the intermediate range) and without making use of a statistical model of ordering which yields the constants in this function (we must clearly use such a model for finding the intensity I_1 for high q).

Depending on the ratio between q and the inverse Debye length $1/r_D$ the diffuse scattering intensity can be found from (3.6.15) (for $qr_D \ll 1$) or (3.6.16) (for $qr_D \gg \sqrt{\varepsilon}$). The first case is realized when the concentration of free charges is large and then the dipole-dipole interaction is screened for large wavelengths of the fluctuation waves ($qr_D \ll 1$). Then the fluctuations can be calculated from the conventional equations of the local theory and according to (3.6.15) the intensity distribution for anomalous scattering near the transition to the ferroelectric state is the same as in the cases discussed in this section. For large Debye lengths and not too small q, however, a contribution must be made by effects typical in ferroelectrics which are caused by the significant influence exerted by the long-range dipole-dipole forces on the polarization fluctuations. This can produce significant changes in the scattering intensity and especially in its angular dependence.

When we approach the transition point in cubic crystals (in the nonferroelectric phase) and take into account the dipole-dipole interaction the scattering intensity grows anomalously for any direction of $\boldsymbol{q}$ if $q \ll d^{-1}$. When we express $\varkappa_{ij}(\boldsymbol{q}) = \varkappa_{\infty ij}(\boldsymbol{q})$ in (3.6.16) in terms of the permittivities ε and ε_∞ for low and high frequencies and use (2.2.8, 9) (here we must make a transition from the special coordinate system chosen in (2.2.9) to the natural coordinate system in which the axes are aligned along the [100]-type directions) we can write an explicit expression for the scattering intensity at $T > T_0$ in the absence of striction for $qr_D \gg \sqrt{\varepsilon}$ and $qd \ll 1$ [1.13]:

$$I_1 = \frac{Nk_B T}{4\pi v}\left[(\varepsilon - \varepsilon_\infty)(\delta_{ij} - n_i n_j) + \frac{\varepsilon - \varepsilon_\infty}{\varepsilon\varepsilon_\infty} n_i n_j\right]\chi_i\chi_j\,, \quad \boldsymbol{n} = \frac{\boldsymbol{q}}{q}\,. \tag{4.2.10}$$

If the dipole-dipole interaction is screened then the term in the square brackets in [4.210] must be replaced with $(\varepsilon - \varepsilon_\infty)\delta_{ij}$, that is, when this interaction has been taken into consideration the angular dependence of the diffuse scattering intensity distribution is, indeed, altered.

Since we have $\varepsilon \to \infty$ in the vicinity of the second-order phase transition point T_0 (because $\varepsilon = a^{-1}(T - T_0)^{-1}$ in the range of applicability of the Landau theory and $\varepsilon \sim \tau^{-\gamma}$ in the fluctuation range) only the first term can be retained in the square brackets in (4.2.10) that is, I_1/T varies with temperature as $\varepsilon - \varepsilon_\infty \approx \varepsilon$. We obtain then

$$I_1 = \frac{Nk_BT}{4\pi v}(\varepsilon - \varepsilon_\infty)\left[\sum_i \chi_i^2 - \left(\sum_i \chi_i n_i\right)^2\right] . \tag{4.2.11}$$

If, for example, in ferroelectrics the displacements χ_i (given by (3.6.43)) are proportional to Q_i then the intensity I_1 depends on $\boldsymbol{Q}$ and $\boldsymbol{q}$ as $Q^2 - (\boldsymbol{Q}\cdot\boldsymbol{n})^2$, it has a peak for $\boldsymbol{Q} \perp \boldsymbol{q}$ and vanishes for $\boldsymbol{Q} \parallel \boldsymbol{q}$ (in contrast to the scattering intensity I_H; see Sect. 4.1.2 in [1.11). The angular dependence of the intensity I_1 in such ferroelectrics is even more complicated (in comparison with the case of $qr_D \ll 1$) below the transition point, that is, in the ferroelectric phase (where the crystal no longer has a cubic symmetry).

The dipole-dipole interaction exerts a particularly strong effect on the anomalous scattering in noncubic crystals (in the nonferroelectric phase). If only one component $\varkappa_{zz}$ of the tensor $\varkappa_{ij}$ tends to infinity at the transition point while $\varkappa_{xx} = \varkappa_{yy}$ and $\varkappa_{ij} = 0$ for $i \neq j$ then we obtain from (3.6.17) the following expression for the anomalous scattering intensity for $qr_D \gg \sqrt{4\pi\varkappa_{zz}}$ and $qd \ll 1$ in the absence of striction if we ignore the terms which remain finite as we approach the transition point [1.13, 16]:

$$\begin{aligned} I_1 &= \frac{Nk_BT}{v}\frac{1}{(\varkappa_{zz})^{-1} + 4\pi\tilde{\varepsilon}_\perp^{-1} n_z^2 + \boldsymbol{q}\hat{\alpha}_{zz}\boldsymbol{q}}\chi_z^2 , \\ \boldsymbol{q}\hat{\alpha}_{zz}\boldsymbol{q} &= \alpha_{zzzz}q_z^2 + \alpha_{zzxx}(q_x^2 + q_y^2) , \\ \tilde{\varepsilon}_\perp &= 1 + 4\pi\varkappa_{xx}(1 - n_z) . \end{aligned} \tag{4.2.12}$$

Here we have also taken into consideration the term $\alpha_{zzij}q_iq_j$ (for the diagonal tensor $\hat{\alpha}_{zz}$ and for $\eta_0 = 0$) besides $(\varkappa_{zz})$ which determines the dependence of the intensity I_1 on q, becoming significant for small $(\varkappa_{zz})^{-1}$. It can be seen from (4.2.12) that under these conditions the anomalous growth of the diffuse scattering intensity occurs only if the vector $\boldsymbol{n}$ is almost perpendicular to the axis Z (if $n_z \ll 1$) and for other directions of $\boldsymbol{q}$ the intensity I_1 grows anomalously only for $qr_D \ll 1$ when the dipole-dipole interaction is no longer effective.

A strong dipole-dipole interaction suppresses the fluctuations and thus enables us to make use of the Landau theory (with the logarithmic additional terms). Using (4.1.37) we can then write the dependence of the intensity I_1 on τ in the explicit form:

$$I_1 = N\frac{\psi(\tau)}{|\tau|}\left[\Gamma_P^{\pm} + \frac{4\pi v}{\tilde{\varepsilon}_\perp k_BT}\frac{\psi(\tau)}{|\tau|}n_z^2 + \frac{v}{k_BT}\frac{\psi(\tau)}{|\tau|}\boldsymbol{q}\hat{\alpha}_{zz}\boldsymbol{q}\right]^{-1}\chi_z^2 . \tag{4.2.13}$$

For ferroelectrics with striction effects, equations (3.6.15–17) must include, besides $-\chi_i$, the terms $f\boldsymbol{Q}\cdot\boldsymbol{A}_q^i$ and $\varkappa_{ij}(\boldsymbol{q})$ or $\varkappa_{ij}$ must be replaced with $\varkappa'_{ij}(\boldsymbol{q})$ or $\varkappa'_{ij}$, respectively. Using the definitions (3.7.5, 13) for these parameters we obtain in the general case, when $q \to 0$ (but $qr_D \gg \sqrt{4\pi \varkappa_{zz}}$) and the small additional terms (for $T \approx T_0$) can be ignored [1.16],

$$I_1 = \frac{Nk_BT}{v} \varkappa'_{ij}(q)(fQ \cdot A^i_q - \chi_i)(fQ \cdot A^j_q - \chi_j),$$

$$\varkappa'^{-1}_{ij}(q) = \varkappa^{-1}_{ij} + 4\pi n_i n_j$$

$$= \varkappa^{-1}_{ij} + c_{lmpr} L^i_{lm}(L^j_{pr} - qn_p A^j_{qr}) + 4\pi n_i n_j \tag{4.2.14}$$

(where $\hat{\varkappa}$ is the dielectric susceptibility under constant stresses) and for uniaxial ferroelectrics we have

$$I_1 = \frac{Nk_BT}{v} \frac{\tilde{\varepsilon}_\perp \varkappa_{zz}}{\tilde{\varepsilon}_\perp + 4\pi \varkappa_{zz} n_z^2}(fQ \cdot A^z_q - \chi_z)^2 . \tag{4.2.15}$$

In these equations all these parameters are also determined from the independent experimental results. According to (4.2.14, 15), the piezoelectric effect causes a decrease in the anomalous scattering intensity for the reflections with $f = 0$ and $\chi_z \neq 0$ and gives rise to hyperbola-shaped intensity distributions for the reflections with $f \neq 0$ (that is, $A^z_q \sim q^{-1}$). The equation (4.2.14) for ferroelectrics with displacements can be derived also with a dynamic model which takes into account the interaction between the polarization fluctuations and the acoustic vibrations [3.223].

The sharp orientational dependence of the anomalous diffuse scattering intensity in ferroelectrics caused by the suppression of the critical fluctuations by the dipole-dipole interaction was analyzed in detail in the case of X-ray diffraction by triglycine sulphate $(NH_2CH_2COOH)_3 \cdot H_2SO_4$ in the vicinity of a second-order phase transition (T_0 = 48°C [4.186]. This monoclinic crystal exhibits no piezoelectric effect so that the intensity distributions for scattering by the polarization fluctuations have inversion centers at the reciprocal lattice points and are described by equations of the type of (4.2.12, 13). The equal-intensity lines around the (340) point shown in Fig. 4.16 are indeed sharply anisotropic and greatly elongated in the direction perpendicular to the axis q_z (we have $q_z \parallel b^*$; the polarization vector is parallel to the monoclinic axis b; a similar pattern was observed in the $(c^* b^*)$ plane; this pattern differs from the almost isotropic scattering intensity distribution shown in Fig. 4.10 for $NaNO_3$ where the fluctuations of the antiferroelectric order parameter have not been suppressed. But the contraction of the equal intensity lines along the q_z axis proved to be weaker than that for the theoretical curves calculated [4.186] from an equation similar to (4.2.12) corresponding to the Ising model. It has been suggested [4.186] that this difference may be attributed to a partial screening of the polarization field by free charges (see (2.2.12)). The behavior of the intensity I_1 as a function of temperature could be described by (4.2.12) with $(\varkappa_{zz})^{-1} \sim T - T_0$ which made it possible to find $\varphi = 1/2$ which agrees with the predicted result for uniaxial ferroelectrics.

The results [4.187] on the energy spectrum of the neutrons inelastically scattered by the uniaxial ferroelectric $Pb_5Ge_3O_{11}$ yielded the same temperature dependence of $(\varkappa_{zz})^{-1} \sim |T - T_0|$ above and below the point T_0 (in agreement with (4.1.3) the coefficient for $T < T_0$ was twice as large).

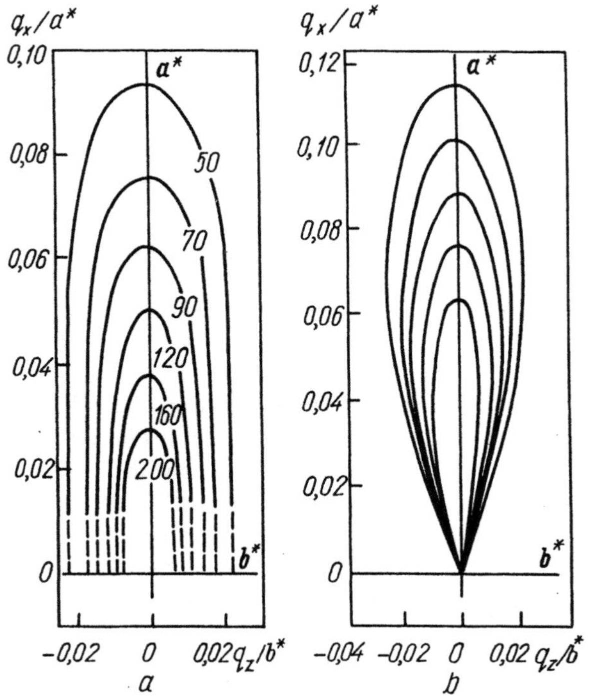

Fig. 4.16a,b. Equal intensity lines in the (a^*b^*) plane near the point (340) of the reciprocal lattice of triglicine sulphate at $T = T_0 + 0.1$ K derived after separation of thermal and incoherent scattering. (**a**) experimental results, (**b**) results calculated from (4.2.12) [4.186]

In neutron scattering experiments the instrumental broadening is usually large. For large q the associated errors are not significant. For small q, however, according to (4.2.12), the intensity I_1 varies rapidly with q_z at ranges which are smaller than the resolution in q_z, the intensity is effectively decreased in the $\boldsymbol{q}$ range under study, and a dip must occur at the plot of I_1 as a function of q_x for $q_z = 0$ (since q_z vanishes only nominally while in fact we have an instrumental spread in q_z which depends on q_x). This effect was observed and explained for neutron scattering in CsD_2PO_4 [4.188, 189].

The crystals of KH_2 PO_4 or KD_2PO_4 which are tetragonal (for $T > T_0$) and exhibit a phase transition of the first order but close to the second order (at 123 K and 221 K, respectively) belong to the class of ferroelectrics possessing a piezoelectric effect in nonferroelectric phase. The critical neutron scattering in KD_2PO_4 is quasielastic in character [3.214; 4.190]. For crystal systems with striction effects the scattering is described by (4.2.14, 15) which take into account the effect of the elastic energy on the fluctuation probabilities and the scattering by distortions produced by the striction strains. The terms $f\boldsymbol{Q} \cdot \boldsymbol{A}^i_{\boldsymbol{q}}$ added to $-\chi_i$ which are associated with this scattering must give rise to violations of the symmetry of the I_1 distribution with respect to the inversion at a reciprocal lattice point $(\boldsymbol{A}^i_{-\boldsymbol{q}} = -\boldsymbol{A}^i_{\boldsymbol{q}})$. These terms complicate further the angular dependence of the scattering intensity distribution (which is cumbersome enough owing to the dependence of $\varkappa'_{ij}(\boldsymbol{q})$ on $\boldsymbol{n}$ determined both by elastic and dipole forces) and also change the behavior of the intensity I_1 as a function of q near the points with

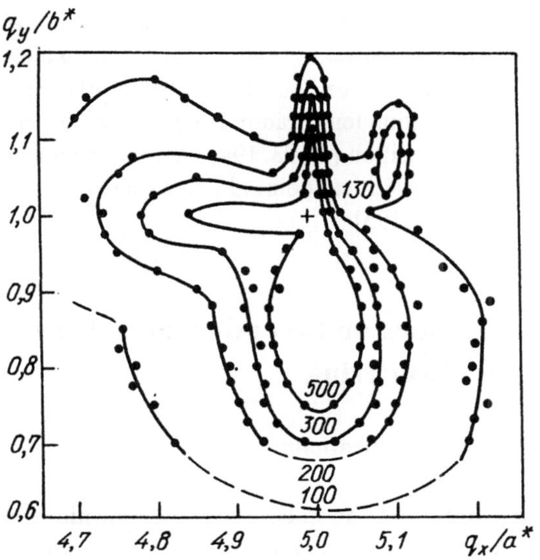

Fig. 4.17. Equal intensity lines in the (001) plane perpendicular to the axis $c*$ near the point (510) of the reciprocal lattice of KD_2PO_4 at 225 K [4.190]

$f \neq 0$ (we have then $I_1 \sim q^{-2}$ for $q \rightarrow 0$ instead of I_1 = const for the crystals with $\boldsymbol{A}_{\boldsymbol{q}}^{i} = 0$).

Figure 4.17 shows the resulting nonsymmetric intensity distribution for neutrons scattered at the (510) point [4.190]. This distribution of a complicated shape is described by the theoretical plots derived [4.191] from an equation similar to (4.2.15) which did not include fitting parameters (in contrast to the above derivation [1.16] of (4.2.15), the respective equation for I_1 in [4.191] was derived within the dynamic model; a detailed microscopic model which makes it possible to calculate the coefficient χ_z in (4.2.15) has been treated in [4.192]). Figure 4.18 illustrates the significant effect exerted by the dipole interaction on the scattering intensity distribution for this crystal. In accordance with (4.2.15), the intensity is high for $\boldsymbol{q} \parallel \boldsymbol{a}^*$ (for $n_z = 0$) and it is low for $\boldsymbol{q} \parallel \boldsymbol{c}^*$ (for $n_z = 1$ (when q are not too small, such anisotropy of the intensity distribution may be partially caused by the sharp anisotropy of the structure and interaction between atoms in KH_2PO_4 [3.216]).

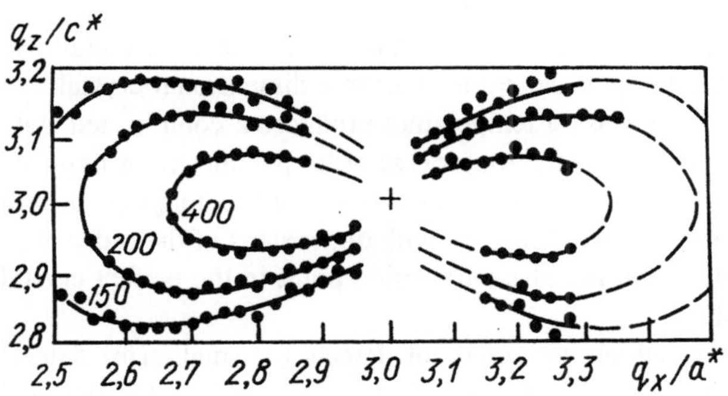

Fig. 4.18. Equal intensity lines in the (010) plane near the point (303) of the reciprocal lattice of KD_2PO_4 at 225 K [4.190]

The dipole forces and striction effects suppress the polarization fluctuations for all $\boldsymbol{q}$ with the exception of two lines in the q_x, q_y plane, that is, for which the transition is described by the Landau theory (see Sect. 4.1). When a relatively low pressure is applied to KH_2PO_4 the transition temperature is decreased and the transition becomes a second-order phase transition [4.193, 194] and, in agreement with theoretical predictions, we have at the tricritical point ($P_c \approx 2$ kbar) $\beta = 1/4$ while for $P = 3.5$ kbar we obtain $\beta = 1/2$ [4.193].

4.3 Effects Produced by Dislocations on Second-Order Phase Transitions and Anomalous Scattering

4.3.1 Phase Transitions in Crystals with Dislocations

Second-Order Phase Transitions in Imperfect Crystals. The phase transitions in imperfect crystals are strongly influenced by the types of defects and their mobility. If the point defects (for instance, impurities) have a high enough mobility and have time to be rearranged during the experiment to adjust to the changed long-range order then the presence of defects (as for nonstoichiometric composition) must cause renormalization of the critical indices [4.37] as discussed in Sect. 4.1.2 and a shifting of the transition point (which remains sharp). Point defects which are practically stationary ("frozen") produce effects on the phase transitions which are determined largely by the symmetry of the defects and the ordered phase [4.195–198]. It is simpler to consider the case of high-symmetry defects which alter only the interaction between atoms, for instance, the energies $\omega(\varrho)$ in the Ising model (see Sect. 1.3). Then we have to consider the random coupling forces described by a term of the type of $c_t\eta^2(\boldsymbol{R}_t)$ in the model Hamiltonian (here the numbers $c_t = 1, 0$ similar to $c_{s\gamma}$ given by (1.1.1) describe the arrangement of defects over the possible positions t). Renormalization group analysis demonstrates that a sharp phase transition exhibiting power-law features is preserved in an imperfect crystal with such defects but the critical indices are changed somewhat [4.199–205]. The "frozen" point defects described by a term similar to $c_t\eta(\boldsymbol{R}_t)$ (or $c_t h(\boldsymbol{r} - \boldsymbol{R}_t)\eta(\boldsymbol{r})$ where h is the effective field) in the Hamiltonian produce a much stronger effect since they give rise to effective fields which are linearly related to the order parameter and violate the crystal symmetry. Under these conditions defects can destroy the long-range order even in a three-dimensional crystal and a continuous transition with decreasing temperature produces a complicated state with alternating regions described by the long-range order parameters of opposite signs [4.206, 217]. Bounded defects of high enough strength belonging to both above types give rise to locally ordered regions with dimensions of the order of r_c at temperatures much higher than the phase transition point in the perfect crystal [4.195, 197, 198, 216–218].

When the crystal imperfections are not point defects (or small-size) defects but large-scale inhomogeneities of a large characteristic size r_1 then the physical picture is different [4.219]. If the size r_1 is sufficiently large we may introduce local transition temperatures $T_0(\boldsymbol{r})$ [4.219, 220] with a characteristic random spread of $\delta T_0 = T_0 \delta\tau_0$. The condition for this is that $r_1 \gg r_c(\delta\tau_0)$ where $r_c(\tau)$ is the correlation length in the perfect crystal. For a homogeneous and isotropic random Gaussian field $\delta T_0(\boldsymbol{r})$ a decrease in temperature at first leads to the production of ordered regions which are defined sharply enough, are not related to each other, and have order parameters of the opposite signs. In this temperature range the long-range order in the entire crystal has not yet been established since the regions with $\pm\eta$ cancel each other out. When these regions account for a significant proportion of the crystal volume at a temperature T_p owing to their growth and aggregation (of the order of 1/6 for $d = 3$) percolation takes place, that is, an infinite cluster with a long-range order parameter of a certain sign is produced throughout the crystal. Then the symmetry of the regions with $\pm\eta$ is lost, that is, a long-range order is established. This means that the temperature T_p is the point of the percolation phase transition [4.219]. This transition is sharp, as is the percolation effect [4.221, 222], and it is described by the critical indices β_p, γ_p, and ν_p though they differ from the conventional indices for the second-order phase transitions in the perfect crystal (according to [4.221, 222], $\beta_p \approx 0.35$, $\gamma_p \approx 1.7$, and $\nu_p \approx 0.9$).

We see that a phase transition in imperfect crystals must also be sharp, with the exception of a crystal containing "frozen" defects which violate the symmetry and produce effective fields. Experimentally observed phase transitions, however, often are spread over a significant temperature range even in crystals where defects have high mobilities or the symmetry conditions do not allow production of effective fields (for instance, for magnetic phase transitions in the absence of relativistic effects). In some cases the transition is spread out because the crystal specimen is not homogeneous enough but in other cases finer physical factors are at work. One of them is a peculiar effect exerted by dislocations on second-order phase transitions which gives rise under certain conditions to a sharp but weakly manifested phase transition at $T \approx T_f$ which is much higher than the transition point T_0^0 for the perfect crystal and a spread-out phase transition near $T \approx T_0^0$ [4.223].

Dislocations exert their characteristic effects through the long-range stress fields produced by them which decrease at a slow rate (as $1/r_\perp$) with increasing distance from the dislocation line. A set of dislocations produce a random field whose effect on an ordering system can be analyzed in the approximation of the local transition temperatures $T_0(\boldsymbol{r})$ owing to smoothness of its variation. In a crystal with dislocations, however, a random field $T_0(\boldsymbol{r})$ is generally more complicated in form than the Gaussian field with a homogeneous and isotropic correlation which has been noted above to give rise to a percolation phase transition. This is why the physical pattern of a phase transition in a crystal with dislocations may differ significantly both from the percolation transition and from the transition in the perfect crystal. This difference is largely determined by a purely geometric factor, namely, by the presence of long dislocation lines extending in given directions around which cylindrical regions with maximal $T_0(\boldsymbol{r})$ are produced. In these regions the ordered domains can appear at temperatures much higher than T_0^0 [4.223–226].

A significant suggestion has been made [4.223] that the ordered regions produced near various dislocation systems can make up a random skeleton extending throughout the crystal but occupying only a small fraction of its volume. The production of such a skeleton qualitatively alters the pattern of the phase transition in crystals with dislocations. At relatively high temperatures the skeleton contains identical fractions of the ordered domains with the opposite signs of the parameter η, that is, the crystal has no long-range order. At a temperature $T_f > T_0^0$ this structure is transformed into a state in which domains with a certain sign of η have a higher content, that is, a true phase transition occurs in the dislocation skeleton which establishes the long-range order parameter $\bar{\eta} \equiv \langle \eta(\boldsymbol{r}) \rangle \neq 0$. This transition involves, however, only a small part of the crystal volume and thus it is difficult to observe in experiment. As the temperature decreases the thickness of the ordered skeleton regions grows and gradually comparatively large ordered clusters are produced around $T \approx T_0^0$ which cover many dislocations and are linked to the skeleton, that is, included in the infinite cluster. They have the same sign of the parameter η as the regions in the skeleton. Thus, in this range of $T \approx T_0^0$ ordering involves the entire crystal and the mean parameter $\bar{\eta}(T)$ varies continuously from very small values to values for the perfect crystal. Experimental data reveal this process as a spread-out second-order phase transition. Thus, the physical factor explaining the spread-out phase transitions found in experiments may be in some circumstances the production of an ordered dislocation skeleton.

The above processes must take place for $T_f - T_0^0 \gg \delta T_0$ where δT_0 is the characteristic spread of $\delta T(\boldsymbol{r})$. For other parameters of the crystal and the dislocation ensemble we have $T_f - T_0^0 \ll \delta T_0$ and large ordered regions can be produced before the phase transition occurs in the skeleton. Then, as discussed above, we obtain the percolation phase transition which involves production of an infinite cluster consisting of randomly arranged ordered regions (which are not linked up to form a skeleton).

Random Field of Transition Temperatures. We shall limit the discussion to crystal systems in which the symmetry conditions allow striction effects which can only be of the second order (not first order) in the parameter η. Then if we consider a sufficiently smoothly varying field the local temperature $T_0(\boldsymbol{r})$ of the phase transition linearly depends on the strain $u_{ij}(\boldsymbol{r})$. To simplify the analysis we shall assume that the temperature $T_0(\boldsymbol{r})$ is determined only by dilatation $u_{ij}(\boldsymbol{r})$, that is, we have

$$\tau_0(\boldsymbol{r}) = \frac{T_0(\boldsymbol{r}) - T_0^0}{T_0^0} = \varkappa_0 u_{ii}(\boldsymbol{r}) \,, \quad \varkappa_0 = -K \frac{1}{T_0^0} \frac{\partial T_0^0}{\partial P} \tag{4.3.1}$$

where P is the pressure. Here we do not take into account the variation of the modulus of dilatation K near T_0^0 which has an effect only in the temperature range described by (4.2.39) which is not significant for our analysis (the same is true for the strains caused by the changes in the orderd. When we discuss solutions we must bear in mind that in the field of a dislocation the concentration is redistributed so that the temperature $T_0(\boldsymbol{r})$ is changed in its turn. In the approximation linear in u_{ij} and δc we may assume that the appropriate renormalization of $\varkappa_0$ takes into account this effect in (4.3.1). We shall assume also that in the range of $\boldsymbol{r}$ under consideration we can ignore the effects of the second order in δc in comparison with the effects linear in u_{ij}.

In crystals which include several systems of straight-line dislocations we can write the dilatation and, hence, $\tau_0(\boldsymbol{r})$ as the superposition of the contributions made by individual defects:

$$\begin{aligned}
\tau_0(\boldsymbol{r}) &= \sum_{\alpha}\sum_{t} c_{t\alpha}\tau_{0\alpha}(\boldsymbol{r}-\boldsymbol{r}_{t\alpha}) , \\
\tau_{0\alpha}(\boldsymbol{r}-\boldsymbol{r}_{t\alpha}) &= \frac{\varkappa_0 b}{|\boldsymbol{r}_{\perp\alpha}-\boldsymbol{r}_{t\alpha}|} f_\alpha(\theta) , \\
f_\alpha(\theta) &= -\frac{1}{2\pi}\frac{1-2\sigma}{1-\sigma}\sin\theta .
\end{aligned} \tag{4.3.2}$$

Here b is the length of the Burgers vector $\boldsymbol{b}_\alpha$, $\boldsymbol{r}_{t\alpha}-\boldsymbol{r}_{\perp\alpha}$ is the vector passing from the point $\boldsymbol{r}$ under consideration to the dislocation line $t\alpha$ in the plane perpendicular to the line ($\boldsymbol{r}_{\perp\alpha}$ is the projection of $\boldsymbol{r}$ in this plane); the equation for $f_\alpha(\theta)$ (where θ is the angle between the vectors $\boldsymbol{r}_{\perp\alpha}-\boldsymbol{r}_{t\alpha}$ and $\boldsymbol{b}_\alpha$) is written for edge dislocations in isotropic crystals but $2\pi|f_\alpha(\theta)| \sim 1$ in other cases, too; and $c_{t\alpha} = 1,0$ are random numbers ($c_{t\alpha}$ is unity if a dislocation line passes through a position t in the plane perpendicular to the lines of the dislocations belonging to the system α and otherwise it vanishes). The distribution of probabilities for the function $\tau_0(\boldsymbol{r})$ given by (4.3.2) can be written as

$$w(\tau_0) = \left\langle \delta\left[\tau_0 - \sum_{\alpha t} c_{t\alpha}\tau_{0\alpha}(\boldsymbol{r}-\boldsymbol{r}_{t\alpha}\right]\right\rangle . \tag{4.3.3}$$

If we use the well-known integral representation for the δ function we can easily perform averaging over $c_{t\alpha}$ for the fully random distribution of dislocations, that is, for statistically independent $c_{t\alpha}$. This averaging is performed in exactly the same way as the calculation of the average intensity of scattering by a crystal with dislocations (see Sect. 5.1 in [1.1]). If the crystal size L is sufficiently large in comparison with the mean distance r_d between dislocations so that $l = \ln(\zeta L/r_d) \gg 1$ (where $\zeta \sim 1$) then, according to the results of Sect. 5.1 in [1.1], $w(\tau_0)$ has a Gaussian shape at the central part and behaves as a power function at the wings:

$$w(\tau_0) = \frac{1}{(2\pi)^{1/2}\delta\tau_0} \exp\left[-\frac{\tau_0^2}{2(\delta\tau_0)^2}\right] \quad \text{for} \quad |\tau_0| \lesssim \delta\tau_0 ,$$
$$w(\tau_0) = \frac{1}{2l\delta\tau_0}\left(\frac{\delta\tau_0}{|\tau_0|}\right)^3 \quad \text{for} \quad |\tau_0| \gg \delta\tau_0 , \tag{4.3.4}$$
$$\delta\tau_0 = \varkappa_0 \overline{f(\theta)}(lnb^2)^{1/2} , \quad \left[\overline{f(\theta)}\right]^2 = \int_0^{2\pi} f_\alpha^2(\theta) d\theta .$$

Here $n = r_d^{-2}$ is the density of dislocations ($n = \sum_\alpha n_\alpha$ where n_α is the density of the dislocation lines of the system α per unit area in the plane perpendicular to these lines).

According to (4.3.4), the mean fluctuations $\delta\tau_0$ of τ_0 are greater by a factor of $\sim \sqrt{l}$ than the inhomogeneities of τ_0 produced by the dislocations which are at a mean distance r_d. This is explained by a slow decrease in $\tau_{0\alpha}(\boldsymbol{r} - \boldsymbol{r}_{t\alpha})$ which gives rise to the conditions under which for $l \gg 1$ the main contribution to the fluctuations of τ_0 in the bulk of the crystal is made by a large number of distant dislocations, rather than several nearest ones. This is why we obtain a Gaussian-type function for $|\tau_0| \lesssim \delta\tau_0$. On the other hand, in a smaller (of the order of l^{-1}) part of the crystal in the regions near the dislocation lines corresponding to $|\boldsymbol{r}_{\perp\alpha} - \boldsymbol{r}_{t\alpha}| < r_1' \sim r_d/\sqrt{l}$ the main contribution to τ_0 is made by the nearest dislocation and here $|\tau_0| > \delta\tau_0$ is sufficiently large. It is these regions that give rise to the power-law variation of $w(\tau_0)$ for large $|\tau_0|$. If the distribution of dislocations is not random the averaging procedure for $w(\tau_0)$ is more complicated and the form of this function in the central part is changed (but its width can still be found from (4.3.4) with a somewhat different l; see Sect. 5.2 in [1.1]). The equation for $w(\tau_0)$ for large $|\tau_0|$ remains valid, though.

Ordered Regions Near Single Dislocations. When the reduced temperature $\tau = (T - T_0^0)/T_0^0$ is comparatively high and $\tau \gg \delta\tau_0$, the strains in the bulk of the crystal are not large enough for causing ordering and only thin cylindrical ordered regions can appear near the dislocation lines. In order to analyze them we must first consider ordering near a single straight-line dislocation. If the fields are smoothly varying we can ignore the gradient terms in the thermodynamic potential density $\varphi(\boldsymbol{r})$, use the approximation of local transition temperatures and find from (4.3.2) the following condition for the boundary $r_0(\theta)$ of the ordered regions, that is, the region where $\tau \leq \tau_0(\boldsymbol{r}) = \tau_{t\alpha}(\boldsymbol{r} - \boldsymbol{r}_{t\alpha})$:

$$0 < r_{\perp\alpha} < r_0(\theta) = \frac{\varkappa_0 b}{\tau} f_\alpha(\theta) \quad \text{fo} \quad \varkappa_0 f_\alpha(\theta) > 0 ,$$
$$r_0(\theta) = 0 \quad \text{for} \quad \varkappa_0 f_\alpha(\theta) < 0 . \tag{4.3.5}$$

Here it is assumed that $\boldsymbol{r}_{t\alpha} = 0$.

For the sake of simplicity we shall consider only systems described by a one-component order parameter. Equations (4.1.32) and (4.3.2,5) yield the following expressions for the order parameter $\eta(\boldsymbol{r})$ and its averaged value $\bar{\eta}$ in the ordered region:

$$\eta(\boldsymbol{r}) = \eta^0[\tau_{0\alpha}(\boldsymbol{r}) - \tau]^\beta = \eta^0\tau^\beta \left[\frac{r_0(\theta)}{r_{\perp\alpha}} - 1\right]^\beta ,$$
$$\tilde{\eta} = \frac{\pi\beta(1-\beta)}{\sin\pi\beta}\eta^0\tau^\beta . \tag{4.3.6}$$

The approximation of local transition temperatures is based on ignoring the dependence of $\varphi(\boldsymbol{r})$ on $\Delta\eta$ but this assumption is valid if $r_c^2[\tau - \tau_0(\boldsymbol{r})]|\Delta\eta| \ll \eta$ where $r_c[\tau - \tau_0(\boldsymbol{r})]$ is the correlation length for the shifted transition (see, for instance, [4.15]). In the main part of the ordered region we have $|\Delta\eta| \sim r_0^{-2}\eta$ where $r_0 = \max r_0(\theta)$, $|\tau - \tau_0(\boldsymbol{r})| \sim \tau$, and $r_c[\tau - \tau_0(\boldsymbol{r})] \sim r_c(\tau) \equiv r_c$, that is, the above condition is satisfied if $r_0 \gg r_c = r_{c0}\tau^{-\nu}$ or

$$\tau^{1-\nu} \ll \tau_1^{1-\nu} , \quad \tau_1^{1-\nu} = \varkappa_0\overline{f(\theta)}b/r_{c0} . \tag{4.3.7}$$

In a layer at the boundary of the ordered region $|\Delta\eta| \sim (r_0 - r_{\perp\alpha})^{-2}\eta$ is large, $r_c|\tau - \tau(\boldsymbol{r})| \gg r_c$ and the above approximation is inapplicable. But the thickness of this layer $\delta r \sim r_0(\tau/\tau_1)^{\xi_1}$ where $\xi_1 = (1-\nu)(1+\nu)^{-1}$ is small in comparison with r_0 if the condition (4.3.7) is satisfied.

If $\varkappa_0 f(\theta)b \gtrsim r_{c0}$ then (4.3.7) yields $\tau_1 \sim 1$, that is, ordered regions with a thickness of $r_0 \gg r_c$ (described in the approximation of local transition temperatures) are produced for all $\tau \ll 1$. In other cases, however, especially in systems with a large interaction range (where $r_{c0} \gg b$ and $\nu = 1/2$) we have $\tau_1 \ll 1$, and the condition (4.3.7) is satisfied only for very small τ. When $\tau > \tau_1$ this condition is not satisfied, we have $r_c > r_0$ and the perturbation produced by the dislocation is not strong enough for distinct ordered regions to be produced. Therefore, τ_1 plays the part of the temperature of production of such regions. The estimate (4.3.7) agrees with the more detailed results obtained for the transition to the superconducting state in the Landau theory ($\nu = 1/2$) with the inclusion of the gradient terms in $\varphi(\boldsymbol{r})$ [4.225].

When we consider the structure of the ordered cylindric region we must bear in mind that in the range of $\tau \sim \tau_1$ the order must be essentially inhomogeneous not only in the transverse direction but also in the longitudinal direction in the cylindric region [4.223]. The explanation is that the boundaries between the antiphase ordered domains with different signs of the order parameter η or between the ordered and disordered regions for $r_0 \sim r_c$ have an interface energy $\sim \gamma r_0^2 \sim \gamma r_c^2$ which is not larger than $k_B T_0^0$ by an order of magnitude (according to the estimate of the interface energy γ per unit area). This inhomogeneous ordered quasi-one-dimensional region consists of domains with sizes $\sim r_0 \sim r_c$ and a formation energy $\sim k_B T_0^0$ and is produced gradually as a result of a significantly spread phase transition (in the range of $\tau \sim \tau_1$).

Even for $\tau \ll \tau_1$ when the ordered region has been produced it can be broken down into antiphase domains. In the Ising model (for instance, for ordering of alloys) the energy of the boundaries between the domains is of the order of $\gamma r_0^2 \sim k_B T_0^0 r_0^2/r_c^2 \gg k_B T_0^0$ and for $\tau \ll \tau_1$ it is a thermodynamically favourable process for the elongated cylindric domains with a mean length of

$$D_0 \sim r_c \exp(\gamma r_0^2/\zeta_1 k_B T_0^0) \sim r_c \exp(r_0^2/\zeta_2 r_c^2) \gg r_c \,, \tag{4.3.8}$$

to be produced (here $\zeta_1 \sim \zeta_2 \sim 1$). In degenerate systems of the Heisenberg type and in ferromagnets with a considerable magnetostatic energy, the equations for the length D_0 have different forms but always $D_0 \gg r_c$ [4.223].

Regions of a changed composition and structure are produced near dislocations also in systems where first-order phase transitions occur [2.28; 4.227–231]. These regions, however, exhibit other types of temperature dependences for the sizes and other parameters.

Phase Transition Associated with the Dislocation Skeleton of Ordered Domains. In crystals with dislocations the character of a second-order phase transition depends significantly on the relation between the temperature τ_1 of the production of ordered regions and the averaged variation $\delta\tau_0$ of the local ordering temperatures. According to (4.3.4, 7), if the condition

$$B_0 = \frac{r_d}{r_{c0}} \left(\frac{\varkappa_0 \overline{f(\theta)} b}{r_{c0}} \right)^{\lambda} \sim \frac{l^{1/2}\tau_1}{\delta\tau_0} \gg l^{1/2} \,, \quad \lambda = \frac{\nu}{1-\nu} \tag{4.3.9}$$

is satisfied, we have $\tau_1 \gg \delta\tau_0$, that is, as the temperature decreases at first at a comparatively high temperature $\tau \sim \tau_1$ cylindric ordered regions are produced near individual dislocations and the thickness r_0 of these regions grows as $1/\tau$ for $\tau \gg \delta\tau_0$.

Since typically a crystal contains several systems of dislocations having different directions of the dislocation lines then purely geometric considerations suggest that the ordered cylindrical regions are not isolated and each of them must sooner or later intersect cylindrical regions belonging to other dislocation systems. As a result, the ordered regions near the randomly distributed dislocations make up a complex irregular skeleton consisting of thin cylinders aligned along several characteristic directions and intersecting cylinders belonging to other systems at randomly distributed positions. The geometric considerations indicate that if the arrangement of the dislocations of different systems is totally uncorrelated the segments between two intersections have the average length

$$D_1 \sim \frac{1}{n r_0} = \frac{r_d^2}{r_0} \sim \frac{r_d^2}{\varkappa_0 \overline{f(\theta)} b} \tau \,. \tag{4.3.10}$$

In the other limiting case when the distribution of dislocations is not entirely random and their lines are linked up forming a Frank network the characteristic segment length is apparently $D_1 \sim r_d$.

The structure of the ordered skeleton is largely determined by the ratio between D_1 and the characteristic domain length D_0 given by (4.3.8). For comparatively large τ when $D_0 \ll D_1$ each part of the skeleton and thus the skeleton as a whole is broken down into fluctuation domains with opposite signs of η. Therefore, the average value $\bar{\eta} = 0$ and no long-range order is established although each domain has a local order. At lower temperatures when (4.3.8) shows D_0 to be large (but

$D_0 \ll L$ where L is the crystal size) the variation of the sign of the parameter η in an isolated ordered cylinder at large enough distances would destroy the long-range order in any case. But for a three-dimensional ($d = 3$) or two-dimensional ($d = 2$) skeleton when $D_0 \gg D_1$ it is unlikely that the order parameter may reverse its sign in individual regions of the skeleton as happens with the ferromagnetic transition. On the other hand, for the production of a region with an opposite sign of the parameter η and a size $\Lambda \gg D_1$ in the skeleton, antiphase boundaries must be produced in $\sim (\Lambda/D_1)^{d-1}$ domains and the probability of this is also exponentially small ($\sim \exp[-(\Lambda/D_1)^{d-1} \ln(D_0/r_c)]$.

Therefore, a phase transition in the dislocation skeleton of the ordered regions must take place at a temperature $T = T_f$ (or $\tau = \tau_f$) for which we have

$$D_0(\tau_f) = \zeta D_1(\tau_f)\,, \quad \zeta \sim 1\,, \tag{4.3.11}$$

For instance, in the Ising model for ferromagnetic-type ordering this is the transition of a skeleton consisting of equal numbers of antiphase domains with opposite signs of the parameter η to a state in which most of the domains have a given sign of the order parameter. The condition (4.3.11) and the expressions (4.3.8, 10) show that in the Ising model with a completely random distribution of dislocations we have

$$r_0^2(\tau_f)/r_c^2(\tau_f) = \zeta_2 \ln(\zeta' r_d^2/r_0 r_c)$$

or

$$\begin{aligned} \tau_f &= \zeta_3 \tau_1 \{\ln[\zeta_4 B_0^2 (\tau_f/\tau_1)^{1+\nu}]\}^{-\lambda_1}\,, \\ \lambda_1 &= \frac{1}{2(1-\nu)}\,, \quad \zeta' \sim \zeta_3 \sim \zeta_4 \sim 1\,. \end{aligned} \tag{4.3.12}$$

Equations (4.3.12) are qualitatively applicable also to the dislocation ensemble with a correlation similar to that discussed in Sect. 5.2 in [1.1]. If the skeleton is based on the Frank network and $D_1 \sim r_d$ then r_d^2/r_c in (4.3.12) must be replaced with r_d and accordingly the arguments of the logarithms must be divided by $B_0\tau_f/\tau_1$ so that τ_f will be increased. (The expressions for τ_f are changed for degenerate ordered systems and ferromagnets with a considerable magnetostatic energy [4.223]. The qualitative estimate (4.3.12) for τ_f derived from the condition (4.3.11) and the respective logarithmic dependence of τ_f on the length D_1 agree with the results of the exact solution for the two-dimensional Ising problem for a regular skeleton made up of one-dimensional chains intersecting at the intervals D_1 [4.232].)

It follows from (4.3.9, 12) that τ_f increases with growing dislocation density. If, for instance, $\varkappa_0 fb - r_{c0}/10$ and $r_d/r_{c0} \sim 10^3 \div 10^4$ then $B_0 \sim 10 \div 100$. Here τ_f may be a fraction of τ_1 and a few times greater (or by a factor of a few tens) than $\delta\tau_0$.

Thus, we may have a considerable temperature range of $\tau_f > \tau \gg \delta\tau_0$ in which the crystal contains a skeleton of ordered regions occupying a small part of its volume. In a considerably smaller range of τ of the order of $\tau_f - \tau \sim$

$\tau_f / \ln(D_1/r_c) \sim \tau_f r_c^2(\tau_f)/r_0^2(\tau_f) \ll \tau_f$ most domains with an order parameter of a given sign disappear and the parameter η averaged over the volume of the domains increases from 0 to $\bar{\eta}$ (see (4.3.6)). Using (4.3.4–6) we obtain the following expression for order parameter $\eta(\boldsymbol{r})$ averaged over the crystal below τ_f (but considerably above $\delta\tau_0$):

$$\bar{\eta} = \frac{\pi\beta(1-\beta)}{4\sin\pi\beta}\eta^0 n b^2 \varkappa_0^2 [\overline{f(\theta)}]^2 \tau^{\beta-2} \quad (\tau_f > \tau \gg \delta\tau_f)\,. \tag{4.3.13}$$

In this range where the skeleton of ordered regions exists in a developed form the averaged parameter $\bar{\eta}$ varies as $\tau^{\beta-2}$ owing to the growth of r_0^2 as τ^{-2} with decreasing τ (although the function $\bar{\eta} \sim \tau^{\beta}$ decreases).

At lower temperatures $\tau \sim \delta\tau_0$, the thicknesses of the ordered regions are large, $r_0 \sim r_d/\sqrt{l}$, and the order is determined by the fields of not only the nearest but also distant dislocations. Superposition of such fields gradually gives rise to large ordered clusters encompassing a large number of dislocations.

The ordered structure in such clusters depends on the type of dislocations and the character of ordering. Under conditions of ferromagnetic ordering the order parameter η (magnetization) in the skeleton and in all ordered clusters has the same sign (since they are linked to the skeleton). Under antiferromagnetic ordering or ordering of alloys, however, the important part is played by the Burgers vectors $\boldsymbol{b}_\alpha$. If dislocations are produced in the ordered phase of the alloy then $\boldsymbol{b}_\alpha$ is a vector of its lattice, dislocations do not prevent long-range order from being established throughout the crystal and the order parameter in the large clusters has the same sign as in the skeleton, as is the case in ferromagnets. But if dislocations are produced in the disordered phase then the vectors $\boldsymbol{b}_\alpha$ do not coincide with the vectors of the lattice of the ordered phase and their very existence disrupts a persistent long-range order. Therefore, at $\tau \lesssim \delta\tau_0$ such dislocations give rise to a complicated field $\eta(\boldsymbol{r})$ with alternating sign consisting of domains with the extent of the order r_d linked with the dislocations which produce a peculiar dislocation glass (it disappears as if dislocations can be paired). Aggregation of such domains gives rise to a smoothly varying field $\eta(\boldsymbol{r})$ of a constant sign; this is found from the condition that the term with $(\nabla\eta)^2$ in $\varphi(\boldsymbol{r})$ has a minimum and the domain boundaries account for a relatively small volume fraction.

In alloys containing dislocations in the disordered phase the "topological" domains discussed above must exist also at temperatures $\tau \gg \delta\tau_0$ when ordering occurs only in the dislocation skeleton near the dislocation lines (such domains are linked to the intersections of the ordered regions near the dislocation lines belonging to different dislocation systems). Under these conditions, therefore, only the "fluctuational" antiphase boundaries disappear at the temperature T_f while the "topological" boundaries are preserved (they do not hinder the formation of the smoothly varying field of $\eta(\boldsymbol{r})$ of a constant sign, that is, the production of a fairly definite order in the dislocation skeleton).

When the order parameter η in the ordered regions has the same sign, the mean value $\bar{\eta}$ can be found as

$$\bar{\eta} = \int_{-\infty}^{\infty} \eta(\tau - \tau_0) w(\tau_0) d\tau_0$$
$$= \eta^0 \int_{\tau}^{\infty} (\tau_O - \tau)^{\beta} w(\tau_0)^{\beta} d\tau_0 \quad (\tau < \tau_f') , \tag{4.3.14}$$

where the probability density $w(\tau_0)$ for the random dislocations is given by (4.3.4). If a structure has emerged which consists of domains with the size of the order of r_d and with opposite signs of η, then (4.3.14) (and (4.3.15)) gives approximately the mean value $\overline{|\eta|}$ of the modulus of $\eta(\boldsymbol{r})$ (under an assumption that the volume of the domain boundaries is small). When $\tau \gg \delta\tau_0$ the equations (4.3.4, 14) lead to (4.3.13) and for $\tau \lesssim \delta\tau_0$ for a random arrangement of dislocation we obtain

$$\bar{\eta} = \eta^0 \frac{\Gamma(1+\beta)}{\sqrt{2\pi}} (\delta\tau_0)^{\beta} \exp\left[-\frac{\tau^2}{4(\delta\tau_0)^2}\right] D_{-1-\beta}\left(\frac{\tau}{\delta\tau_0}\right) \quad (\tau \lesssim \delta\tau_0) , \tag{4.3.15}$$

where $D_{-1-\beta(x)}$ is a parabolic cylindrical function. When τ is negative and $-\tau \gg \delta\tau_0$ equations (4, 3.14, 15) describe a continuous transition from the asymptotic function (4.3.13) corresponding to the "skeleton ordering" to the function $\eta^0(-\tau)^{\beta}$ corresponding to ordering of almost the entire crystal.

In the range of $\tau \gg \delta\tau_0$ the averaged order parameter $\bar{\eta}$ is very small and thus the transition to the skeleton ordered state at $\tau = \tau_f$ can remain unnoticed in experiments. Then the fact that the order parameter dependence $\eta \sim (-\tau)^{\beta}$ (and similar dependences for other parameters) has been established in the range of $|\tau| \sim \delta\tau_0$ will be interpreted as a phase transition and in experiments such dependences correspond to a second-order phase transition spread over the range $\delta\tau_0$.

We see that the existence of the dislocation skeleton causes a spread-out transition near $\tau = 0$ in addition to the true transition at $\tau = \tau_f$ which is difficult to notice. To find the range of spreading of the transition assume, for instance, that $\varkappa_0 f(\theta)\sqrt{1} \sim 1$ and $n \sim 10^9\,\mathrm{cm}^{-2}$. Then (4.3.4) yields $\delta\tau_0 \sim 10^{-3}$. The behavior of the averaged order parameter as a function of temperature in the transition range is illustrated by the plot of $y = (\bar{\eta}/\eta^0)\sqrt{2}\delta\tau_0)^{-1/2}$ given in Fig. 4.19 for $\beta = 1/3$ and $l = 8.5$. The plot of $\tilde{\beta} = d(\ln\bar{\eta})/d(\ln|\tau|)$ also shown in Fig. 4.19 illustrates the approximation function $\bar{\eta}(\tau) \sim (-\tau)^{\beta}$. It can be seen that $\tilde{\beta} > \beta$ in a certain temperature range and at the peak $\tilde{\beta} \approx 0.5$. The experimental results for spread-out phase transitions observed in various τ ranges in some solids are reported, for instance in [4.233–237]. For example, the Mössbauer results for the magnetic transition in Ni demonstrated a dependence of $\tilde{\beta}$ on τ similar to that shown in Fig. 4.19 (with $\tilde{\beta} \approx 0.5 > \beta$ for very small τ) [4.233]. Similar spreading of the peak is observed for the plots of the specific heat and other properties. Since it is assumed that the range $\delta\tau$ given by (4.1.39) where $\delta K \gtrsim K$ is small in comparison with $\delta\tau_0$ the first order striction phase transition discussed in Sect. 4.1.2 must not occur in this case.

Percolation Phase Transition in Crystals with Dislocations. In systems with small $\varkappa_0$ or $r_{c0} \gg b$ when the dislocation density is high the condition $B_0 \gg l^{1/2}$,

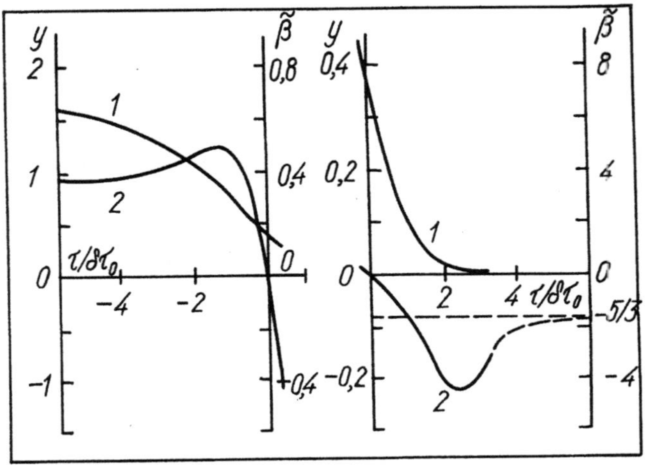

Fig. 4.19. The reduced averaged order parameter y (*curve 1*) and the effective critical index β (*curve 2*) as functions of the reduced temperature $\tau/\delta\tau_0$ for second-order phase transitions in crystals with dislocations. The vertical axes have different scales in two segments of the figure

which is the reverse of (4.3.9) is satisfied. Then $\tau_1 \ll \delta\tau_0$ and the ordered regions are not produced near individual dislocations at $\tau \sim \delta\tau_0$. They are produced only at $\tau \sim \delta\tau_0$ and involve many dislocations at once. At first, these regions are isolated and the order parameters η have opposite signs in various regions. As temperature decreases the size and the number of the ordered regions grow until at a temperature $\tau = \tau_p$ an infinite cluster is produced with a certain sign of η, that is, percolation takes place (in the case of ferromagnetic-type ordering or when dislocations are produced in the ordered phase). This temperature corresponds to a percolation phase transition of the second order in an inhomogeneous system. In some cases $B_0 \gg l^{1/2}$ and the skeleton of ordered regions (which have order parameters of opposite signs) is produced at $\tau \sim \tau_1 \gg \delta\tau_0$ but no phase transitions occur in the skeleton at temperatures significantly higher than $\delta\tau_0$ so that the aggregation of the ordered regions with a size $\gtrsim r_d$ at $\tau = \tau_p \sim \delta\tau_0$ causes a percolation phase transition.

At $\tau \approx \tau_p \sim \delta\tau_0$ the strain in the ordered regions is primarily determined not by the nearest dislocation but rather by a large number of distant dislocations. Therefore, the distribution of τ_0 probabilities for $|\tau_0| \gtrsim \delta\tau_0$ is close to a Gaussian distribution and the percolation transition in a crystal with dislocations may be described with the results calculated from the phenomenological theory with a Gaussian probability distribution [4.219] (with the parameters selected for the above problem). In the three-dimensional model the percolation phase transition occurs when the total content of the ordered regions amounts to 0.17 of the crystal volume [4.221, 222]. For the distribution (4.3.4) this content is obtained at

$$\tau = \tau_p \approx 0.95\delta\tau_0 = 0.95\varkappa_0\overline{f(\theta)}(\ln b^2)^{1/2} . \tag{4.3.16}$$

According to (4.3.16), at $\tau_f \lesssim \delta\tau_0$ the dislocations cause a shift of the phase transition point which is proportional to $\sqrt{n}$. For instance, if $\varkappa_0 \overline{f(\theta)}\sqrt{l} \sim 1$ then $\tau_p \sim \delta\tau_0 \sim 10^{-4}$ for $n = 10^7\,\mathrm{cm}^{-2}$ and $\tau_p \sim 10^{-2}$ for $n = 10^{11}\,\mathrm{cm}^{-2}$. In the range of $|\tau - \tau_p| \ll \delta\tau_0$ a significant part is played by the processes of the percolation aggregation of the ordered regions and in the approximation of local transition temperatures (the conditions for its applicability are discussed in [4.219–223]) the behavior of $\bar{\eta}$, $\varkappa$, and r_c as functions of $\tau - \tau_p$ is described by the percolation critical indices $\beta_p \approx 0.35$, $\gamma_p \approx 1.7$, and $\nu_{pp} \approx 0.9$. The percolation dependences for $\varkappa$ and $\bar{\eta}$ are replaced with the functions (4.1.32) for the perfect crystals only when $|\tau| \gg \delta\tau_0$ (the behavior of $\bar{\eta}$ and $\varkappa$ in the intermediate range is analyzed in [4.219]).

In a crystal with uniformly and randomly distributed dislocations a sharp percolation phase transition can be spread over a range $\sim \delta\tau_0/l$ owing to the effects produced by the crystal boundaries on the dislocation fields. They give rise to additional terms in $\delta\tau_0$ which are smaller by a factor of l than the conventional terms but depend on $\boldsymbol{r}$ and vary at distances of the order of L. The spreading of this transition can also be caused by a macroscopically inhomogeneous plastic strain (and inhomogeneity in the composition of the crystal).

The above results are applicable not only to the straight-line dislocations but also to dislocation loops of a large radius $R_0 \gg r_d$. We must only replace n with $N_L R_0 \sim n$, l with $\sim \ln(R_0/r_d)$, and $\overline{|f(\theta)|^2}$ with $2\pi\langle f_\alpha^2(\theta)\rangle$ where the angle brackets denote averaging over θ and over the perimeter of the dislocation loop. If $R_0 \ll r_d$ then a linked skeleton is not produced and a spread of τ_0 is mainly determined by the effect of the nearest loop. The percolation phase transitions for the isolated ordered regions have been analyzed in [4.223, 238].

4.3.2 Scattering by Long-Range Order Inhomogeneities Produced by Dislocations Near a Second-Order Phase Transition

Near a second-order phase transition diffuse scattering by defects exhibits some characteristic features. They are explained by the fact that a defect produces a region where the order parameter is changed and the size of such regions increases rapidly as the temperature approaches the transition point. For point defects or bounded defects of a small radius the region where the order parameter is perturbed has a size of the order of the correlation length r_c which is large near T_0 so that the intensity of X-ray or light scattering increases considerably near this point [2.23; 4.76, 87, 195, 196, 198, 239]. (For instance, the "frozen" defects, which violate the symmetry and are related to the order parameter through a term similar to $c_t h(\boldsymbol{r}-\boldsymbol{R}_t)\eta(\boldsymbol{r})$ in the Hamiltonian, in the case of normal concentrations produce static waves of variation of the order parameter, $\eta_k \sim \varkappa(\boldsymbol{k})h_k$ and give rise to diffuse scattering intensity $\sim cN\varkappa^2(\boldsymbol{k})|h(\boldsymbol{k})|^2$ which is proportional to $\varkappa^2(\boldsymbol{k})$ and thus to $\tau^{-2\gamma}$ for $k \to 0$ [4.76, 196, 239].)

The defects which generate slowly decreasing stress fields can produce larger regions of slowly varying order parameter $\eta(\boldsymbol{r})$ with the size $r_0 \gg r_c$ which can be analyzed in the approximation of local transition temperatures. These regions must produce high-intensity diffuse scattering. Under certain conditions this scattering has a higher intensity than the conventional scattering by critical fluctuations in perfect crystals. Since the regions where the order parameter is changed are static (or quasistatic if, for instance, the point defects producing them can migrate between two equivalent positions reversing the sign of the order parameter η), they cause elastic (or quasielastic) scattering. If the critical fluctuations in perfect crystals have high characteristic relaxation frequencies, the intensity of scattering by defects can be identified by analyzing the energy distribution of the scattered neutrons or Mössbauer photons. The defect-produced inhomogeneities must give rise to a narrow central peak while the critical fluctuations produce a wider distribution [2.23; 4.76, 196, 239, 240] (apart from the defects, other factors, however, may give rise to the central peak in the energy distribution [2.23; 4.76, 241–245]).

The defects producing slowly decreasing stress fields are illustrated by dislocations which give rise to a skeleton of thin ordered regions at relatively high temperatures $\tau \sim \tau_1$ and a complicated inhomogeneous distribution of $\eta(\boldsymbol{r})$ at $|\tau| \lesssim \delta\tau_0$ (see Sect. 4.3). Disk-shaped or more isotropic distributions of the intensity of scattering by these inhomogeneities must accordingly be produced near the superlattice points of the reciprocal lattice. We shall discuss them below when we consider crystal systems described by a one-component order parameter η [4.246].

Denote the time-averaged order parameter (dependent on $\boldsymbol{r}$ in a crystal with dislocations) by $\eta(\boldsymbol{r}) = \bar{\eta} + \Delta\eta(\boldsymbol{r})$ and its fluctuation by $\delta\eta(\boldsymbol{r})$. As was done for (3.6.3), we may assume that the structure amplitude of the sth lattice cell $f_s = \chi[\eta(\boldsymbol{R}_s)+\delta\eta(\boldsymbol{R}_s)]$ is proportional to $\eta+\delta\eta$ in the vicinities of the superlattice reflections ($\boldsymbol{Q} \approx \boldsymbol{G}'$) (scattering in the vicinities of the lattice reflection ($\boldsymbol{Q} \approx \boldsymbol{G}$) and neutron scattering near the magnetic transition point have also been analyzed in [4.246]). When we substitute this expression for the structure amplitude into the general equation (3.6.2) we can write the scattering intensity in the vicinity of these reflections as a sum of the term $I^{(1)}$ describing the scattering by the inhomogeneous time-averaged distribution of the order parameter and the term $I^{(2)}$ describing the critical scattering by the order fluctuations produced by dislocations in the crystal:

$$I = I^{(1)} + I^{(2)}\,, \quad I^{(1)} = \frac{|\chi|^2}{v^2}\int\!\!\int \eta(\boldsymbol{R}_s)\eta(\boldsymbol{R}_{s'})e^{i\boldsymbol{q}\cdot\boldsymbol{R}_{ss'}}e^{i\boldsymbol{Q}\cdot\boldsymbol{u}_{ss'}}\,d\boldsymbol{R}_s d\boldsymbol{R}_{s'}\,,$$
$$I^{(2)} = \frac{|\chi|^2}{v^2}\int\!\!\int \delta\eta(\boldsymbol{R}_s)\delta\eta(\boldsymbol{R}_{s'})e^{i\boldsymbol{q}\cdot\boldsymbol{R}_{ss'}}e^{i\boldsymbol{Q}\cdot\boldsymbol{u}_{ss'}}\,d\boldsymbol{R}_s d\boldsymbol{R}_{s'}\,, \quad (\boldsymbol{Q} \approx \boldsymbol{G}')\,. \tag{4.3.17}$$

If we consider only the reflections with large Miller indices for which $bQ \sim (\sum h_i^2)^{1/2} \gg 1$ we can find the integral for the term $I^{(1)}$ in the approximation of smoothly varying distortions (see Sect. 1.5 in [1.1]). The main contribution to the integral is made by the range of small $R_{ss'}$, where $\eta(\boldsymbol{R}_{s'}) \approx \eta(\boldsymbol{R}_s)$ and $u_{ss'i} \approx u_{i,j}(\boldsymbol{R}_s)R_{ss'j}$ (here $u_{i,j}$ is the distortion tensor). We obtain in this approximation

$$I^{(1)} = \frac{(2\pi)^3}{v^2}|\chi|^2 \int \eta^2(\boldsymbol{r})\delta[\boldsymbol{q} + \nabla(\boldsymbol{Q}\cdot\boldsymbol{u}(\boldsymbol{r}))]d\boldsymbol{r} \quad (\boldsymbol{Q} \approx \boldsymbol{G}')\,. \tag{4.3.18}$$

This equation gives the intensity of scattering for any smoothly varying distribution of both the order parameter and the distortions.

Scattering by a Skeleton of Ordered Regions. Let us consider first the scattering intensity $I^{(1)}$ at relatively high temperatures when a skeleton of thin cylindric ordered regions (with the thicknesses $r_0 \ll r_d$) are produced around the dislocation lines of various types α. Since the arrangement of dislocations is random the scattering by the ordered regions at different dislocations occurs incoherently and $I^{(1)}$ may be found simply as the sum of intensities of scattering by various regions. Using (4.3.1, 6, 18) we can write

$$\begin{aligned} I^{(1)} &= C \sum_\alpha n_\alpha E_\alpha \,, \quad C = \frac{8\pi^3}{v^2}|\chi|^2(\eta^0)^2 V \,, \\ E_\alpha &= \frac{1}{\mathcal{L}} \int_0^{\mathcal{L}} dz_\alpha \int_{S_\alpha} d\boldsymbol{r}_{\perp\alpha} \left\langle [\varkappa_0 u_{ii}(\boldsymbol{r}) - \tau]^{2\beta} \delta\left(\boldsymbol{q}_{\perp\alpha} + \frac{\partial}{\partial \boldsymbol{r}_{\perp\alpha}}(\boldsymbol{Q}\cdot\boldsymbol{u}(\boldsymbol{r}))\right]\right. \\ &\quad \times \delta\left.\left[q_{z\alpha} + \frac{\partial}{\partial z_\alpha}(\boldsymbol{Q}\cdot\boldsymbol{u}(\boldsymbol{r}))\right]\right\rangle \quad (r_0 \ll r_\alpha) \,. \end{aligned} \tag{4.3.19}$$

Here the density n_α of dislocations of the type α is defined as the total length of the lines per unit volume, z_α and $\boldsymbol{r}_{\perp\alpha}$ are the coordinates along the dislocation line and in the perpendicular plane, and integration over $\boldsymbol{r}_{\perp\alpha}$ is performed over the surface area S_α of the ordered cylindric region while integration over z_α is performed along some length $\mathcal{L}$ which is eliminated from the final results.

In the dislocation skeleton of the ordered regions the main contribution to $u_{ii}(\boldsymbol{r})$ and $\partial(\boldsymbol{Q}\cdot\boldsymbol{u}(\boldsymbol{r}))/\partial\boldsymbol{r}_{\perp\alpha}$ is made by the distortion field $u_{i,j} = bf_{\alpha ij}(\theta)/r_{\perp\alpha}$ generated by the nearest dislocation (the functions $f_{\alpha ij}(\theta)$ for the specific types of dislocations are known and we have $|f_{\alpha ij}(\theta)| \sim 10^{-1}$). But the nearest dislocation does not affect $\partial(\boldsymbol{Q}\cdot\boldsymbol{u}(\boldsymbol{r}))/\partial z_\alpha$ since this function depends on a complicated random field generated by many distant dislocations. The averaging of the last factor in (4.3.19) over the positions of the randomly arranged dislocations can be performed in the same way as in the analysis of the dislocation broadening of the regular reflections (see Sect. 5.1.1 in [1.1]). Then the equation for E_α in (4.3.19) is broken down into factors which depend only on the components $q_{z\alpha}$ and $\boldsymbol{q}_{\perp\alpha}$ of the vector $\boldsymbol{q}$ along the axis Z_α and in the perpendicular plane:

$$\begin{aligned} &E_\alpha(\boldsymbol{q}) = E_\alpha(\boldsymbol{q}_{\perp\alpha})E_\alpha(q_{z\alpha}) \,, \\ &E_\alpha(q_{z\alpha}) = \frac{1}{\sigma_\alpha}\exp\left(-\frac{\pi q_{z\alpha}^2}{\sigma_\alpha^2}\right) \,, \\ &\sigma_\alpha^2 = \sigma^2\Phi_\alpha \,, \quad \Phi_\alpha = \frac{1}{4\pi}\sum_{\alpha'}\frac{n_{\alpha'}}{n}\int_0^{2\pi}(m_j f_{\alpha' j3})^2 d\theta \sim 1 \,, \\ &\boldsymbol{m} = \frac{\boldsymbol{Q}}{Q} \,, \quad \sigma^2 = \frac{1}{2}(Qb)^2 nl \,, \quad l = \ln\frac{\sqrt{\pi}\sigma L}{2Qb} \approx \ln(\sqrt{nl}L) \,. \end{aligned} \tag{4.3.20}$$

Here L is the crystal size and $bf_{\alpha' j3}(\theta) r_{\perp\alpha}$ are the components of the distortion tensor for dislocations of the type α' in the reference frame linked to the dislocations α.

Equation (4.3.20) describes the variation of $E_\alpha(\boldsymbol{q})$ as a function of $q_{z\alpha}$ in the case when the dislocation broadening σ_α is fairly large so that $1/\sigma_\alpha$ is smaller than other characteristic dimensions of the coherence range along the axis Z_α, and, in particular, than the domain size D_0. In the reversed case of small $D_0 \ll 1/\sigma$ (for instance, for $\tau \sim \tau_1$) the field of $\eta(\boldsymbol{r})$ along the axis Z_α cannot be regarded as varying smoothly and (4.3.20) is, strictly speaking, inapplicable (though it may still be used for calculating $E_\alpha(\boldsymbol{q}_{\perp\alpha})$). Then the dependence on $q_{z\alpha}$ is approximately described by the following function (see, for instance, Sect. 5.1.4 in [1.1]):

$$E_\alpha(q_{z\alpha}) \approx \frac{1}{\pi} \frac{\zeta_\| D_0}{1 + (\zeta_\| D_0 q_{z\alpha})^2}\,, \quad \zeta_\| \sim 1\,. \tag{4.3.21}$$

It can be seen from (4.3.19, 20) that the function $E_\alpha(\boldsymbol{q}_{\perp\alpha})$ is determined by integration of $|\varkappa_0 u_{ii}(\boldsymbol{r}) - \tau|^{2\beta} \delta(\boldsymbol{q}_{\perp\alpha} + \partial(\boldsymbol{Q}\cdot\boldsymbol{u})/\partial \boldsymbol{r}_{\perp\alpha})$ over $d\boldsymbol{r}_{\perp\alpha}$. We shall use below the notation

$$\begin{gathered}\left|\frac{\partial}{\partial \boldsymbol{r}_{\perp\alpha}}(\boldsymbol{Q}\cdot\boldsymbol{u}(\boldsymbol{r}))\right| = \frac{Q_{\perp\alpha} b}{r_{\perp\alpha}} F(\theta_{Q\alpha}, \theta_\alpha)\,, \\ \xi(\theta_\alpha) = \frac{f_\alpha(\theta_\alpha)}{F(\theta_{Q\alpha}, \theta_\alpha)}\,, \quad f_\alpha(\theta_\alpha) = \sum_\alpha f_{\alpha ii}(\theta_\alpha)\,,\end{gathered} \tag{4.3.22}$$

where $\boldsymbol{Q}_{\perp\alpha}$ is the component of $\boldsymbol{Q}$ perpendicular to the axis Z_α (the dislocation line), and $\theta_{Q\alpha}$ and θ_α are the polar angles for $\boldsymbol{Q}_{\perp\alpha}$ and $\boldsymbol{r}_{\perp\alpha}$. Since $u_{ii} = bf_\alpha(\theta)/r_{\perp\alpha}$ the integration over $d\boldsymbol{r}_{\perp\alpha}$ gives

$$\begin{aligned}E_\alpha(\boldsymbol{q}_{\perp\alpha}) &= \frac{Q_{\perp\alpha}^2 b^2}{q_{\perp\alpha}^4}\left[\xi(\theta_0)\varkappa_0 \frac{q_{\perp\alpha}}{Q_{\perp\alpha}} - \tau\right]^{2\beta} Y\left[\xi(\theta_0)\varkappa_0 \frac{q_{\perp\alpha}}{Q_{\perp\alpha}} - \tau\right] \\ &\quad \times F^2(\theta_{Q\alpha}, \theta_0)\left|\frac{d\chi_\alpha}{d\theta}\right|^{-1}_{\theta=\theta_0}\,.\end{aligned} \tag{4.3.23}$$

Here $Y(x) = 1$ for $x > 0$ and $Y(x) = 0$ for $x < 0$ while θ_0 is determined by the equation $\chi_\alpha(\theta_{Q\alpha}, \theta_0) = \theta_{q\alpha}$ where $\theta_{q\alpha}$ and $\chi_\alpha(\theta_{Q\alpha}, \theta)$ are the polar angles for the vectors $\boldsymbol{q}_{\perp\alpha}$ and $\partial(\boldsymbol{Q}\cdot\boldsymbol{u})/\partial \boldsymbol{r}_{\perp\alpha}$.

It can be seen from (4.3.19–23) that the breadth δq_z of the intensity distribution for scattering by the ordered regions around dislocations near the axis $q_{z\alpha}$ is of the order of the largest of the two parameters, $\sigma \sim bQ/r_d$ or π/D_0, while its width in the perpendicular plane, $\delta q_\perp \sim Q\tau/\varkappa_0 \sim Qb/r_0$ (see (4.3.5)), is considerably larger than in the region where the skeleton of ordered regions is sharply defined (where $r_0 \ll r_d$). This means that each system of dislocation lines gives rise to a slab-shaped scattering intensity distribution $I_\alpha^{(1)}$ in the reciprocal lattice space while the total intensity distribution $I^{(1)} = \sum_\alpha I_\alpha^{(1)}$ is the superposition of such slab-shaped intensity distributions with thicknesses δq_z and widths $\delta q_\perp$ which are perpendicular to various possible directions of the dislocation lines.

The δ functions in (4.3.19) enable us to calculate immediately the integrated scattering intensity $I_i^{(1)}$:

$$I_i^{(1)} = \frac{\pi}{2} C \frac{\beta(1-2\beta)}{\sin 2\pi\beta} n(\varkappa_0 b)^2 [\overline{f(\theta)}]^2 \tau^{2\beta-2} , \quad (4.3.24)$$

Here $\overline{f(\theta)}$ is given by (4.3.4). The calculated dependence for the intensity, $I_i^{(1)} \sim \tau^{2\beta-2}$, is explained by the fact that the volume of the ordered regions is proportional to $r_0^2 \sim \tau^{-2}$ while the mean value of η^2 in these regions is proportional to $\tau^{2\beta}$.

Simple symmetric functions (4.3.20) or (4.3.21) (or their convolution product) describe the intensity distribution $I_\alpha^{(1)}$ along the axis $q_{z\alpha}$. The thickness of this slab-shaped distribution $\delta q_z \sim (\sigma + \pi D_0^{-1})$ at the temperatures $\tau \sim \tau_1$ of production of the ordered regions (where $D_0 \sim r_0 \ll r_d$ and $\sigma \ll \pi D_0^{-1}$) is related to the length D_0 of the bounded ordered regions (for $\tau \sim \tau_1$ we have $\delta q_z \delta q_\perp$). In the range of $\tau \ll \tau_1$ where $D_0 \gg r_0$ and $D_0 \gg r_d/bQ$, however, the main part is played by the deformation broadening $\delta q_z \sim \sigma \sim bQ/r_d$. As noted in Sect. 4.3.1, at temperatures just below the phase transition point τ_f for the dislocation skeleton a single-domain state is produced in the bulk of the skeleton under certain conditions and thus the broadening due to finiteness of D_0 disappears. According to (4.3.10, 11), for the random skeleton we have $D_0(\tau_f) - r_d^2/r_0(\tau_f)$ and for the skeleton produced by the Frank network we have $D_0(\tau_f) \sim r_d$. In the first case the relative variation of δq_z is small ($\sim 2\pi r_0/bQr_d$) but in the second case it is of the order of $2\pi/bQ$ and can be considerable.

The distribution of the intensity $I_\alpha^{(1)}$ in the plane of $\boldsymbol{q}_{\perp\alpha}$ (of the slab) exhibits a complicated dependence on the angles $\theta_{Q\alpha}$ and θ_0 and on the distance $\boldsymbol{q}_{\perp\alpha}$ to the reciprocal lattice point. Since the ordered domains are produced in the crystal regions distorted by dislocations with a changed lattice parameter in each direction the intensity peak exhibits a certain displacement q_m from the reciprocal lattice point and the entire intensity distribution is to one side of this point. As illustrated in Fig. 4.20, the dependence of $E_\alpha(\boldsymbol{q}_{\perp\alpha})$ on $q_{\perp\alpha}$ for the given direction of $\boldsymbol{q}_{\perp\alpha}$ (in the sector where $E_\alpha(\boldsymbol{q}_{\perp\alpha}) \neq 0$) is described by an asymmetric plot calculated from (4.3.23). It does not vanish for $q_{\perp\alpha} > q_\alpha^0 = Q_{\perp\alpha}\tau/\varkappa_0\xi$ (that is, it does not approach the reciprocal lattice point) and it reaches a maximum for $q_{\perp\alpha} = q_m$ where

$$\begin{aligned} q_m &= \frac{2}{2-\beta}\frac{F(\theta_{Q\alpha},\theta_0)}{\varkappa_0 f(\theta_0)} Q_{\perp\alpha}\tau \sim \frac{Q_{\perp\alpha}\tau}{\varkappa_0} \sim \frac{1}{r_0} , \\ E_\alpha(q_m) &= \frac{(2-\beta)^{4-2\beta}\beta^{2\beta} f_\alpha^4(\theta_0)\varkappa_0^4 b^2}{16F^2(\theta_{Q\alpha},\theta_0)Q_{\perp\alpha}^2\tau^{4-2\beta}\left|\frac{d\chi_\alpha}{d\theta}\right|} . \end{aligned} \quad (4.3.25)$$

For large $q_{\perp\alpha} \gg q_m$ we have $E_\alpha(\boldsymbol{q}_{1\alpha}) \approx (2/\beta)^{2\beta} E_\alpha(q_m)(q_m/q_{\perp\alpha})^{4-2\beta}$ which decreases as $q_{\perp\alpha}^{-(4-2\beta)}$. The half-width $\delta q_\perp$ of the plot of $E_\alpha(\boldsymbol{q}_{\perp\alpha})$ is of the order of the shift of the maximum q_m with respect to the lattice point and it equals $0.51q_m$ for $\beta = 1/3$ and $0.68q_m$ for $\beta = 1/2$. Since the thickness of the ordered

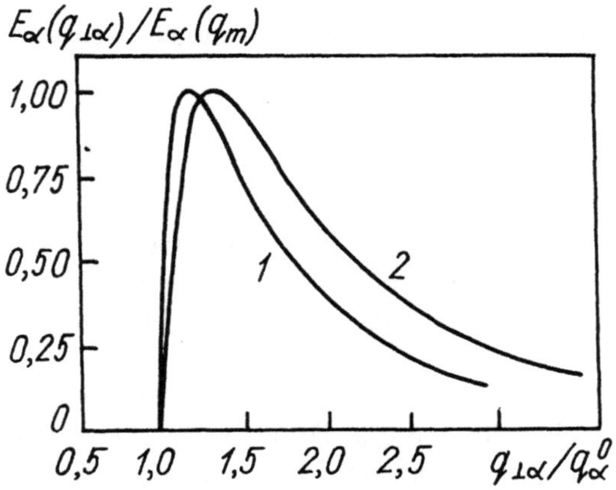

Fig. 4.20. The ratio of the intensity of scattering by the ordered region near a single dislocation α to its maximum value as a function of the distance to the reciprocal lattice point (in terms of q_α^0 units). *Curve 1* for $\beta = 1/3$, *curve 2* for $\beta = 1/2$

region increases with decreasing temperature, $r_0 \sim \tau^{-1}$, we find that q_m and $\delta q_\perp$ are proportional to τ. According to (4.3.25), for lattice points with large Miller indices q_m varies as $Q_{\perp\alpha}$ and $E_\alpha(\boldsymbol{q}_m)$ varies as $Q_{\perp\alpha}^{-2}$.

In addition to the intensity $I^{(1)}$ of scattering by the long-range order inhomogeneities a significant contribution near the transition is made by the intensity $I^{(2)}$ of critical scattering by the fluctuations of the order parameter (see (4.3.17)) whose distribution has a considerably larger width (of the order of r_c^{-1}). We can easily distinguish $I^{(1)}$ against the background of $I^{(2)}$ in X-ray diffraction studies if $I^{(1)}(\boldsymbol{q}_{\perp\alpha} = \boldsymbol{q}_m)$ is larger than max $I^{(2)}$. In an inhomogeneously distorted crystal $I^{(2)}$ generally differs from the scattering intensity found for perfect crystals. If we ignore this difference (and thus overestimate $I^{(2)}$) we can take the conventional estimate of $I^{(2)} \sim |\chi|^2 N\tau^{-\gamma}$ (see (4.2.3)). Then we obtain from (4.3.19, 20, 25) that the production of ordered regions at $\tau \sim \tau_1$ (when $r_0 \sim r_c$) in a small crystal volume at dislocations gives rise only to a small additional term $I^{(1)}$ for the critical scattering intensity $I^{(2)}$. But as the temperature decreases the intensity $I^{(1)}$ grows at a much faster rate than $I^{(2)}$ (and it is concentrated in a smaller volume of the reciprocal lattice space) so that for conventional dislocation densities of the order of $10^8\,\mathrm{cm}^{-2}$ the intensity $I^{(1)}$ can be much greater than $I^{(2)}$ even in the range in which the skeleton of ordered regions exists in the crystal if the following condition is satisfied:

$$\frac{\sqrt{n}b\varkappa_0^4}{Q^3 v} \gg \tau^{4-2\beta-\gamma} . \tag{4.3.26}$$

For instance, for $\varkappa_0 = 1$, $n = 10^9\,\mathrm{cm}^{-2}$ (that is, $\sqrt{n}b = 10^{-3}$), and $Q^3 v = 3 \cdot 10^2$ this condition is satisfied for $\tau < 10^{-3}$ and for $n = 10^7\,\mathrm{cm}^{-2}$ for $\tau < 10^{-4}$. In fact, the strong condition (4.3.26) must not necessarily be satisfied for making possible an identification of $I^{(1)}$. For scattering of neutrons or Mössbauer photons the intensity $I^{(1)}$ can be identified from the analysis of the energy distribution associated with $I^{(1)}$ near the central peak even if $I^{(1)} < I^{(2)}$.

Scattering Near a Spread-Out Phase Transition. As the temperature decreases $r_0(\tau)$ grows, the widths $\delta q_\perp$ decrease and the $I^{(1)}$ distributions gradually approach the reciprocal lattice points while their slab-shaped structure is disappearing. For $\tau \lesssim \delta\tau_0$ the dislocation skeleton of ordered regions is replaced with a complicated random arrangement of ordered domains of a large size ($\gg r_d$) which are produced by the effects of the fields of distant dislocations (in addition to the field of the nearest dislocation) and occupy a large part of the crystal volume. For a crystal containing randomly distributed dislocations the intensity of scattering in this temperature range can be calculated by averaging (4.3.18) over the dislocation positions. This averaging is performed as for a crystal with dislocations far from the phase transition (see Sect. 5.1 in [1.1]) but we must take into account not only the random strains here but also the essentially nonhomogeneous distribution $\eta(\boldsymbol{r})$.

Appropriate averaging procedures were performed in [4.246] yielding both the function $I^{(1)}(\boldsymbol{q})$ and its moments. For $\tau \lesssim \delta\tau_0$ the zeroth moment, that is, the integrated scattering intensity in the vicinity of a superlattice point, is given by

$$M_0 I_i^{(1)} = \int I^{(1)}(\boldsymbol{q})d\boldsymbol{q} = C\frac{\Gamma(1+2\beta)}{\sqrt{2\pi}}(\delta\tau_0)^{2\beta}\exp\left[-\frac{\tau^2}{4(\delta\tau_0)^2}\right]D_{-(I-2\beta)}\left(\frac{\tau}{\delta\tau_0}\right). \tag{4.3.27}$$

Replacing the parabolic cylinder functions $D_\varsigma(z)$ with asymptotic expressions we can easily see that at $-\tau \gg \delta\tau_0$, that is, at temperatures considerably lower than the point of the spread-out transition, M_0 is proportional to $\eta^3(\tau)$ in a perfect crystal (as should be the case for the intensity of a superlattice reflection). In the range of $|\tau| \sim \delta\tau_0$ the intensity $I_i^{(1)}$ decreases rapidly and for $\tau \gg \delta\tau_0$ its variation is described by (4.3.24). Figure 4.21 presents a plot of $I_i^{(1)}$ for $\beta = 1/3$ (curve 1; the dashed line shows the part of the curve where the approximations (4.3.24, 27) are not accurate and $I_i^{(1)}$ is obtained by interpolation) and the plot of $2\tilde{\beta} = d(\ln I_i^{(1)})/d(\ln|\tau|)$ which is obtained when we approximate the experimentally recorded plot of $I^{(1)}$ with the function $(-\tau)^{2\beta}$ (see curve 2). In the range of $-\tau \sim 2\delta\tau_0$ the exponent $\tilde{\beta}$ grows (reaching its upper limit of 0.362) but the growth rate is appreciably lower than for the similar approximation of $\tilde{\eta} \sim (-\tau)^\beta$ (see Fig. 4.19).

The median point and the width of the intensity distribution are described by its first and second reduced moments m_i and m_{ij}. In the range of $|\tau| \lesssim \delta\tau_0$ we have [4.246]

$$\begin{aligned} m_i &= \frac{1}{M_0}\int q_i I^{(1)}(\boldsymbol{q})d\boldsymbol{q} = P_i\sigma\frac{D_{-2\beta}(t)}{D_{-(1+2\beta)}(t)}, \quad t = \frac{\tau}{\delta\tau_0}, \\ m_{ij} &= \frac{1}{M_0}\int q_i q_j I^{(1)}(\boldsymbol{q})d\boldsymbol{q} - m_i m_j \\ &= \sigma^2 P_{ij} + m_i m_j\left[\frac{D_{1-2\beta}(t)D_{-(1+2\beta)}(t)}{D^2_{-2\beta}(t)} - 1\right], \end{aligned} \tag{4.3.28}$$

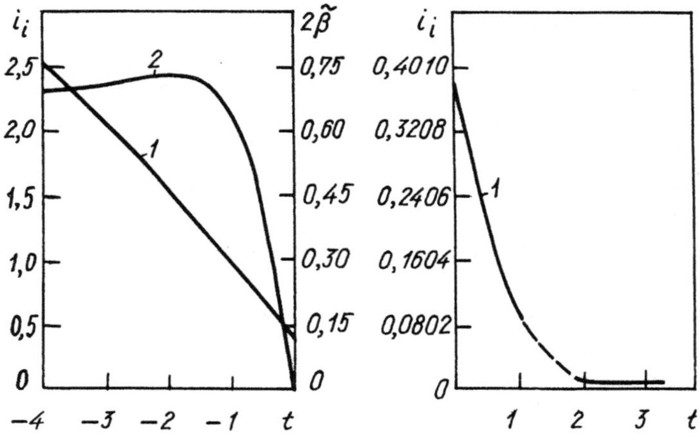

Fig. 4.21. Behavior of $i_i = I_i^{(1)}C^{-1}(\delta\tau_0)^{-2\beta}$ and 2β as functions of the reduced temperature $t = \tau/\delta\tau_0$. Different scales are used for the plots at $t > 0$ and $t < 0$

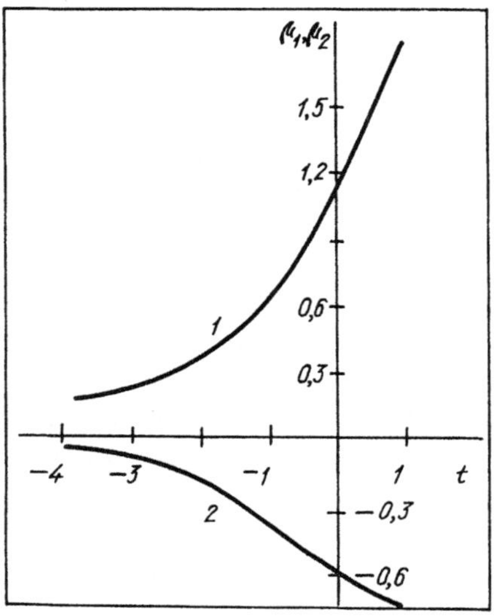

Fig. 4.22. Moments of the scattering intensity distribution as functions of reduced temperature $t = \tau/\delta\tau_0$

where $P_i \sim 1$ and $P_{ij} \sim 1$ are complicated functions of the direction of Q/Q. Figure 4.22 presents the plots of $\mu_1 = m_i/\sigma P_i$ (curve 1) and $\mu_2 = (m_{ij} - P_{ij}\sigma^2)/\sigma^2 P_i P_j$ (curve 2) as functions of t which demonstrate that with a decrease in temperature the width of the distribution $I^{(1)}(q)$ decreases and its median point approaches the reciprocal lattice point. In the limit of $-\tau \gg \delta\tau_0$ we obtain a Gaussian intensity distribution with a characteristic dislocation broadening $\sim \sigma$ (as for the lattice reflections) and with a median point which is almost at the reciprocal lattice point. Figure 4.23 illustrates the transition from the symmetric Gaussian distribution around the reciprocal lattice point at $t \gg 1$ to the asymmetric shifted distribution at $|\tau| \sim 1$ calculated from the equations derived

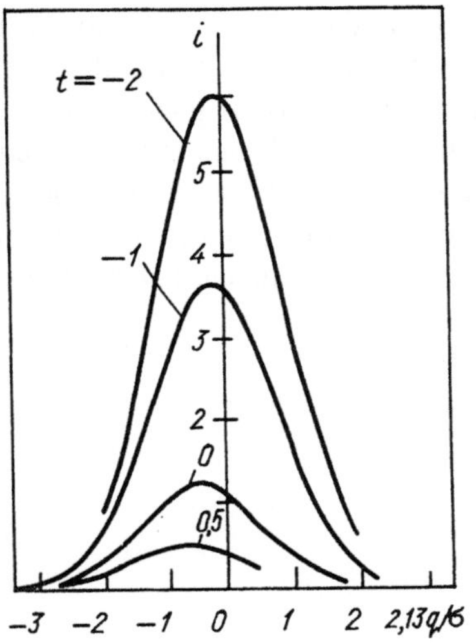

Fig. 4.23. The scattering intensity distribution at various temperatures in the case of dislocations in an elastically isotropic cubic crystal with the Poisson ratio for $\beta = 1/3$, $i = I^{(1)}/I_0$, $I_0 = 8.63 \cdot 10^{-2} C(\delta\tau_0)^{2\beta}\sigma^{-3}$

in [4.246] for $I^{(1)}(\boldsymbol{q})$ for an elastically isotropic crystal with dislocation lines and the Burgers vectors parallel to $\langle 100\rangle$ for $\boldsymbol{q} \parallel \boldsymbol{Q} \parallel \langle 100\rangle$.

The above effects produced by dislocations on diffraction in the critical range must manifest themselves even for normal dislocation densities $n \sim 10^7 \div 10^{10}\,\mathrm{cm}^{-2}$. In experiments these effects must be studied for various controlled dislocation densities (and for various types and arrangements of dislocations).

Appendix: Calculation of the Mean Squares of the Fourier Components in the Microscopic Theory

Let us consider the Fourier components of the fluctuations corresponding to the given vector $\boldsymbol{k}$ and find the averages $\langle c_{k\gamma} c^*_{k\gamma}\rangle$ in the crystal broken down into sublattices as described in Sect. 1.3 [1.4]. Denote the atomic planes perpendicular to $\boldsymbol{k}$ by the superscript δ. If $\boldsymbol{k}$ is not perpendicular to any system of atomic planes then the crystal can be broken down into "infinitesimally thin" layers perpendicular to $\boldsymbol{k}$ which can also be denoted by δ. The number of points belonging to one of the sublattices γ is the same for different parallel planes δ_γ which contain the points of this sublattice. Denote this number by $\mathfrak{N}_0$.When we perform summation over the points in one plane (atomic layer) the factor $\exp(i\boldsymbol{k} \cdot \boldsymbol{E}_{s\gamma})$ in (1.3.3) remains constant. Therefore, we can replace $c_{s\gamma}$ in the sum (1.3.3) over s with the mean concentration c^δ_γ of the atoms A on the points of the sublattice γ in the plane δ_γ. If we then align the axis Z with the vector $\boldsymbol{k}$ then (1.3.3) can be rewritten as

$$c_{k\gamma} = \frac{\mathfrak{N}_0}{N^0} \sum_{\delta_\gamma=1}^{N^0/\mathfrak{N}_0} (c^\delta_\gamma - c_\gamma) \exp(ikZ_{\delta\gamma}) \,, \qquad \text{(A.1)}$$

where $Z_{\delta\gamma}$ is the segment on the axis Z from the origin to the intersection with the plane δ_γ. Equations (A.1) and (1.3.3) show that the fluctuations of c_k are determined by the deviations of the concentrations c_γ^δ in various planes from the mean concentrations, c_γ. The probability of a fluctuational change in the concentrations c_γ^δ can be found from the general equation (1.2.1) of the thermodynamic theory of fluctuations. The least work required for reversible production of such a fluctuation can be written as the expansion of the thermodynamic potential Φ of the entire crystal in the powers of $c_\gamma^\delta - c_\gamma$:

$$\mathfrak{R} = \frac{1}{2}\sum_\gamma\sum_\delta \frac{\partial^2\Phi}{\partial c_\gamma^{\delta 2}}(c_\gamma^\delta - c_\gamma)^2 + \sum_{\substack{\gamma,\gamma' \\ (\gamma<\gamma')}} \frac{\partial^2\Phi}{\partial c_\gamma^\delta \partial c_{\gamma'}^{\delta'}}(c_\gamma^\delta - c_\gamma)(c_{\gamma'}^{\delta'} - c_{\gamma'})$$
$$+\sum_\gamma \sum_{\substack{\delta_\gamma\delta'_\gamma \\ (\delta_\gamma<\delta'_\gamma)}} \frac{\partial^2\Phi}{\partial c_\gamma^\delta \partial c_\gamma^{\delta'}}(c_\gamma^\delta - c_\gamma)(c_\gamma^{\delta'} - c_\gamma)\,. \tag{A.2}$$

If we choose a sufficiently large ν' then for $\delta \neq \delta'$, the planes with the same γ are far from each other, the derivatives $\partial^2\Phi/\partial c_\gamma^\delta \partial c_\gamma^{\delta'}$ are small and the last term in (A.2) can be omitted. In the case under consideration when we can ignore the direct and indirect interaction between atoms in different lattice cells (containing ν' points; see Sect. 1.3) the derivatives of Φ with respect to the concentrations in the atomic planes which appear in the first two terms in (A.2) are related to the derivatives with respect to the conventional concentration for the entire crystal by the following formulas:

$$\frac{\partial^2\Phi}{\partial c_\gamma^{\delta 2}} = \frac{\mathfrak{R}_0}{N^0}\frac{\partial^2\Phi}{\partial c_\gamma^2}\,,$$
$$\frac{\partial^2\Phi}{\partial c_\gamma^\delta \partial c_{\gamma'}^{\delta'}} = \frac{\mathfrak{R}_0 z''}{N^0 z'}\frac{\partial^2\Phi}{\partial c_\gamma \partial c_{\gamma'}}\,. \tag{A.3}$$

Here z' is the total number of the points in the sublattice γ' which are nearest to the given point of the sublattice γ, and z'' is the number of points of the sublattice γ' which are in the plane δ' (if there are no such neighbouring γ' points in the plane δ' then the respective derivative $\partial^2\Phi/\partial c_\gamma^\delta \partial c_\gamma^{\delta'}$ vanishes).

Let us write for the wave-like variation of the concentration c_γ^δ,

$$c_\gamma^\delta - c_\gamma = c_{k\gamma}\exp(-ikZ_{\gamma\delta}) + c_{k\gamma}^*\exp(ikZ_{\gamma\delta})\,. \tag{A.4}$$

Substituting (A.3,4) into (A.2) and using the equation $k(Z_{\gamma'\delta'} - Z_{\gamma\delta}) = \boldsymbol{k}\cdot\boldsymbol{\varrho}_{\mathfrak{z}\gamma\gamma'}$, where the vector $\boldsymbol{k}$ is parallel to the axis Z and $\boldsymbol{\varrho}_{\mathfrak{z}\gamma\gamma'}$ is the vector passing from a point of the sublattice γ in the plane δ to a point (with the number $\mathfrak{z}$) of the sublattice γ' in the plane δ', we obtain the following distribution of probabilities for the fluctuations of the kth Fourier components of the concentrations:

$$w \sim \exp\left(-\frac{\mathfrak{R}_k}{k_B T}\right) = \exp\left[-\frac{1}{2}\sum_{\gamma,\gamma'=1}^{\nu} a_{\gamma\gamma'}(c_{k\gamma}c_{k\gamma'}^* + c_{k\gamma}^* c_{k\gamma'})\right]\,, \tag{A.5}$$

where

$$a_{\gamma\gamma'} = \frac{1}{k_{\mathrm{B}}T}\frac{\partial^2 \Phi}{\partial c_\gamma^2} \; ;$$

$$a_{\gamma\gamma'} = \frac{1}{z' k_{\mathrm{B}}T}\frac{\partial^2 \Phi}{\partial c_\gamma \partial c_{\gamma'}} \sum_{\jmath=1}^{z'} \cos \boldsymbol{k}\cdot\boldsymbol{\varrho}_{\jmath\gamma\gamma'} \quad (\gamma \neq \gamma') \, . \tag{A.6}$$

The averaged squares and products of the Fourier components can be calculated from (A.5) in the usual way (as it was done in Sect. 1.2):

$$\tfrac{1}{2}\langle (c_{\boldsymbol{k}\gamma} c^*_{\boldsymbol{k}\gamma'} + c^*_{\boldsymbol{k}\gamma} c_{\boldsymbol{k}\gamma'}) \rangle = b_{\gamma\gamma'} \, . \tag{A.7}$$

Here $b_{\gamma\gamma'}$ make up a matrix which is the inverted matrix $\|a_{\gamma\gamma'}\|$ given by (A.6). Equations (1.3.3) and (A.7) yield

$$\langle |c_{\boldsymbol{k}}|^2 \rangle = \frac{1}{\nu'^2} \sum_{\gamma,\gamma'=1}^{\nu'} b_{\gamma\gamma'} \, , \quad \|b\| = \|a^{-1}\| \, . \tag{A.8}$$

We see that in order to find the mean squared Fourier components for the concentration fluctuations we must know the thermodynamic potential of the solution as a function of the variables c_γ, then calculate the matrix $\|a_{\gamma\gamma'}\|$ from (A.6) and find the sum of all the matrix elements of the inverse matrix $\|b_{\gamma\gamma'}\|$.

In some instances we can directly calculate the sum of all matrix elements of the inverse matrix $\|b_{\gamma\gamma'}\|$ without inverting the matrix $\|a_{\gamma\gamma'}\|$. This can be done if the sum A of the matrix elements of each row in the matrix is independent of the row number. It can be readily seen then that

$$\langle |c_{\boldsymbol{k}}|^2 \rangle = \frac{1}{\nu'^2} \sum_{\gamma,\gamma'=1}^{\nu'} b_{\gamma\gamma'} = \frac{1}{\nu' A} \, , \tag{A.9}$$

where

$$A = \frac{1}{k_{\mathrm{B}}T}\frac{\partial^2 \Phi}{\partial c_\gamma^2} + \frac{1}{z' k_{\mathrm{B}}T} \sum_{\substack{\gamma'=1 \\ (\gamma\neq\gamma')}}^{\nu'} \frac{\partial^2 \Phi}{\partial c_\gamma \partial c_{\gamma'}} \sum_{\jmath=1}^{z'} \cos \boldsymbol{k}\cdot\boldsymbol{\varrho}_{\jmath\gamma\gamma'} \, . \tag{A.10}$$

(Indeed, summation of the equations $\sum_{\gamma''=1}^{\nu'} b_{\gamma'\gamma''} a_{\gamma''\gamma} = \delta_{\gamma\gamma'}$, which relate the matrix elements of the two matrices, over all γ and γ' yields $\sum^{\nu'} b_{\gamma'\gamma''} a_{\gamma''\gamma} = \nu'$. Taking the sum over γ and using $\sum_{\gamma=1}^{\nu'} a_{\gamma''\gamma} = A$ we obtain $\sum_{\gamma,\gamma'=1}^{\nu'} b_{\gamma\gamma'} = \nu'/A$, that is, (A.9)).

References

Chapter 1

1.1 M.A. Krivoglaz: *Diffraction of X-Rays and Neutrons in Imperfect Crystals* (Springer, Berlin, Heidelberg 1991)
1.2 L.D. Landau: Zh. Eksp. Teor. Fiz. **7**, 1232–1241 (1937)
1.3 M.A. Krivoglaz: Zh. Eksp. Teor. Fiz. **31**, 625–635 (1956)
1.4 M.A. Krivoglaz: Zh. Eksp. Teor. Fiz. **32**, 1368–1381 (1957)
1.5 P.C. Clapp, S.C. Moss: Phys. Rev. **142**, 418–427 (1966)
1.6 S.C. Moss, P.C. Gehlen: in *Critical Phenomena in Alloys, Magnets and Superconductors*, ed. by R.E. Mills, E. Ascher, R.I. Jaffee (McGraw-Hill, New York, 1971) pp. 363–374
1.7 H. Kanzaki: J. Phys. Chem. Solids **2**, 24–36 (1957)
1.8 M.A. Krivoglaz: Zh. Eksp. Teor. Fiz. **34**, 204–218 (1958)
1.9 M.A. Krivoglaz, O.O. Tikhonova: Ukr. Fiz. Zh. **3**, 297–312 (1958)
1.10 M.A. Krivoglaz: Zh. Eksp. Teor. Fiz. **34**, 405–411 (1958)
1.11 M.A. Krivoglaz, O.O. Tikhonova: Ukr. Fiz. Zh. **5**, 174–189 (1960)
1.12 M.A. Krivoglaz: Kristallographiya **5**, 24–31 (1960)
1.13 M.A. Krivoglaz: *Theory of X-Ray and Thermal-Neutron Scattering by Real Cristals* (Plenum Press, New York 1969)
1.14 H.E. Cook: J. Phys. Chem. Solids **30**, 1097–1112 (1969)
1.15 M.A. Krivoglaz: Fiz. Tverd. Tela **3**, 3682–3690 (1961)
1.16 M.A. Krivoglaz: Fiz. Tverd. Tela **5**, 3439–3452 (1963)
1.17 M.A. Krivoglaz, Hao Thyu: in *Defects and Properties of Crystal Lattice*, ser. Metallofizika (Nauk. Dumka, Kiev, 1968) pp. 84–96 (in Russian).
1.18 A.G. Khachaturyan, T.A. Onisimova: Fiz. Met. i Metalloved. **26**, 973–980 (1968)
1.19 G.V. Kurdyumov, A.G. Khachaturyan: Metallurgical Transactions **3**, 1069–1076 (1972)
1.20 M.A. Krivoglaz: Fiz. Met. i Metalloved. **8**, 648–666 (1959)
1.21 K. Binder: Phys. Rev. Lett. **45**, 811–814 (1980)
1.22 L. Guttmann: in *Solid State Physics*, ed. by F. Seitz, D. Turnbull, vol. 3 (Academic Press, New York 1956) pp. 145–223
1.23 M.A. Krivoglaz, A.A. Smirnov: *The Theory of Order-Disorder in Alloys* (Macdonald, London 1964)
1.24 T. Muto, Y. Takagi: *The Theory of Order-Disorder Transitions* (Academic Press, New York 1955)
1.25 A.G. Khachaturyan: *Theory of Structural Transformations in Solids* (Wiley, New York 1983)
1.26 D. de Fontaine: in *Solid State Physics*, ed. by H. Ehrenreich, F. Seitz, D. Turnbull, vol. 34 (Academic Press, New York 1979) pp. 73–274
1.27 V.I. Iveronova, A.A. Katsnelson: *Short-Range Order* (Nauka, Moscow 1977) (in Russian)
1.28 J.M. Cowley: J. Appl. Phys. **21**, 24–30 (1950)
1.29 B.E. Warren: *X-Ray Diffraction* (Addison-Wesley, Reading, Mass. 1969); reprinted by Dover Publ., New York 1990
1.30 A.A. Katsnelson, Sh.A. Alimov, P.Sh. Dazhaev et al: Fiz. Met. i Metalloved. **26**, 987–995 (1968)

1.31 L.D. Landau, E.M. Lifshits: *Statistical Physics*, vol. 5 (Pergamon Press, Oxford 1980)
1.32 M.A. Leontovich: *Statistical Physics* (Gostechizdat, Moscow 1944) (in Russian)
1.33 I.E. Dzyaloshinskii: Zh. Eksp. Teor. Fiz. **46**, 1420–1437 (1964)
1.34 V.L. Indenbom: Kristallographya **5**, 115–125 (1960)
1.35 A.P. Levaneuk, D.G. Sannikov: Usp. Fiz. Nauk **112**, 562–589 (1974)
1.36 M.A. Krivoglaz, S.A. Rybak: Zh. Eksp. Teor. Fiz. **33**, 139–150 (1957)
1.37 M.A. Leontovich: *Introduction to Thermodynamics* (GITTL, Mowcow 1950)
1.38 L.D. Landau, E.M. Lifshitz and L. Pitaevskii: *Electrodynamics of Continuous Media*, vol. 6 (Pergamon Press, Oxford 1984)
1.39 M.A. Krivoglaz: Dokl. Akad. Nauk SSSR **117**, 213–216 (1957)
1.40 M.A. Krivoglaz: Zh. Eksp. Teor. Fiz. **34**, 355–370 (1958)
1.41 W. Harrison: *Pseudopotentials in the Theory of Metals* (N.A. Benjamin Inc., New York, Amsterdam 1966)
1.42 V. Heine: The Pseudopotential Concept in *Solid State Physics*, vol. 24 (Academic Press, New York, London 1970)
1.43 C.M. van Baal: Physica **64**, 571–586 (1973)
1.44 P.G. De Gennes, J. Friedel: J. Phys. Chem. Solids **4**, 71–77 (1958)
1.45 R. Brout: *Phase Transitions* (New York 1965)
1.46 B.L. Gyorffy, G.M. Stocks: Phys. Rev. Lett. **50**, 374–377 (1983)
1.47 M.A. Krivoglaz: Fiz. Met. i. Metalloved. **5**, 203–211 (1957)
1.48 D. de Fontaine: J. Phys. Chem. Solids **34**, 1285–1304 (1973)
1.49 D.A. Badalyan: Fiz. Tverd. Tela **21**, 2017–2021 (1979)
1.50 M.A. Krivoglaz: Zh. Fiz. Khimii **31**, 1930–1942 (1957)
1.51 V.G. Vaks, A.I. Larkin. S.A. Pikin: Zh. Eksp. Teor. Fiz. **51**, 361–375 (1966)
1.52 T.S. Chang: J. Chem. Phys. **9**, 169–174 (1941)
1.53 R.A. Tahir-Kheli: Phys. Rev. **188**, 1142–1153 (1969)
1.54 M.E. Fischer, R.J. Burford: Phys. Rev. **156**, 583–622 (1967)
1.55 D.W. Hoffman: Metallurgical Transactions **3**, 3231–3238 (1972)
1.56 C.G. Shirley, S. Wilkins: Phys. Rev. B. **6**, 1252–1263 (1972)
1.57 S. Wilkins: Phys. Rev. B. **2**, 3935–3942 (1970)
1.58 R. Kikuchi: Phys. Rev. **81**, 988–1003 (1951)
1.59 P.C. Clapp: J. Phys. Chem. Solids **30**, 2589–2598 (1969)
1.60 N.S. Golosov, L.E. Popov, L.Ya. Pudan: J. Phys. Chem. Solids **34**, 1149–1164 (1973)
1.61 R. Kikuchi, H. Sato: Acta met. **22**, 1099–1112 (1974)
1.62 R. Kikuchi: J. Chem. Phys. **60**, 1071–1080 (1974)
1.63 J.M. Sanchez, D. de Fontaine: Phys. Rev. B. **17**, 2926–2936 (1978)
1.64 D.A. Badalyan, A.G. Khachaturyan: Fiz. Tverd. Tela **14**, 2625–2630 (1972)
1.65 L. Van Hove: Phys. Rev. **89**, 1189–1193 (1953)
1.66 J.C. Phillips: Phys. Rev. **104**, 1263–1277 (1956)
1.67 L.D. Landau: Zh. Eksp. Teor. Fiz. **7**, 19–32 (1937)
1.68 L.D. Landau: Zh. Eksp. Teor. Fiz. **7**, 627–632 (1937)
1.69 E.M. Lifshits: Zh. Eksp. Teor. Fiz. **11**, 255–268 (1941)
1.70 E.M. Lifshits: Zh. Eksp. Teor. Fiz. **11**, 269–281 (1941)
1.71 A.G. Khachaturyan: Fiz. Tverd. Tela **5**, 26–35 (1963)
1.72 P.C. Clapp, S.C. Moss: Phys. Rev. **171**, 754–763 (1968)
1.73 S.C. Moss, P.C. Clapp: Phys. Rev. **171**, 764–777 (1968)
1.74 D. de Fontaine: Acta met. **23**, 553–571 (1975)
1.75 Ya.I. Frenkel: Zh. Eksp. Teor. Fiz. **9**, 952–962 (1939)
1.76 Ya.I. Frenkel: *Kinetic Theory of Liquids* (Izd. AN SSSR, Moscow 1959)
1.77 M.A. Krivoglaz: Fiz. Met. i. Metalloved. **8**, 514–530 (1959)
1.78 C.G. Shirley: Phys. Rev. **10**, 1149–1159 (1974)
1.79 D.A. Vul, M.A. Krivoglaz: Metallofizika **7**, 3–12 (1985)
1.80 D.A. Vul: Fiz. Met. i. Metalloved. **65**, 66–73 (1988)

Chapter 2

2.1 M.A. Krivoglaz: Usp. Fiz. Nauk **8**, 162–170 (1963)
2.2 J.W. Cahn: Acta Met. **10**, 907–913 (1962)
2.3 H. Wagner, H. Horner: Adv. Phys. **23**, 587–637 (1974)
2.4 H. Wagner: in *Hydrogen in Metals*, vol. 1 Basic Research, ed. by Alefeld and J. Völkl, Topics in Applied Physics, vol. 28 (Springer, Berlin, Heidelberg, New York 1978)
2.5 J.W. Cahn, J.E. Hillard: J. Chem. Phys. **31**, 688–699 (1958)
2.6 J.W. Cahn: Acta met. **10**, 179–183 (1962)
2.7 J.W. Cahn: Transactions AIME **242**, 166–180 (1968)
2.8 K.V. Chuistov: *Modulated Structures in Ageing Alloys* (Nauk, Dumka, Kiev 1975) (in Russian)
2.9 M.A. Krivoglaz: Fiz. Met. i. Metalloved. **17**, 161–167 (1964)
2.10 G. Alefeld, J. Völkl, G. Schaumann: Phys. Stat. Sol. **37**, 337–351 (1970)
2.11 C. Zener: Phys. Rev. **74**, 639–647 (1948)
2.12 J.D. Eshelby: Proc. Roy. Soc. **A241**, 376 (1957) **A252**, 561 (1959)
2.13 A.G. Khachaturyan: Fiz. Tverd. Tela **4**, 2840–2844 (1962)
2.14 D. de Fontaine, H.E. Cook: in *Critical Phenomena in Alloys, Magnets and Superconductors*, ed. by R.E. Mills, E. Ascher, R.I. Jaffee (McGraw-Hill, New York 1971) pp. 257–275
2.15 G.V. Kurdyumov, A.G. Khachaturyan: Acta Met. **23**, 1077–1087 (1975)
2.16 T. Kajitani, H.E. Cook: Acta Met. **26**, 1371–1378 (1978)
2.17 A.P. Levanyuk, A.A. Sobyanin: Pis'ma v Zh. Eksp. Teor. Fiz. **11**, 540–543 (1970)
2.18 J. Villain: Sol. State Commun. **8**, 295–297 (1970)
2.19 D.E. Khmelnitskiy: Fiz. Tverd. Tela **16**, 3188–3190 (1974)
2.20 R.A. Cowley: Phys. Rev. **13**, 4877–4885 (1976)
2.21 R. Folk, H. Iro, F. Schwabl: Zs. Phys. B **25**, 69–81 (1976)
2.22 V.L. Ginzburg, A.P. Levanyuk: Phys. Lett. A **47**, 345–346 (1974)
2.23 V.L. Ginzburg, A.P. Levanyuk, A.A. Sobyanin: Usp. Fiz. Nauk **130**, 615–673 (1980)
2.24 F.R.N. Nabarro: Proc. Roy. Soc. **175**, 519–538 (1940)
2.25 J.D. Eshelby: Proc. Roy. Soc. **241**, 376–396 (1957)
2.26 A.L. Roitburd: Usp. Fiz. Nauk **113**, 69–104 (1974)
2.27 A.L. Roitburd: in *Solid State Physics*, ed. by D. Turnbull, H. Ehrenreich, vol. 33 (Academic Press, New York 1978) pp. 317–390
2.28 J.W. Christian: *Transormations in Metals and Alloys* (Pergamon Press 1978)
2.29 A.G. Khachaturyan: Fiz. Tverd. Tela **8**, 2709–2717 (1966)
2.30 H. Yamauchi, D. de Fontaine: Acta Met. **27**, 763–776 (1979)
2.31 M.A. Krivoglaz: Vopr. Fiz. Met. i. Metalloved. **13**, 17–34 (1961)
2.32 A.G. Khachaturyan, G.A. Shatalov: Fiz. Tverd. Tela **11**, 159–166 (1969)
2.33 A.L. Roitburd, A.G. Khachaturyan: Probl. Metalloved. i Fiz. Metallov **9**, 78–87 (1969)
2.34 A.G. Khachaturyan: Phys. Stat. Sol. **35**, 119–132 (1969)
2.35 J.H. Evans: Nature **229**, 403–404 (1971)
2.36 G. Martin: J. Phys. **38**, 419–429 (1977)
2.37 V.K. Tewary, R. Bullough: J. Phys. **2**, L69–L72 (1972)
2.38 L.A. Maksimov, A.I. Ryazanov: Fiz. Met. i. Metalloved. **41**, 284–291 (1976); Zh. Eksp. Teor. Fiz. **79**, 2311–2327 (1980)
2.39 A.L. Roitburd: Fiz. Tverd. Tela **10**, 3619–3627 (1968)
2.40 A.G. Khachaturyan, G.A. Shatalov: Zh. Eksp. Teor. Fiz. **56**, 1037–1045 (1969)
2.41 J.S. Bowles, J.K. Mackenzie: Acta Met. **2**, 129–147 (1954)
2.42 M.S. Wechsler: Acta Met. **7**, 793–802 (1959)
2.43 B.A. Bilby, I.V. Khristyan: Usp. Fiz. Nauk **70**, 515–564 (1960)
2.44 A.M. Stoneham, P.J. Durham: J. Phys. Chem. Solids **34**, 2127–2135 (1973)
2.45 E. Iguchi: J. Phys. Chem. Solids **38**, 1039–1045 (1977)
2.46 C. Kittel: Solid State Commun. **25**, 519–520 (1978)

2.47 C.R.A. Catlow: in *Modulated Structures*, Proc. AIP Conf., vol. 53 (AIP, New York 1979) pp. 149–161
2.48 N. Daumas, A. Herold: Comp. Rend. Acad. Sci. C **268**, 373–375 (1969)
2.49 R. Clarke, N. Caswell, S.A. Solin, P.M. Horn: Phys. Rev. Lett. **42**, 61–64 (1979)
2.50 S.A. Safran, D.R. Hamann: Phys. Rev. Lett. **42**, 1410–1413 (1979)
2.51 A.P. Levanyuk, K.A. Minaeva, B.A. Strukov: Fiz. Tverd. Tela **9**, 3095–3103 (1967)
2.52 B.V. Egorov: Fiz. Tverd. Tela **25**, 3450–3451 (1983)
2.53 S.Ya. Geguzina, M.A. Krivoglaz: Fiz. Tverd. Tela **9**, 3095–3103 (1967)
2.54 A.P. Levanyuk, N.V. Shchedrina: Fiz. Tverd. Tela **14**, 3012–3016 (1972)
2.55 S.V. Maleev: Zh. Eksp. Teor. Fiz. **66**, 1809–1822 (1974)
2.56 S.V. Maleev: Zh. Eksp. Teor. Fiz. **69**, 1398–1414 (1975)
2.57 A.P. Levanyuk: Izv. Akad. Nauk. Ser. Fiz. **29**, 879–881 (1965)
2.58 A.I. Larkin, D.I. Khmelnitskyi: Zh. Eksp. Teor. Fiz. **56**, 2087–2098 (1969)
2.59 A. Aharony: Phys. Rev. B. **8**, 3363–3370 (1973)
2.60 C. Kittel: in *Solid State Physics*, ed. by F. Seitz, D. Turnbull, H. Ehrenreich, vol. 22 (Academic Press, New York 1968) pp. 1–26
2.61 S.V. Vonsovskii: *Magnetizm* (Wiley, New York 1974)
2.62 J.M. Ziman: *Principles of the Theory of Solids* (University Press, Cambridge 1964)
2.63 M.A. Krivoglaz: *Theory of X-Ray and Thermal-Neutron Scattering by Real Cristals* (Plenum, New York 1969)
2.64 S.C. Moss, R.H. Walker: J. Appl. Cristallogr. **8**, 96–107 (1974)
2.65 W. Kohn: Phys. Rev. Lett. **2**, 393–394 (1959)
2.66 A.B. Migdal: Zh. Eksp. Teor. Fiz. **34**, 1438–1446 (1958)
2.67 T.M. Hayes, H. Brooks, A. Bienenstock: Phys. Rev. **175**, 699–710 (1968)
2.68 J.E. Ingesfield: J. Phys. C. **2**, 1285–1292 (1969)
2.69 M. Kogachi: J. Phys. Chem. Solids **35**, 109–115 (1974)
2.70 G.L. Krasko, A.B. Makhnovetskii: Phys. Stat. Sol. b **65**, 869–878 (1974)
2.71 A.A. Katsnelson, V.M. Silonov, F.A. Khwaja: Phys. Stat. Sol. b **91**, 11–33 (1979)
2.72 A.A. Katsnelson, L.I. Yastrebov: *Pseudopotential Theory of Cristal Structures* (MGU, Moscow 1981) (in Russian)
2.73 J. Lindhard: Kgl. Danske Videnskab. Selskab. Mat. in *Fiz. Med.* **8**, 8–36 (1954)
2.74 P.L. Taylor: Phys. Rev. **131**, 1995–1999 (1963)
2.75 L.M. Roth, H.J. Zeiger, T.A. Kaplan: Phys. Rev. **149**, 519–525 (1966)
2.76 A.M. Afanasyev, Yu. Kogan: Zh. Eksp. Teor. Fiz. **43**, 1456–1463 (1962)
2.77 T.M. Rice, G.K. Scott: Phys. Rev. Lett. **35**, 120–123 (1975)
2.78 P. de Gennes: J. Phys. Radium **23**, 630–636 (1962)
2.79 P.B. Littlewood, V. Heine: J. Phys. C **14**, 2943–2949 (1981)
2.80 S.K. Sinha, B.N. Harmon: Phys. Rev. Lett **35**, 1515–1518 (1975)
2.81 J. Hafner, W. Hanke, H. Bilz: in *Electron- Phonon Interactions and Phase Transitions*, ed. by T. Riste (Plenum, New York 1977) pp. 200–208
2.82 E.V. Kozlov, V.M. Dementyev: Izv. Vyssh. Uchebn. Zaved. Fiz. **12**, 63–78 (1982)
2.83 H. Sato, R.S. Toth: in *Alloying Behavior and Effects in Concentrated Solid Solutions*, ed. by T.B. Massalski (Gordon and Breach, New York 1965) pp. 295–419
2.84 M. Tachiki, K. Teramoto: J. Phys. Chem. Solids **27**, 335–348 (1966)
2.85 D. Watanabe: In *Modulated Structures*, Proc. AIP Conf., vol. 53 (AIP, New York 1979) pp. 229–239
2.86 S.-K. Chan, V. Heine: J. Phys. F. **3**, 795–809 (1973)
2.87 J. Friedel: in *Electron-Phonon Interactions and Phase Transitions*, ed. by T. Riste (Plenum, New York 1977) pp. 1–49
2.88 W.M. Lomer: Proc. Phys. Soc. **80**, 489–496 (1962)
2.89 A.W. Overhauser: Phys. Rev. **128**, 1437–1452 (1962)
2.90 K. Yosida, A. Watabe: Progr. Theor. Phys. **28**, 361–370 (1962)

2.91 W.C. Koehler: in *Magnetic Properties of Rare Earth Metals*, ed. by R.J. Elliot (Plenum, London, New York 1972) pp. 81–128
2.92 R. Peierls: *Quantum Theroy of Solids* (Oxford 1955)
2.93 H. Frohlich: Proc. Roy. Soc. A **223**, 296–305 (1954)
2.94 L.N. Bulaevskii: Usp. Fiz. Nauk **115**, 263–300 (1975)
2.95 P. Bak: In *Electron-Phonon Interactions and Phase Transitions*, ed. by T. Riste (Plenum, New York 1977) pp. 66–87
2.96 R. Comes: in *Modulated Structures*, Proc. AIP Conf., vol. 53 (AIP, New York 1979) pp. 31–42
2.97 J.M. Williams, A.J. Schultz: in *Modulated Structures*, Proc. AIP Conf., vol. 53 (AIP, New York 1979) pp. 187–192
2.98 R.H. Friend, D. Jerome: J. Phys. C **12**, 1441–1477 (1979)
2.99 L.P. Gor'kov, I.E. Dzyaloshynskii: Zh. Eksp. Teor. Fiz. **67**, 397–417 (1974)
2.100 B. Horovitz, H. Gutfreund, M. Weger: Phys. Rev. B **12**, 3174–3185 (1975)
2.101 K. Saub, S. Barisic, J. Friedel: Phys. Lett. A **56**, 302–304 (1976)
2.102 L.N. Bulaevskii, D.I. Khomskii: Zh. Eksp. Teor. Fiz. **73**, 1148–1162 (1977)
2.103 J.A. Wilson, F.J. Di Salvo, S. Mahajan: Adv. Phys. **24**, 117–201 (1975)
2.104 L.N. Bulaevskii: Usp. Fiz. Nauk **120**, 259–271 (1976)
2.105 F.D. Di Salvo: in *ELectron-Phonon Interactions and Phase Transitions*, ed. by T. Riste (Plenum, New York 1977) pp. 107–136
2.106 R.J. Elliott, F.A. Wedgwood: Proc. Phys. Soc. **81**, 846–855 (1963)
2.107 I.E. Dzyaloshinskii: Zh. Eksp. Teor. Fiz. **47**, 336–348 (1964)
2.108 D.A. Vul, M.A. Krivoglaz: Fiz. Met. i. Metalloved. **52**, 1127–1140 (1981)
2.109 D.A. Vul, M.A. Krivoglaz: Metallofizika **4**, 3 (1982)
2.110 W.L. McMillan: Phys. Rev. B **12** 1187–1196 (1975)
2.111 I.E. Dzyaloshinskii: Zh. Eksp. Teor. Fiz. **47**, 992–1002 (1964)
2.112 P. Bak, J. von Boehm: Phys. Rev. B **21**, 5297–5308 (1964)
2.113 S. Aubry: in *Solitons and Condensed Matter Physics*, ed. by A.R. Bishop, T. Schneider (Springer, Berlin, Heidelberg 1978) pp. 264–277
2.114 D.E. Moncton, J.D. Axe, F.J. Di Salvo: Phys. Rev. Lett. **34**, 734–737 (1975)
2.115 W.L. McMillan: Phys. Rev. B **14**, 1496–1502 (1976)
2.116 H.E. Cook: Acta Met. **22**, 239–248 (1974)
2.117 H.E. Cook: J. Appl. Crystallogr. **8**, 132–140 (1975)
2.118 A.W. Overhauser: Phys. Rev. B **3**, 3173–3182 (1971)
2.119 Ya.I. Frenkel, T.M. Kontorova: Zh. Eksp. Teor. Fiz. **8**, 1340–1348 (1938)
2.120 F.C. Frank, J.H. Van-der-Merwe: Proc. Roy. Soc. A **198**, 205–225 (1949)
2.121 V.L. Indenbom: Kristallografiya **3**, 197–205 (1958)
2.122 V.L. Pokrovskii: Solid State Commun. **26**, 77–79 (1978)
2.123 L.N. Bulaevskii, D.I. Khomskii: Zh. Eksp. Teor. Fiz. **74**, 1863–1871 (1978)
2.124 V.L. Prokrovskii, A.L. Talapov: Zh. Eksp. Teor. Fiz. **75**, 1151–1157 (1978)
2.125 G. Theodorou, T.M. Rice: Phys. Rev. B **18**, 2840–2856 (1978)
2.126 V.L. Pokrovskii, A.L. Talapov: Zh. Eksp. Teor. Fiz. **78**, 269–295 (1980)
2.127 P. Bak, D. Mukamel, J. Villain, K. Wentowska: Phys. Rev. B **19**, 1610–1613 (1979)
2.128 A.D. Novaco: Phys. Rev. B **22**, 1645–1659 (1980)
2.129 C.H. Chen, J.M. Gibson, R.M. Fleming: Phys. Rev. B **26**, 184–205 (1982)
2.130 S. Ogawa: in *Order-Disorder Transformation in Alloys*, ed. by H. Warlimont (Springer, Berlin, Heidelberg 1974) pp. 240–264
2.131 M. Tachiki, S. Maekawa: J. Phys. Soc. Jap. **28**, 375–379 (1970)
2.132 S. Kubo, K. Adachi: J. Phys. Soc. Jap. **35**, 776–783 (1973)
2.133 M. Kataoka: J. Phys. Soc. Jap. **46**, 1176–1184 (1979)
2.134 D.A. Vul, M.A. Krivoglaz: Fiz. Met. i. Metalloved. **51**, 231–245 (1981)
2.135 D.A. Vul, M.A. Krivoglaz: Fiz. Met. i. Metalloved. **55**, 868–881 (1983)
2.136 K. Fujiwara: J. Phys. Soc. Jap. **12**, 7–13 (1957)
2.137 J. Kakinoki, T. Minagava: Acta Crystallogr. A **27**, 647–659 (1971)

2.138 J. Kakinoki, T. Minagava: Acta Crystallogr. A **28**, 120–133 (1972)
2.139 T. Minagava: Acta Crystallogr. A **28**, 308–318 (1972)
2.140 M. Guymont, D. Gratias, R. Portier: in *Modulated Structures*, Proc. AIP Conf., vol. 53 (AIP, New York 1979) pp. 244–246
2.141 M. Kataoka, H. Iwasaki: J. Phys. F. **11**, 1545–1556 (1981)
2.142 D.A. Vul: Fiz. Met. i. Metalloved. **65**, 66–73 (1988)
2.143 M. Greenholz, A. Kidron: Acta Crystallogr. A **26**, 311–314 (1970)
2.144 B. Borie, S.L. Sass, A. Andreassen: Acta Crystallogr. A **29**, 585–593; 594–602 (1973)
2.145 M. Sauvage, E. Parthe: Acta Crystallogr. A **30**, 239–246 (1974)
2.146 S. Hashimoto: Acta Crystallogr. A **30**, 792–798 (1974)
2.147 R. Di Ridder, G. van Tendeloo, D. van Dyck, S. Amelinckx: J. Phys. **38**, 178–186 (1977)
2.148 B. Borie: Acta Crystallogr. A **38**, 438–442 (1982): J. Phys. F **12**, 1285–1292 (1982)
2.149 J.M. Sanchez, D. de Fontaine: J. Phys. **38**, 444–452 (1977)
2.150 J.M. Sanchez, D. de Fontaine: J. Appl. Crystallogr. **10**, 220–227 (1977)
2.151 B. Horovitz, J.L. Murray, J.A. Krumhansl: Phys. Rev. B **18**, 3549–3558 (1978)
2.152 R. Pynn: J. Phys. F **8**, 1–13 (1978)
2.153 M.A. Krivoglaz: Pis'ma Zh. Eksp. Teor. Fiz. **24**, 403–406 (1976)
2.154 M.A. Krivoglaz: Zh. Eksp. Teor. Fiz. **83**, 355–369 (1983)
2.155 M.A. Krivoglaz: Metallofizika **6**, 3 (1984)
2.156 M.A. Krivoglaz: Fiz. Met. i. Metalloved. **66**, 1045–1072 (1988)
2.157 M.A. Krivoglaz: Electronic Theory of Heterogeneous States in Solids in *Electron Structure and Electron Properties of Metals and Alloys* (Nauk. Dumka. Kiev 1988)
2.158 D.A. Vul, M.A. Krivoglaz, O.V. Martynenko: Fiz. Tverd. Tela **30**, 3417–3425 (1988)
2.159 E.G. Brovman, Yu. M. Kagan: Usp. Fiz. Nauk **112**, 369–426 (1974)
2.160 L.D. Landau, E.M. Lifshitz: *Quantum Mechanics* (Pergamon Press, Oxford 1977)
2.161 R. Hosemann, S.N. Bagchi: *Direct Analysis of Matter by Diffraction* (North-Holland Publ., Amsterdam 1962)
2.162 L.E. Tanner: Phys. Stat. Sol. **30**, 685–701 (1968)
2.163 S.L. Sass: J. Less-Common Metals **28**, 157–173 (1972)
2.164 Yu. A. Bagaryatzkii, G.I. Nosova, T.V. Tagunova: Dokl. Akad. Nauk SSSR **105**, 1225–1228 (1955)
2.165 J.M. Silcock: Acta Met. **6**, 481–493 (1958)
2.166 D. de Fonatine: Acta Met. **18**, 275–279 (1970)
2.167 H.E. Cook: Acta Met. **21**, 1445–1449 (1973)
2.168 H.W. Myron, A.J. Freeman, S.C. Moss: Solid State Commun. **17**, 1467–1470 (1975)
2.169 A.L. Simons, C.M. Varma: Solid State Commun. **35**, 317–319 (1980)
2.170 M.A. Krivoglaz: Zh. Eksp. Teor. Fiz. **63**, 670–683 (1972)
2.171 M.A. Krivoglaz, A.I. Karasevskii: Fiz. Tverd. Tela **16**, 1458–1471 (1974)
2.172 M.A. Krivoglaz, A.I. Karasevskii: Zh. Eksp. Teor. Fiz. **69**, 297–310 (1975)
2.173 M.A. Krivoglaz, A.I. Karasevskii: Fiz. Tverd. Tela **17**, 2565–2577 (1975)
2.174 V.M. Nabutovskii, N.A. Nemov, Yu.G. Peisahovitch: Zh. Eksp. Teor. Fiz. **79**, 2196–2205 (1980)
2.175 A.I. Larkin, D.E. Khmelnitzkii: Zh. Eksp. Teor. Fiz. **55**, 2345–2354 (1968)
2.176 Yu.I. Balkarey, V.G. Baru, A.A. Suhanov: Fiz. Tverd. Tela **15**, 2223–2225 (1973)
2.177 S.I. Pekar, V.I. Pipa, V.N. Piskovoy: Pis'ma Zh. Eksp. Teor. Fiz. **12**, 338–340 (1970)

Chapter 3

3.1 M.A. Krivoglaz: Fiz. Met. i. Metalloved. **12**, 465–475 (1961)
3.2 W.H. Zachariasen: *Theory of X-Ray Diffraction in Crystals* (John Wiley & Sons, New York 1945)
3.3 M. Laue: *Röntgenstrahlinterferenzen* (Academische Verlagsgesellschaft, Leipzig 1948)
3.4 R.N. James: *The Optical Principles of the Diffraction of X-Rays* (New York 1948) reprinted.

3.5 V.I. Iveronova, G.P. Revkevich: *Theory of X-Rays Scattering* (Izd. MGu Moscow 1978) (in Russian)
3.6 B.K. Vainshtein: *Symmetry of Crystals. Methods of Structural Crystallography* (Plenum, New York, London 1981)
3.7 J.M. Cowley: *Diffraction Physics* (North-Holland 1975) 2nd Ed.
3.8 A. Ahiezer, I. Pomeranchuk: *Problems of Nuclear Theory* (Gostehteorizdat, Moscow 1950) (in Russian)
3.9 W. Marshall, S.W. Lovesey: *Theory of Thermal Neutron Scattering* (Clarendon Press, Oxford 1971)
3.10 G.E. Bacon: *Neutron Diffraction* (Clarendon Press, Oxford 1975)
3.11 G. Kostorz, S.W. Lovesey: in *Treatise on Materials Science and Technology*, ed. by G. Kostorz (Academic Press, New York 1979) pp. 1–67
3.12 Yu.A. Izyumov, V.E. Naish, R.P. Ozerov: (Atomizcat, Moscow 1981) (in Russian)
3.13 B. Borie, C.J. Sparks Jr: Acta Crystallogr. A **27**, 198–201 (1971)
3.14 J.E. Tibbals: J. Appl. Crystallogr. **8**, 111–114 (1975)
3.15 P. Georgopoulos, J.B. Cohen: J. Phys. **38**, 191–196 (1977)
3.16 M. Hayakawa, P. Bardhan, J.B. Cohen: J. Appl. Crystallogr. **8**, 87–95 (1975)
3.17 H. Chen, R.J. Comstock, J.B. Cohen: Ann. Rev. Mater. Sci. **9**, 51–86 (1979)
3.18 M. Laue: Ann. Phys. **78**, 167–176 (1925)
3.19 I. Pomeranchuk: Zh. Eksp. Teor. Fiz. **8**, 894–906 (1938)
3.20 Yu.N. Obraztzov: Zh. Eksp. Teor. Fiz. **8**, 593–598 (1938)
3.21 I. Lifshitz: Zh. Eksp. Teor. Fiz. **9**, 500–511 (1939)
3.22 A.A. Smirnov, S.V. Vonsovskii: J. Phys. USSR **5**, 263–266 (1941)
3.23 P.A. Flinn, B.L. Averbach, P.S. Rudman: Acta Crystallogr. **7**, 153–158 (1954)
3.24 V.M. Danilenko, M.A. Krivoglaz, Z.A. Matysina, A.A. Smirnov: Fiz. Met. i. Metalloved. **4**, 28–34 (1957)
3.25 W. Cohran: Acta Crystallogr. **9**, 259–262 (1956)
3.26 W. Cohran, G. Kartha: Acta Crystallogr. **9**, 941–943 (1956)
3.27 B.E. Warren, B.L. Averbach, B.W. Roberts: J. Appl. Phys. **22**, 1493–1496 (1951)
3.28 B. Borie: Acta Crystallogr. **10**, 89–96 (1957)
3.29 B. Borie: Acta Crystallogr. **12**, 280–282 (1959)
3.30 A.A. Smirnov, E.A. Tihonova: Fiz. Tverd. Tela **3**, 1238–1248 (1961)
3.31 H. Kanzaki: J. Phys. Chem. Solids **2**, 107–114 (1957)
3.32 A.B. Bhatia, D.E. Thornton: Phys. Rev. B **4**, 2325–2328 (1971)
3.33 J.E. Gragg Jr.: J. Phys. Chem. Solids **32**, 1195–1203 (1971)
3.34 M.A. Krivoglaz: Fiz. Tverd. Tela **9**, 641–656 (1960)
3.35 D.T. Keating: J. Phys. Chem. Solids **32**, 2693–2698 (1971)
3.36 H. Behr, H. Metzger, J. Peisl: J. Phys. F **13**, L75–L76 (1983)
3.37 H. Metzger, H. Behr, J. Peisl: Z. Phys. B **4**, 295–299 (1982)
3.38 H. Metzger, H. Behr, G. Steyrer, J. Peisl: Phys. Rev. Lett. **50**, 843–845 (1983)
3.39 N. Norman, B.E. Warren: J. Appl. Phys. **22**, 483–486 (1951)
3.40 A.A. Katznelson, A.M. Silonov, V.M. Silonov: Fiz. Met. i. Metalloved. **51**, 455–462 (1981)
3.41 A.N. Tihonov, V.Ya. Arsenin: *Solutions of Incorrect Problems* (Nauka, Moscow 1974) (in Russian)
3.42 M. Morinaga: Acta Met. **25**, 957–962 (1977)
3.43 P. Bardhan, J.B. Cohen: Acta Crystallogr. **32**, 597–614 (1976)
3.44 K. Ohshima, D. Watanabe, J. Harada: Acta Crystallogr. A **32**, 883–892 (1976)
3.45 S.C. Moss: Phys. Rev. Lett. **22**, 1108–1111 (1969)
3.46 M.A. Krivoglaz, Khao Thyu: Fiz. Met. i. Metalloved. **21**, 817–823 (1966)
3.47 H.E. Cook: J. Phys. Chem. Solids **30**, 1097–1112 (1969)
3.48 S.C. Moss: J. Appl. Phys. **35**, 3547–3553 (1964)
3.49 H. Conrad, G. Bauer, G. Alefeld et al.: Zs. Phys. **266**, 239–244 (1974)
3.50 W. Munzing, N. Stump, G. Goltz: J. Appl. Crystallogr. **11**, 588–591 (1978)

3.51 V.I. Izotov, L.M. Utevskii: Fiz. Met. i. Metalloved. **25**, 98–110 (1968)
3.52 M.A. Krivoglaz: Fiz. Met. i. Metalloved. **8**, 514–530 (1959)
3.53 V.I. Iveronova, A.A. Katznelson: Fiz. Met. i. Metalloved. **11**, 40–45 (1961)
3.54 H. Kubo, I. Cornelis, C.M. Wayman: Acta Met. **28**, 405–416 (1980)
3.55 S.C. Moss: Ann. Rev. in Mater. Sci. **3**, 293–316 (1973)
3.56 W. Schmatz: in *Treatise on Materials Science and Technology*, ed. by H.H. Herman, vol. 2 (Academic Press, New York 1973) pp. 105–229
3.57 W. Schmatz: in *Neutron Diffraction*, ed. by H. Dachs (Springer, Berlin, Heidelberg, New York 1978) pp. 151–196
3.58 G.S. Bauer: in *Treatise on Materials Science and Technology*, ed. by G. Kostorz, vol. 15 (Academic Press, New York 1979) pp. 291–336
3.59 J.E. Gragg Jr., M. Hayakawa, J.B. Cohen: J. Appl. Crystallogr. **6**, 59–66 (1973)
3.60 J.E. Epperson, P. Furnrohr, C. Ortiz: Acta Crystallogr. A **34**, 667–681 (1978)
3.61 S. Lefevre, F. Bley, M. Bessiere et al.: Acta Crystallogr. A **36**, 1–7 (1980)
3.62 Z. Wilchinsky: J. Appl. Phys. **15**, 806–812 (1944)
3.63 P. Furnrohr, J. Epperson, V. Gerold: Zs. Metallkunde **71**, 403–409 (1980)
3.64 H. Suzuki, J. Harada, T. Nakashima, K. Adachi: Acta Crystallogr. **38**, 522–529 (1982)
3.65 B. Mozer, D. Keating, S.C. Moss: Phys. Rev. **175**, 868–876 (1968)
3.66 J. Vrijen, S. Radelaar: Phys. Rev. B **17**, 409–421 (1978)
3.67 W. Wagner, R. Poerschke, A. Axmann: Phys. Rev. B **21**, 3087–3099 (1980)
3.68 M. Morinaga, J.B. Cohen: Acta Crystallogr. A **35**, 745–756 (1979)
3.69 M. Morinaga, J.B. Cohen: Acta Crystallogr. A **35**, 975–989 (1979)
3.70 S.V. Semenovskaya, A.I. Kitaygorodskii, A.A. Smirnova: Kristallografiya **23**, 299–308 (1978)
3.71 A.S. Lashko: Vopr. Fiz. Met. i. Metalloved. **15**, 80–86 (1962)
3.72 D.-H. Wu, R.A. Tahir-Kheli: J. Phys. Soc. Jap. **31**, 641–648 (1971)
3.73 E. Metcalfe, J.A. Leake: Acta Met. **23**, 1135–1143 (1975)
3.74 A.A. Katznelson, P.P. Safonov, V.G. Moiseenko, V.M. Silonov: Fiz. Met. i. Metalloved. **43**, 110–115 (1977)
3.75 L.A. Safronova, A.A. Katznelson, S.V. Sveshnikov, Yu.M. Lvov: Fiz. Met. i. Metalloved. **43**, 76–80 (1977)
3.76 Z. Racz, M.F. Collins: Phys. Rev. B **21**, 229–237 (1980)
3.77 J.W. Cahn, R. Kikuchi: Acta Met. **27**, 1329–1336 (1979)
3.78 J.M. Sanchez, D. de Fontaine: in *Modulated Structures*, Proc. AIP Conf., vol. 53 (AIP, New York 1979) pp. 133–145
3.79 M.M. Naumova, S.V. Semenovskaya: Fiz. Tverd. Tela **13**, 381–387 (1971)
3.80 S.V. Semenovskaya: Fiz. Tverd. Tela **8**, 2841–2847 (1966)
3.81 S.V. Semenovskaya: Fiz. Tverd. Tela **9**, 3408–3414 (1967)
3.82 S.V. Semenovskaya: Phys. Stat. Sol. b **64**, 291–303 (1974)
3.83 S.V. Semenovskaya, D.M. Umidov: Phys. Stat. Sol. b **64**, 627–633 (1974)
3.84 V.T. Bublik, S.S. Gorelik, A.A. Zaitsev, A.Y. Polyakov: Phys. Stat. Sol. b **66**, 427–432 (1974)
3.85 V.T. Bublik, M.V. Karman, V.F. Kleptsin, V.N. Leikin: Phys. Stat. Sol. a **32**, 631–638 (1975)
3.86 V.T. Bublik, A.A. Zaitsev: Phys. Stat. Sol. a **39**, 345–349 (1977)
3.87 V.T. Bublik: Kristall und Technik **12**, 849–857 (1977)
3.88 S.V. Semenovskaya, A.A. Smirnova, A.I. Kitaygorodskii: Kristallografiya **23**, 99–106 (1978)
3.89 A.I. Kitaygorodskii: *Molecular Crystals* (Nauka, Moscow 1971) (in Russian)
3.90 K.S. Singwi, A. Sjolander: Phys. Rev. **120**, 1093–1102 (1960)
3.91 M.A. Krivoglaz: Zh. Eksp. Teor. Fiz. **40**, 1812–1824 (1961)
3.92 C.T. Chudley, R.J. Elliott: Proc. Phys. Soc. **77**, 353–361 (1961)
3.93 M.A. Krivoglaz, S.P. Repetskii: Fiz. Tverd. Tela **8**, 2908–2918 (1966)
3.94 M.A. Krivoglaz, S.P. Repetskii: Ukr. Fiz. Zh. **11**, 1215–1221 (1966)
3.95 S. Asenov, T. Ruskov, T. Tomov, I. Spirov: Phys. Lett. A **79**, 349–352 (1980)
3.96 J.G. Mullen: Phys. Lett. A **79**, 457–460 (1980)
3.97 W. Petry, G. Vogl, W. Mansel: Phys. Rev. Lett. **45**, 1862–1865 (1980)

3.98 M. Ait-Salem, T. Springer, A. Heidemann, B. Alefeld: Phyl. Mag. A **39**, 797–814 (1979)
3.99 F. Probst, F.E. Wagner, M. Karger: J. Phys. F **10**, 2081–2091 (1980)
3.100 G. Goltz, A. Heidemann, H. Mehrer et al.: Phyl. Mag. A **41**, 723–744 (1980)
3.101 T. Springer: in *Dynamics of Solids and Liquids by Neutron Scattering*, ed. by S.W. Lovesay, T. Springer. Topics in Current Physics, vol. 3 (Springer, Berlin, Heidelberg 1977)
3.102 S. Mantl, W. Petry, K. Schroeder, C. Vogl: Phys. Rev. B **27**, 5313–5331 (1983)
3.103 J. Bardeen, C. Herring: in *Imperfections in Nearly Perfect Cristals*, (John Wiley, London 1952) pp. 261–288
3.104 M.C.D. Ure, P.A. Flinn: Phys. Rev. B **15**, 1261–1270 (1977)
3.105 M.A. Krivoglaz, S.P. Repetzkii: Fiz. Met. i. Metalloved. **32**, 899–910 (1971)
3.106 A.G. Khachaturyan: Fiz. Tverd. Tela **9**, 2594–2603 (1967)
3.107 A.G. Khachaturyan: Fiz. Tverd. Tela **11**, 3534–3537 (1969)
3.108 H.E. Cook, D. de Fontaine, J.E. Hilliard: Acta Met. **17**, 765–773 (1969)
3.109 H.E. Cook: J. Phys. Chem. Solids **30**, 2427–2437 (1969)
3.110 M.M. Naumova, S.V. Semenovskaya, Ya.S. Umanskii: Fiz. Tverd. Tela **12**, 75–982 (1970)
3.111 M.M. Naumova, S.V. Semenovskaya: Fiz. Tverd. Tela **12**, 3632–3633 (1970)
3.112 H. Chen, J.B. Cohen: J. Phys. **38**, 314–327 (1977)
3.113 J. Vrijen, J. Aalders, C. Van Dijk: Phys. Rev. B **22**, 1503–1514 (1980)
3.114 W. Wagner, R. Poerschke, A. Axmann, D. Schwahn: Phys. Rev. B **21**, 3087–3099 (1980)
3.115 W. Wagner, R. Poerschke, H. Wollenberger: J. Phys. F **12**, 405–424 (1982)
3.116 V.I. Iveronova, A.A. Katznelson: Kristallografiya **5**, 71–78 (1960)
3.117 V.I. Iveronova, A.A. Katznelson, G.P. Revkevich: Fiz. Met. i. Metalloved. **26**, 1064–1068 (1968)
3.118 P. Sh. Dazhaev, A.A. Katznelson: Fiz. Met. i. Metalloved. **30**, 663–665 (1970)
3.119 V.I. Iveronova, A.A. Katznelson, M.D. Kondratyeva, G.P. Revkevich: Fiz. Met. i. Metalloved. **35**, 355–362 (1973)
3.120 A.A. Katznelson: Izv. Vyssh. Uchbn. Zaved. Fiz. **10**, 17–24 (1971)
3.121 V.I. Iveronova, A.A. Katznelson: Kristallografiya **16**, 1259–1263 (1971)
3.122 H.E. Cook: Mater. Sci. Eng. **25**, 127–134 (1976)
3.123 H. Chen, J.B. Cohen: Acta Met. **27**, 603–611 (1979)
3.124 M.A. Krivoglaz: *Theory of X-Ray and Thermal-Neutron Scattering by Real Crystals* (Plenum, New York 1969)
3.125 R. Raether: Z. Angew. Phys. **4**, 53–59 (1952)
3.126 M.J. Marcinowski, L. Zwell: Acta Met. **11**, 373–390 (1963)
3.127 S.C. Moss: in *Local Atomic Arrangements Studied by X-Ray Diffractions*, ed. by J.B. Cohen, J.E. Hillard (Gordon & Breach, New York 1965) pp. 95–122
3.128 D. Watanabe, P.M.J. Fisher: J. Phys. Soc. Jap. **20**, 2170–2179 (1965)
3.129 S. Hashimoto, S. Ogawa: J. Phys. Soc. Jap. **29**, 710–721 (1970)
3.130 R.O. Scattergood, S.C. Moss, M.B. Bever: Acta Met. **18**, 1087–1098 (1970)
3.131 K. Ohshima, D. Watanabe: Acta Crystallogr. A **29**, 520–526 (1973)
3.132 K. Ohshima, D. Watanabe: Acta Crystallogr. A **33**, 784–788 (1977)
3.133 M.A. Krivoglaz, A.A. Smirnov: Fiz. Met. i. Metalloved. **7**, 151–152 (1977)
3.134 I.Ya. Dekhtyar: Phys. Rep. **9**, 242–253 (1974)
3.135 K. Fujiwara, O. Sucoka, T. Imura: J. Phys. Soc. Jap. **24**, 467–475 (1968)
3.136 M. Hasegawa, T. Suzuki, M. Hirabayashi: J. Phys. Soc. Jap. **43**, 89–96 (1977)
3.137 G. Nilsson, S. Rolandson: Phys. Rev. **9**, 3278–3284 (1974)
3.138 S.V. Semenovskaya: Fiz. Tverd. Tela **20**, 1360–1364 (1978)
3.139 S.V. Sudareva, E.N. Zhuravleva: Fiz. Met. i. Metalloved. **50**, 130–141 (1980)
3.140 J.B. Goodenough: Phys. Rev. B **5**, 2764–2774 (1972)
3.141 L.M. Huisman, A.E. Carlsson, C.D. Gelatt Jr.: Phys. Rev. B **22**, 991–1006 (1980)
3.142 J.R. Castles, J.M. Cowley, A.E. Spargo: Acta Crystallogr. A **27**, 376–383 (1971)
3.143 H. Terauchi, J.B. Cohen: J. Phys. Chem. Solids **39**, 681–686 (1978)
3.144 M.B. Stearns: Phys. Rev. **129**, 1136–1144 (1963)
3.145 A.A. Katznelson, Yu.I. Ustinovshchikov: Fiz. Met. i. Metalloved. **46**, 649–650 (1978)

3.146 F.A. Sidorenko, V.Ya. Elner, B.V. Ryzhenko, P.V. Geld: Fiz. Tverd. Tela **21**, 1216–1218 (1979)
3.147 P. Gehlen, J.B. Cohen: Phys. Rev. **139**, 844–855 (1965)
3.148 J.E. Gragg Jr., P. Bardhan, J.B. Cohen: in *Critical Phenomena in Alloys, Magnets and Superconductors*, ed. by R.E. Mills, E. Ascher, R.I. Jaffee (McGraw-Hill, New York 1971) pp. 309–337
3.149 J.E. Epperson: J. Appl. Crystallogr. **12**, 351–356 (1979)
3.150 J.E. Epperson, P. Furnrohr: Acta Crystallogr. A **36**, 372–378 (1980)
3.151 B. Borie, C.J. Sparks Jr.: Acta Crystallogr. **17**, 827–835 (1964)
3.152 J.E. Epperson, G. Kostorz, C. Ortiz et al.: Acta Met. **27**, 1363–1372 (1979)
3.153 R.O. Williams: Metallurgical Transactions **5**, 1843–1850 (1974)
3.154 W. Gaudig, H. Warlimont: Acta Met. **26**, 709–724 (1978)
3.155 H. Chen, J.B. Cohen, R. Ghosh: J. Phys. Chem. Solids **38**, 855–857 (1977)
3.156 P. Bardhan, H. Chen, J.B. Cohen: Phyl. Mag. **35**, 1653–1666 (1977)
3.157 G.C. Kuczynski, M. Doyama, M.E. Fine: J. Appl. Phys. **27**, 651–655 (1956)
3.158 R. Sinclair, K. Schneider, G. Thomas: Acta Met. **23**, 873–883 (1975)
3.159 R. Sinclair, G. Thomas: J. Appl. Crystallogr. **8**, 206–210 (1975)
3.160 R. Sinclair, G. Thomas: J. Phys. **38**, 165–171 (1977)
3.161 J. Vrijen, C. van Dijk, E.W. van Royen, S. Radelaar: J. Phys. **38**, 341–346 (1977)
3.162 W. Lin, J.E. Spruiell, R.O. Williams: J. Appl. Crystallogr. **3**, 297–305 (1970)
3.163 W.M. Stobbs, J.-P. Chevalier; Acta Met. **26**, 233–240 (1978)
3.164 H. Warlimont, H.P. Aubauer: Zs. Metallkunde **64**, 484–491 (1973)
3.165 H. Berg Jr., J.B. Cohen: Acta Met. **21**, 1579–1588 (1973)
3.166 H. Berg. Jr., T.T. Tsong, J.B. Cohen: Acta Met. **21**, 1589–1599 (1973)
3.167 P.C. Clapp: J. Phys. Chem. Solids **30**, 2589–2598 (1969)
3.168 P.C. Clapp: Phys. Rev. B **4**, 255–270 (1971)
3.169 K. Ohshima, J. Harada: Sci. Repts. Inst. Tohoki Univ. **A29**, 123–128 (1981)
3.170 N. Tanaka, K. Ohshima, J. Harada, K. Mihama: in *Modulated Structures*, Proc. AIP Conf., vol. 53 (AIP, New York 1979) pp. 292–294
3.171 S. Hashimoto, H. Iwasaki: in *Modulated Structures*, Proc. AIP Conf., vol. 53 (AIP, New York 1979) pp. 283–285
3.172 S. Hashimoto, H. Iwasaki: Phys. Stat. Sol. a **51**, 673–682 (1979)
3.173 J.-P. Chevalier, A.J. Craven: Phyl. Mag. **36**, 67–79 (1977)
3.174 H. Sato, S.A. Werner, R. Kikuchi: J. Phys. **35**, 23–26 (1974)
3.175 N. Ahmed, T.J. Hicks: Solid State Commun. **15**, 415–419 (1974)
3.176 H. Bouchiai, E. Dartyge, P. Monod, M. Lambert: Phys. Rev. B **23**, 1375–1383 (1981)
3.177 E. Dartyge, H. Bouchiat, P. Monod: Phys. Rev. B **25**, 6995–7002 (1982)
3.178 J.P. Boilot, A. Kahn, J. Thery et al.: J. Phys. **38**, 145–150 (1977)
3.179 W.L. Roth: J. Phys. **38**, 151–158 (1977)
3.180 D.B. McWhan, P.D. Dernier, C. Vettier et al.: Phys. Rev. B **17**, 4043–4059 (1978)
3.181 R.J. Cava, D.B. McWhan: Phys. Rev. Lett. **45**, 2046–2050 (1980)
3.182 J.P. Boilot, G. Collin, Ph. Colomban, R. Comes: Phys. Rev. B **22**, 5912–5923 (1980)
3.183 M. Brunel, F. de Bergevin, M. Gondrand: J. Phys. Chem. Solids **33**, 1927–1941 (1972)
3.184 M. Morinaga: in *Modulated Structures*, Proc. AIP Conf., vol. 53 (AIP Conf., New York 1979) pp. 370–372
3.185 D. van Dyck, R. de Ridder, S. Amelinckx: Phys. Stat. Sol. a **59**, 513–530 (1980)
3.186 S. Ijima, J.M. Cowley: J. Phys. **38**, 135–144 (1977)
3.187 B.S. Hickman: J. Mater. Sci. **4**, 554–563 (1969)
3.188 D. de Fontaine, N.E. Paton, J.C. Williams: Acta Met. **19**, 1153–1162 (1971)
3.189 K.K. McCabe, S.L. Sass: Phyl. Mag. **23**, 957–970 (1971)
3.190 I.V. Lyasotzkii, Yu.D. Tyapkin: Fiz. Met. i. Metalloved. **36**, 1260–1270 (1973)
3.191 I.V. Lyasotzkii, N.B. Dyakonova, Yu.D. Tyapkin: Dokl. Akad. Nauk SSSR **237**, 577–580 (1977)
3.192 N.B. Dyakonova, I.V. Lyasotzkii: Fiz. Met. i. Metalloved. **53**, 525–532 (1982)
3.193 I.V. Lyasotzkii, N.V. Dyakonova: Fiz. Met. i. Metalloved. **53**, 1161–1168 (1982)

3.194 S.C. Moss, D.T. Keating, J.D. Axe: in *Conf. on Phase Transformations and Applications in Material Science*, ed. by L.E. Cross, (Pergamon, New York 1973) pp. 179–188
3.195 D.T. Keating, S.J. LaPlaca: J. Phys. Chem. Solids **35**, 879–891 (1974)
3.196 J.D. Axe, D.T. Keating, S.C. Moss: Phys. Rev. Lett. **35**, 530–533 (1975)
3.197 W. Lin, H. Spalt, B.W. Batterman: Phys. Rev. B **13**, 5158–5169 (1976)
3.198 J.D. Axe, A. Heidemann, W.S. Howells et al.: Solid State Commun. **24**, 743–745 (1977)
3.199 C. Stassis, J. Zarestky, N. Wakabayashi: Phys. Rev. Lett. **41**, 1726–1729 (1978)
3.200 I.V. Lyasotzkii, N.B. Dyakonova: Dokl. Akad. Nauk SSSR **256**, 388–392 (1981)
3.201 C.B. Walker: Phys. Rev. B **28**, 674–688 (1983)
3.202 H. Dosch, A. Schwerin, J. Peisl: Phys. Rev. B **34**, 1654–1661 (1986)
3.203 T.S. Kuan, S.L. Sass: Acta Met. **24**, 1053–1059 (1976)
3.204 T.S. Kuan, S.L. Sass: Phil. Mag. **36**, 1473–1498 (1977)
3.205 C.B. Walker: J. Phys. Chem. **44**, 167–169 (1983); **28**, 674–688 (1983)
3.206 Yu.D. Tyapkin, I.V. Lyasotzkii, E.I. Malienko: Fiz. Met. i. Metalloved. **37**, 107–118 (1974)
3.207 I.V. Lyasotzkii, N.B. Dyakonova: Fiz. Met. i. Metalloved. **50**, 142–150 (1980)
3.208 N.B. Dyakonova, I.V. Lyasotzkii: Fiz. Met. i. Metalloved. **52**, 119–127 (1981)
3.209 R.S. Seymour: Acta Crystallogr. **27**, 348–353 (1971)
3.210 G. Coulon, M. Descamps: J. Phys. C **13**, 945–962 (1980); **13**, 2847–2856 (1980)
3.211 J. Villain, J. Schneider: in *Physics and Chemistry of Ice* (Royal Society of Canada, Ottawa 1973) pp. 285–290
3.212 M. Descamps, G. Coulon: Chem. Phys. **25**, 117–130 (1977)
3.213 J. Schneider, C. Zeyen: J. Phys. C **13**, 4121–4126 (1980)
3.214 G.L. Paul, W. Cochran, W.J.L. Buyers, R.A. Cowley: Phys. Rev. B **2**, 4603–4612 (1970)
3.215 V.G. Vaks: *Introduction to Microscopic Theory of Ferroelectrics* (Nauka, Moscow 1973) (in Russian)
3.216 S. Havlin, H. Sompolinzky: Phys. Rev. B **21**, 5212–5220 (1977)
3.217 S. Halvin, R.A. Corvley, H. Sompolinsky: J. Phys. C **15**, 6057–6066 (1982)
3.218 R. Youngblood, J.D. Axe, B.M. McCoy: Phys. Rev. B **21**, 5212–5220 (1980)
3.219 D.E. Moncton, J.D. Axe, F.J. Di Salvo: Phys. Rev. B **16**, 801–819 (1977)
3.220 J.D. Axe: Pys. Rev. B **21**, 4181–4190 (1980)
3.221 G.F. Giuliani, A.W. Overhauser: Phys. Rev. B **23**, 3737–3743 (1981)
3.222 A.D. Bruce, R.A. Cowley; J. Phys. C **11**, 3609–3630 (1978)
3.223 R.A. Cowley: Adv. Phys. **29**, 1–110 (1980)
3.224 V.A. Golovko, A.P. Levanyuk: Zh. Eksp. Teor. Fiz. **81**, 2296–2313 (1981)
3.225 P.A. Lee, T.M. Rice, P.W. Anderson: Solid State Commun. **14**, 703–709 (1974)

Chapter 4

4.1 L.S. Ornstein, F. Zernike: Phys. Zs. **19**, 134–158 (1918)
4.2 M.A. Krivoglaz: Ukr. Fiz. Zh. **2**, 114–121 (1957)
4.3 G.Ya. Lubarsky: *The Application of Group Theory in Physics* (Pergamon Press, New York 1966)
4.4 V.L. Ginzburg: Usp. Fiz. Nauk **38**, 490–525 (1949)
4.5 A.F. Devonshire: Adv. Phys. **3**, 85-130 (1954)
4.6 A.P. Levanuk: Zh. Eksp. Teor. Fiz. **36**, 810–818 (1959)
4.7 V.L. Ginzburg: Fiz. Tverd. Tela **2**, 2031–2043 (1960)
4.8 J. Als-Nielsen, R.J. Birgeneau: Amer. J. Phys. **45**, 554–560 (1977)
4.9 F.J. Wagner, E.K. Riedel: Phys. Rev. B **7**, 248–256 (1973)
4.10 L.P. Kadanov, W. Gotze, D. Hambleuetal: Rev. Modern Phys. **39**, 395–431 (1967)
4.11 A.Z. Patashinskii, V.L. Pokrovskii: Usp. Fiz. Nauk **121**, 55–96 (1977)
4.12 P.-A. Lingard: in *Neutron Diffraction*, ed. by H. Dachs (Springer, Berlin, Heidelberg 1978) pp. 197–242
4.13 Shang-Keng Ma: *Modern Theory of Critical Phenomena* (W.A. Benjamin Inc., London, Amsterdam 1978)

4.14 A.D. Bruce: Adv. Phys. **29**, 111–217 (1980)
4.15 A.Z. Patashinskii, V.L. Pokrovskii: *Fluctuational Theory of Phase Transitions* (Nauka, Moscow 1982) (in Russian)
4.16 A.Z. Patashinskii, V.L. Pokrovskii: Zh. Eksp. Teor. Fiz. **50**, 439–447 (1966)
4.17 L.P. Kadanoff: Physics **2**, 263–272 (1966)
4.18 M.E. Fisher, R.J. Burford: Phys. Rev. **156**, 583–622 (1967)
4.19 D.S. Ritchie, M.E. Fisher: Phys. Rev. **5**, 2668–2692 (1972)
4.20 M.E. Fisher, A. Aharony: Phys. Rev. B **10**, 2818–2833 (1974)
4.21 D.R. Nelson: Phys. Rev. B **14**, 1123–1127 (1976)
4.22 E. Brezin, J.C. Le Guillou, J. Zinn-Justin: Phys. Rev. Lett. **32**, 473–475 (1974)
4.23 Y. Achiam, J.M. Kosterlitz: J. Phys. C **10**, 4559–4569 (1977)
4.24 J.F. Nicoll: Phys. Rev. B **21**, 1124–1132 (1980)
4.25 K.G. Wilson: Phys. Rev. B **4**, 3174–3205 (1971)
4.26 K.G. Wilson, M.E. Fisher: Phys. Rev. Lett. **28**, 240–243 (1972)
4.27 K.G. Wilson, J. Kogut: Phys. Reports **12c**, 75–199 (1974)
4.28 M.E. Fisher: Rev. Modern Phys. **46**, 595–616 (1974)
4.29 I.R. Yuhnovskii: Teor. i Mat. Fiz. **36**, 3773–399 (1978)
4.30 W.J. Camp, D.M. Saul, J.P. Van Dyke, M. Wortis: Phys. Rev. B **14**, 3990–4001 (1976)
4.31 J.C. Le Guillou, J. Zinn-Justin: Phys. Rev. B **21**, 3976–3998 (1980)
4.32 D.Z. Albert: Phys. Rev. B **25**, 4810–4818 (1982)
4.33 A.A. Vladimirov, D.I. Kazakov, O.V. Tarasov: Zh. Eksp. Teor. Fiz. **77**, 1035–1045 (1979)
4.34 C. Domb: in *Phase Transitions and Critical Phenomena*, ed. by C. Domb, M.S. Green, vol. 3 (Academic Press, New York 1974) pp. 357–484
4.35 W.J. Camp, J.P. Van Dyke: J. Phys. A **9**, 731–749 (1976)
4.36 L.N. Lipatov: Zh. Eksp. Teor. Fiz. **72**, 411–427 (1977)
4.37 M.E. Fisher: Phys. Rev. **176**, 257–272 (1968)
4.38 A. Aharony: in *Phase Transitions and Critical Phenomena*, ed. by C. Domb, M.S. Green, vol. 6 (Academic Press, New York 1976) pp. 357–424
4.39 T. Nattermann: J. Phys. C **9**, 3337–3354 (1976)
4.40 R.A. Cowley, A.D. Bruce: J. Phys. C **11**, 3577–3590 (1978)
4.41 I.F. Lyuksyutov, V.L. Pokrovskii: Pis'ma v Zh. Eksp. Teor. Fiz. **21**, 22–25 (1975)
4.42 I.F. Lyuksyutov, V.L. Pokrovskii, D.E. Khmelnitzkii: Zh. Eksp. Teor. Fiz. **69**, 1817–1824 (1975)
4.43 S.A. Brazovskii, I.E. Dzyaloshinskii: Pis'ma v Zh. Eksp. Teor. Fiz. **21**, 360–364 (1975)
4.44 D. Mukamel, S. Krinsky: Phys. Rev. B **13**, 5065–5085 (1976)
4.45 P. Bak, D. Mukamel: Phys. Rev. **13**, 5086–5094 (1976)
4.46 S.A. Brazovskii, I.E. Dzyaloshinskii, B.G. Kuharenko: Zh. Eksp. Teor. Fiz. **70**, 2257–2267 (1976)
4.47 I.E. Dzyaloshinskii: Zh. Eksp. Teor. Fiz. **72**, 1930–1945 (1977)
4.48 E. Domany, D. Mukamel, M.E. Fisher: Phys. Rev. B **15**, 5432–5441 (1977)
4.49 A. Aharony, A.D. Bruce: Phys. Rev. Lett. **42**, 462–464 (1979)
4.50 A.L. Korzhenevskii: Zh. Eksp. Teor. Fiz. **81**, 1071–1087 (1981)
4.51 D. Blankschtein, A. Aharony: Phys. Rev. B **26**, 415–419 (1982)
4.52 D. Blankschtein, D. Mukamel: Phys. Rev. B **25**, 6939–6951 (1982)
4.53 D. Blankschtein, A. Aharony: Phys. Rev. B **28**, 386–401 (1983)
4.54 A. Aharony: Phys. Rev. B **8**, 3363-3370 (1973)
4.55 E. Brezin, J. Zinn-Justin: Phys. Rev. **13**, 251–254 (1976)
4.56 E.K. Riedel, F. Wegner: Phys. Rev. Lett. **24**, 730–733 (1970)
4.57 R.B. Griffiths: Phys. Rev. Lett. **24**, 1479–1482 (1970)
4.58 D. Bailin, A. Love: J. Phys. C **10**, 3633–3639 (1977)
4.59 A. Aharony, M.E. Fisher: Phys. Rev. **8**, 3323–3342 (1973)
4.60 A.I. Sokolov, A.K. Tagantzev: Zh. Eksp. Teor. Fiz. **76**, 181–193 (1979)
4.61 A.L. Korzhenevskii: Fiz. Tverd. Tela **23**, 77–85 (1981)
4.62 O.K. Rice: J. Chem. Phys. **22**, 1535–1544 (1954)

4.63 A.I. Larkin, S.A. Pikin: Zh. Eksp. Teor. Fiz. **56**, 1664–1674 (1969)
4.64 J. Sak: Phys. Rev. B **10**, 3957–3960 (1974)
4.65 J. Bruno, J. Sak: Phys. Rev. B **22**, 3302–3324 (1980)
4.66 D.J. Bergman, B.I. Halperin: Phys. Rev. B **13**, 2145–2175 (1976)
4.67 K.K. Murata: Phys. Rev. B **15**, 4328–4335 (1977)
4.68 T. Nattermann: J. Phys. A **10**, 1757–1772 (1977)
4.69 A. Munster, K. Sagel: Zeit. Physik. Chem. **12**, 145–167 (1957)
4.70 A. Munster: in *Fluctuation Phenomena on Solids*, ed. by R.E. Burgess (Academic Press, New York 1965) pp. 180–266
4.71 J. Rudnick, D.R. Nelson: Phys. Rev. B **13**, 2208–2221 (1976)
4.72 H. Unoki, T. Sakudo: J. Phys. Soc. Jap. **23**, 546–552 (1967)
4.73 G. Shirane, Y. Yamada: Phys. Rev. **177**, 858–863 (1969)
4.74 G. Shirane: Rev. Modern Phys. **46**, 437–449 (1974)
4.75 B. Dorner, R. Comes: in *Dynamics of Solids and Liquids by Neutron Scattering*, ed. by S.W. Lovesay, T. Springer, Topics in Current Physics, vol. 3 (Springer, Berlin, Heidelberg 1977)
4.76 A.D. Bruce, R.A. Cowley: Adv. Phys. **29**, 219–321 (1980)
4.77 M.E. Lines, A.M. Glass: *Principles and Application of Ferroelectrics and Related Materials* (Clarendon Press, Oxford 1977)
4.78 C.B. Walker, D.T. Keating: Phys. Rev. **130**, 1726–1734 (1963)
4.79 J. Als-Nielsen, O.W. Dietrich: Phys. Rev. **153**, 706–710 (1967)
4.80 O.W. Dietrich, J. Als-Nielsen: Phys. Rev. **153**, 711–717 (1967)
4.81 J. Als-Nielsen: Phys. Rev. **185**, 664–666 (1969)
4.82 L. Guttman, H.C. Schnyders: Phys. Rev. Lett. **22**, 520–522 (1969)
4.83 J. Als-Nielsen, O.W. Dietrich: Phys. Rev. **153**, 717–721 (1967)
4.84 D.R. Chipman, C.B. Walker: Phys. Rev. B **5**, 3823–3831 (1972)
4.85 O. Rathmann, J. Als-Nielsen: Phys. Rev. B **9**, 3921–3926 (1974)
4.86 L. Guttman, H.C. Schnyders, G.J. Arai: Phys. Rev. Lett. **22**, 517–519 (1969)
4.87 A.D. Bruce: J. Phys. C **14**, 193–210 (1981)
4.88 M.D. Rechtin, S.C. Moss, B.L. Averbach: Phys. Rev. Lett. **24**, 1485–1489 (1970)
4.89 J.A. Ouedele, M.F. Collins: Phys. Rev. B **16**, 3208–3212 (1977)
4.90 J.W. Essam, D.L. Hunter: J. Phys. C **1**, 392–407 (1968)
4.91 P. Schofield: Phys. Rev. Lett. **22**, 606–608 (1969)
4.92 L.B. Ebert: Ann. Rev. Mater. Sci. **6**, 181–211 (1976)
4.93 J.E. Fischer, T.E. Thompson: Phys. Today **31**, 36–40 (1978)
4.94 G.S. Parry: Mater. Sci. Eng. **31**, 99–106 (1977)
4.95 J.B. Hastings, W.D. Ellenson, J.E. Fischer: Phys. Rev. Lett. **42**, 1552–1556 (1979)
4.96 H. Zabel, S.C. Moss, N. Caswell, S.A. Solin: Phys. Rev. Lett. **43**, 2022–2025 (1979)
4.97 H. Zabel, Y.M. Jan, S.C. Moss: Physica B **99**, 453–456 (1980)
4.98 S. Ohnishi, S. Sugano: Solid State Commun. **36**, 823–826 (1980)
4.99 J.R. Dahn, D.C. Dahn, R.R. Haering: Solid State Commun. **42**, 179–183 (1982)
4.100 R. Clarke, N. Caswell, S.A. Solin, P.M. Horn: Phys. Rev. Lett. **43**, 2018–2022 (1979)
4.101 M. Suzuki, H. Ikeda, H. Suematsu et al.: J. Phys. Soc. Jap. **49**, 671–677 (1980)
4.102 M. Mori, S.C. Moss, Y.M. Jan, H. Zabel: Phys. Rev. B **25**, 1287–1296 (1982)
4.103 B. Bak, E. Domany: Phys. Rev. B **23**, 1320–1324 (1981)
4.104 B.I. Halperin: in *Physics of Defects*, ed. by R. Balian, M. Kleman, J.-P. Poirier (North-Holland Publ., Amsterdam 1981) pp. 813–857
4.105 I. Naiki, Y. Yamada: J. Phys. Soc. Jap. **51**, 257 (1982)
4.106 M. Mori, S.C. Moss, Y.M. Jan: Phys. Rev. B **27**, 6385–6394 (1983)
4.107 C. Horie, S.A. Solin, H. Miyazaki et al.: Phys. Rev. B **27**, 3796–3805 (1983)
4.108 H. Terauchi, Y. Yamada: J. Phys. Soc. Jap. **33**, 446–454 (1972)
4.109 A. Hüller, W. Press: Phys. Rev. Lett. **29**, 266–269 (1972)
4.110 W.B. Yelon, D.E. Cox, P.J. Kortman: Phys. Rev. **9**, 4843–4856 (1974)
4.111 H. Terauchi, Y. Noda, Y. Yamada: J. Phys. Soc. Jap. **32**, 1560–1564 (1972)

4.112 Y. Yamada, M. Mori, Y. Noda. J. Phys. Soc. Jap. **32**, 1565–1576 (1972)
4.113 Y. Yamada, Y. Noda, J.D. Axe, G. Shirane: Phys. Rev. B **9**, 4429–4438 (1974)
4.114 R.A. Cowley: Phys. Rev. **134**, 981–997 (1964)
4.115 T. Riste, E.J. Samuelsen, K. Otnes, J. Feder: Solid State Commun. **9**, 1455–1458 (1971)
4.116 S.M. Shapiro, J.D. Axe, G. Shirane, R. Riste: Phys. Rev. B **6**, 4332–4341 (1972)
4.117 J. Töpler, B. Alefeld, A. Heidemann: J. Phys. C **10**, 635–643 (1977)
4.118 D.A. Bruce, W.G. Stirling: J. Phys. C **16**, 841–854 (1983)
4.119 J. Hutton, R.J. Nelmes, G.M. Meyer, V.R. Eiriksson: J. Phys. C **12**, 5393–5410 (1979)
4.120 J. Hutton, R.J. Nelmes: J. Phys. C **14**, 1713–1736 (1981)
4.121 K.A. Müller: in *Dynamical Critical Phenomena and Related Topics*, ed. by J. Ehlers, K. Hepp, R. Kippenhahn et al. (Springer, Berlin, Heidelberg 1979) pp. 210–250
4.122 H. Jex, J. Maetz, M. Mullner: Phys. Rev. B **21**, 1209–1218 (1980)
4.123 J.Y. Bizare, J.C. Fayet, W. Berlinger, K.A. Muller: Phys. Rev. Lett. **42**, 465–468 (1979)
4.124 R. Comes, M. Lambert, A. Guinier: Solid State Commun. **6**, 715–719 (1968)
4.125 R. Comes, M. Lambert, A. Guinier: Acta Crystallogr. A **26**, 244–254 (1970)
4.126 F. Denoyer, R. Comes, M. Lambert: Solid State Commun. **8**, 1979–1081 (1970)
4.127 F. Denoyer, R. Comes, M. Lambert: Acta Crystallogr. A **27**, 414–420 (1971)
4.128 R. Comes, F. Denoyer, L. Deschamps, M. Lambert: Phys. Lett. A **34**, 65–66 (1971)
4.129 K. Gesi, J.D. Axe, G. Shirane, A. Linz: Phys. Rev. B **5**, 1933–1941 (1972)
4.130 J. Harada, J.D. Axe, G. Shirane: Phys. Rev. B **4**, 155–162 (1971)
4.131 K. Ishida, G. Honjo: J. Phys. Soc. Jap. **34**, 1279–1288 (1973)
4.132 R. Comes, G. Shirane: Phys. Rev. B **5**, 1886–1891 (1972)
4.133 M.D. Fontana, G. Dolling, G.E. Kugel, C. Carabatos: Phys. Rev. B **20**, 3850–3858 (1979)
4.134 J. Slater: Phys. Rev. **78**, 748–761 (1950)
4.135 A. Hüller: Zs. Phys. **220**, 145–158 (1969)
4.136 R. Clarke: Phys. Rev. Lett. **39**, 1550–1553 (1977)
4.137 M. Sato, B.H. Grier, G. Shirane, T. Akahane: Phys. Rev. B **25**, 6876–6885 (1982)
4.138 Y. Yamada, T. Yamada: J. Phys. Soc. Jap. **21**, 2167–2177 (1966)
4.139 J. Sakurai, R.A. Cowley, G. Dolling: J. Phys. Soc. Jap. **28**, 1426–1445 (1970)
4.140 K.D. Erhardt, K.H. Michel: Zs. Phys. B **41**, 329–339 (1981)
4.141 K.H. Michel: Phys. Rev. B **24**, 3998–4007 (1981)
4.142 A.P. Levanyuk, D.G. Sannikov: Fiz. Tverd. Tela **18**, 1927–1932 (1976)
4.143 J.D.C. McConnell, V. Heine: Phys. Rev. Lett. **B46**, 1092–1095 (1981)
4.144 J.D.C. McConnell, V. Heine: J. Phys. C **15**, 2387–2402 (1982)
4.145 I. Shibuya: J. Phys. Soc. Jap. **16**, 490–494 (1961)
4.146 M.L. Canut, J. Mendiola: Phys. Stat. Sol. **5**, 313–327 (1964)
4.147 J.L. Amoros, M. Amoros: *Molecular Crystals, Their Transforms and Diffuse Scattering* (John Wiley, New York 1968)
4.148 C.J. Majkrzak, J.D. Axe, A.D. Bruce: Phys. Rev. B **22**, 5278–5283 (1980)
4.149 H. Mashiyama: J. Phys. Soc. Jap. **50**, 2655–2659 (1981)
4.150 M. Iizumi, J.D. Axe, G. Shirane, K. Shimaoka: Phys. Rev. B **15**, 4392–4411 (1981)
4.151 C.-E. Chen, Y. Schlesinger, A.J. Heeger: Phys. Rev. B **24**, 5139–5142 (1981)
4.152 R. Pynn, J.D. Axe: J. Phys. C **9**, L199–L202 (1976)
4.153 D. Mukamel: Phys. Rev. Lett. **34**, 481–483 (1975)
4.154 J. Schneck, J.C. Toledano, C. Joffrin et al.: Phys. Rev. B **25**, 1766–1785 (1982)
4.155 P.W. Anderson, E.I. Blount: Phys. Rev. Lett. **14**, 217–219 (1965)
4.156 J.B. Goodenough, A.L. Loeb: Phys. Rev. **98**, 391–408 (1955)
4.157 M. Kataoka, J. Kanamori: J. Phys. Soc. Jap. **32**, 113–134 (1972)
4.158 N.N. Kristofel: *Theory of Small-Radius Impurities in Ionic Crystals* (Nauka, Moscow 1974) (in Russian)
4.159 K.I. Kugel, D.I. Khomskii: Usp. Fiz. Nauk **136**, 621–664 (1982)
4.160 H. Terauchi, M. Mori, Y. Yamada: J. Phys. Soc. Jap. **32**, 1049–1058 (1972)
4.161 R.J. Birgeneau, J.K. Kjems, G. Shirane, L.G. Van Uitert: Phys. Rev. B **10**, 2512–2534 (1974)

4.162 R.J. Elliott, R.T. Harley, W. Hayes, S.R.P. Smith: Proc. Roy. Soc. A **328**, 217–266 (1972)
4.163 G. Shirane, J.D. Axe, R.J. Birgeneau: Solid State Commun. **9**, 397–400 (1971)
4.164 G. Shirane, J.D. Axe: Phys. Rev. B **4**, 2957–2963 (1971)
4.165 J.D. Axe, G. Shirane: Phys. Rev. B **8**, 1965–1977 (1973)
4.166 L. Testardi, M. Veger, I. Goldberg: *Superconducting Compounds with Tungsten β Structure* (Mir, Moscow 1977) (in Russian)
4.167 L.P. Gorkov: Appendix II in *Ref. 4.166.*
4.168 Y. Noda, R. Youngblood, G. Shirane, Y. Yamada: J. Phys. Soc. Jap. **48**, 1576–1587 (1980)
4.169 Y. Noda, M. Mori Y. Yamada: J. Phys. Soc. Jap. **45**, 954–966 (1978)
4.170 M. Mori, Y. Noda, Y. Yamada: J. Phys. Soc. Jap. **48**, 1288–1297 (1980)
4.171 M. Mori, Y. Noda, Y. Yamada: in *Modulated Structures*, Proc. AIP Conf., vol. 53 (AIP, New York 1979) pp. 214–216
4.172 Y. Yamada, M. Mori, Y. Noda, M. Iizumi: Solid State Commun. **32**, 827–830 (1979)
4.173 Y. Yamada, N. Wakabayashi, R.M. Nicklow: Phys. Rev. **21**, 4642–4648 (1980)
4.174 Y. Yamada: Phil. Mag. **42**, 377–385 (1980)
4.175 Y. Fujii, G. Shirane, Y. Yamada: Phys. Rev. B **11**, 2036–2041 (1975)
4.176 K. Chiba, K. Suzuki, S. Chikazumi: J. Phys. Soc. Jap. **39**, 839–840 (1975)
4.177 S.M. Shapiro, M. Iizumi, G. Shirane: Phys. Rev. B **14**, 200–207 (1976)
4.178 H. Zabel, G. Alefeld, H. Peisl: Scripta Met. **9**, 1345–1349 (1975)
4.179 W. Fenzl, H. Zabel, H. Peisl: J. Phys. F. **12**, 1897–1906 (1982)
4.180 J.W. Cahn: J. Chem. Phys. **66**, 3667–3672 (1977)
4.181 H. Zabel, B. Schönfeld, S.C. Moss: J. Phys. Chem. Solids **42**, 897–900 (1981)
4.182 D. Schwahn, W. Schmatz: Acta Met. **26**, 1571–1578 (1978)
4.183 E.G. Ponyatovskii: Dokl. Acad. Nauk SSSR **120**, 1021–1023 (1958)
4.184 H. Zabel: Phys. Stat. Sol. a **37**, K67–K70 (1976)
4.185 H.S.U. Jo, S.C. Moss: Solid State Commun. **30**, 365–368 (1979)
4.186 Y. Fujii, Y. Yamada: J. Phys. Soc. Jap. **30**, 1676–1685 (1971)
4.187 S. Satija, R.A. Cowley: Phys. Rev. B **25**, 6765–6770 (1982)
4.188 B.C. Frazer, D. Semmingsen, W.D. Ellenson, G. Shirane: Phys. Rev. B **20**, 2745–2754 (1979)
4.189 R. Youngblood, B.C. Frazer, J. Eckert, G. Shirane: Phys. Rev. B **22**, 228–235 (1980)
4.190 J. Skalyo Jr., B.C. Frazer, G. Shirane: Phys. Rev. B **1**, 278–286 (1970)
4.191 R.A. Cowley: Phys. Rev. Lett. **36**, 744–747 (1976)
4.192 H.-J. Bleif, R.A. Cowley, R.J. Nelmes: J. Phys. C **15**, L201–L205 (1982)
4.193 P. Bastie, M. Vallade, C. Vettier, C.M.E. Zeyen: Phys. Rev. Lett. **40**, 337–340 (1978)
4.194 G.M. Meyer, R.J. Nelmes, C. Vettier: J. Phys. C **13**, 4035–4051 (1980)
4.195 A.P. Levanyuk, V.V. Osipov, A.A. Sobyanin: in *The Theory of Light Scattering in Solids*, ed. by V.M. Agranovich, J.L. Birman, vol. 2 (Nauka, Moscow 1976) pp. 628–642
4.196 B.I. Halperin, C.M. Varma: Phys. Rev. B **14**, 4030–4044 (1976)
4.197 K.N. Hock, H. Thomas: Z. Phys. B **27**, 267–271 (1977)
4.198 A.P. Levanyuk, V.V. Osipov, A.S. Sigov, A.A. Sobyanin: Zh. Eksp. Teor. Fiz. **76**, 345–368 (1979)
4.199 A.B. Harris: J. Phys. C **7**, 1671–1692 (1974)
4.200 A.B. Harris, T.C. Lubensky: Phys. Rev. Lett. **33**, 1540–1543 (1974)
4.201 T.C. Lubensky: Phys. Rev. B **11**, 3573–3580 (1975)
4.202 D.E. Khmelnitzkii: Zh. Eksp. Teor. Fiz. **68**, 1960–1968 (1975)
4.203 A. Aharony: Phys. Rev. B **12**, 1038–1056 (1975)
4.204 G. Grinstein, A. Luther: Phys. Rev. B **13**, 1329–1343 (1976)
4.205 K.E. Newman, E.K. Riedel: Phys. Rev. B **25**, 264–280 (1982)
4.206 Y. Imry, S. Ma: Phys. Rev. Lett. **35**, 1399–1401 (1975)
4.207 G. Grinstein: Phys. Rev. Lett. **37**, 944–947 (1976)
4.208 G. Parisi, N. Sourlas: Phys. Rev. Lett. **43**, 744–745 (1979)
4.209 S. Fishman, A. Aharony: J. Phys. C **12**, L729–L733 (1979)
4.210 E. Pytte, Y. Imry, D. Mukamel: Phys. Rev. Lett. **46**, 1173–1177 (1981)

4.211 K. Binder, Y. Imry, E. Pytte: Phys. Rev. B **24**, 6736–6739 (1981)
4.212 H. Yoshizawa, R.A. Cowley, G. Shirane et al.: Phys. Rev. Lett. **48**, 438–441 (1982)
4.213 D. Mukamel, E. Pytte: Phys. Rev. B **25**, 4779–4786 (1982)
4.214 Y. Shapira, A. Aharony: J. Phys. C **15**, 1361–1380 (1982)
4.215 R.J. Birgeneau, A.N. Berker: Phys. Rev. B **26**, 3751–3757 (1982)
4.216 V. Wildpaner, H. Rauch, K. Binder: J. Phys. Chem. Solids **34**, 925–935 (1973)
4.217 H. Schmidt, F. Schwabl: Phys. Lett. A **61**, 476–478 (1977)
4.218 K.H. Hock, R. Schaffer, H. Thomas: Z. Phys. B **36**, 151–160 (1979)
4.219 S.L. Ginzburg: Zh. Eksp. Teor. Fiz. **73**, 1961–1966 (1977)
4.220 M.A. Mikulinskii: Zh. Eksp. Teor. Fiz. **55**, 2376–2386 (1968)
4.221 S. Kirkpatrick: Rev. Modern Phys. **45**, 574–588 (1973)
4.222 B.I. Shklovskii, A.L. Efros: Usp. Fiz. Nauk **117**, 401–436 (1975)
4.223 I.M. Dubrovskii, M.A. Krivoglaz: Zh. Eksp. Teor. Fiz. **77**, 1017–1031 (1979)
4.224 F.J. Bartis: Phys. Lett. A **43**, 61–62 (1973)
4.225 V.M. Nabutovskii, B.Ya. Shapiro: Zh. Eksp. Teor. Fiz. **75**, 948–959 (1978)
4.226 M.V. Belousov, B.E. Volf: Pis'ma v Zh. Eksp. Teor. Fiz. **31**, 348–352 (1980)
4.227 J.W. Cahn: Acta Met. **5**, 169–172 (1957)
4.228 C.C. Dollins: Acta Met. **18**, 1209–1215 (1970)
4.229 D.M. Barnett: Scripta Met. **5**, 261–266 (1971)
4.230 M. Suezawa, H.E. Cook: Acta Met. **26**, 1205–1213 (1978)
4.231 M. Suezawa, H.E. Cook: Acta Met. **26**, 1665–1670 (1978)
4.232 A.M. Kosevich, V.A. Shklovskii: Zh. Eksp. Teor. Fiz. **55**, 1131–1141 (1968)
4.233 D.G. Howard, B.D. Dunlap, J.G. Dash: Phys. Rev. Lett. **15**, 628–632 (1965)
4.234 S. Arajs, B.L. Tehan, E.E. Anderson, A.A. Stelmach: Phys. Stat. Sol. **41**, 639–648 (1970)
4.235 E.E. Anderson, H.J. Munson, S. Arajs et al.: J. Appl. Phys. **41**, 1274–1276 (1970)
4.236 G.M. Drabkin, A.I. Okorokov, V.I. Volkov, A.F. Shchebetov: Pis'ma v. Zh. Eksp. Teor. Fiz. **13**, 3–7 (1971)
4.237 A.I. Okorokov, Ya.A. Kasman: Fiz. Tverd. Tela **14**, 3065–3068 (1972)
4.238 I.M. Dubrovskii, M.A. Krivoglaz: Fiz. Met. i. Metalloved. **49**, 1134–1143 (1980)
4.239 J.D. Axe, S.M. Shapiro, G. Shirane: in *Anharmonic Lattices, Structural Transitions and Melting*, ed. by T. Riste (Noordholf, Leiden 1974) pp. 23–37
4.240 E. Courtens: Phys. Rev. Lett. **41**, 1171–1174 (1978)
4.241 T. Schneider, E. Stoll: Phys. Rev. Lett. **31**, 1254–1258 (1973)
4.242 J.A. Krumhansl, J.R. Schreiffer: Phys. Rev. B **11**, 3535–3545 (1975)
4.243 S. Aubry: J. Chem. Phys. **64**, 3392–3402 (1976)
4.244 C.M. Varma: Phys. Rev. B **14**, 244–254 (1976)
4.245 H.E. Cook: Phys. Rev. B **15**, 1477–1488 (1977)
4.246 I.M. Dubrovskii, M.A. Krivoglaz: Fiz. Tverd. Tela **23**, 762–773 (1981)

Springer-Verlag and the Environment

We at Springer-Verlag firmly believe that an international science publisher has a special obligation to the environment, and our corporate policies consistently reflect this conviction.

We also expect our business partners – paper mills, printers, packaging manufacturers, etc. – to commit themselves to using environmentally friendly materials and production processes.

The paper in this book is made from low- or no-chlorine pulp and is acid free, in conformance with international standards for paper permanency.

MIX
Papier aus verantwortungsvollen Quellen
Paper from responsible sources
FSC® C105338
FSC
www.fsc.org

If you have any concerns about our products,
you can contact us on
ProductSafety@springernature.com

In case Publisher is established outside the EU,
the EU authorized representative is:
Springer Nature Customer Service Center GmbH
Europaplatz 3, 69115 Heidelberg, Germany

Printed by Libri Plureos GmbH
in Hamburg, Germany